STUDENT'S SOLUTIONS MANUAL

JUDITH A. PENNA

PRECALCULUS
GRAPHS & MODELS
A UNIT CIRCLE APPROACH

ALGEBRA AND TRIGONOMETRY
GRAPHS & MODELS
A UNIT CIRCLE APPROACH

Marvin L. Bittinger
Indiana University—Purdue University at Indianapolis

Judith A. Beecher
Indiana University—Purdue University at Indianapolis

David J. Ellenbogen
Community College of Vermont

Judith A. Penna
Indiana University—Purdue University at Indianapolis

Boston San Francisco New York
London Toronto Sydney Tokyo Singapore Madrid
Mexico City Munich Paris Cape Town Hong Kong Montreal

Reproduced by Addison Wesley Longman from camera-ready copy supplied by the author.

Copyright © 2001 Addison Wesley Longman

ISBN   0-201-71349-7

1 2 3 4 5 6 7 8 9 10 VG 04 03 02 01 00

Contents

Chapter G

Introduction to Graphs and the Graphing Calculator

1. To graph $(4, 0)$ we move from the origin 4 units to the right of the y-axis. Since the second coordinate is 0, we do not move up or down from the x-axis.

   To graph $(-3, -5)$ we move from the origin 3 units to the left of the y-axis. Then we move 5 units down from the x-axis.

   To graph $(-1, 4)$ we move from the origin 1 unit to the left of the y-axis. Then we move 4 units up from the x-axis.

   To graph $(0, 2)$ we do not move to the right or the left of the y-axis since the first coordinate is 0. From the origin we move 2 units up.

   To graph $(2, -2)$ we move from the origin 2 units to the right of the y-axis. Then we move 2 units down from the x-axis.

   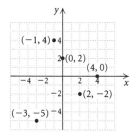

3. To graph $(-5, 1)$ we move from the origin 5 units to the left of the y-axis. Then we move 1 unit up from the x-axis.

   To graph $(5, 1)$ we move from the origin 5 units to the right of the y-axis. Then we move 1 unit up from the x-axis.

   To graph $(2, 3)$ we move from the origin 2 units to the right of the y-axis. Then we move 3 units up from the x-axis.

   To graph $(2, -1)$ we move from the origin 2 units to the right of the y-axis. Then we move 1 unit down from the x-axis.

   To graph $(0, 1)$ we do not move to the right or the left of the y-axis since the first coordinate is 0. From the origin we move 1 unit up.

   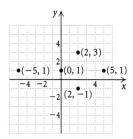

5. Enter the coordinates of the points in lists, set up a STAT PLOT, and then graph the points in the standard window. See the Graphing Calculator Manual that accompanies the text for the procedure.

   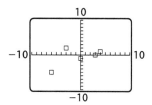

7. Enter the coordinates of the points in lists, set up a STAT PLOT, and then graph the points in the standard window.

   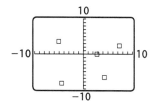

9. To determine whether $(1, -1)$ is a solution, substitute 1 for x and -1 for y.

$$\frac{y = 2x - 3}{-1 \; ? \; 2 \cdot 1 - 3}$$
$$\begin{array}{c|c} & 2 - 3 \\ -1 & -1 \qquad \text{TRUE} \end{array}$$

   The equation $-1 = -1$ is true, so $(1, -1)$ is a solution.

   To determine whether $(0, 3)$ is a solution, substitute 0 for x and -3 for y.

$$\frac{y = 2x - 3}{3 \ ? \ 2 \cdot 0 - 3}$$
$$0 - 3$$
$$3 \ \big| \ -3 \qquad \text{FALSE}$$

The equation $3 = -3$ is false, so $(0, 3)$ is not a solution.

11. To determine whether $\left(\frac{2}{3}, \frac{3}{4}\right)$ is a solution, substitute $\frac{2}{3}$ for x and $\frac{3}{4}$ for y.

$$\frac{6x - 4y = 1}{6 \cdot \frac{2}{3} - 4 \cdot \frac{3}{4} \ ? \ 1}$$
$$4 - 3$$
$$1 \ \big| \ 1 \quad \text{TRUE}$$

The equation $1 = 1$ is true, so $\left(\frac{2}{3}, \frac{3}{4}\right)$ is a solution.

To determine whether $\left(1, \frac{3}{2}\right)$ is a solution, substitute 1 for x and $\frac{3}{2}$ for y.

$$\frac{6x - 4y = 1}{6 \cdot 1 - 4 \cdot \frac{3}{2} \ ? \ 1}$$
$$6 - 6$$
$$0 \ \big| \ 1 \quad \text{FALSE}$$

The equation $0 = 1$ is false, so $\left(1, \frac{3}{2}\right)$ is not a solution.

13. To determine whether $\left(-\frac{1}{2}, -\frac{4}{5}\right)$ is a solution, substitute $-\frac{1}{2}$ for a and $-\frac{4}{5}$ for b.

$$\frac{2a + 5b = 3}{2\left(-\frac{1}{2}\right) + 5\left(-\frac{4}{5}\right) \ ? \ 3}$$
$$-1 - 4$$
$$-5 \ \big| \ 3 \quad \text{FALSE}$$

The equation $-5 = 3$ is false, so $\left(-\frac{1}{2}, -\frac{4}{5}\right)$ is not a solution.

To determine whether $\left(0, \frac{3}{5}\right)$ is a solution, substitute 0 for a and $\frac{3}{5}$ for b.

$$\frac{2a + 5b = 3}{2 \cdot 0 + 5 \cdot \frac{3}{5} \ ? \ 3}$$
$$0 + 3$$
$$3 \ \big| \ 3 \quad \text{TRUE}$$

The equation $3 = 3$ is true, so $\left(0, \frac{3}{5}\right)$ is a solution.

15. To determine whether $(-0.75, 2.75)$ is a solution, substitute -0.75 for x and 2.75 for y.

$$\frac{x^2 - y^2 = 3}{(-0.75)^2 - (2.75)^2 \ ? \ 3}$$
$$0.5625 - 7.5625$$
$$-7 \ \big| \ 3 \quad \text{FALSE}$$

The equation $-7 = 3$ is false, so $(-0.75, 2.75)$ is not a solution.

To determine whether $(2, -1)$ is a solution, substitute 2 for x and -1 for y.

$$\frac{x^2 - y^2 = 3}{2^2 - (-1)^2 \ ? \ 3}$$
$$4 - 1$$
$$3 \ \big| \ 3 \quad \text{TRUE}$$

The equation $3 = 3$ is true, so $(2, -1)$ is a solution.

17. First create a table of values as described in the Graphing Calculator Manual that accompanies the text.

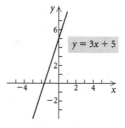

Then plot the points in the table and draw the graph.

$y = 3x + 5$

19. First create a table of values as described in the Graphing Calculator Manual that accompanies the text. Then plot the points in the table and draw the graph.

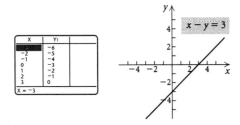

21. First create a table of values as described in the Graphing Calculator Manual that accompanies the text. Then plot the points in the table and draw the graph.

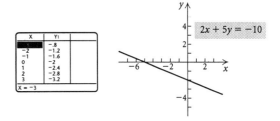

29. First create a table of values as described in the Graphing Calculator Manual that accompanies the text. Then plot the points in the table and draw the graph.

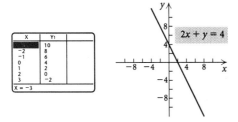

23. First create a table of values as described in the Graphing Calculator Manual that accompanies the text. Then plot the points in the table and draw the graph.

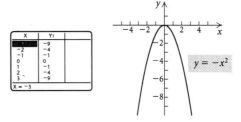

31. First create a table of values as described in the Graphing Calculator Manual that accompanies the text. Then plot the points in the table and draw the graph.

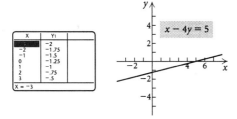

25. First create a table of values as described in the Graphing Calculator Manual that accompanies the text. Then plot the points in the table and draw the graph.

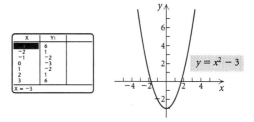

33. First create a table of values as described in the Graphing Calculator Manual that accompanies the text. Then plot the points in the table and draw the graph.

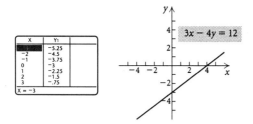

27. First create a table of values as described in the Graphing Calculator Manual that accompanies the text. Then plot the points in the table and draw the graph.

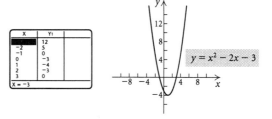

35. Graph (b) is the graph of $y = 3 - x$.

37. Graph (a) is the graph of $y = x^2 + 2x + 1$.

39. Enter the equation, select the standard window, and graph the equation as described in the Graphing Calculator Manual that accompanies the text.

$$y = 2x + 1$$

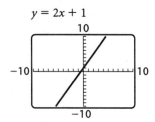

41. First solve the equation for y: $y = -4x+7$. Enter the equation in this form, select the standard window, and graph the equation as described in the Graphing Calculator Manual that accompanies the text.

$$4x + y = 7$$

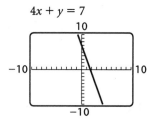

43. Enter the equation, select the standard window, and graph the equation as described in the Graphing Calculator Manual that accompanies the text.

$$y = \frac{1}{3}x + 2$$

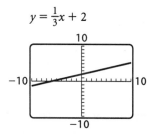

45. First solve the equation for y.

$$2x + 3y = -5$$
$$3y = -2x - 5$$
$$y = \frac{-2x - 5}{3}, \text{ or } \frac{1}{3}(-2x - 5)$$

Enter the equation in "$y =$" form, select the standard window, and graph the equation as described in the Graphing Calculator Manual that accompanies the text.

$$2x + 3y = -5$$

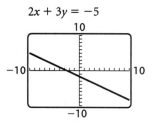

47. Enter the equation, select the standard window, and graph the equation as described in the Graphing Calculator Manual that accompanies the text.

$$y = x^2 + 6$$

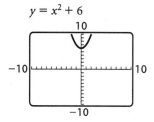

49. Enter the equation, select the standard window, and graph the equation as described in the Graphing Calculator Manual that accompanies the text.

$$y = 2 - x^2$$

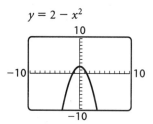

51. Enter the equation, select the standard window, and graph the equation as described in the Graphing Calculator Manual that accompanies the text.

$$y = x^2 + 4x - 2$$

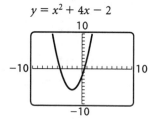

53. Standard window:

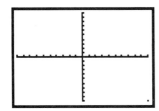

$[-25, 25, -25, 25]$, Xscl = 5, Yscl = 5

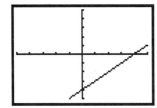

We see that $[-25, 25, -25, 25]$ is a better choice for this graph.

55. Standard window:

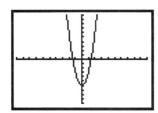

$[-4, 4, -4, 4]$

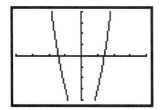

We see that the standard window is a better choice for this graph.

57. Standard window:

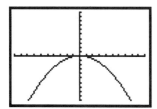

$[-1, 1, -0.3, 0.3]$, Xscl = 0.1, Yscl = 0.1

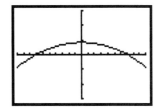

We see that $[-1, 1, -0.3, 0.3]$ is a better choice for this graph.

59. Graph the equations and use the INTERSECT feature as described in the Graphing Calculator Manual that accompanies the text.

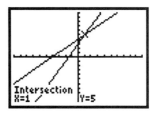

The point of intersection is $(1, 5)$.

61. Graph the equations and use the INTERSECT feature as described in the Graphing Calculator Manual that accompanies the text. The coordinates are rational numbers, so we can use the ▷Frac feature to find their exact values.

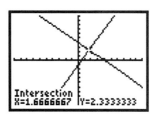

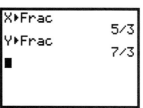

The point of intersection is $\left(\dfrac{5}{3}, \dfrac{7}{3}\right)$.

63. Graph the equations and use the INTERSECT feature as described in the Graphing Calculator Manual that accompanies the text. The coordinates are rational numbers, so we can use the ▷Frac feature to find their exact values.

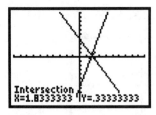

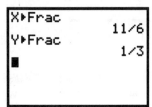

The point of intersection is $\left(\dfrac{11}{6}, \dfrac{1}{3}\right)$.

65. Graph the equations and use the INTERSECT feature as described in the Graphing Calculator Manual that accompanies the text.

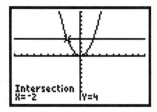

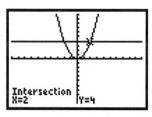

The points of intersection are $(-2, 4)$ and $(2, 4)$.

67. Graph the equations and use the INTERSECT feature as described in the Graphing Calculator Manual that accompanies the text.

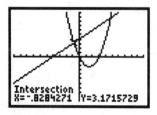

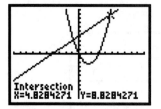

The coordinates are not rational numbers. We give their decimal approximations. The points of intersection are approximately $(-0.8284271, 3.1715729)$ and $(4.8284271, 8.8284271)$.

Chapter R

Basic Concepts of Algebra

Exercise Set R.1

1. Whole numbers: $\sqrt[3]{8}$, 0, 9, $\sqrt{25}$ ($\sqrt[3]{8} = 2$, $\sqrt{25} = 5$)

3. Irrational numbers: $\sqrt{7}$, $5.242242224\ldots$, $-\sqrt{14}$, $\sqrt[5]{5}$, $\sqrt[3]{4}$

   (Although there is a pattern in $5.242242224\ldots$, there is no repeating block of digits.)

5. Rational numbers: -12, $5.\overline{3}$, $-\dfrac{7}{3}$, $\sqrt[3]{8}$, 0, -1.96, 9, $4\dfrac{2}{3}$, $\sqrt{25}$, $\dfrac{5}{7}$

7. Answers may vary. Some examples are $-\dfrac{3}{4}$, $5.7\overline{6}$, $9\dfrac{1}{8}$, -1.067.

9. Answers may vary. Some examples are -1, -5, -352.

11. This is a closed interval, so we use brackets. Interval notation is $[-3, 3]$.

13. This is a half-open interval. We use a bracket on the left and a parenthesis on the right. Interval notation is $[-14, -11)$.

15. This interval is of unlimited extent in the negative direction, and the endpoint -4 is included. Interval notation is $(-\infty, -4]$.

17. This interval is of unlimited extent in the negative direction, and the endpoint 3.8 is not included. Interval notation is $(-\infty, 3.8)$.

19. $\{x | x \neq 7\}$ is equivalent to $\{x | x < 7\} \cup \{x | x > 7\}$. Thus, interval notation is $(-\infty, 7) \cup (7, \infty)$.

21. The endpoints 0 and 5 are not included in the interval, so we use parentheses. Interval notation is $(0, 5)$.

23. The endpoint -9 is included in the interval, so we use a bracket before the -9. The endpoint -4 is not included, so we use a parenthesis after the -4. Interval notation is $[-9, -4)$.

25. Both endpoints are included in the interval, so we use brackets. Interval notation is $[x, x + h]$.

27. The endpoint p is not included in the interval, so we use a parenthesis before the p. The interval is of unlimited extent in the positive direction, so we use the infinity symbol ∞. Interval notation is (p, ∞).

29. Since 6 is an element of the set of natural numbers, the statement is true.

31. Since 3.2 is not an element of the set of integers, the statement is false.

33. Since $-\dfrac{11}{5}$ is an element of the set of rational numbers, the statement is true.

35. Since $\sqrt{11}$ is an element of the set of real numbers, the statement is false.

37. Since 24 is an element of the set of whole numbers, the statement is false.

39. Since 1.089 is not an element of the set of irrational numbers, the statement is true.

41. Since every whole number is an integer, the statement is true.

43. Since every rational number is a real number, the statement is true.

45. Since there are real numbers that are not integers, the statement is false.

47. The sentence $6x = x6$ illustrates the commutative property of multiplication.

49. The sentence $-3 \cdot 1 = -3$ illustrates the multiplicative identity property.

51. The sentence $5(ab) = (5a)b$ illustrates the associative property of multiplication.

53. The sentence $2(a + b) = (a + b)2$ illustrates the commutative property of multiplication.

55. The sentence $-6(m + n) = -6(n + m)$ illustrates the commutative property of addition.

57. The sentence $8 \cdot \dfrac{1}{8} = 1$ illustrates the multiplicative inverse property.

59. The distance of -7.1 from 0 is 7.1, so $|-7.1| = 7.1$.

61. The distance of $\frac{5}{4}$ from 0 is $\frac{5}{4}$, so $\left|\frac{5}{4}\right| = \frac{5}{4}$.

63. $|-5-6| = |-11| = 11$, or

$|6-(-5)| = |6+5| = |11| = 11$

65. $|-2-(-8)| = |-2+8| = |6| = 6$, or

$|-8-(-2)| = |-8+2| = |-6| = 6$

67. $|12.1 - 6.7| = |5.4| = 5.4$, or

$|6.7 - 12.1| = |-5.4| = 5.4$

69. Discussion and Writing

71. Answers may vary. One such number is $0.124124412444\ldots$.

73. Answers may vary. Since $-\frac{1}{101} = 0.\overline{0099}$ and $-\frac{1}{100} = -0.01$, one such number is -0.00999.

75. Since $1^2 + 3^2 = 10$, the hypotenuse of a right triangle with legs of lengths 1 unit and 3 units has a length of $\sqrt{10}$ units.

$c^2 = 1^2 + 3^2$
$c^2 = 10$
$c = \sqrt{10}$

Exercise Set R.2

1. $18° = 1$ (For any nonzero real number, $a^0 = 1$.)

3. $5^8 \cdot 5^{-6} = 5^{8+(-6)} = 5^2$, or 25

5. $m^{-5} \cdot m^5 = m^{-5+5} = m^0 = 1$

7. $7^3 \cdot 7^{-5} \cdot 7 = 7^{3+(-5)+1} = 7^{-1}$, or $\frac{1}{7}$

9. $2x^3 \cdot 3x^2 = 2 \cdot 3 \cdot x^{3+2} = 6x^5$

11. $(5a^2b)(3a^{-3}b^4) = 5 \cdot 3 \cdot a^{2+(-3)} \cdot b^{1+4} = 15a^{-1}b^5$, or $\frac{15b^5}{a}$

13. $(2x)^3(3x)^2 = 2^3x^3 \cdot 3^2x^2 = 8 \cdot 9 \cdot x^{3+2} = 72x^5$

15. $\frac{b^{40}}{b^{37}} = b^{40-37} = b^3$

17. $\frac{x^2y^{-2}}{x^{-1}y} = x^{2-(-1)}y^{-2-1} = x^3y^{-3}$, or $\frac{x^3}{y^3}$

19. $\frac{32x^{-4}y^3}{4x^{-5}y^8} = \frac{32}{4}x^{-4-(-5)}y^{3-8} = 8xy^{-5}$, or $\frac{8x}{y^5}$

21. $(2ab^2)^3 = 2^3a^3(b^2)^3 = 2^3a^3b^{2\cdot3} = 8a^3b^6$

23. $(-2x^3)^4 = (-2)^4(x^3)^4 = (-2)^4x^{3\cdot4} = 16x^{12}$

25. $(-5c^{-1}d^{-2})^{-2} = (-5)^{-2}c^{-1(-2)}d^{-2(-2)} = \frac{c^2d^4}{(-5)^2} = \frac{c^2d^4}{25}$

27. $\left(\frac{24a^{10}b^{-8}c^7}{3a^6b^{-3}b^5}\right)^5 = (8a^4b^{-5}c^2)^5 = 8^5a^{20}b^{-25}c^{10}$, or $\frac{8^5a^{20}c^{10}}{b^{25}}$

29. Convert 405,000 to scientific notation.

We want the decimal point to be positioned between the 4 and the first 0, so we move it 5 places to the left. Since 405,000 is greater than 10, the exponent must be positive.

$$405,000 = 4.05 \times 10^5$$

31. Convert 0.00000039 to scientific notation.

We want the decimal point to be positioned between the 3 and the 9, so we move it 7 places to the right. Since 0.00000039 is a number between 0 and 1 the exponent must be negative.

$$0.00000039 = 3.9 \times 10^{-7}$$

33. Convert 0.000016 to scientific notation.

We want the decimal point to be positioned between the 1 and the 6, so we move it 5 places to the right. Since 0.000016 is a number between 0 and 1, the exponent must be negative.

$$0.000016 \text{ m}^3 = 1.6 \times 10^{-5} \text{ m}^3$$

35. Convert 8.3×10^{-5} to decimal notation.

The exponent is negative, so the number is between 0 and 1. We move the decimal point 5 places to the left.

$$8.3 \times 10^{-5} = 0.000083$$

37. Convert 2.07×10^7 to decimal notation.

The exponent is positive, so the number is greater than 10. We move the decimal point 7 places to the right.

$$2.07 \times 10^7 = 20,700,000$$

39. Convert $\$1.1358 \times 10^{10}$ to decimal notation.

The exponent is positive, so the number is greater than 10. We move the decimal point 10 places to the right.

$$\$1.1358 \times 10^{10} = \$11,358,000,000$$

41. $(3.1 \times 10^5)(4.5 \times 10^{-3})$

$\cdot = (3.1 \times 4.5) \times (10^5 \times 10^{-3})$

$= 13.95 \times 10^2$ This is not scientific notation.

$= (1.395 \times 10) \times 10^2$

$= 1.395 \times 10^3$ Writing scientific notation

43. $\dfrac{6.4 \times 10^{-7}}{8.0 \times 10^6} = \dfrac{6.4}{8.0} \times \dfrac{10^{-7}}{10^6}$

$= 0.8 \times 10^{-13}$ This is not scientific notation.

$= (8 \times 10^{-1}) \times 10^{-13}$

$= 8 \times 10^{-14}$ Writing scientific notation

45. The average cost per mile is the total cost divided by the number of miles.

$$\dfrac{\$210 \times 10^6}{17.6}$$

$$= \dfrac{\$210 \times 10^6}{1.76 \times 10}$$

$$\approx \$119 \times 10^5$$

$$\approx (\$1.19 \times 10^2) \times 10^5$$

$$\approx \$1.19 \times 10^7$$

The average cost per mile was about $\$1.19 \times 10^7$.

47. First find the number of seconds in 1 hour:

1 hour $= 1 \; \text{hr} \times \dfrac{60 \; \text{min}}{1 \; \text{hr}} \times \dfrac{60 \; \text{sec}}{1 \; \text{min}} = 3600$ sec

The number of disintegrations produced in 1 hour is the number of disintegrations per second times the number of seconds in 1 hour.

37 billion $\times 3600$

$= 37,000,000,000 \times 3600$

$= 3.7 \times 10^{10} \times 3.6 \times 10^3$ Writing scientific notation

$= (3.7 \times 3.6) \times (10^{10} \times 10^3)$

$= 13.32 \times 10^{13}$ Multiplying

$= (1.332 \times 10) \times 10^{13}$

$= 1.332 \times 10^{14}$

One gram of radium produces 1.332×10^{14} disintegrations in 1 hour.

49. $3 \cdot 2 - 4 \cdot 2^2 + 6(3 - 1)$

$= 3 \cdot 2 - 4 \cdot 2^2 + 6 \cdot 2$ Working inside parentheses

$= 3 \cdot 2 - 4 \cdot 4 + 6 \cdot 2$ Evaluating 2^2

$= 6 - 16 + 12$ Multiplying

$= -10 + 12$ Adding in order

$= 2$ from left to right

51. $16 \div 4 \cdot 4 \div 2 \cdot 256$

$= 4 \cdot 4 \div 2 \cdot 256$ Multiplying and dividing in order from left to right

$= 16 \div 2 \cdot 256$

$= 8 \cdot 256$

$= 2048$

53. $\dfrac{4(8 - 6)^2 - 4 \cdot 3 + 2 \cdot 8}{3^1 + 19^0}$

$= \dfrac{4 \cdot 2^2 - 4 \cdot 3 + 2 \cdot 8}{3 + 1}$ Calculating in the numerator and in the denominator

$= \dfrac{4 \cdot 4 - 4 \cdot 3 + 2 \cdot 8}{4}$

$= \dfrac{16 - 12 + 16}{4}$

$= \dfrac{4 + 16}{4}$

$= \dfrac{20}{4}$

$= 5$

55. Since interest is compounded semiannually, $n = 2$. Substitute $\$2125$ for P, 6.2% or 0.062 for i, 2 for n, and 5 for t in the compound interest formula.

$A = P\left(1 + \dfrac{i}{n}\right)^{nt}$

$= \$2125\left(1 + \dfrac{0.062}{2}\right)^{2 \cdot 5}$ Substituting

$= \$2125(1 + 0.031)^{2 \cdot 5}$ Dividing

$= \$2125(1.031)^{2 \cdot 5}$ Adding

$= \$2125(1.031)^{10}$ Multiplying 2 and 5

$\approx \$2125(1.357021264)$ Evaluating the exponential expression

$\approx \$2883.670185$ Multiplying

$\approx \$2883.67$ Rounding to the nearest cent

57. Since interest is compounded quarterly, $n = 4$. Substitute $\$6700$ for P, 4.5% or 0.045 for i, 4 for n, and 6 for t in the compound interest formula.

$A = P\left(1 + \dfrac{i}{n}\right)^{nt}$

$= \$6700\left(1 + \dfrac{0.045}{4}\right)^{4 \cdot 6}$ Substituting

$= \$6700(1 + 0.01125)^{4 \cdot 6}$ Dividing

$= \$6700(1.01125)^{4 \cdot 6}$ Adding

$= \$6700(1.01125)^{24}$ Multiplying 4 and 6

$\approx \$6700(1.307991226)$ Evaluating the exponential expression

$\approx \$8763.541217$ Multiplying

$\approx \$8763.54$ Rounding to the nearest cent

59. Discussion and Writing

61. $(x^t \cdot x^{3t})^2 = (x^{4t})^2 = x^{4t \cdot 2} = x^{8t}$

63. $(t^{a+x} \cdot t^{x-a})^4 = (t^{2x})^4 = t^{2x \cdot 4} = t^{8x}$

65. $\left[\dfrac{(3x^a y^b)^3}{(-3x^a y^b)^2}\right]^2 = \left[\dfrac{27x^{3a} y^{3b}}{9x^{2a} y^{2b}}\right]^2$

$= \left[3x^a y^b\right]^2$

$= 9x^{2a} y^{2b}$

67. $P = \$98,000 - \$16,000 = \$82,000$;

$i = 8\dfrac{1}{2}\%,$ or 0.085; $n = 12 \cdot 25 = 300$

$$M = P\left[\dfrac{\dfrac{i}{12}\left(1+\dfrac{i}{12}\right)^n}{\left(1+\dfrac{i}{12}\right)^n - 1}\right]$$

$$M = \$82,000\left[\dfrac{\dfrac{0.085}{12}\left(1+\dfrac{0.085}{12}\right)^{300}}{\left(1+\dfrac{0.085}{12}\right)^{300} - 1}\right]$$

$\approx \$660.29$ Using a calculator

69. $P = \$135,000 - \$18,000 = \$117,000$;

$i = 7\dfrac{1}{2}\%,$ or 0.075; $n = 12 \cdot 20 = 240$

$$M = P\left[\dfrac{\dfrac{i}{12}\left(1+\dfrac{i}{12}\right)^n}{\left(1+\dfrac{i}{12}\right)^n - 1}\right]$$

$$M = \$117,000\left[\dfrac{\dfrac{0.075}{12}\left(1+\dfrac{0.075}{12}\right)^{240}}{\left(1+\dfrac{0.075}{12}\right)^{240} - 1}\right]$$

$\approx \$924.54$ Using a calculator

71. Deselect the graph of y_3 and inspect the graphs of y_1 and y_2 in a suitable window such as $[-2, 2, -2, 2]$. The graph of y_1 lies on or above the graph of y_2 for $x \le 1$.

Exercise Set R.3

1. $-5y^4 + 3y^3 + 7y^2 - y - 4 =$
$-5y^4 + 3y^3 + 7y^2 + (-y) + (-4)$

Terms: $-5y^4$, $3y^3$, $7y^2$, $-y$, -4

The degree of the term of highest degree, $-5y^4$, is 4. Thus, the degree of the polynomial is 4.

3. $3a^4 b - 7a^3 b^3 + 5ab - 2 = 3a^4 b + (-7a^3 b^3) + 5ab + (-2)$

Terms: $3a^4 b$, $-7a^3 b^3$, $5ab$, -2

The degrees of the terms are 5, 6, 2, and, 0, respectively, so the degree of the polynomial is 6.

5. $(5x^2 y - 2xy^2 + 3xy - 5) +$
$\quad\quad (-2x^2 y - 3xy^2 + 4xy + 7)$
$= (5 - 2)x^2 y + (-2 - 3)xy^2 + (3 + 4)xy +$
$\quad\quad (-5 + 7)$
$= 3x^2 y - 5xy^2 + 7xy + 2$

7. $(2x + 3y + z - 7) + (4x - 2y - z + 8) +$
$\quad\quad (-3x + y - 2z - 4)$
$= (2 + 4 - 3)x + (3 - 2 + 1)y + (1 - 1 - 2)z +$
$\quad\quad (-7 + 8 - 4)$
$= 3x + 2y - 2z - 3$

9. $(3x^2 - 2x - x^3 + 2) - (5x^2 - 8x - x^3 + 4)$
$= (3x^2 - 2x - x^3 + 2) + (-5x^2 + 8x + x^3 - 4)$
$= (3 - 5)x^2 + (-2 + 8)x + (-1 + 1)x^3 + (2 - 4)$
$= -2x^2 + 6x - 2$

11. $(x^4 - 3x^2 + 4x) - (3x^3 + x^2 - 5x + 3)$
$= (x^4 - 3x^2 + 4x) + (-3x^3 - x^2 + 5x - 3)$
$= x^4 - 3x^3 + (-3 - 1)x^2 + (4 + 5)x - 3$
$= x^4 - 3x^3 - 4x^2 + 9x - 3$

13. $(a - b)(2a^3 - ab + 3b^2)$
$= (a - b)(2a^3) + (a - b)(-ab) + (a - b)(3b^2)$

      Using the distributive property
$= 2a^4 - 2a^3 b - a^2 b + ab^2 + 3ab^2 - 3b^3$

      Using the distributive property
      three more times
$= 2a^4 - 2a^3 b - a^2 b + 4ab^2 - 3b^3$ Collecting like
      terms

15. $(x + 5)(x - 3)$
$= x^2 - 3x + 5x - 15$ Using FOIL
$= x^2 + 2x - 15$ Collecting like terms

17. $(2a + 3)(a + 5)$
$= 2a^2 + 10a + 3a + 15$ Using FOIL
$= 2a^2 + 13a + 15$ Collecting like terms

19. Construct a table of values for $y_1 = (2x + 3)(x + 5)$ and $y_2 = 2x^2 + 13x + 15$. The values of y_1 and y_2 are the same for each given x-value, so the product appears to be correct.

21. $(2x + 3y)(2x + y)$
$= 4x^2 + 2xy + 6xy + 3y^2$ Using FOIL
$= 4x^2 + 8xy + 3y^2$

23. $(y+5)^2$

$= y^2 + 2\cdot y \cdot 5 + 5^2$

$\qquad [(A+B)^2 = A^2 + 2AB + B^2]$

$= y^2 + 10y + 25$

25. $(5x-3)^2$

$= (5x)^2 - 2\cdot 5x \cdot 3 + 3^2$

$\qquad [(A-B)^2 = A^2 - 2AB + B^2]$

$= 25x^2 - 30x + 9$

27. Construct a table of values for $y_1 = (5x-3)^2$ and $y_2 = 25x^2 - 30x + 9$. The values of y_1 and y_2 are the same for each given x-value, so the product appears to be correct.

29. $(2x+3y)^2$

$= (2x)^2 + 2(2x)(3y) + (3y)^2$

$\qquad [(A+B)^2 = A^2 + 2AB + B^2]$

$= 4x^2 + 12xy + 9y^2$

31. $(2x^2 - 3y)^2$

$= (2x^2)^2 - 2(2x^2)(3y) + (3y)^2$

$\qquad [(A-B)^2 = A^2 - 2AB + B^2]$

$= 4x^4 - 12x^2 y + 9y^2$

33. $(a+3)(a-3)$

$= a^2 - 3^2 \qquad [(A+B)(A-B) = A^2 - B^2]$

$= a^2 - 9$

35. Construct a table of values for $y_1 = (x+3)(x-3)$ and $y_2 = x^2 - 9$. The values of y_1 and y_2 are the same for each given x-value, so the product appears to be correct.

37. $(3x-2y)(3x+2y)$

$= (3x)^2 - (2y)^2 \qquad [(A-B)(A+B) = A^2 - B^2]$

$= 9x^2 - 4y^2$

39. $(2x+3y+4)(2x+3y-4)$

$= [(2x+3y)+4][(2x+3y)-4]$

$= (2x+3y)^2 - 4^2$

$= 4x^2 + 12xy + 9y^2 - 16$

41. $(x+1)(x-1)(x^2+1)$

$= (x^2-1)(x^2+1)$

$= x^4 - 1$

43. Discussion and Writing

45. $(a^n + b^n)(a^n - b^n) = (a^n)^2 - (b^n)^2$

$= a^{2n} - b^{2n}$

47. $(a^n + b^n)^2 = (a^n)^2 + 2\cdot a^n \cdot b^n + (b^n)^2$

$= a^{2n} + 2a^n b^n + b^{2n}$

49. $(x-1)(x^2+x+1)(x^3+1)$

$= [(x-1)x^2 + (x-1)x + (x-1)\cdot 1](x^3+1)$

$= (x^3 - x^2 + x^2 - x + x - 1)(x^3+1)$

$= (x^3-1)(x^3+1)$

$= (x^3)^2 - 1^2$

$= x^6 - 1$

51. $(x^{a-b})^{a+b}$

$= x^{(a-b)(a+b)}$

$= x^{a^2 - b^2}$

53. $(a+b+c)^2$

$= (a+b+c)(a+b+c)$

$= (a+b+c)(a) + (a+b+c)(b) + (a+b+c)(c)$

$= a^2 + ab + ac + ab + b^2 + bc + ac + bc + c^2$

$= a^2 + b^2 + c^2 + 2ab + 2ac + 2bc$

Exercise Set R.4

1. $2x - 10 = 2\cdot x - 2\cdot 5 = 2(x-5)$

3. $3x^4 - 9x^2 = 3x^2 \cdot x^2 - 3x^2 \cdot 3 = 3x^2(x^2-3)$

5. $4a^2 - 12a + 16 = 4\cdot a^2 - 4\cdot 3a + 4\cdot 4 = 4(a^2 - 3a + 4)$

7. $a(b-2) + c(b-2) = (b-2)(a+c)$

9. $x^3 + 3x^2 + 6x + 18$

$= x^2(x+3) + 6(x+3)$

$= (x+3)(x^2+6)$

11. $y^3 - 3y^2 - 4y + 12$

$= y^2(y-3) - 4(y-3)$

$= (y-3)(y^2-4)$

$= (y-3)(y+2)(y-2)$

13. The graphs of $y_1 = x^3 + 3x^2 + 6x + 18$ and $y_2 = (x+3)(x^2+6)$ appear to coincide. A table of values also shows that the values of y_1 and y_2 are the same for the given x-values.

15. $p^2 + 6p + 8$

We look for two numbers with a product of 8 and a sum of 6. By trial, we determine that they are 2 and 4.

$$p^2 + 6p + 8 = (p+2)(p+4)$$

17. $2n^2 + 9n - 56$

We look for factors $(pn+q)(rn+s)$ for which $pn \cdot rn = 2n^2$ and $q \cdot s = -56$. When we multiply the inside terms, then the outside terms, and add, we must have $9n$. By trial, we determine the factorization:

$$2n^2 + 9n - 56 = (2n - 7)(n + 8)$$

19. $y^4 - 4y^2 - 21$

Think of this polynomial as $u^2 - 4u - 21$, where we have mentally substituted u for y^2. Then we look for factors of -21 whose sum is -4. By trial we determine the factors to be -7 and 3, so

$$u^2 - 4u - 21 = (u - 7)(u + 3).$$

Then, substituting y^2 for u, we obtain the factorization of the original trinomial.

$$y^4 - 4y^2 - 21 = (y^2 - 7)(y^2 + 3)$$

Neither of these factors can be factored further, so the factorization is complete.

21. The graphs of $y_1 = x^2 + 6x + 8$ and $y_2 = (x+2)(x+4)$ appear to coincide. A table of values also shows that the values of y_1 and y_2 are the same for the given x-values.

23. $\begin{aligned} 9x^2 - 25 &= (3x)^2 - 5^2 \\ &= (3x + 5)(3x - 5) \end{aligned}$

25. $\begin{aligned} 4xy^4 - 4xz^2 &= 4x(y^4 - z^2) \\ &= 4x[(y^2)^2 - z^2] \\ &= 4x(y^2 + z)(y^2 - z) \end{aligned}$

27. $\begin{aligned} y^2 - 6y + 9 &= y^2 - 2 \cdot y \cdot 3 + 3^2 \\ &= (y - 3)^2 \end{aligned}$

29. $\begin{aligned} 1 - 8x + 16x^2 &= 1^2 - 2 \cdot 1 \cdot 4x + (4x)^2 \\ &= (1 - 4x)^2 \end{aligned}$

31. The graphs of $y_1 = 1 - 8x + 16x^2$ and $y_2 = (1 - 4x)^2$ appear to coincide. A table of values also shows that the values of y_1 and y_2 are the same for the given x-values.

33. $\begin{aligned} x^3 + 8 &= x^3 + 2^3 \\ &= (x + 2)(x^2 - 2x + 4) \end{aligned}$

35. $\begin{aligned} m^3 - 1 &= m^3 - 1^3 \\ &= (m - 1)(m^2 + m + 1) \end{aligned}$

37. The graphs of $y_1 = x^3 - 1$ and $y_2 = (x - 1)(x^2 + x + 1)$ appear to coincide. A table of values also shows that the values of y_1 and y_2 are the same for the given x-values.

39. $\begin{aligned} 18a^2b - 15ab^2 &= 3ab \cdot 6a - 3ab \cdot 5b \\ &= 3ab(6a - 5b) \end{aligned}$

41. $\begin{aligned} x^3 - 4x^2 + 5x - 20 &= x^2(x - 4) + 5(x - 4) \\ &= (x - 4)(x^2 + 5) \end{aligned}$

43. $\begin{aligned} 8x^2 - 32 &= 8(x^2 - 4) \\ &= 8(x + 2)(x - 2) \end{aligned}$

45. $4y^2 - 5$

There are no common factors. We might try to factor this polynomial as a difference of squares, but there is no integer which yields 5 when squared. Thus, the polynomial is prime.

47. The graphs of $y_1 = x^3 - 4x^2 + 5x - 20$ and $y_2 = (x - 4)(x^2 + 5)$ appear to coincide. A table of values also shows that the values of y_1 and y_2 are the same for the given x-values.

49. $x^2 + 9x + 20$

We look for two numbers with a product of 20 and a sum of 9. They are 4 and 5.

$$x^2 + 9x + 20 = (x + 4)(x + 5)$$

51. $y^2 - 6y + 5$

We look for two numbers with a product of 5 and a sum of -6. They are -5 and -1.

$$y^2 - 6y + 5 = (y - 5)(y - 1)$$

53. $2a^2 + 9a + 4$

We look for factors $(pa+q)(ra+s)$ for which $pa \cdot qa = 2a^2$ and $q \cdot s = 4$. When we multiply the inside terms, then the outside terms, and add, we must have $9a$. By trial, we determine the factorization:

$$2a^2 + 9a + 4 = (2a + 1)(a + 4)$$

55. $6x^2 + 7x - 3$

We look for factors $(px+q)(rx+s)$ for which $px \cdot rx = 6$ and $q \cdot s = -3$. When we multiply the inside terms, then the outside terms, and add, we must have $7x$. By trial, we determine the factorization:

$$6x^2 + 7x - 3 = (3x - 1)(2x + 3)$$

57. $\begin{aligned} y^2 - 18y + 81 &= y^2 - 2 \cdot y \cdot 9 + 9^2 \\ &= (y - 9)^2 \end{aligned}$

59. $\begin{aligned} x^2y^2 - 14xy + 49 &= (xy)^2 - 2 \cdot xy \cdot 7 + 7^2 \\ &= (xy - 7)^2 \end{aligned}$

61. $\begin{aligned} 4ax^2 + 20ax - 56a &= 4a(x^2 + 5x - 14) \\ &= 4a(x + 7)(x - 2) \end{aligned}$

63. $3z^3 - 24 = 3(z^3 - 8)$
$$= 3(z^3 - 2^3)$$
$$= 3(z - 2)(z^2 + 2z + 4)$$

65. $16a^7b + 54ab^7$
$$= 2ab(8a^6 + 27b^6)$$
$$= 2ab[(2a^2)^3 + (3b^2)^3]$$
$$= 2ab(2a^2 + 3b^2)(4a^4 - 6a^2b^2 + 9b^4)$$

67. The graphs of $y_1 = 6x^2 + 7x - 3$ and $y_2 = (3x - 1)(2x + 3)$ appear to coincide. A table of values also shows that the values of y_1 and y_2 are the same for the given x-values.

69. Discussion and Writing

71. $y^4 - 84 + 5y^2$
$$= y^4 + 5y^2 - 84$$
$$= u^2 + 5u - 84 \qquad \text{Substituting } u \text{ for } y^2$$
$$= (u + 12)(u - 7)$$
$$= (y^2 + 12)(y^2 - 7) \quad \text{Substituting } y^2 \text{ for } u$$

73. $y^2 - \dfrac{8}{49} + \dfrac{2}{7}y = y^2 + \dfrac{2}{7}y - \dfrac{8}{49}$
$$= \left(y + \frac{4}{7}\right)\left(y - \frac{2}{7}\right)$$

75. $(x + h)^3 - x^3$
$$= [(x + h) - x][(x + h)^2 + x(x + h) + x^2]$$
$$= (x + h - x)(x^2 + 2xh + h^2 + x^2 + xh + x^2)$$
$$= h(3x^2 + 3xh + h^2)$$

77. $(y - 4)^2 + 5(y - 4) - 24$
$$= u^2 + 5u - 24 \qquad \text{Substituting } u \text{ for } y - 4$$
$$= (u + 8)(u - 3)$$
$$= (y - 4 + 8)(y - 4 - 3) \quad \text{Substituting } y - 4$$
$$\text{for } u$$
$$= (y + 4)(y - 7)$$

79. $x^{2n} + 5x^n - 24 = (x^n)^2 + 5x^n - 24$
$$= (x^n + 8)(x^n - 3)$$

81. $x^2 + ax + bx + ab = x(x + a) + b(x + a)$
$$= (x + a)(x + b)$$

83. $25y^{2m} - (x^{2n} - 2x^n + 1)$
$$= (5y^m)^2 - (x^n - 1)^2$$
$$= [5y^m + (x^n - 1)][5y^m - (x^n - 1)]$$
$$= (5y^m + x^n - 1)(5y^m - x^n + 1)$$

85. $(y - 1)^4 - (y - 1)^2$
$$= (y - 1)^2[(y - 1)^2 - 1]$$
$$= (y - 1)^2[y^2 - 2y + 1 - 1]$$
$$= (y - 1)^2(y^2 - 2y)$$
$$= y(y - 1)^2(y - 2)$$

Exercise Set R.5

1. Since $-\dfrac{3}{4}$ is defined for all real numbers, the domain is $\{x | x \text{ is a real number}\}$.

3. $\dfrac{3x - 3}{x(x - 1)}$

The denominator is 0 when the factor $x = 0$ and also when $x - 1 = 0$, or $x = 1$. The domain is $\{x | x \text{ is a real number } and \ x \neq 0 \ and \ x \neq 1\}$.

5. We first factor the denominator completely.
$$\frac{7x^2 - 28x + 28}{(x^2 - 4)(x^2 + 3x - 10)} = \frac{7x^2 - 28x + 28}{(x + 2)(x - 2)(x + 5)(x - 2)}$$
We see that $x + 2 = 0$ when $x = -2$, $x - 2 = 0$ when $x = 2$, and $x + 5 = 0$ when $x = -5$. Thus, the domain is $\{x | x \text{ is a real number } and \ x \neq -2 \ and \ x \neq 2 \ and \ x \neq -5\}$.

7. Let $y_1 = \dfrac{3x - 3}{x(x - 1)}$ and find the values of x for which the table entry is "ERROR."

X	Y₁	Y₂
−1	−3	
0	ERROR	
1	ERROR	
2	1.5	
3	1	
4	.75	
5	.6	
X = 0		

9. $\dfrac{x^2 - y^2}{(x - y)^2} \cdot \dfrac{1}{x + y}$
$$= \frac{(x^2 - y^2) \cdot 1}{(x - y)^2(x + y)}$$
$$= \frac{(x + y)(x - y) \cdot 1}{(x - y)(x - y)(x + y)}$$
$$= \frac{1}{x - y}$$

11. $\dfrac{x^2 - 2x - 35}{2x^3 - 3x^2} \cdot \dfrac{4x^3 - 9x}{7x - 49}$
$$= \frac{(x - 7)(x + 5)(x)(2x + 3)(2x - 3)}{x \cdot x(2x - 3)(7)(x - 7)}$$
$$= \frac{(x + 5)(2x + 3)}{7x}$$

13. $\dfrac{a^2 - a - 6}{a^2 - 7a + 12} \cdot \dfrac{a^2 - 2a - 8}{a^2 - 3a - 10}$

$$= \dfrac{(a-3)(a+2)(a-4)(a+2)}{(a-4)(a-3)(a-5)(a+2)}$$

$$= \dfrac{a+2}{a-5}$$

15. $\dfrac{m^2 - n^2}{r+s} \div \dfrac{m-n}{r+s}$

$$= \dfrac{m^2 - n^2}{r+s} \cdot \dfrac{r+s}{m-n}$$

$$= \dfrac{(m+n)(m-n)(r+s)}{(r+s)(m-n)}$$

$$= m+n$$

17. $\dfrac{3x+12}{2x-8} \div \dfrac{(x+4)^2}{(x-4)^2}$

$$= \dfrac{3x+12}{2x-8} \cdot \dfrac{(x-4)^2}{(x+4)^2}$$

$$= \dfrac{3(x+4)(x-4)(x-4)}{2(x-4)(x+4)(x+4)}$$

$$= \dfrac{3(x-4)}{2(x+4)}$$

19. $\dfrac{x^2 - y^2}{x^3 - y^3} \cdot \dfrac{x^2 + xy + y^2}{x^2 + 2xy + y^2}$

$$= \dfrac{(x+y)(x-y)(x^2 + xy + y^2)}{(x-y)(x^2 + xy + y^2)(x+y)(x+y)}$$

$$= \dfrac{1}{x+y} \cdot \dfrac{(x+y)(x-y)(x^2 + xy + y^2)}{(x+y)(x-y)(x^2 + xy + y^2)}$$

$$= \dfrac{1}{x+y} \cdot 1 \quad \text{Removing a factor of 1}$$

$$= \dfrac{1}{x+y}$$

21. $\dfrac{(x-y)^2 - z^2}{(x+y)^2 - z^2} \div \dfrac{x-y+z}{x+y-z}$

$$= \dfrac{(x-y)^2 - z^2}{(x+y)^2 - z^2} \cdot \dfrac{x+y-z}{x-y+z}$$

$$= \dfrac{(x-y+z)(x-y-z)(x+y-z)}{(x+y+z)(x+y-z)(x-y+z)}$$

$$= \dfrac{(x-y+z)(x+y-z)}{(x-y+z)(x+y-z)} \cdot \dfrac{x-y-z}{x+y+z}$$

$$= 1 \cdot \dfrac{x-y-z}{x+y+z} \quad \text{Removing a factor of 1}$$

$$= \dfrac{x-y-z}{x+y+z}$$

23. The graphs of $y_1 = \dfrac{x^2 - 2x - 35}{2x^3 - 3x^2} \cdot \dfrac{4x^3 - 9x}{7x - 49}$ and

$y_2 = \dfrac{(x+5)(2x+3)}{7x}$ appear to coincide. A table of

values also shows that the values of y_1 and y_2 are the same for the given x-values.

25. $\dfrac{3}{2a+3} + \dfrac{2a}{2a+3}$

$$= \dfrac{3+2a}{2a+3}$$

$$= 1$$

27. $\dfrac{y}{y-1} + \dfrac{2}{1-y}$

$$= \dfrac{y}{y-1} + \dfrac{-1}{-1} \cdot \dfrac{2}{1-y}$$

$$= \dfrac{y}{y-1} + \dfrac{-2}{y-1}$$

$$= \dfrac{y-2}{y-1}$$

29. $\dfrac{x}{2x-3y} - \dfrac{y}{3y-2x}$

$$= \dfrac{x}{2x-3y} - \dfrac{-1}{-1} \cdot \dfrac{y}{3y-2x}$$

$$= \dfrac{x}{2x-3y} - \dfrac{-y}{2x-3y}$$

$$= \dfrac{x+y}{2x-3y} \qquad [x - (-y) = x+y]$$

31. $\dfrac{3}{x+2} + \dfrac{2}{x^2-4}$

$$= \dfrac{3}{x+2} + \dfrac{2}{(x+2)(x-2)}, \quad \text{LCD is } (x+2)(x-2)$$

$$= \dfrac{3}{x+2} \cdot \dfrac{x-2}{x-2} + \dfrac{2}{(x+2)(x-2)}$$

$$= \dfrac{3x-6}{(x+2)(x-2)} + \dfrac{2}{(x+2)(x-2)}$$

$$= \dfrac{3x-4}{(x+2)(x-2)}$$

33. $\dfrac{y}{y^2 - y - 20} - \dfrac{2}{y+4}$

$$= \dfrac{y}{(y+4)(y-5)} - \dfrac{2}{y+4}, \quad \text{LCD is } (y+4)(y-5)$$

$$= \dfrac{y}{(y+4)(y-5)} - \dfrac{2}{y+4} \cdot \dfrac{y-5}{y-5}$$

$$= \dfrac{y}{(y+4)(y-5)} - \dfrac{2y-10}{(y+4)(y-5)}$$

$$= \dfrac{y - (2y-10)}{(y+4)(y-5)}$$

$$= \dfrac{y - 2y + 10}{(y+4)(y-5)}$$

$$= \dfrac{-y + 10}{(y+4)(y-5)}$$

35.
$$\frac{3}{x+y} + \frac{x-5y}{x^2-y^2}$$
$$= \frac{3}{x+y} + \frac{x-5y}{(x+y)(x-y)}, \text{ LCD is } (x+y)(x-y)$$
$$= \frac{3}{x+y} \cdot \frac{x-y}{x-y} + \frac{x-5y}{(x+y)(x-y)}$$
$$= \frac{3x-3y}{(x+y)(x-y)} + \frac{x-5y}{(x+y)(x-y)}$$
$$= \frac{4x-8y}{(x+y)(x-y)}$$

37.
$$\frac{9x+2}{3x^2-2x-8} + \frac{7}{3x^2+x-4}$$
$$= \frac{9x+2}{(3x+4)(x-2)} + \frac{7}{(3x+4)(x-1)},$$
$$\text{LCD is } (3x+4)(x-2)(x-1)$$
$$= \frac{9x+2}{(3x+4)(x-2)} \cdot \frac{x-1}{x-1} + \frac{7}{(3x+4)(x-1)} \cdot \frac{x-2}{x-2}$$
$$= \frac{9x^2-7x-2}{(3x+4)(x-2)(x-1)} + \frac{7x-14}{(3x+4)(x-1)(x-2)}$$
$$= \frac{9x^2-16}{(3x+4)(x-2)(x-1)}$$
$$= \frac{(3x+4)(3x-4)}{(3x+4)(x-2)(x-1)}$$
$$= \frac{3x-4}{(x-2)(x-1)}$$

39.
$$\frac{5a}{a-b} + \frac{ab}{a^2-b^2} + \frac{4b}{a+b}$$
$$= \frac{5a}{a-b} + \frac{ab}{(a+b)(a-b)} + \frac{4b}{a+b},$$
$$\text{LCD is } (a+b)(a-b)$$
$$= \frac{5a}{a-b} \cdot \frac{a+b}{a+b} + \frac{ab}{(a+b)(a-b)} + \frac{4b}{a+b} \cdot \frac{a-b}{a-b}$$
$$= \frac{5a^2+5ab}{(a+b)(a-b)} + \frac{ab}{(a+b)(a-b)} + \frac{4ab-4b^2}{(a+b)(a-b)}$$
$$= \frac{5a^2+10ab-4b^2}{(a+b)(a-b)}$$

41.
$$\frac{7}{x+2} - \frac{x+8}{4-x^2} + \frac{3x-2}{4-4x+x^2}$$
$$= \frac{7}{x+2} - \frac{x+8}{(2+x)(2-x)} + \frac{3x-2}{(2-x)^2},$$
$$\text{LCD is } (2+x)(2-x)^2$$
$$= \frac{7}{2+x} \cdot \frac{(2-x)^2}{(2-x)^2} - \frac{x+8}{(2+x)(2-x)} \cdot \frac{2-x}{2-x} +$$
$$\frac{3x-2}{(2-x)^2} \cdot \frac{2+x}{2+x}$$
$$= \frac{28-28x+7x^2-(16-6x-x^2)+3x^2+4x-4}{(2+x)(2-x)^2}$$
$$= \frac{28-28x+7x^2-16+6x+x^2+3x^2+4x-4}{(2+x)(2-x)^2}$$
$$= \frac{11x^2-18x+8}{(2+x)(2-x)^2}, \text{ or } \frac{11x^2-18x+8}{(x+2)(x-2)^2}$$

43.
$$\frac{1}{x+1} + \frac{x}{2-x} + \frac{x^2+2}{x^2-x-2}$$
$$= \frac{1}{x+1} + \frac{x}{2-x} + \frac{x^2+2}{(x+1)(x-2)}$$
$$= \frac{1}{x+1} + \frac{-1}{-1} \cdot \frac{x}{2-x} + \frac{x^2+2}{(x+1)(x-2)}$$
$$= \frac{1}{x+1} + \frac{-x}{x-2} + \frac{x^2+2}{(x+1)(x-2)},$$
$$\text{LCD is } (x+1)(x-2)$$
$$= \frac{1}{x+1} \cdot \frac{x-2}{x-2} + \frac{-x}{x-2} \cdot \frac{x+1}{x+1} + \frac{x^2+2}{(x+1)(x-2)}$$
$$= \frac{x-2}{(x+1)(x-2)} + \frac{-x^2-x}{(x+1)(x-2)} + \frac{x^2+2}{(x+1)(x-2)}$$
$$= \frac{x-2-x^2-x+x^2+2}{(x+1)(x-2)}$$
$$= \frac{0}{(x+1)(x-2)}$$
$$= 0$$

45. The graphs of $y_1 = \frac{3}{x+2} + \frac{2}{x^2-4}$ and
$y_2 = \frac{3x-4}{(x+2)(x-2)}$ appear to coincide. A table of
values also shows that the values of y_1 and y_2 are the
same for the given x-values.

47.
$$\frac{\frac{x^2-y^2}{xy}}{\frac{x-y}{y}} = \frac{x^2-y^2}{xy} \cdot \frac{y}{x-y}$$
$$= \frac{(x+y)(x-y)y}{xy(x-y)}$$
$$= \frac{x+y}{x}$$

49. $\dfrac{a - a^{-1}}{a + a^{-1}} = \dfrac{a - \dfrac{1}{a}}{a + \dfrac{1}{a}} = \dfrac{a \cdot \dfrac{a}{a} - \dfrac{1}{a}}{a \cdot \dfrac{a}{a} + \dfrac{1}{a}}$

$$= \dfrac{\dfrac{a^2 - 1}{a}}{\dfrac{a^2 + 1}{a}}$$

$$= \dfrac{a^2 - 1}{a} \cdot \dfrac{a}{a^2 + 1}$$

$$= \dfrac{a^2 - 1}{a^2 + 1}$$

51. $\dfrac{c + \dfrac{8}{c^2}}{1 + \dfrac{2}{c}} = \dfrac{c \cdot \dfrac{c^2}{c^2} + \dfrac{8}{c^2}}{1 \cdot \dfrac{c}{c} + \dfrac{2}{c}}$

$$= \dfrac{\dfrac{c^3 + 8}{c^2}}{\dfrac{c + 2}{c}}$$

$$= \dfrac{c^3 + 8}{c^2} \cdot \dfrac{c}{c + 2}$$

$$= \dfrac{(c + 2)(c^2 - 2c + 4)\cancel{c}}{\cancel{c} \cdot c(c + 2)}$$

$$= \dfrac{c^2 - 2c + 4}{c}$$

53. $\dfrac{x^2 + xy + y^2}{\dfrac{x^2}{y} - \dfrac{y^2}{x}} = \dfrac{x^2 + xy + y^2}{\dfrac{x^2}{y} \cdot \dfrac{x}{x} - \dfrac{y^2}{x} \cdot \dfrac{y}{y}}$

$$= \dfrac{x^2 + xy + y^2}{\dfrac{x^3 - y^3}{xy}}$$

$$= (x^2 + xy + y^2) \cdot \dfrac{xy}{x^3 - y^3}$$

$$= \dfrac{(x^2 + xy + y^2)(xy)}{(x - y)(x^2 + xy + y^2)}$$

$$= \dfrac{x^2 + xy + y^2}{x^2 + xy + y^2} \cdot \dfrac{xy}{x - y}$$

$$= 1 \cdot \dfrac{xy}{x - y}$$

$$= \dfrac{xy}{x - y}$$

55. $\dfrac{\dfrac{x}{y} - \dfrac{y}{x}}{\dfrac{1}{y} + \dfrac{1}{x}} = \dfrac{\dfrac{x}{y} - \dfrac{y}{x}}{\dfrac{1}{y} + \dfrac{1}{x}} \cdot \dfrac{xy}{xy}$, LCM is xy

$$= \dfrac{\left(\dfrac{x}{y} - \dfrac{y}{x}\right)(xy)}{\left(\dfrac{1}{y} + \dfrac{1}{x}\right)(xy)}$$

$$= \dfrac{x^2 - y^2}{x + y}$$

$$= \dfrac{(\cancel{x + y})(x - y)}{(\cancel{x + y}) \cdot 1}$$

$$= x - y$$

57. $\dfrac{\dfrac{1}{x - 3} + \dfrac{2}{x + 3}}{\dfrac{3}{x - 1} - \dfrac{4}{x + 2}} = \dfrac{\dfrac{1}{x - 3} \cdot \dfrac{x + 3}{x + 3} + \dfrac{2}{x + 3} \cdot \dfrac{x - 3}{x - 3}}{\dfrac{3}{x - 1} \cdot \dfrac{x + 2}{x + 2} - \dfrac{4}{x + 2} \cdot \dfrac{x - 1}{x - 1}}$

$$= \dfrac{\dfrac{x + 3 + 2(x - 3)}{(x - 3)(x + 3)}}{\dfrac{3(x + 2) - 4(x - 1)}{(x - 1)(x + 2)}}$$

$$= \dfrac{\dfrac{x + 3 + 2x - 6}{(x - 3)(x + 3)}}{\dfrac{3x + 6 - 4x + 4}{(x - 1)(x + 2)}}$$

$$= \dfrac{\dfrac{3x - 3}{(x - 3)(x + 3)}}{\dfrac{-x + 10}{(x - 1)(x + 2)}}$$

$$= \dfrac{3x - 3}{(x - 3)(x + 3)} \cdot \dfrac{(x - 1)(x + 2)}{-x + 10}$$

$$= \dfrac{(3x - 3)(x - 1)(x + 2)}{(x - 3)(x + 3)(-x + 10)}, \text{ or}$$

$$\dfrac{3(x - 1)^2(x + 2)}{(x - 3)(x + 3)(-x + 10)}$$

59. $\dfrac{\dfrac{a}{1 - a} + \dfrac{1 + a}{a}}{\dfrac{1 - a}{a} + \dfrac{a}{1 + a}} = \dfrac{\dfrac{a}{1 - a} \cdot \dfrac{a}{a} + \dfrac{1 + a}{a} \cdot \dfrac{1 - a}{1 - a}}{\dfrac{1 - a}{a} \cdot \dfrac{1 + a}{1 + a} + \dfrac{a}{1 + a} \cdot \dfrac{a}{a}}$

$$= \dfrac{\dfrac{a^2 + (1 - a^2)}{a(1 - a)}}{\dfrac{(1 - a^2) + a^2}{a(1 + a)}}$$

$$= \dfrac{1}{\cancel{a}(1 - a)} \cdot \dfrac{\cancel{a}(1 + a)}{1}$$

$$= \dfrac{1 + a}{1 - a}$$

61. $\dfrac{\dfrac{1}{a^2} + \dfrac{2}{ab} + \dfrac{1}{b^2}}{\dfrac{1}{a^2} - \dfrac{1}{b^2}} = \dfrac{\dfrac{1}{a^2} + \dfrac{2}{ab} + \dfrac{1}{b^2}}{\dfrac{1}{a^2} - \dfrac{1}{b^2}} \cdot \dfrac{a^2 b^2}{a^2 b^2},$

$$\text{LCM is } a^2 b^2$$

$$= \frac{b^2 + 2ab + a^2}{b^2 - a^2}$$

$$= \frac{(\cancel{b+a})(b+a)}{(\cancel{b+a})(b-a)}$$

$$= \frac{b+a}{b-a}$$

63. Discussion and Writing

65. $\dfrac{(x+h)^2 - x^2}{h} = \dfrac{x^2 + 2xh + h^2 - x^2}{h}$

$$= \frac{2xh + h^2}{h}$$

$$= \frac{\cancel{h}(2x+h)}{\cancel{h} \cdot 1}$$

$$= 2x + h$$

67. $\dfrac{(x+h)^3 - x^3}{h} = \dfrac{x^3 + 3x^2 h + 3xh^2 + h^3 - x^3}{h}$

$$= \frac{3x^2 h + 3xh^2 + h^3}{h}$$

$$= \frac{\cancel{h}(3x^2 + 3xh + h^2)}{\cancel{h} \cdot 1}$$

$$= 3x^2 + 3xh + h^2$$

69. $\left[\dfrac{\dfrac{x+1}{x-1} + 1}{\dfrac{x+1}{x-1} - 1} \right]^5 = \left[\dfrac{\dfrac{(x+1) + (x-1)}{x-1}}{\dfrac{(x+1) - (x-1)}{x-1}} \right]^5$

$$= \left[\frac{2x}{x-1} \cdot \frac{x-1}{2} \right]^5$$

$$= \left[\frac{\cancel{2}x(\cancel{x-1})}{1 \cdot \cancel{2}(\cancel{x-1})} \right]^5$$

$$= x^5$$

71. $\dfrac{n(n+1)(n+2)}{2 \cdot 3} + \dfrac{(n+1)(n+2)}{2}$

$$= \frac{n(n+1)(n+2)}{2 \cdot 3} + \frac{(n+1)(n+2)}{2} \cdot \frac{3}{3},$$

$$\text{LCD is } 2 \cdot 3$$

$$= \frac{n(n+1)(n+2) + 3(n+1)(n+2)}{2 \cdot 3}$$

$$= \frac{(n+1)(n+2)(n+3)}{2 \cdot 3} \quad \begin{array}{l}\text{Factoring the num-} \\ \text{erator by grouping}\end{array}$$

73. $\dfrac{x^2 - 9}{x^3 + 27} \cdot \dfrac{5x^2 - 15x + 45}{x^2 - 2x - 3} + \dfrac{x^2 + x}{4 + 2x}$

$$= \frac{(x+3)(x-3)(5)(x^2 - 3x + 9)}{(x+3)(x^2 - 3x + 9)(x-3)(x+1)} + \frac{x^2 + x}{4 + 2x}$$

$$= \frac{(x+3)(x-3)(x^2 - 3x + 9)}{(x+3)(x-3)(x^2 - 3x + 9)} \cdot \frac{5}{x+1} + \frac{x^2 + x}{4 + 2x}$$

$$= 1 \cdot \frac{5}{x+1} + \frac{x^2 + x}{4 + 2x}$$

$$= \frac{5}{x+1} + \frac{x^2 + x}{2(2+x)}$$

$$= \frac{5 \cdot 2(2+x) + (x^2 + x)(x+1)}{2(x+1)(2+x)}$$

$$= \frac{20 + 10x + x^3 + 2x^2 + x}{2(x+1)(2+x)}$$

$$= \frac{x^3 + 2x^2 + 11x + 20}{2(x+1)(2+x)}$$

Exercise Set R.6

1. $\sqrt{(-11)^2} = |-11| = 11$

3. $\sqrt{16y^2} = \sqrt{(4y)^2} = |4y| = 4|y|$

5. $\sqrt{(b+1)^2} = |b+1|$

7. $\sqrt[3]{-27x^3} = \sqrt[3]{(-3x)^3} = -3x$

9. $\sqrt{x^2 - 4x + 4} = \sqrt{(x-2)^2} = |x-2|$

11. $\sqrt[5]{32} = \sqrt[5]{2^5} = 2$

13. $\sqrt{180} = \sqrt{36 \cdot 5} = \sqrt{36} \cdot \sqrt{5} = 6\sqrt{5}$

15. $\sqrt[3]{54} = \sqrt[3]{27 \cdot 2} = \sqrt[3]{27} \cdot \sqrt[3]{2} = 3\sqrt[3]{2}$

17. $\sqrt{128c^2 d^4} = \sqrt{64c^2 d^4 \cdot 2} = |8cd^2|\sqrt{2} = 8\sqrt{2}\,|c|d^2$

19. The values of $y_1 = \sqrt{x^2 - 4x + 4}$ and $y_2 = |x-2|$ appear to be the same for any value of x.

21. $\sqrt{2x^3 y}\sqrt{12xy} = \sqrt{24x^4 y^2} = \sqrt{4x^4 y^2 \cdot 6} = 2x^2 y\sqrt{6}$

23. $\sqrt[3]{3x^2 y}\sqrt[3]{36x} = \sqrt[3]{108x^3 y} = \sqrt[3]{27x^3 \cdot 4y} = 3x\sqrt[3]{4y}$

25. $\sqrt[3]{2(x+4)}\sqrt[3]{4(x+4)^4} = \sqrt[3]{8(x+4)^5}$

$$= \sqrt[3]{8(x+4)^3 \cdot (x+4)^2}$$

$$= 2(x+4)\sqrt[3]{(x+4)^2}$$

27. $\sqrt[6]{\dfrac{m^{12} n^{24}}{64}} = \sqrt[6]{\left(\dfrac{m^2 n^4}{2}\right)^6} = \dfrac{m^2 n^4}{2}$

29. $\dfrac{\sqrt[3]{40m}}{\sqrt[3]{5m}} = \sqrt[3]{\dfrac{40m}{5m}} = \sqrt[3]{8} = 2$

31. $\dfrac{\sqrt[3]{3x^2}}{\sqrt[3]{24x^5}} = \sqrt[3]{\dfrac{3x^2}{24x^5}} = \sqrt[3]{\dfrac{1}{8x^3}} = \dfrac{1}{2x}$

33. $\sqrt[3]{\dfrac{64a^4}{27b^3}} = \sqrt[3]{\dfrac{64 \cdot a^3 \cdot a}{27 \cdot b^3}}$

$\qquad = \dfrac{\sqrt[3]{64a^3}\,\sqrt[3]{a}}{\sqrt[3]{27b^3}}$

$\qquad = \dfrac{4a\sqrt[3]{a}}{3b}$

35. $\sqrt{\dfrac{7x^3}{36y^6}} = \sqrt{\dfrac{7 \cdot x^2 \cdot x}{36 \cdot y^6}}$

$\qquad = \dfrac{\sqrt{x^2}\sqrt{7x}}{\sqrt{36y^6}}$

$\qquad = \dfrac{x\sqrt{7x}}{6y^3}$

37. $9\sqrt{50} + 6\sqrt{2} = 9\sqrt{25 \cdot 2} + 6\sqrt{2}$

$\qquad = 9 \cdot 5\sqrt{2} + 6\sqrt{2}$

$\qquad = 45\sqrt{2} + 6\sqrt{2}$

$\qquad = (45 + 6)\sqrt{2}$

$\qquad = 51\sqrt{2}$

39. $\quad 8\sqrt{2x^2} - 6\sqrt{20x} - 5\sqrt{8x^2}$

$= 8x\sqrt{2} - 6\sqrt{4 \cdot 5x} - 5\sqrt{4x^2 \cdot 2}$

$= 8x\sqrt{2} - 6 \cdot 2\sqrt{5x} - 5 \cdot 2x\sqrt{2}$

$= 8x\sqrt{2} - 12\sqrt{5x} - 10x\sqrt{2}$

$= -2x\sqrt{2} - 12\sqrt{5x}$

41. $\quad \left(\sqrt{3} - \sqrt{2}\right)\left(\sqrt{3} + \sqrt{2}\right)$

$= \left(\sqrt{3}\right)^2 - \left(\sqrt{2}\right)^2$

$= 3 - 2$

$= 1$

43. $(1 + \sqrt{3})^2 = 1^2 + 2 \cdot 1 \cdot \sqrt{3} + (\sqrt{3})^2$

$\qquad = 1 + 2\sqrt{3} + 3$

$\qquad = 4 + 2\sqrt{3}$

45. The graphs of $y_1 = 8\sqrt{2x^2} - 6\sqrt{20x} - 5\sqrt{8x^2}$ and $y_2 = -2x\sqrt{2} - 12\sqrt{5x}$ appear to coincide. A table of values also shows that the values of y_1 and y_2 are the same for the given x-values.

47. We use the Pythagorean theorem to find b, the airplane's horizontal distance from the airport. We have $a = 3700$ and $c = 14,200$.

$$c^2 = a^2 + b^2$$
$$14,200^2 = 3700^2 + b^2$$
$$201,640,000 = 13,690,000 + b^2$$
$$187,950,000 = b^2$$
$$13,709.5 \approx b$$

The airplane is about $13,709.5$ ft horizontally from the airport.

49. a) $\quad h^2 + \left(\dfrac{a}{2}\right)^2 = a^2 \qquad$ Pythagorean theorem

$\qquad h^2 + \dfrac{a^2}{4} = a^2$

$\qquad h^2 = \dfrac{3a^2}{4}$

$\qquad h = \sqrt{\dfrac{3a^2}{4}}$

$\qquad h = \dfrac{a}{2}\sqrt{3}$

b) Using the result of part (a) we have

$A = \dfrac{1}{2} \cdot \text{base} \cdot \text{height}$

$A = \dfrac{1}{2}a \cdot \dfrac{a}{2}\sqrt{3} \quad \left(\dfrac{a}{2} + \dfrac{a}{2} = a\right)$

$A = \dfrac{a^2}{4}\sqrt{3}$

51.

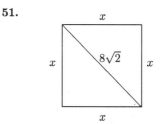

$x^2 + x^2 = (8\sqrt{2})^2 \quad$ Pythagorean theorem

$\qquad 2x^2 = 128$

$\qquad x^2 = 64$

$\qquad x = 8$

53. $\sqrt{\dfrac{2}{3}} = \sqrt{\dfrac{2}{3} \cdot \dfrac{3}{3}} = \sqrt{\dfrac{6}{9}} = \dfrac{\sqrt{6}}{\sqrt{9}} = \dfrac{\sqrt{6}}{3}$

55. $\dfrac{\sqrt[3]{5}}{\sqrt[3]{4}} = \dfrac{\sqrt[3]{5}}{\sqrt[3]{4}} \cdot \dfrac{\sqrt[3]{2}}{\sqrt[3]{2}} = \dfrac{\sqrt[3]{10}}{\sqrt[3]{8}} = \dfrac{\sqrt[3]{10}}{2}$

57. $\sqrt[3]{\dfrac{16}{9}} = \sqrt[3]{\dfrac{16}{9} \cdot \dfrac{3}{3}} = \sqrt[3]{\dfrac{48}{27}} = \dfrac{\sqrt[3]{48}}{\sqrt[3]{27}} =$

$\dfrac{\sqrt[3]{8 \cdot 6}}{3} = \dfrac{2\sqrt[3]{6}}{3}$

59.
$$\frac{6}{3+\sqrt{5}} = \frac{6}{3+\sqrt{5}} \cdot \frac{3-\sqrt{5}}{3-\sqrt{5}}$$
$$= \frac{6(3-\sqrt{5})}{9-5}$$
$$= \frac{6(3-\sqrt{5})}{4}$$
$$= \frac{3(3-\sqrt{5})}{2} = \frac{9-3\sqrt{5}}{2}$$

61.
$$\frac{6}{\sqrt{m}-\sqrt{n}} = \frac{6}{\sqrt{m}-\sqrt{n}} \cdot \frac{\sqrt{m}+\sqrt{n}}{\sqrt{m}+\sqrt{n}}$$
$$= \frac{6(\sqrt{m}+\sqrt{n})}{(\sqrt{m})^2 - (\sqrt{n})^2}$$
$$= \frac{6\sqrt{m}+6\sqrt{n}}{m-n}$$

63. $\dfrac{\sqrt{12}}{5} = \dfrac{\sqrt{12}}{5} \cdot \dfrac{\sqrt{3}}{\sqrt{3}} = \dfrac{\sqrt{36}}{5\sqrt{3}} = \dfrac{6}{5\sqrt{3}}$

65. $\sqrt[3]{\dfrac{7}{2}} = \sqrt[3]{\dfrac{7}{2} \cdot \dfrac{49}{49}} = \sqrt[3]{\dfrac{343}{98}} = \dfrac{\sqrt[3]{343}}{\sqrt[3]{98}} = \dfrac{7}{\sqrt[3]{98}}$

67. $\dfrac{\sqrt{11}}{\sqrt{3}} = \dfrac{\sqrt{11}}{\sqrt{3}} \cdot \dfrac{\sqrt{11}}{\sqrt{11}} = \dfrac{\sqrt{121}}{\sqrt{33}} = \dfrac{11}{\sqrt{33}}$

69.
$$\frac{9-\sqrt{5}}{3-\sqrt{3}} = \frac{9-\sqrt{5}}{3-\sqrt{3}} \cdot \frac{9+\sqrt{5}}{9+\sqrt{5}}$$
$$= \frac{9^2 - (\sqrt{5})^2}{27+3\sqrt{5}-9\sqrt{3}-\sqrt{15}}$$
$$= \frac{81-5}{27+3\sqrt{5}-9\sqrt{3}-\sqrt{15}}$$
$$= \frac{76}{27+3\sqrt{5}-9\sqrt{3}-\sqrt{15}}$$

71.
$$\frac{\sqrt{a}+\sqrt{b}}{3a} = \frac{\sqrt{a}+\sqrt{b}}{3a} \cdot \frac{\sqrt{a}-\sqrt{b}}{\sqrt{a}-\sqrt{b}}$$
$$= \frac{(\sqrt{a})^2 - (\sqrt{b})^2}{3a(\sqrt{a}-\sqrt{b})}$$
$$= \frac{a-b}{3a\sqrt{a}-3a\sqrt{b}}$$

73. $x^{3/4} = \sqrt[4]{x^3}$

75. $16^{3/4} = (16^{1/4})^3 = (\sqrt[4]{16})^3 = 2^3 = 8$

77. $125^{-1/3} = \dfrac{1}{125^{1/3}} = \dfrac{1}{\sqrt[3]{125}} = \dfrac{1}{5}$

79. $a^{5/4}b^{-3/4} = \dfrac{a^{5/4}}{b^{3/4}} = \dfrac{\sqrt[4]{a^5}}{\sqrt[4]{b^3}} = \dfrac{a\sqrt[4]{a}}{\sqrt[4]{b^3}}$, or $a\sqrt[4]{\dfrac{a}{b^3}}$

81. $\left(\sqrt[4]{13}\right)^5 = \sqrt[4]{13^5} = 13^{5/4}$, or $\sqrt[4]{13^5} = 13\sqrt[4]{13}$

83. $\sqrt[3]{20^2} = 20^{2/3}$

85. $\sqrt[3]{\sqrt{11}} = \left(\sqrt{11}\right)^{1/3} = (11^{1/2})^{1/3} = 11^{1/6}$

87. $\sqrt{5}\sqrt[3]{5} = 5^{1/2} \cdot 5^{1/3} = 5^{1/2+1/3} = 5^{5/6}$

89. $\sqrt[5]{32^2} = 32^{2/5} = (32^{1/5})^2 = 2^2 = 4$

91. $(2a^{3/2})(4a^{1/2}) = 8a^{3/2+1/2} = 8a^2$

93. $\left(\dfrac{x^6}{9b^{-4}}\right)^{1/2} = \left(\dfrac{x^6}{3^2 b^{-4}}\right)^{1/2} = \dfrac{x^3}{3b^{-2}}$, or $\dfrac{x^3 b^2}{3}$

95. $\dfrac{x^{2/3}y^{5/6}}{x^{-1/3}y^{1/2}} = x^{2/3-(-1/3)}y^{5/6-1/2} = xy^{1/3} = x\sqrt[3]{y}$

97. The graphs of $y_1 = (2x^{3/2})(4x^{1/2})$ and $y_2 = 8x^2$ appear to coincide. A table of values also shows that the values of y_1 and y_2 are the same for the given x-values.

99.
$$\sqrt[3]{6}\sqrt{2} = 6^{1/3}2^{1/2} = 6^{2/6}2^{3/6}$$
$$= (6^2 2^3)^{1/6}$$
$$= \sqrt[6]{36 \cdot 8}$$
$$= \sqrt[6]{288}$$

101.
$$\sqrt[4]{xy}\sqrt[3]{x^2y} = (xy)^{1/4}(x^2y)^{1/3} = (xy)^{3/12}(x^2y)^{4/12}$$
$$= \left[(xy)^3(x^2y)^4\right]^{1/12}$$
$$= \left[x^3y^3x^8y^4\right]^{1/12}$$
$$= \sqrt[12]{x^{11}y^7}$$

103.
$$\sqrt[3]{a^4\sqrt{a^3}} = \left(a^4\sqrt{a^3}\right)^{1/3} = (a^4a^{3/2})^{1/3}$$
$$= (a^{11/2})^{1/3}$$
$$= a^{11/6}$$
$$= \sqrt[6]{a^{11}}$$
$$= a\sqrt[6]{a^5}$$

105.
$$\frac{\sqrt{(a+x)^3}\sqrt[3]{(a+x)^2}}{\sqrt[4]{a+x}} = \frac{(a+x)^{3/2}(a+x)^{2/3}}{(a+x)^{1/4}}$$
$$= \frac{(a+x)^{26/12}}{(a+x)^{3/12}}$$
$$= (a+x)^{23/12}$$
$$= \sqrt[12]{(a+x)^{23}}$$
$$= (a+x)\sqrt[12]{(a+x)^{11}}$$

107. Discussion and Writing

109. $\sqrt{1+x^2} + \dfrac{1}{\sqrt{1+x^2}}$

$= \sqrt{1+x^2} \cdot \dfrac{1+x^2}{1+x^2} + \dfrac{1}{\sqrt{1+x^2}} \cdot \dfrac{\sqrt{1+x^2}}{\sqrt{1+x^2}}$

$= \dfrac{(1+x^2)\sqrt{1+x^2}}{1+x^2} + \dfrac{\sqrt{1+x^2}}{1+x^2}$

$= \dfrac{(2+x^2)\sqrt{1+x^2}}{1+x^2}$

111. $\left(\sqrt{a^{\sqrt{a}}}\right)^{\sqrt{a}} = \left(a^{\sqrt{a}/2}\right)^{\sqrt{a}} = a^{a/2}$

113. Graph $y_1 = x^{1/2}$ and $y_2 = x^{1/3}$ in a window that shows the relative positions of the graphs. One suitable choice is $[-1, 4, -1, 3]$.

a) Find the first coordinates of the points of intersection of the graphs. They are 0 and 1, so $x^{1/2} = x^{1/3}$ for $x = 0$ or $x = 1$.

b) Find the values of x for which the graph of y_1 lies above the graph of y_2. They are $\{x | x > 1\}$.

c) Find the values of x for which the graph of y_1 lies below the graph of y_2. They are $\{x | 0 < x < 1\}$.

Exercise Set R.7

---

1. $4x + 5 = 21$

$\quad\quad 4x = 16$ Subtracting 5 on both sides

$\quad\quad\ \ x = 4$ Dividing by 4 on both sides

We can also graph $y_1 = 4x + 5$ and $y_2 = 21$ and find the first coordinate of the point of intersection.

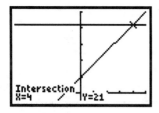

The solution is 4.

3. $y + 1 = 2y - 7$

$\quad\quad 1 = y - 7$ Subtracting y on both sides

$\quad\quad 8 = y$ Adding 7 on both sides

We can also graph $y_1 = x + 1$ and $y_2 = 2x - 7$ and find the first coordinate of the point of intersection.

⊞

The solution is 8.

5. $5x - 2 + 3x = 2x + 6 - 4x$

$\quad\quad 8x - 2 = 6 - 2x$ Collecting like terms

$\quad\quad 8x + 2x = 6 + 2$ Adding $2x$ and 2 on both sides

$\quad\quad\quad 10x = 8$ Collecting like terms

$\quad\quad\quad\ \ x = \dfrac{8}{10}$ Dividing by 10 on both sides

$\quad\quad\quad\ \ x = \dfrac{4}{5}$, or 0.8 Simplifying

We can also graph $y_1 = 5x - 2 + 3x$ and $y_2 = 2x + 6 - 4x$ and find the first coordinate of the point of intersection.

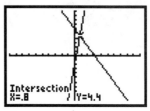

The solution is $\dfrac{4}{5}$, or 0.8.

7. $7(3x + 6) = 11 - (x + 2)$

$\quad 21x + 42 = 11 - x - 2$ Using the distributive property

$\quad 21x + 42 = 9 - x$ Collecting like terms

$\quad 21x + x = 9 - 42$ Adding x and subtracting 42 on both sides

$\quad\quad\ \ 22x = -33$ Collecting like terms

$\quad\quad\quad\ x = -\dfrac{33}{22}$ Dividing by 22 on both sides

$\quad\quad\quad\ x = -\dfrac{3}{2}$, or -1.5 Simplifying

We can also graph $y_1 = 7(3x + 6)$ and $y_2 = 11 - (x + 2)$ and find the first coordinate of the point of intersection.

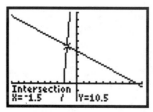

The solution is $-\dfrac{3}{2}$, or -1.5.

9. $(2x - 3)(3x - 2) = 0$

$\quad 2x - 3 = 0 \ \ or \ \ 3x - 2 = 0$ Using the principle of zero products

$\quad\quad 2x = 3 \ \ or \quad\ \ 3x = 2$

$\quad\quad\ \ x = \dfrac{3}{2} \ \ or \quad\ \ x = \dfrac{2}{3}$

We can also graph $y_1 = (2x - 3)(3x - 2)$ and $y_2 = 0$ and find the first coordinates of the points of intersection.

The solutions are $\frac{3}{2}$ and $\frac{2}{3}$.

11. $3x^2 + x - 2 = 0$

$(3x - 2)(x + 1) = 0$ Factoring

$3x - 2 = 0 \;\; or \;\; x + 1 = 0$ Using the principle
                                                       of zero products

$x = \frac{2}{3} \;\; or \;\;\;\;\;\; x = -1$

We can also graph $y_1 = 3x^2 + x - 2$ and $y_2 = 0$ and find the first coordinates of the points of intersection.

The solutions are $\frac{2}{3}$ and -1.

13. $4x^2 - 12 = 0$

$4x^2 = 12$

$x^2 = 3$

$x = \sqrt{3} \; or \; x = -\sqrt{3}$ Using the principle
                                                    of square roots

We can also graph $y_1 = 4x^2 - 12$ and $y_2 = 0$ and find the first coordinates of the points of intersection.

The solutions are $\sqrt{3}$ and $-\sqrt{3}$, or approximately 1.732 and -1.732.

15. $2x^2 = 6x$

$2x^2 - 6x = 0$ Subtracting $6x$ on both sides

$2x(x - 3) = 0$

$2x = 0 \;\; or \;\; x - 3 = 0$

$x = 0 \;\; or \;\;\;\;\;\; x = 3$

We can also graph $y_1 = 2x^2$ and $y_2 = 6x$ and find the first coordinates of the points of intersection.

The solutions are 0 and 3.

17. $3y^3 - 5y^2 - 2y = 0$

$y(3y^2 - 5y - 2) = 0$

$y(3y + 1)(y - 2) = 0$

$y = 0 \;\; or \;\; 3y + 1 = 0 \;\;\; or \;\; y - 2 = 0$

$y = 0 \;\; or \;\;\;\;\;\; y = -\frac{1}{3} \;\; or \;\;\;\;\;\; y = 2$

We can also graph $y_1 = 3x^3 - 5x^2 - 2x$ and $y_2 = 0$ and find the first coordinates of the points of intersection.

The solutions are $-\frac{1}{3}$, 0 and 2.

19. $7x^3 + x^2 - 7x - 1 = 0$

$x^2(7x + 1) - (7x + 1) = 0$

$(x^2 - 1)(7x + 1) = 0$

$(x + 1)(x - 1)(7x + 1) = 0$

$x + 1 = 0 \;\;\; or \;\; x - 1 = 0 \;\; or \;\; 7x + 1 = 0$

$x = -1 \;\; or \;\;\;\;\;\; x = 1 \;\; or \;\;\;\;\;\; x = -\frac{1}{7}$

We can also graph $y_1 = 7x^3 + x^2 - 7x - 1$ and $y_2 = 0$ and find the first coordinates of the points of intersection.

The solutions are -1, $-\frac{1}{7}$, and 1.

21. $A = \frac{1}{2}bh$

$2A = bh$ Multiplying by 2 on both sides

$\frac{2A}{h} = b$ Dividing by h on both sides

23. $P = 2l + 2w$

$P - 2l = 2w$ Subtracting $2l$ on both sides

$\frac{P - 2l}{2} = w$ Dividing by 2 on both sides

25. $A = \frac{1}{2}h(b_1 + b_2)$

$2A = h(b_1 + b_2)$ Multiplying by 2 on
                                          both sides

$\frac{2A}{b_1 + b_2} = h$ Dividing by $b_1 + b_2$ on both sides

27. $V = \frac{4}{3}\pi r^3$

$3V = 4\pi r^3$ Multiplying by 3 on both sides

$\frac{3V}{4r^3} = \pi$ Dividing by $4r^3$ on both sides

29. $F = \frac{9}{5}C + 32$

$F - 32 = \frac{9}{5}C$ Subtracting 32 on both sides

$\frac{5}{9}(F - 32) = C$ Multiplying by $\frac{5}{9}$ on both sides

31. Discussion and Writing

33. $2x - \{x - [3x - (6x + 5)]\} = 4x - 1$

$2x - \{x - [3x - 6x - 5]\} = 4x - 1$

$2x - \{x - [-3x - 5]\} = 4x - 1$

$2x - \{x + 3x + 5\} = 4x - 1$

$2x - \{4x + 5\} = 4x - 1$

$2x - 4x - 5 = 4x - 1$

$-2x - 5 = 4x - 1$

$-6x - 5 = -1$

$-6x = 4$

$x = -\frac{2}{3}$

We can also graph $y_1 = 2x - (x - (3x - (6x + 5)))$ and $y_2 = 4x - 1$ and find the first coordinate of the point of intersection.

The solution is $-\dfrac{2}{3}$.

35.
$$(x - 2)^3 = x^3 - 2$$
$$x^3 - 6x^2 + 12x - 8 = x^3 - 2$$
$$0 = 6x^2 - 12x + 6$$
$$0 = 6(x^2 - 2x + 1)$$
$$0 = 6(x - 1)(x - 1)$$
$$x - 1 = 0 \ \ or \ \ x - 1 = 0$$
$$x = 1 \ \ or \ \ \ \ \ x = 1$$

We can also graph $y_1 = (x - 2)^3$ and $y_2 = x^3 - 2$ and find the first coordinate of the point of intersection.

The solution is 1.

37.
$$(6x^3 + 7x^2 - 3x)(x^2 - 7) = 0$$
$$x(6x^2 + 7x - 3)(x^2 - 7) = 0$$
$$x(3x - 1)(2x + 3)(x^2 - 7) = 0$$
$$x{=}0 \ or \ 3x - 1{=}0 \ \ or \ 2x + 3{=}0 \ \ \ or \ x^2 - 7 = 0$$
$$x{=}0 \ or \ \ \ \ \ \ x{=}\frac{1}{3} \ or \ \ \ \ \ \ x{=}{-}\frac{3}{2} \ or \ x = \sqrt{7} \ or$$
$$x = -\sqrt{7}$$

We can also graph $y_1 = (6x^3 + 7x^2 - 3x)(x^2 - 7)$ and $y_2 = 0$ and find the first coordinates of the points of intersection.

The exact solutions are $-\sqrt{7}$, $-\dfrac{3}{2}$, 0, $\dfrac{1}{3}$, and $\sqrt{7}$.

Chapter 1

Graphs, Functions, and Models

Exercise Set 1.1

1. This correspondence is a function, because each member of the domain corresponds to exactly one member of the range.

3. This correspondence is a function, because each member of the domain corresponds to exactly one member of the range.

5. This correspondence is not a function, because there is a member of the domain (m) that corresponds to more than one member of the range (A and B).

7. This correspondence is a function, because each member of the domain corresponds to exactly one member of the range.

9. This correspondence is a function, because each car has exactly one license number.

11. This correspondence is a function, because each member of the family has exactly one eye color.

13. This correspondence is not a function, because at least one student will have more than one neighboring seat occupied by another student.

15. The relation is a function, because no two ordered pairs have the same first coordinate and different second coordinates.

The domain is the set of all first coordinates: $\{2, 3, 4\}$.

The range is the set of all second coordinates: $\{10, 15, 20\}$.

17. The relation is not a function, because the ordered pairs $(-2, 1)$ and $(-2, 4)$ have the same first coordinate and different second coordinates.

The domain is the set of all first coordinates: $\{-7, -2, 0\}$.

The range is the set of all second coordinates: $\{3, 1, 4, 7\}$.

19. The relation is a function, because no two ordered pairs have the same first coordinate and different second coordinates.

The domain is the set of all first coordinates: $\{-2, 0, 2, 4, -3\}$.

The range is the set of all second coordinates: $\{1\}$.

21. The point $(-1, 2)$ is on the graph, so $f(-1) = 2$; the point $(0, 0)$ is on the graph, so $f(0) = 0$; the point $(1, -2)$ is on the graph, so $f(1) = -2$.

23. $g(x) = 3x^2 - 2x + 1$

a) $g(0) = 3 \cdot 0^2 - 2 \cdot 0 + 1 = 1$

b) $g(-1) = 3(-1)^2 - 2(-1) + 1 = 6$

c) $g(3) = 3 \cdot 3^2 - 2 \cdot 3 + 1 = 22$

d) $g(-x) = 3(-x)^2 - 2(-x) + 1 = 3x^2 + 2x + 1$

e) $g(1 - t) = 3(1 - t)^2 - 2(1 - t) + 1 =$
$3(1 - 2t + t^2) - 2(1 - t) + 1 = 3 - 6t + 3t^2 - 2 + 2t + 1 =$
$3t^2 - 4t + 2$

25. $g(x) = x^3$

a) $g(2) = 2^3 = 8$

b) $g(-2) = (-2)^3 = -8$

c) $g(-x) = (-x)^3 = -x^3$

d) $g(3y) = (3y)^3 = 27y^3$

e) $g(2 + h) = (2 + h)^3 = 8 + 12h + 6h^2 + h^3$

27. $g(x) = \dfrac{x - 4}{x + 3}$

a) $g(5) = \dfrac{5 - 4}{5 + 3} = \dfrac{1}{8}$

b) $g(4) = \dfrac{4 - 4}{4 + 7} = 0$

c) $g(-3) = \dfrac{-3 - 4}{-3 + 3} = \dfrac{-7}{0}$

Since division by 0 is not defined, $g(-3)$ does not exist.

d) $g(-16.25) = \dfrac{-16.25 - 4}{-16.25 + 3} = \dfrac{-20.25}{-13.25} = \dfrac{81}{53}$

e) $g(x + h) = \dfrac{x + h - 4}{x + h + 3}$

29. $g(x) = \dfrac{x}{\sqrt{1 - x^2}}$

$g(0) = \dfrac{0}{\sqrt{1 - 0^2}} = \dfrac{0}{\sqrt{1}} = \dfrac{0}{1} = 0$

$g(-1) = \dfrac{-1}{\sqrt{1 - (-1)^2}} = \dfrac{-1}{\sqrt{1 - 1}} = \dfrac{-1}{\sqrt{0}} = \dfrac{-1}{0}$

Since division by 0 is not defined, $g(-1)$ does not exist.

$g(5) = \dfrac{5}{\sqrt{1 - 5^2}} = \dfrac{5}{\sqrt{1 - 25}} = \dfrac{5}{\sqrt{-24}}$

Since $\sqrt{-24}$ is not defined as a real number, $g(5)$ does not exist as a real number.

$$g\left(\frac{1}{2}\right) = \frac{\frac{1}{2}}{\sqrt{1 - \left(\frac{1}{2}\right)^2}} = \frac{\frac{1}{2}}{\sqrt{1 - \frac{1}{4}}} = \frac{\frac{1}{2}}{\sqrt{\frac{3}{4}}} =$$

$$\frac{\frac{1}{2}}{\frac{\sqrt{3}}{2}} = \frac{1}{2} \cdot \frac{2}{\sqrt{3}} = \frac{1 \cdot 2}{2\sqrt{3}} = \frac{1}{\sqrt{3}}, \text{ or } \frac{\sqrt{3}}{3}$$

31. We can substitute any real number for x. Thus, the domain is the set of all real numbers, or $(-\infty, \infty)$.

33. The input 0 results in a denominator of 0. Thus, the domain is $\{x | x \neq 0\}$, or $(-\infty, 0) \cup (0, \infty)$.

35. We can substitute any real number in the numerator, but we must avoid inputs that make the denominator 0. We find these inputs.

$$2 - x = 0$$
$$2 = x$$

The domain is $\{x | x \neq 2\}$, or $(-\infty, 2) \cup (2, \infty)$.

37. We find the inputs that make the denominator 0:

$$x^2 - 4x - 5 = 0$$
$$(x - 5)(x + 1) = 0$$
$$x - 5 = 0 \ \ or \ \ x + 1 = 0$$
$$x = 5 \ \ or \ \ \ \ \ \ x = -1$$

The domain is $\{x | x \neq 5 \ and \ x \neq -1\}$, or $(-\infty, -1) \cup (-1, 5) \cup (5, \infty)$.

39.

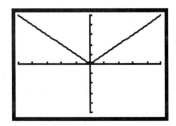

Domain: all real numbers, or $(-\infty, \infty)$

Range: $[0, \infty)$

41.

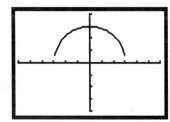

Domain: $[-3, 3]$

Range: $[0, 3]$

43.

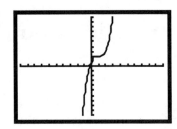

Domain: all real numbers, or $(-\infty, \infty)$

Range: all real numbers, or $(-\infty, \infty)$

45.

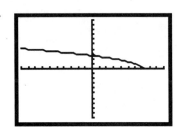

Domain: $(-\infty, 7]$

Range: $[0, \infty)$

47.

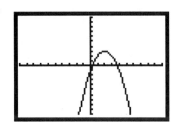

Domain: all real numbers, or $(-\infty, \infty)$

Range: $(-\infty, 3]$

49. This is not the graph of a function, because we can find a vertical line that crosses the graph more than once.

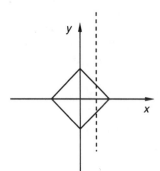

51. This is the graph of a function, because there is no vertical line that crosses the graph more than once.

53. This is the graph of a function, because there is no vertical line that crosses the graph more than once.

55. This is not the graph of a function, because we can find a vertical line that crosses the graph more than once.

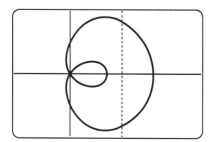

57. This is the graph of a function, because there is no vertical line that crosses the graph more than once.

The inputs on the x-axis that correspond to points on the graph extend from 0 to 5, inclusive. Thus, the domain is $\{x|0 \le x \le 5\}$, or $[0, 5]$.

The outputs on the y-axis extend from 0 to 3, inclusive. Thus, the range is $\{y|0 \le y \le 3\}$, or $[0, 3]$.

59. This is the graph of a function, because there is no vertical line that crosses the graph more than once.

The inputs on the x-axis that correspond to points on the graph extend from -2π to 2π inclusive. Thus, the domain is $\{x| - 2\pi \le x \le 2\pi\}$, or $[-2\pi, 2\pi]$.

The outputs on the y-axis extend from -1 to 1, inclusive. Thus, the range is $\{y| - 1 \le y \le 1\}$, or $[-1, 1]$.

61. $E(t) = 1000(100 - t) + 580(100 - t)^2$

a) $E(99.5) = 1000(100-99.5)+580(100-99.5)^2$
$= 1000(0.5) + 580(0.5)^2$
$= 500 + 580(0.25) = 500 + 145$
$= 645$ m above sea level

b) $E(100) = 1000(100 - 100) + 580(100 - 100)^2$
$= 1000 \cdot 0 + 580(0)^2 = 0 + 0$
$= 0$ m above sea level, or at sea level

63. $T(0.5) = 0.5^{1.31} \approx 0.4$ acres
$T(10) = 10^{1.31} \approx 20.4$ acres
$T(20) = 20^{1.31} \approx 50.6$ acres
$T(100) = 100^{1.31} \approx 416.9$ acres
$T(200) = 200^{1.31} \approx 1033.6$ acres

65. Discussion and Writing

67. Answers may vary. Two possibilities are $f(x) = x$, $g(x) = x + 1$ and $f(x) = x^2$ and $g(x) = x^2 - 4$.

69. $f(x - 1) = 5x$
$f(6) = f(7 - 1) = 5 \cdot 7 = 35$

Exercise Set 1.2

1. a) Yes. Each input is 1 more than the one that precedes it.

b) Yes. Each output is 3 more than the one that precedes it.

c) Yes. Constant changes in inputs result in constant changes in outputs.

3. a) Yes. Each input is 15 more than the one that precedes it.

b) No. The change in the outputs varies.

c) No. Constant changes in inputs do not result in constant changes in outputs.

5. Two points on the line are $(0, 3)$ and $(5, 0)$.
$$m = \frac{y_2 - y_1}{x_2 - x_1} = \frac{0 - 3}{5 - 0} = \frac{-3}{5}, \text{ or } -\frac{3}{5}$$

7. $m = \dfrac{y_2 - y_1}{x_2 - x_1} = \dfrac{3 - 3}{3 - 0} = \dfrac{0}{3} = 0$

9. $m = \dfrac{y_2 - y_1}{x_2 - x_1} = \dfrac{2 - 4}{-1 - 9} = \dfrac{-2}{-10} = \dfrac{1}{5}$

11. $m = \dfrac{y_2 - y_1}{x_2 - x_1} = \dfrac{6 - (-9)}{-5 - 4} = \dfrac{15}{-9} = -\dfrac{5}{3}$

13. $m = \dfrac{y_2 - y_1}{x_2 - x_1} = \dfrac{2 - (-3)}{\pi - \pi} = \dfrac{5}{0}$

Since division by 0 is not defined, the slope is not defined.

15. $m = \dfrac{y_2 - y_1}{x_2 - x_1} = \dfrac{(a+h)^2 - a^2}{a + h - a} = \dfrac{a^2 + 2ah + h^2 - a^2}{h} =$
$\dfrac{2ah + h^2}{h} = \dfrac{h(2a + h)}{h} = 2a + h$

17. $m = \dfrac{920.58}{13,740} = 0.067$

The road grade is 6.7%.

We find an equation of the line with slope 0.067 and containing the point $(13,740, 920.58)$:
$$y - 920.58 = 0.067(x - 13,740)$$
$$y - 920.58 = 0.067x - 920.58$$
$$y = 0.067x$$

19. We can use any two points on the graph to find the rate of change. We will use $(1991, 800)$ and $(1995, 1200)$.

$$\text{Rate of change} = \frac{\text{change in } y}{\text{change in } x}$$
$$= \frac{1200 - 800}{1995 - 1991}$$
$$= \frac{400}{4}$$
$$= 100$$

The rate of change is $100 per year.

21. First express $1\frac{1}{2}$ hr as 90 min. Then we have the points $(50, 10)$ and $(50 + 90, 25)$, or $(50, 10)$ and $(140, 25)$.

$$\text{Speed} = \text{average rate of change}$$
$$= \frac{25 - 10}{140 - 50}$$
$$= \frac{15}{90}$$
$$= \frac{1}{6}$$

The speed is $\frac{1}{6}$ km per minute.

23. $y = \frac{3}{5}x - 7$

The equation is in the form $y = mx + b$ where $m = \frac{3}{5}$ and $b = -7$. Thus, the slope is $\frac{3}{5}$, and the y-intercept is $(0, -7)$.

25. $f(x) = 5 - \frac{1}{2}x$, or $f(x) = -\frac{1}{2}x + 5$

The second equation is in the form $y = mx + b$ where $m = -\frac{1}{2}$ and $b = 5$. Thus, the slope is $-\frac{1}{2}$ and the y-intercept is $(0, 5)$.

27. Solve the equation for y.

$$3x + 2y = 10$$
$$2y = -3x + 10$$
$$y = -\frac{3}{2}x + 5$$

Slope: $-\frac{3}{2}$; y-intercept: $(0, 5)$

29. Solve the equation for $f(x)$.

$$4x - 3f(x) - 15 = 6$$
$$-3f(x) = -4x + 21$$
$$f(x) = \frac{4}{3}x - 7$$

Slope: $\frac{4}{3}$; y-intercept: $(0, -7)$

31. Substitute $\frac{2}{9}$ for m, 0 for x_1, and 4 for y_1 in the point-slope equation.

$$y - y_1 = m(x - x_1)$$
$$y - 4 = \frac{2}{9}(x - 0)$$
$$y - 4 = \frac{2}{9}x$$
$$y = \frac{2}{9}x + 4 \qquad \text{Slope-intercept equation}$$

33. Substitute -4 for m, 0 for x_1, and -7 for y_1 in the point-slope equation.

$$y - y_1 = m(x - x_1)$$
$$y - (-7) = -4(x - 0)$$
$$y + 7 = -4x$$
$$y = -4x - 7 \qquad \text{Slope-intercept equation}$$

35.
$$y - y_1 = m(x - x_1)$$
$$y - \frac{3}{4} = -4.2(x - 0) \quad \text{Substituting}$$
$$y - \frac{3}{4} = -4.2x$$
$$y = -4.2x + \frac{3}{4} \quad \text{Slope-intercept equation}$$

37.
$$y - y_1 = m(x - x_1)$$
$$y - 7 = \frac{2}{9}(x - 3) \quad \text{Substituting}$$
$$y - 7 = \frac{2}{9}x - \frac{2}{3}$$
$$y = \frac{2}{9}x + \frac{19}{3} \quad \text{Slope-intercept equation}$$

39.
$$y - y_1 = m(x - x_1)$$
$$y - (-2) = 3(x - 1)$$
$$y + 2 = 3x - 3$$
$$y = 3x - 5 \qquad \text{Slope-intercept equation}$$

41.
$$y - y_1 = m(x - x_1)$$
$$y - (-1) = -\frac{3}{5}(x - (-4))$$
$$y + 1 = -\frac{3}{5}(x + 4)$$
$$y + 1 = -\frac{3}{5}x - \frac{12}{5}$$
$$y = -\frac{3}{5}x - \frac{17}{5} \qquad \text{Slope-intercept equation}$$

43. $m = \dfrac{-4-5}{2-(-1)} = \dfrac{-9}{3} = -3$

Using the point $(-1,5)$, we get
$$y - 5 = -3(x-(-1)), \text{ or } y - 5 = -3(x+1).$$
Using the point $(2,-4)$, we get
$$y - (-4) = -3(x-2), \text{ or } y + 4 = -3(x-2).$$
In either case, the slope-intercept equation is $y = -3x + 2$.

45. $m = \dfrac{4-0}{-1-7} = \dfrac{4}{-8} = -\dfrac{1}{2}$

Using the point $(7,0)$, we get
$$y - 0 = -\frac{1}{2}(x-7).$$
Using the point $(-1,4)$, we get
$$y - 4 = -\frac{1}{2}(x-(-1)), \text{ or}$$
$$y - 4 = -\frac{1}{2}(x+1).$$
In either case, the slope-intercept equation is $y = -\dfrac{1}{2}x + \dfrac{7}{2}$.

47. We solve each equation for y.
$$
\begin{array}{ll}
x + 2y = 5 & 2x + 4y = 8 \\[4pt]
y = -\dfrac{1}{2}x + \dfrac{5}{2} & y = -\dfrac{1}{2}x + 2
\end{array}
$$
We see that $m_1 = -\dfrac{1}{2}$ and $m_2 = -\dfrac{1}{2}$. Since the slopes are the same and the y-intercepts, $\dfrac{5}{2}$ and 2, are different, the lines are parallel.

49. We solve each equation for y.
$$
\begin{array}{ll}
y = 4x - 5 & 4y = 8 - x \\[4pt]
 & y = -\dfrac{1}{4}x + 2
\end{array}
$$
We see that $m_1 = 4$ and $m_2 = -\dfrac{1}{4}$. Since $m_1 m_2 = 4\left(-\dfrac{1}{4}\right) = -1$, the lines are perpendicular.

51. $y = \dfrac{2}{7}x + 1; \ m = \dfrac{2}{7}$

The line parallel to the given line will have slope $\dfrac{2}{7}$. We use the point-slope equation for a line with slope $\dfrac{2}{7}$ and containing the point $(3,5)$:
$$y - y_1 = m(x - x_1)$$
$$y - 5 = \frac{2}{7}(x - 3)$$
$$y - 5 = \frac{2}{7}x - \frac{6}{7}$$
$$y = \frac{2}{7}x + \frac{29}{7} \quad \text{Slope-intercept form}$$

The slope of the line perpendicular to the given line is the opposite of the reciprocal of $\dfrac{2}{7}$, or $-\dfrac{7}{2}$. We use the point-slope equation for a line with slope $-\dfrac{7}{2}$ and containing the point $(3,5)$:
$$y - y_1 = m(x - x_1)$$
$$y - 5 = -\frac{7}{2}(x - 3)$$
$$y - 5 = -\frac{7}{2}x + \frac{21}{2}$$
$$y = -\frac{7}{2}x + \frac{31}{2} \quad \text{Slope-intercept form}$$

53. $y = -0.3x + 4.3; \ m = -0.3$

The line parallel to the given line will have slope -0.3. We use the point-slope equation for a line with slope -0.3 and containing the point $(-7,0)$:
$$y - y_1 = m(x - x_1)$$
$$y - 0 = -0.3(x - (-7))$$
$$y = -0.3x - 2.1 \quad \text{Slope-intercept form}$$

The slope of the line perpendicular to the given line is the opposite of the reciprocal of -0.3, or $\dfrac{1}{0.3} = \dfrac{10}{3}$.
We use the point-slope equation for a line with slope $\dfrac{10}{3}$ and containing the point $(-7,0)$:
$$y - y_1 = m(x - x_1)$$
$$y - 0 = \frac{10}{3}(x - (-7))$$
$$y = \frac{10}{3}x + \frac{70}{3} \quad \text{Slope-intercept form}$$

55.
$$3x + 4y = 5$$
$$4y = -3x + 5$$
$$y = -\frac{3}{4}x + \frac{5}{4}; \ m = -\frac{3}{4}$$

The line parallel to the given line will have slope $-\dfrac{3}{4}$. We use the point-slope equation for a line with slope $-\dfrac{3}{4}$ and containing the point $(3,-2)$:
$$y - y_1 = m(x - x_1)$$
$$y - (-2) = -\frac{3}{4}(x - 3)$$
$$y + 2 = -\frac{3}{4}x + \frac{9}{4}$$
$$y = -\frac{3}{4}x + \frac{1}{4} \quad \text{Slope-intercept form}$$

The slope of the line perpendicular to the given line is the opposite of the reciprocal of $-\dfrac{3}{4}$, or $\dfrac{4}{3}$. We use the point-slope equation for a line with slope $\frac{4}{3}$ and containing the point $(3, -2)$:

$$y - y_1 = m(x - x_1)$$

$$y - (-2) = \frac{4}{3}(x - 3)$$

$$y + 2 = \frac{4}{3}x - 4$$

$$y = \frac{4}{3}x - 6 \quad \text{Slope-intercept form}$$

57. $x = -1$ is the equation of a vertical line. The line parallel to the given line is a vertical line containing the point $(3, -3)$, or $x = 3$.

The line perpendicular to the given line is a horizontal line containing the point $(3, -3)$, or $y = -3$.

59. a) $W(h) = 3.5h - 110$

b) $y = 3.5x - 110$

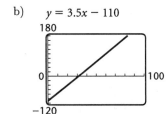

c) $W(62) = 3.5(62) - 110 = 217 - 110 = 107$ lb

d) Both the height and weight must be positive. Solving $h > 0$ and $3.5h - 110 > 0$, we find that the domain of the function is $\{h | h > 31.43\}$, or $(31.43, \infty)$.

61. $D(F) = 2F + 115$

a) $y = 2x + 115$

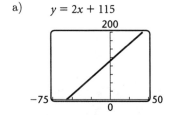

b) $D(0) = 2 \cdot 0 + 115 = 115$ ft

$D(-20) = 2(-20) + 115 = -40 + 115 = 75$ ft

$D(10) = 2 \cdot 10 + 115 = 20 + 115 = 135$ ft

$D(32) = 2 \cdot 32 + 115 = 64 + 115 = 179$ ft

c) For $F < -57.5°$, $D(F)$ is negative; above $32°$, ice doesn't form

63. a) $D(r) = \dfrac{11r + 5}{10} = \dfrac{11}{10}r + \dfrac{5}{10}$

The slope is $\dfrac{11}{10}$.

For each mph faster the car travels, it takes $\dfrac{11}{10}$ ft longer to stop.

b) $y = \dfrac{11x + 5}{10}$

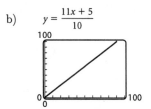

c) $D(5) = \dfrac{11 \cdot 5 + 5}{10} = \dfrac{60}{10} = 6$ ft

$D(10) = \dfrac{11 \cdot 10 + 5}{10} = \dfrac{115}{10} = 11.5$ ft

$D(20) = \dfrac{11 \cdot 20 + 5}{10} = \dfrac{225}{10} = 22.5$ ft

$D(50) = \dfrac{11 \cdot 50 + 5}{10} = \dfrac{555}{10} = 55.5$ ft

$D(65) = \dfrac{11 \cdot 65 + 5}{10} = \dfrac{720}{10} = 72$ ft

d) The speed cannot be negative. $D(0) = \dfrac{1}{2}$ which says that a stopped car travels $\dfrac{1}{2}$ ft before stopping. Thus, 0 is not in the domain. The speed can be positive, so the domain is $\{r | r > 0\}$, or $(0, \infty)$.

65. The rise represents "Height of corn in feet," and the run represents "Number of weeks after planting."

$$\text{Slope} = \frac{\text{rise}}{\text{run}} = \frac{\text{Height of corn in feet}}{\text{Number of weeks after planting}}$$

67. $C(t) = 60 + 40t$

$y = 60 + 40x$

$C(6) = 60 + 40 \cdot 6 = \$300$

69. Let x = the number of shirts produced.

$$C(x) = 800 + 3x$$

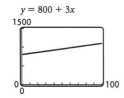

$y = 800 + 3x$

$$C(75) = 800 + 3 \cdot 75 = \$1025$$

71. Discussion and Writing

73. $f(x) = x^2 - 3x$

$$f(-5) = (-5)^2 - 3(-5) = 25 + 15 = 40$$

75. $f(x) = x^2 - 3x$

$$f(-a) = (-a)^2 - 3(-a) = a^2 + 3a$$

77. The slope of the line containing $(-3, k)$ and $(4, 8)$ is

$$\frac{8 - k}{4 - (-3)} = \frac{8 - k}{7}.$$

The slope of the line containing $(5, 3)$ and $(1, -6)$ is

$$\frac{-6 - 3}{1 - 5} = \frac{-9}{-4} = \frac{9}{4}.$$

The slopes must be equal in order for the lines to be parallel:

$$\frac{8 - k}{7} = \frac{9}{4}$$

$$32 - 4k = 63 \quad \text{Multiplying by 28}$$

$$-4k = 31$$

$$k = -\frac{31}{4}, \text{ or } -7.75$$

79. To express F as a function of C, we consider C to be the independent variable. First find the slope of the function containing the points $(0, 32)$ and $(100, 212)$.

$$\frac{32 - 212}{0 - 100} = \frac{-180}{-100} = \frac{9}{5}$$

Now find the function with slope $\frac{9}{5}$ and containing the point $(0, 32)$.

$$F - 32 = \frac{9}{5}(C - 0)$$

$$F - 32 = \frac{9}{5}C$$

$$F(C) = \frac{9}{5}C + 32$$

To express C as a function of F, we consider F to be the independent variable. First find the slope of the function containing the points $(32, 0)$ and $(212, 100)$.

$$\frac{100 - 0}{212 - 32} = \frac{100}{180} = \frac{5}{9}$$

Now find the function with slope $\frac{5}{9}$ and containing the point $(32, 0)$.

$$C - 0 = \frac{5}{9}(F - 32)$$

$$C(F) = \frac{5}{9}(F - 32)$$

81. False. For example, let $f(x) = x + 1$. Then $f(cd) = cd + 1$, but $f(c)f(d) = (c+1)(d+1) = cd + c + d + 1 \neq cd + 1$ for $c \neq -d$.

83. False. For example, let $f(x) = x + 1$. Then $f(c-d) = c - d + 1$, but $f(c) - f(d) = c + 1 - (d + 1) = c - d$.

85.

$$f(x) = mx + b$$

$$f(x + 2) = f(x) + 2$$

$$m(x + 2) + b = mx + b + 2$$

$$mx + 2m + b = mx + b + 2$$

$$2m = 2$$

$$m = 1$$

Thus, $f(x) = 1 \cdot x + b$, or $f(x) = x + b$.

Exercise Set 1.3

1. Yes. The rate of change of the occupancy rates generally seems to be constant, so the data might be modeled by a linear function.

3. No. The population figures seem to be rising more rapidly after 1993 than before (that is, the slope is not constant), so this data cannot be modeled by a linear function.

5. No. The data points fall faster from 0 to 2 than after 2 (that is, the rate of change is not constant), so they cannot be modeled by a linear function.

7. Yes. The rate of change seems to be constant, so the scatterplot might be modeled by a linear function.

9. a) Using the linear regression feature on a grapher we get $y = 1304.844444x + 21,950.01111$, where x = the number of years after 1980.

In 2005, $x = 2005 - 1980 = 25$. Find y when $x = 25$.

$y = 1304.844444(25) + 21,950.01111 \approx 54,571$

There will be approximately 54,571 shopping centers in 2005.

b) $r \approx 9890$; since r is close to 1, the regression line is a good fit.

11. a) Using the linear regression feature on a grapher, we get $y = 645.7x + 9799.2$.

In 2004-2005, $x = 7$:
$$y = 645.7(7) + 9799.2 = \$14,319.10$$

In 2006-2007, $x = 9$:
$$y = 645.7(9) + 9799.2 = \$15,610.50$$

In 2010-2011, $x = 13$:
$$y = 645.7(13) + 9799.2 = \$18,193.30$$

b) $r = 0.9994$; since this is close to 1, the regression line is a good fit.

c) The president's assertion is about \$1000, or 6% higher than the estimate from the linear model.

13. a) Using the linear regression feature on a grapher, we get $M = 0.2H + 156$.

b) For $H = 40$: $M = 0.2(40) + 156 = 164$ beats per minute

For $H = 65$: $M = 0.2(65) + 156 = 169$ beats per minute

For $H = 76$: $M = 0.2(76) + 156 \approx 171$ beats per minute

For $H = 84$: $M = 0.2(84) + 156 \approx 173$ beats per minute

c) $r = 1$; all the data points are on the regression line so it should be a good predictor.

15. Discussion and Writing

17. $m = \dfrac{y_2 - y_1}{x_2 - x_1}$
$$= \frac{-1 - (-8)}{-5 - 2} = \frac{-1 + 8}{-7}$$
$$= \frac{7}{-7} = -1$$

19. Substitute $\dfrac{3}{4}$ for m and -5 for b in $y = mx + b$. We have $y = \dfrac{3}{4}x - 5$.

21. a) Using the home run data from the years 1939 through 1942, 1946 through 1951, and 1954 through 1960, we get $H = -0.4020073552x + 813.785167$. (Rounding the constants reduces the accuracy of any values calculated using this function.)

In 1943,
$$y = -0.4020073552(1943) + 813.785167 \approx 33.$$

In 1944,
$$y = -0.4020073552(1944) + 813.785167 \approx 32.$$

In 1945,
$$y = -0.4020073552(1945) + 813.785167 \approx 32.$$

In 1952,
$$y = -0.4020073552(1952) + 813.785167 \approx 29.$$

In 1953,
$$y = -0.4020073552(1953) + 813.785167 \approx 29.$$

The correlation coefficient is -0.3668. Since this is low, the function would not be a good predictor.

b) Adding the number of home runs predicted for the war years and the data in the table for the other years, we get 662 home runs. Thus, Ted Williams would not have broken Hank Aaron's home run record according to this function.

c) Using the RBI data for the years listed in part (a), we get $R = -3.353892124x + 6646.484217$. (Rounding the constants greatly reduces the accuracy of values found using this function.)

In 1943,
$$R = -3.353892124(1943) + 6646.484217 \approx 130.$$

In 1944,
$$R = -3.353892124(1944) + 6646.484217 \approx 127.$$

In 1945,
$$R = -3.353892124(1945) + 6646.484217 \approx 123.$$

In 1952,
$$R = -3.353892124(1952) + 6646.484217 \approx 100.$$

In 1953,
$$R = -3.353892124(1953) + 6646.484217 \approx 96.$$

The correlation coefficient is -0.7817. Since this is fairly high, the function would be an above-average predictor but not good enough to provide a high degree of confidence in its ability to predict.

d) Adding the number of RBIs predicted for the war years and the data in the table for the other years, we get 2378 RBIs. Thus, Ted Williams would have broken Hank Aaron's RBI record according to this function.

Exercise Set 1.4

1. a) The graph rises from left to right on $(-5, 1)$.

b) The graph drops from left to right on $(3, 5)$.

c) The graph neither rises nor drops on $(1, 3)$.

3. a) The graph rises from left to right on $(-3, -1)$ and on $(3,5)$.

b) The graph drops from left to right on $(1, 3)$.

c) The graph neither rises nor drops on $(-5, 3)$.

5.

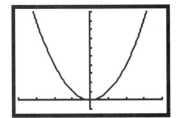

The graph is increasing on $(0, \infty)$ and decreasing on $(-\infty, 0)$. Using the MINIMUM feature we find that the relative minimum is 0 at $x = 0$. There are no relative maxima.

7.

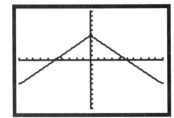

The graph is increasing on $(-\infty, 0)$ and decreasing on $(0, \infty)$. Using the MAXIMUM feature we find that the relative maximum is 5 at $x = 0$. There are no relative minima.

9.

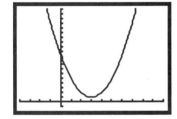

Beginning at the left side of the window, the graph first drops as we move to the right. We see that the function is decreasing on $(-\infty, 3)$. We then observe that the function is increasing on $(3, \infty)$. The MINIMUM feature also shows that the relative minimum is 1 at $x = 3$. There are no relative maxima.

11.

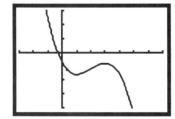

Beginning at the left side of the window, the graph first drops as we move to the right. We see that the function is decreasing on $(-\infty, 1)$. We then find that the function is increasing on $(1, 3)$ and decreasing again on $(3, \infty)$. The MAXIMUM and MINIMUM

features also show that the relative maximum is -4 at $x = 3$ and the relative minimum is -8 at $x = 1$.

13.

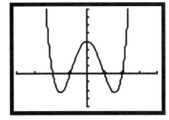

We find that the function is increasing on $(-1.552.0)$ and on $(1.552, \infty)$ and decreasing on $(-\infty, -1.552)$ and on $(0, 1.552)$. The relative maximum is 4.07 at $x = 0$ and the relative minima are -2.314 at $x = -1.552$ and -2.314 at $x = 1.552$.

15. a)

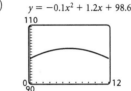

$y = -0.1x^2 + 1.2x + 98.6$

b) Using the MAXIMUM feature we find that the relative maximum is 102.2 at $t = 6$.

c) Using the result in part (b), we know that the patient's temperature was the highest at $t = 6$, or 6 days after the onset of the illness and that the highest temperature was 102.2°F.

17. Graph $y = \dfrac{8x}{x^2 + 1}$.

Increasing: $(-1, 1)$

Decreasing: $(-\infty, -1)$, $(1, \infty)$

19. Graph $y = x\sqrt{4 - x^2}$, for $-2 \le x \le 2$.

Increasing: $(-1.414, 1.414)$

Decreasing: $(-2, -1.414)$, $(1.414, 2)$

21. After t minutes, the balloon has risen $120t$ ft. We use the Pythagorean theorem.

$$[d(t)]^2 = (120t)^2 + (400)^2$$
$$d(t) = \sqrt{(120t)^2 + (400)^2}$$

We only considered the positive square root since distance must be nonnegative.

23. If $x =$ the length of the rectangle, in meters, then the width is $\dfrac{48 - 2x}{2}$, or $24 - x$. We use the formula Area = length × width:

$$A(x) = x(24 - x)$$
$$A(x) = 24x - x^2$$

25. Let w = the width of the rectangle. Then the length $= \dfrac{40 - 2w}{2}$, or $20 - w$. Divide the rectangle into quadrants as shown below.

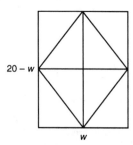

In each quadrant there are two congruent triangles. One triangle is part of the rhombus and both are part of the rectangle. Thus, in each quadrant the area of the rhombus is one-half the area of the rectangle. Then, in total, the area of the rhombus is one-half the area of the rectangle.

$$A(w) = \frac{1}{2}(20 - w)(w)$$

$$A(w) = 10w - \frac{w^2}{2}$$

27. We will use similar triangles, expressing all distances in feet. $\left(6 \text{ in.} = \dfrac{1}{2} \text{ ft}, \ s \text{ in.} = \dfrac{s}{12} \text{ ft, and } d \text{ yd} = 3d \text{ ft} \right)$ We have

$$\frac{3d}{7} = \frac{\frac{1}{2}}{\frac{s}{12}}$$

$$\frac{s}{12} \cdot 3d = 7 \cdot \frac{1}{2}$$

$$\frac{sd}{4} = \frac{7}{2}$$

$$d = \frac{4}{s} \cdot \frac{7}{2}, \text{ so}$$

$$d(s) = \frac{14}{s}.$$

29. a) If the length $= x$ feet, then the width $= 30 - x$ feet.

$$A(x) = x(30 - x)$$
$$A(x) = 30x - x^2$$

b) The length of the rectangle must be positive and less than 30 ft, so the domain of the function is $\{x | 0 < x < 30\}$, or $(0, 30)$.

c) $y = 30x - x^2$

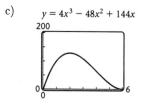

d) Using the MAXIMUM feature, we find that the maximum area occurs when $x = 15$. Then the dimensions that yield the maximum area are length = 15 ft and width = $30 - 15$, or 15 ft.

31. a) When a square with sides of length x are cut from each corner, the length of each of the remaining sides of the piece of cardboard is $12 - 2x$. Then the dimensions of the box are x by $12 - 2x$ by $12 - 2x$. We use the formula Volume = length × width × height to find the volume of the box:

$$V(x) = (12 - 2x)(12 - 2x)(x)$$
$$V(x) = (144 - 48x + 4x^2)(x)$$
$$V(x) = 144x - 48x^2 + 4x^3$$

This can also be expressed as $V(x) = 4x(x-6)^2$.

b) The length of the sides of the square corners that are cut out must be positive and less than half the length of a side of the piece of cardboard. Thus, the domain of the function is $\{x | 0 < x < 6\}$, or $(0, 6)$.

c) $y = 4x^3 - 48x^2 + 144x$

d) Using the MAXIMUM feature, we find that the maximum value of the volume occurs when $x = 2$. When $x = 2$, $12 - 2x = 12 - 2 \cdot 2 = 8$, so the dimensions that yield the maximum volume are 8 cm by 8 cm by 2 cm.

33. a) The length of a diameter of the circle (and a diagonal of the rectangle) is $2 \cdot 8$, or 16 ft. Let l = the length of the rectangle. Use the Pythagorean theorem to write l as a function of x.

$$x^2 + l^2 = 16^2$$
$$x^2 + l^2 = 256$$
$$l^2 = 256 - x^2$$
$$l = \sqrt{256 - x^2}$$

Since the length must be positive, we considered only the positive square root.

Use the formula Area = length × width to find the area of the rectangle:

$$A(x) = x\sqrt{256 - x^2}$$

b) The width of the rectangle must be positive and less than the diameter of the circle. Thus, the domain of the function is $\{x | 0 < x < 16\}$, or $(0, 16)$.

c) $y = x\sqrt{256 - x^2}$

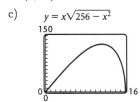

d) Using the MAXIMUM feature, we find that the maximum area occurs when x is about 11.314. When $x \approx 11.314$, $\sqrt{256 - x^2} \approx \sqrt{256 - (11.314)^2} \approx 11.313$. Thus, the dimensions that maximize the area are about 11.314 ft by 11.313 ft. (Answers may vary slightly due to rounding differences.)

35. $f(x) = \begin{cases} \dfrac{1}{2}x, & \text{for } x < 0, \\ x + 3, & \text{for } x \geq 0 \end{cases}$

We create the graph in two parts. Graph $f(x) = \dfrac{1}{2}x$ for inputs x less than 0. The graph $f(x) = x + 3$ for inputs x greater than or equal to 0.

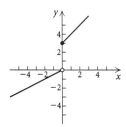

37. $f(x) = \begin{cases} -\dfrac{3}{4}x + 2, & \text{for } x < 4, \\ -1, & \text{for } x \geq 4 \end{cases}$

We create the graph in two parts. Graph $f(x) = -\dfrac{3}{4}x + 2$ for inputs x less than 4. The graph $f(x) = -1$ for inputs x greater than or equal to 4.

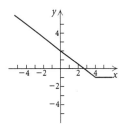

39. $f(x) = \begin{cases} x + 1, & \text{for } x \leq -3, \\ -1, & \text{for } -3 < x < 4, \\ \dfrac{1}{2}x, & \text{for } x \geq 4 \end{cases}$

We create the graph in three parts. Graph $f(x) = x + 1$ for inputs x less than or equal to -3. Graph $f(x) = -1$ for inputs greater than -3 and less than 4. Then graph $f(x) = \dfrac{1}{2}x$ for inputs greater than or equal to 4.

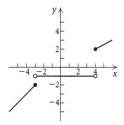

41. $f(x) = \begin{cases} 2, & \text{for } x = 5, \\ \dfrac{x^2 - 25}{x - 5}, & \text{for } x \neq 5 \end{cases}$

When $x \neq 5$, the denominator of $(x^2 - 25)/(x - 5)$ is nonzero so we can simplify:

$$\frac{x^2 - 25}{x - 5} = \frac{(x + 5)(x - 5)}{x - 5} = x + 5.$$

Thus, $f(x) = x + 5$, for $x \neq 5$.

The graph of this part of the function consists of a line with a "hole" at the point $(5, 10)$, indicated by an open dot. At $x = 5$, we have $f(5) = 2$, so the point $(5, 2)$ is plotted below the open dot.

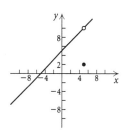

43. $f(x) = \text{int}(x)$

See Example 7.

45. $f(x) = 1 + \text{int}(x)$

This function can be defined by a piecewise function with an infinite number of statements:

$$f(x) = \begin{cases} \vdots \\ \vdots \\ -1, & \text{for } -2 \le x < -1, \\ 0, & \text{for } -1 \le x < -0, \\ 1, & \text{for } 0 \le x < 1, \\ 2, & \text{for } 1 \le x < 2, \\ \vdots \\ \vdots \end{cases}$$

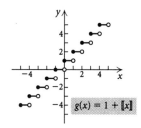

$g(x) = 1 + [\![x]\!]$

47. Use DOT mode.

$$y = \begin{cases} \sqrt[3]{x}, & \text{for } x \le -1, \\ x^2 - 3x, & \text{for } -1 < x < 4, \\ \sqrt{x-4}, & \text{for } x \ge 4 \end{cases}$$

49. $(f+g)(5) = f(5) + g(5)$

$$= (5^2 - 3) + (2 \cdot 5 + 1)$$
$$= 25 - 3 + 10 + 1$$
$$= 33$$

51. $(f-g)(-1) = f(-1) - g(-1)$

$$= ((-1)^2 - 3) - (2(-1) + 1)$$
$$= -2 - (-1) = -2 + 1$$
$$= -1$$

53. $(fg)(2) = f(2) \cdot g(2)$

$$= (2^2 - 3)(2 \cdot 2 + 1)$$
$$= 1 \cdot 5 = 5$$

55. $(f/g)(-1) = \dfrac{f(-1)}{g(-1)}$

$$= \frac{(-1)^2 - 3}{2(-1) + 1}$$
$$= \frac{-2}{-1} = 2$$

57. $(fg)\left(-\dfrac{1}{2}\right) = f\left(-\dfrac{1}{2}\right) \cdot g\left(-\dfrac{1}{2}\right)$

$$= \left[\left(-\frac{1}{2}\right)^2 - 3\right]\left[2\left(-\frac{1}{2}\right) + 1\right]$$
$$= -\frac{11}{4} \cdot 0 = 0$$

59. $(f/g)(\sqrt{3}) = \dfrac{f(\sqrt{3})}{g(\sqrt{3})}$

$$= \frac{(\sqrt{3})^2 - 3}{2\sqrt{3} + 1}$$
$$= \frac{3 - 3}{2\sqrt{3} + 1}$$
$$= \frac{0}{2\sqrt{3} + 1} = 0$$

61. $f(x) = x - 3$, $g(x) = \sqrt{x+4}$

a) Any number can be an input in f, so the domain of f is the set of all real numbers.

The inputs of g must be nonnegative, so we have $x + 4 \ge 0$, or $x \ge -4$. Thus, the domain of g is $[-4, \infty)$.

The domain of $f + g$, $f - g$, and fg is the set of all numbers in the domains of both f and g. This is $[-4, \infty)$.

The domain of ff is the domain of f, or the set of all real numbers.

The domain of f/g is the set of all numbers in the domains of f and g, excluding those for which $g(x) = 0$. Since $g(-4) = 0$, the domain of f/g is $(-4, \infty)$.

The domain of g/f is the set of all numbers in the domains of g and f, excluding those for which $f(x) = 0$. Since $f(3) = 0$, the domain of g/f is $[-4, 3) \cup (3, \infty)$.

b) $(f+g)(x) = f(x) + g(x) = x - 3 + \sqrt{x+4}$

$(f-g)(x) = f(x) - g(x) = x - 3 - \sqrt{x+4}$

$(fg)(x) = f(x) \cdot g(x) = (x-3)\sqrt{x+4}$

$(ff)(x) = [f(x)]^2 = (x-3)^2 = x^2 - 6x + 9$

$(f/g)(x) = \dfrac{f(x)}{g(x)} = \dfrac{x-3}{\sqrt{x+4}}$

$(g/f)(x) = \dfrac{g(x)}{f(x)} = \dfrac{\sqrt{x+4}}{x-3}$

63. $f(x) = x^3$, $g(x) = 2x^2 + 5x - 3$

a) Since any number can be an input for either f or g, the domain of f, g, $f + g$, $f - g$, fg, and ff is the set of all real numbers.

Since $g(-3) = 0$ and $g\left(\dfrac{1}{2}\right) = 0$, the domain of f/g is the set of all real numbers except -3 and $\dfrac{1}{2}$.

Since $f(0) = 0$, the domain of g/f is the set of all real numbers except 0.

b) $(f + g)(x) = f(x) + g(x) = x^3 + 2x^2 + 5x - 3$

$(f - g)(x) = f(x) - g(x) = x^3 - (2x^2 + 5x - 3) = x^3 - 2x^2 - 5x + 3$

$(fg)(x) = f(x) \cdot g(x) = x^3(2x^2 + 5x - 3) = 2x^5 + 5x^4 - 3x^3$

$(ff)(x) = \left[f(x)\right]^2 = (x^3)^2 = x^6$

$(f/g)(x) = \dfrac{f(x)}{g(x)} = \dfrac{x^3}{2x^2 + 5x - 3}$

$(g/f)(x) = \dfrac{g(x)}{f(x)} = \dfrac{2x^2 + 5x - 3}{x^3}$

65. $f(x) = x^2 - 4$, $g(x) = x^2 + 2$

$(f + g)(x) = f(x) + g(x) = x^2 - 4 + x^2 + 2 = 2x^2 - 2$

$(f - g)(x) = f(x) - g(x) = x^2 - 4 - (x^2 + 2) = x^2 - 4 - x^2 - 2 = -6$

$(fg)(x) = f(x) \cdot g(x) = (x^2 - 4)(x^2 + 2) = x^4 - 2x^2 - 8$

$(f/g)(x) = \dfrac{f(x)}{g(x)} = \dfrac{x^2 - 4}{x^2 + 2}$

67. $f(x) = \sqrt{x - 7}$, $g(x) = x^2 - 25$

$(f + g)(x) = f(x) + g(x) = \sqrt{x - 7} + x^2 - 25$

$(f - g)(x) = f(x) - g(x) = \sqrt{x - 7} - (x^2 - 25) = \sqrt{x - 7} - x^2 + 25$

$(fg)(x) = f(x) \cdot g(x) = \sqrt{x - 7}(x^2 - 25)$

$(f/g)(x) = \dfrac{f(x)}{g(x)} = \dfrac{\sqrt{x - 7}}{x^2 - 25}$

69. From the graph, we see that the domain of F is $[0, 9]$ and the domain of G is $[3, 10]$. The domain of $F + G$ is the set of numbers in the domains of both F and G. This is $[3, 9]$.

71. The domain of G/F is the set of numbers in the domains of both F and G (See Exercise 69.), excluding those for which $F = 0$. Since $F(6) = 0$ and $F(8) = 0$, the domain of G/F is $[3, 6) \cup (6, 8) \cup (8, 9]$.

73.

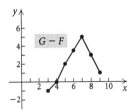

75. a) $P(x) = R(x) - C(x) = 60x - 0.4x^2 - (3x + 13) = 60x - 0.4x^2 - 3x - 13 = -0.4x^2 + 57x - 13$

b) $R(100) = 60 \cdot 100 - 0.4(100)^2 = 6000 - 0.4(10,000) = 6000 - 4000 = 2000$

$C(100) = 3 \cdot 100 + 13 = 300 + 13 = 313$

$P(100) = R(100) - C(100) = 2000 - 313 = 1687$

c)

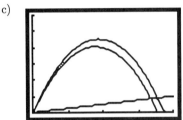

77. $f(x) = 3x^2 - 2x + 1$

$f(x + h) = 3(x + h)^2 - 2(x + h) + 1 = 3(x^2 + 2xh + h^2) - 2(x + h) + 1 = 3x^2 + 6xh + 3h^2 - 2x - 2h + 1$

$f(x) = 3x^2 - 2x + 1$

$\dfrac{f(x + h) - f(x)}{h} =$

$\dfrac{(3x^2 + 6xh + 3h^2 - 2x - 2h + 1) - (3x^2 - 2x + 1)}{h} =$

$\dfrac{3x^2 + 6xh + 3h^2 - 2x - 2h + 1 - 3x^2 + 2x - 1}{h} =$

$\dfrac{6xh + 3h^2 - 2h}{h} = \dfrac{h(6x + 3h - 2)}{h \cdot 1} =$

$\dfrac{h}{h} \cdot \dfrac{6x + 3h - 2}{1} = 6x + 3h - 2$

79. $f(x) = x^3$

$f(x + h) = (x + h)^3 = x^3 + 3x^2h + 3xh^2 + h^3$

$f(x) = x^3$

$\dfrac{f(x + h) - f(x)}{h} = \dfrac{x^3 + 3x^2h + 3xh^2 + h^3 - x^3}{h} =$

$\dfrac{3x^2h + 3xh^2 + h^3}{h} = \dfrac{h(3x^2 + 3xh + h^2)}{h \cdot 1} =$

$\dfrac{h}{h} \cdot \dfrac{3x^2 + 3xh + h^2}{1} = 3x^2 + 3xh + h^2$

81. $f(x) = \dfrac{x-4}{x+3}$

$$\frac{f(x+h)-f(x)}{h} = \frac{\dfrac{x+h-4}{x+h+3}-\dfrac{x-4}{x+3}}{h} =$$

$$\frac{\dfrac{x+h-4}{x+h+3}-\dfrac{x-4}{x+3}}{h} \cdot \frac{(x+h+3)(x+3)}{(x+h+3)(x+3)} =$$

$$\frac{(x+h-4)(x+3)-(x-4)(x+h+3)}{h(x+h+3)(x+3)} =$$

$$\frac{x^2+hx-4x+3x+3h-12-(x^2+hx+3x-4x-4h-12)}{h(x+h+3)(x+3)} =$$

$$\frac{x^2+hx-x+3h-12-x^2-hx+x+4h+12}{h(x+h+3)(x+3)} =$$

$$\frac{7h}{h(x+h+3)(x+3)} = \frac{h}{h} \cdot \frac{7}{(x+h+3)(x+3)} =$$

$$\frac{7}{(x+h+3)(x+3)}$$

83. Discussion and writing

85. $f(x) = -2x - 3$

We can substitute any real number for x, so the domain is the set of all real numbers, or $(-\infty, \infty)$.

87. $h(x) = \dfrac{x}{|x+5|}$

We can substitute any real number in the numerator, but the input -5 results in a denominator of 0. Thus, the domain is $\{x | x \neq -5\}$, or $(-\infty, -5) \cup (-5, \infty)$.

89. Graph $y = x^4 + 4x^3 - 36x^2 - 160x + 400$

Increasing: $(-5, -2)$, $(4, \infty)$

Decreasing: $(-\infty, -5)$, $(-2, 4)$

Relative maximum: 560 at $x = -2$

Relative minima: 425 at $x = -5$, -304 at $x = 4$

91. a) The function $C(t)$ can be defined piecewise.

$$C(t) = \begin{cases} 2, & \text{for } 0 < t < 1, \\ 4, & \text{for } 1 \leq t < 2, \\ 6, & \text{for } 2 \leq t < 3, \\ \cdot \\ \cdot \\ \cdot \end{cases}$$

We graph this function.

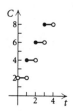

b) From the definition of the function in part (a), we see that it can be written as

$$C(t) = 2[\text{int}(t) + 1], \ t > 0.$$

93. If $[\text{int}(x)]^2 = 25$, then $\text{int}(x) = -5$ or $\text{int}(x) = 5$. For $-5 \leq x < -4$, $\text{int}(x) = -5$. For $5 \leq x < 6$, $\text{int}(x) = 5$. Thus, the possible inputs for x are $\{x | -5 \leq x < -4 \text{ or } 5 \leq x < 6\}$.

95. a) We add labels to the drawing in the text.

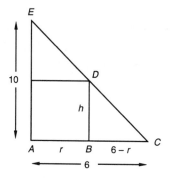

We write a proportion involving the lengths of the sides of the similar triangles BCD and ACE. Then we solve it for h.

$$\frac{h}{6-r} = \frac{10}{6}$$

$$h = \frac{10}{6}(6-r) = \frac{5}{3}(6-r)$$

$$h = \frac{30-5r}{3}$$

Thus, $h(r) = \dfrac{30-5r}{3}$.

b) $\quad V = \pi r^2 h$

$$V(r) = \pi r^2 \left(\frac{30-5r}{3}\right) \quad \text{Substituting for } h$$

c) We first express r in terms of h.

$$h = \frac{30-5r}{3}$$

$$3h = 30 - 5r$$

$$5r = 30 - 3h$$

$$r = \frac{30-3h}{5}$$

$$V = \pi r^2 h$$

$$V(h) = \pi \left(\frac{30-3h}{5}\right)^2 h$$

$$\text{Substituting for } r$$

We can also write $V(h) = \pi h \left(\dfrac{30-3h}{5}\right)^2$.

Exercise Set 1.5

1. If the graph were folded on the x-axis, the parts above and below the x-axis would not coincide, so the graph is not symmetric with respect to the x-axis.

If the graph were folded on the y-axis, the parts to the left and right of the y-axis would coincide, so the graph is symmetric with respect to the y-axis.

If the graph were rotated $180°$, the resulting graph would not coincide with the original graph, so it is not symmetric with respect to the origin.

3. If the graph were folded on the x-axis, the parts above and below the x-axis would coincide, so the graph is symmetric with respect to the x-axis.

If the graph were folded on the y-axis, the parts to the left and right of the y-axis would not coincide, so the graph is not symmetric with respect to the y-axis.

If the graph were rotated $180°$, the resulting graph would not coincide with the original graph, so it is not symmetric with respect to the origin.

5. If the graph were folded on the x-axis, the parts above and below the x-axis would not coincide, so the graph is not symmetric with respect to the x-axis.

If the graph were folded on the y-axis, the parts to the left and right of the y-axis would not coincide, so the graph is not symmetric with respect to the y-axis.

If the graph were rotated $180°$, the resulting graph would coincide with the original graph, so it is symmetric with respect to the origin.

7.

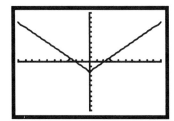

The graph is symmetric with respect to the y-axis. It is not symmetric with respect to the x-axis or the origin.

Test algebraically for symmetry with respect to the x-axis:

$$y = |x| - 2 \qquad \text{Original equation}$$
$$-y = |x| - 2 \qquad \text{Replacing } y \text{ by } -y$$
$$y = -|x| + 2 \qquad \text{Simplifying}$$

The last equation is not equivalent to the original equation, so the graph is not symmetric with respect to the x-axis.

Test algebraically for symmetry with respect to the y-axis:

$$y = |x| - 2 \qquad \text{Original equation}$$
$$y = |-x| - 2 \qquad \text{Replacing } x \text{ by } -x$$
$$y = |x| - 2 \qquad \text{Simplifying}$$

The last equation is equivalent to the original equation, so the graph is symmetric with respect to the y-axis.

Test algebraically for symmetry with respect to the origin:

$$y = |x| - 2 \qquad \text{Original equation}$$
$$-y = |-x| - 2 \qquad \text{Replacing } x \text{ by } -x \text{ and } y \text{ by } -y$$
$$-y = |x| - 2 \qquad \text{Simplifying}$$
$$y = -|x| + 2$$

The last equation is not equivalent to the original equation, so the graph is not symmetric with respect to the origin.

9.

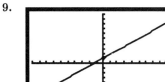

The graph is not symmetric with respect to the x-axis, the y-axis, or the origin.

Test algebraically for symmetry with respect to the x-axis:

$$5y = 4x + 5 \qquad \text{Original equation}$$
$$5(-y) = 4x + 5 \qquad \text{Replacing } y \text{ by } -y$$
$$-5y = 4x + 5 \qquad \text{Simplifying}$$
$$5y = -4x - 5$$

The last equation is not equivalent to the original equation, so the graph is not symmetric with respect to the x-axis.

Test algebraically for symmetry with respect to the y-axis:

$$5y = 4x + 5 \qquad \text{Original equation}$$
$$5y = 4(-x) + 5 \qquad \text{Replacing } x \text{ by } -x$$
$$5y = -4x + 5 \qquad \text{Simplifying}$$

The last equation is not equivalent to the original equation, so the graph is not symmetric with respect to the y-axis.

Test algebraically for symmetry with respect to the origin:

$$5y = 4x + 5 \quad \text{Original equation}$$

$$5(-y) = 4(-x) + 5 \quad \text{Replacing } x \text{ by } -x$$
$$\text{and}$$
$$y \text{ by } -y$$

$$-5y = -4x + 5 \quad \text{Simplifying}$$

$$5y = 4x - 5$$

The last equation is not equivalent to the original equation, so the graph is not symmetric with respect to the origin.

11.

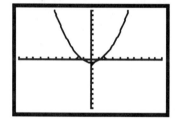

The graph is symmetric with respect to the y-axis. It is not symmetric with respect to the x-axis or the origin.

Test algebraically for symmetry with respect to the x-axis:

$$5y = 2x^2 - 3 \quad \text{Original equation}$$

$$5(-y) = 2x^2 - 3 \quad \text{Replacing } y \text{ by } -y$$

$$-5y = 2x^2 - 3 \quad \text{Simplifying}$$

$$5y = -2x^2 + 3$$

The last equation is not equivalent to the original equation, so the graph is not symmetric with respect to the x-axis.

Test algebraically for symmetry with respect to the y-axis:

$$5y = 2x^2 - 3 \quad \text{Original equation}$$

$$5y = 2(-x)^2 - 3 \quad \text{Replacing } x \text{ by } -x$$

$$5y = 2x^2 - 3$$

The last equation is equivalent to the original equation, so the graph is symmetric with respect to the y-axis.

Test algebraically for symmetry with respect to the origin:

$$5y = 2x^2 - 3 \quad \text{Original equation}$$

$$5(-y) = 2(-x)^2 - 3 \quad \text{Replacing } x \text{ by } -x \text{ and}$$
$$y \text{ by } -y$$

$$-5y = 2x^2 - 3 \quad \text{Simplifying}$$

$$5y = -2x^2 + 3$$

The last equation is not equivalent to the original equation, so the graph is not symmetric with respect to the origin.

13.

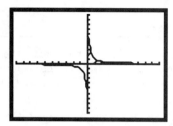

The graph is not symmetric with respect to the x-axis or the y-axis. It is symmetric with respect to the origin.

Test algebraically for symmetry with respect to the x-axis:

$$y = \frac{1}{x} \quad \text{Original equation}$$

$$-y = \frac{1}{x} \quad \text{Replacing } y \text{ by } -y$$

$$y = -\frac{1}{x} \quad \text{Simplifying}$$

The last equation is not equivalent to the original equation, so the graph is not symmetric with respect to the x-axis.

Test algebraically for symmetry with respect to the y-axis:

$$y = \frac{1}{x} \quad \text{Original equation}$$

$$y = \frac{1}{-x} \quad \text{Replacing } x \text{ by } -x$$

$$y = -\frac{1}{x} \quad \text{Simplifying}$$

The last equation is not equivalent to the original equation, so the graph is symmetric with respect to the y-axis.

Test algebraically for symmetry with respect to the origin:

$$y = \frac{1}{x} \quad \text{Original equation}$$

$$-y = \frac{1}{-x} \quad \text{Replacing } x \text{ by } -x \text{ and } y \text{ by } -y$$

$$y = \frac{1}{x} \quad \text{Simplifying}$$

The last equation is equivalent to the original equation, so the graph is symmetric with respect to the origin.

15. Test for symmetry with respect to the x-axis:

$$5x - 5y = 0 \quad \text{Original equation}$$

$$5x - 5(-y) = 0 \quad \text{Replacing } y \text{ by } -y$$

$$5x + 5y = 0 \quad \text{Simplifying}$$

The last equation is not equivalent to the original equation, so the graph is not symmetric with respect to the x-axis.

Test for symmetry with respect to the y-axis:

$$5x - 5y = 0 \quad \text{Original equation}$$
$$5(-x) - 5y = 0 \quad \text{Replacing } x \text{ by } -x$$
$$-5x - 5y = 0 \quad \text{Simplifying}$$
$$5x + 5y = 0$$

The last equation is not equivalent to the original equation, so the graph is not symmetric with respect to the y-axis.

Test for symmetry with respect to the origin:

$$5x - 5y = 0 \quad \text{Original equation}$$
$$5(-x) - 5(-y) = 0 \quad \text{Replacing } x \text{ by } -x \text{ and } y \text{ by } -y$$
$$-5x + 5y = 0 \quad \text{Simplifying}$$
$$5x - 5y = 0$$

The last equation is equivalent to the original equation, so the graph is symmetric with respect to the origin.

17. Test for symmetry with respect to the x-axis:

$$3x^2 - 2y^2 = 3 \quad \text{Original equation}$$
$$3x^2 - 2(-y)^2 = 3 \quad \text{Replacing } y \text{ by } -y$$
$$3x^2 - 2y^2 = 3 \quad \text{Simplifying}$$

The last equation is equivalent to the original equation, so the graph is symmetric with respect to the x-axis.

Test for symmetry with respect to the y-axis:

$$3x^2 - 2y^2 = 3 \quad \text{Original equation}$$
$$3(-x)^2 - 2y^2 = 3 \quad \text{Replacing } x \text{ by } -x$$
$$3x^2 - 2y^2 = 3 \quad \text{Simplifying}$$

The last equation is equivalent to the original equation, so the graph is symmetric with respect to the y-axis.

Test for symmetry with respect to the origin:

$$3x^2 - 2y^2 = 3 \quad \text{Original equation}$$
$$3(-x)^2 - 2(-y)^2 = 3 \quad \text{Replacing } x \text{ by } -x \text{ and } y \text{ by } -y$$
$$3x^2 - 2y^2 = 3 \quad \text{Simplifying}$$

The last equation is equivalent to the original equation, so the graph is symmetric with respect to the origin.

19. Test for symmetry with respect to the x-axis:

$$y = |2x| \quad \text{Original equation}$$
$$-y = |2x| \quad \text{Replacing } y \text{ by } -y$$
$$y = -|2x| \quad \text{Simplifying}$$

The last equation is not equivalent to the original equation, so the graph is not symmetric with respect to the x-axis.

Test for symmetry with respect to the y-axis:

$$y = |2x| \quad \text{Original equation}$$
$$y = |2(-x)| \quad \text{Replacing } x \text{ by } -x$$
$$y = |-2x| \quad \text{Simplifying}$$
$$y = |2x|$$

The last equation is equivalent to the original equation, so the graph is symmetric with respect to the y-axis.

Test for symmetry with respect to the origin:

$$y = |2x| \quad \text{Original equation}$$
$$-y = |2(-x)| \quad \text{Replacing } x \text{ by } -x \text{ and } y \text{ by } -y$$
$$-y = |-2x| \quad \text{Simplifying}$$
$$-y = |2x|$$
$$y = -|2x|$$

The last equation is not equivalent to the original equation, so the graph is not symmetric with respect to the origin.

21. Test for symmetry with respect to the x-axis:

$$2x^4 + 3 = y^2 \quad \text{Original equation}$$
$$2x^4 + 3 = (-y)^2 \quad \text{Replacing } y \text{ by } -y$$
$$2x^4 + 3 = y^2 \quad \text{Simplifying}$$

The last equation is equivalent to the original equation, so the graph is symmetric with respect to the x-axis.

Test for symmetry with respect to the y-axis:

$$2x^4 + 3 = y^2 \quad \text{Original equation}$$
$$2(-x)^4 + 3 = y^2 \quad \text{Replacing } x \text{ by } -x$$
$$2x^4 + 3 = y^2 \quad \text{Simplifying}$$

The last equation is equivalent to the original equation, so the graph is symmetric with respect to the y-axis.

Test for symmetry with respect to the origin:

$$2x^4 + 3 = y^2 \quad \text{Original equation}$$
$$2(-x)^4 + 3 = (-y)^2 \quad \text{Replacing } x \text{ by } -x \text{ and } y \text{ by } -y$$
$$2x^4 + 3 = y^2 \quad \text{Simplifying}$$

The last equation is equivalent to the original equation, so the graph is symmetric with respect to the origin.

23. Test for symmetry with respect to the x-axis:

$$3y^3 = 4x^3 + 2 \quad \text{Original equation}$$
$$3(-y)^3 = 4x^3 + 2 \quad \text{Replacing } y \text{ by } -y$$
$$-3y^3 = 4x^3 + 2 \quad \text{Simplifying}$$
$$3y^3 = -4x^3 - 2$$

The last equation is not equivalent to the original equation, so the graph is not symmetric with respect to the x-axis.

Test for symmetry with respect to the y-axis:

$$3y^3 = 4x^3 + 2 \qquad \text{Original equation}$$
$$3y^3 = 4(-x)^3 + 2 \quad \text{Replacing } x \text{ by } -x$$
$$3y^3 = -4x^3 + 2 \qquad \text{Simplifying}$$

The last equation is not equivalent to the original equation, so the graph is not symmetric with respect to the y-axis.

Test for symmetry with respect to the origin:

$$3y^3 = 4x^3 + 2 \qquad \text{Original equation}$$
$$3(-y)^3 = 4(-x)^3 + 2 \quad \text{Replacing } x \text{ by } -x$$
$$\text{and } y \text{ by } -y$$
$$-3y^3 = -4x^3 + 2 \qquad \text{Simplifying}$$
$$3y^3 = 4x^3 - 2$$

The last equation is not equivalent to the original equation, so the graph is not symmetric with respect to the origin.

25. Test for symmetry with respect to the x-axis:

$$xy = 12 \qquad \text{Original equation}$$
$$x(-y) = 12 \qquad \text{Replacing } y \text{ by } -y$$
$$-xy = 12 \qquad \text{Simplifying}$$
$$xy = -12$$

The last equation is not equivalent to the original equation, so the graph is not symmetric with respect to the x-axis.

Test for symmetry with respect to the y-axis:

$$xy = 12 \qquad \text{Original equation}$$
$$-xy = 12 \qquad \text{Replacing } x \text{ by } -x$$
$$xy = -12 \quad \text{Simplifying}$$

The last equation is not equivalent to the original equation, so the graph is not symmetric with respect to the y-axis.

Test for symmetry with respect to the origin:

$$xy = 12 \qquad \text{Original equation}$$
$$-x(-y) = 12 \qquad \text{Replacing } x \text{ by } -x \text{ and } y \text{ by } -y$$
$$xy = 12 \quad \text{Simplifying}$$

The last equation is equivalent to the original equation, so the graph is symmetric with respect to the origin.

27. The graph is symmetric with respect to the y-axis, so the function is even.

29. The graph is symmetric with respect to the origin, so the function is odd.

31. The graph is not symmetric with respect to either the y-axis or the origin, so the function is neither even nor odd.

33. $\quad f(x) = -3x^3 + 2x$

$$f(-x) = -3(-x)^3 + 2(-x) = 3x^3 - 2x$$
$$-f(x) = -(-3x^3 + 2x) = 3x^3 - 2x$$
$$f(-x) = -f(x), \text{ so } f \text{ is odd.}$$

35. $\quad f(x) = 5x^2 + 2x^4 - 1$

$$f(-x) = 5(-x)^2 + 2(-x)^4 - 1 = 5x^2 + 2x^4 - 1$$
$$f(x) = f(-x), \text{ so } f \text{ is even.}$$

37. $\quad f(x) = x^{17}$

$$f(-x) = (-x)^{17} = -x^{17}$$
$$-f(x) = -x^{17}$$
$$f(-x) = -f(x), \text{ so } f \text{ is odd.}$$

39. $\quad f(x) = \dfrac{1}{x^2}$

$$f(-x) = \frac{1}{(-x)^2} = \frac{1}{x^2}$$
$$f(x) = f(-x), \text{ so } f \text{ is even.}$$

41. $\quad f(x) = 8$

$$f(-x) = 8$$
$$f(x) = f(-x), \text{ so } f \text{ is even.}$$

43. Think of the graph of $g(x) = x^2$. Since $f(x) = g(x) + 1$, the graph of $f(x) = x^2 + 1$ is the graph of $g(x) = x^2$ shifted up 1 unit.

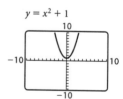

$y = x^2 + 1$

45. Think of the graph of $f(x) = x^3$. Since $g(x) = f(x + 5)$, the graph of $g(x) = (x + 5)^3$ is the graph of $f(x) = x^3$ shifted left 5 units.

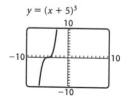

$y = (x + 5)^3$

47. Think of the graph of $g(x) = x^2$. Since $f(x) = -g(x)$, the graph of $f(x) = -x^2$ is the graph of $g(x) = x^2$ reflected across the x-axis.

$y = -x^2$

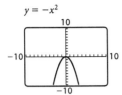

49. Think of the graph of $g(x) = \sqrt{x}$. Since $f(x) = 3g(x) - 5$, the graph of $f(x) = 3\sqrt{x} - 5$ is the graph of $g(x) = \sqrt{x}$ stretched vertically by multiplying each y-coordinate by 3 and then shifted down 5 units.

$y = 3\sqrt{x} - 5$

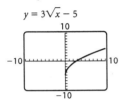

51. Think of the graph of $f(x) = |x|$. Since

$g(x) = f\left(\dfrac{1}{3}x\right) - 4$, the graph of $g(x) = \left|\dfrac{1}{3}x\right| - 4$ is

the graph of $f(x) = |x|$ stretched horizontally by multiplying each x-coordinate by 3 and then shifted down 4 units.

$y = \left|\dfrac{1}{3}x\right| - 4$

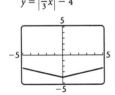

53. Think of the graph of $g(x) = x^2$. Since $f(x) = g(x+5) - 4$, the graph of $f(x) = (x+5)^2 - 4$ is the graph of $g(x) = x^2$ shifted left 5 units and then down 4 units.

$y = (x + 5)^2 - 4$

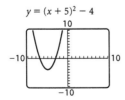

55. Think of the graph of $g(x) = x^2$. Since $f(x) = -\dfrac{1}{4}g(x - 5)$, the graph of $f(x) = -\dfrac{1}{4}(x - 5)^2$ is the graph of $g(x) = x^2$ shifted right 5 units, shrunk vertically by multiplying each y-coordinate by $\dfrac{1}{4}$, and reflected across the x-axis.

$y = -\dfrac{1}{4}(x - 5)^2$

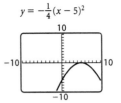

57. Think of the graph of $g(x) = \dfrac{1}{x}$. Since $f(x) = g(x + 3) + 2$, the graph of $f(x) = \dfrac{1}{x + 3} + 2$ is the graph of $g(x) = \dfrac{1}{x}$ shifted left 3 units and up 2 units.

$y = \dfrac{1}{x + 3} + 2$

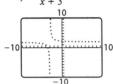

59. Shape: $h(x) = x^2$

Turn $h(x)$ upside-down (that is, reflect it across the x-axis): $g(x) = -h(x) = -x^2$

Shift $g(x)$ right 8 units: $f(x) = g(x - 8) = -(x - 8)^2$

61. Shape: $h(x) = |x|$

Shift $h(x)$ left 7 units: $g(x) = h(x + 7) = |x + 7|$

Shift $g(x)$ up 2 units: $f(x) = g(x) + 2 = |x + 7| + 2$

63. Shape: $h(x) = \dfrac{1}{x}$

Shrink $h(x)$ vertically by a factor of $\dfrac{1}{2}$ $\left(\text{that is,}\right.$ multiply each function value by $\left.\dfrac{1}{2}\right)$:

$g(x) = \dfrac{1}{2}h(x) = \dfrac{1}{2} \cdot \dfrac{1}{x}$, or $\dfrac{1}{2x}$

Shift $g(x)$ down 3 units: $f(x) = g(x) - 3 = \dfrac{1}{2x} - 3$

65. Shape: $m(x) = x^2$

Turn $m(x)$ upside-down (that is, reflect it across the x-axis): $h(x) = -m(x) = -x^2$

Shift $h(x)$ right 3 units: $g(x) = h(x - 3) = -(x - 3)^2$

Shift $g(x)$ up 4 units: $f(x) = g(x) + 4 = -(x - 3)^2 + 4$

67. Shape: $m(x) = \sqrt{x}$

Reflect $m(x)$ across the y-axis: $h(x) = m(-x) = \sqrt{-x}$

Shift $h(x)$ left 2 units: $g(x) = h(x + 2) = \sqrt{-(x + 2)}$

Shift $g(x)$ down 1 unit: $f(x) = g(x) - 1 = \sqrt{-(x + 2)} - 1$

69. Shape: $h(x) = x^3$

Shift $h(x)$ left 4 units: $g(x) = h(x + 4) = (x + 4)^3$

Shrink $g(x)$ vertically by a factor of 0.83 (that is, multiply each function value by 0.83): $f(x) = 0.83g(x) = 0.83(x + 4)^3$

71. Each y-coordinate is multiplied by -2. We plot and connect $(-4, 0)$, $(-3, 4)$, $(-1, 4)$, $(2, -6)$, and $(5, 0)$.

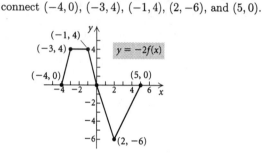

73. The graph is reflected across the y-axis and stretched horizontally by a factor of 2. That is, each x-coordinate is multiplied by -2 $\left(\text{or divided by } -\dfrac{1}{2}\right)$. We plot and connect $(8, 0)$, $(6, -2)$, $(2, -2)$, $(-4, 3)$, and $(-10, 0)$.

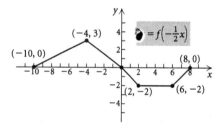

75. The graph is shifted right 1 unit so each x-coordinate is increased by 1. The graph is also reflected across the x-axis, shrunk vertically by a factor of 2, and shifted up 3 units. Thus, each y-coordinate is multiplied by $-\dfrac{1}{2}$ and then increased by 3. We plot and connect $(-3, 3)$, $(-2, 4)$, $(0, 4)$, $(3, 1.5)$, and $(6, 3)$.

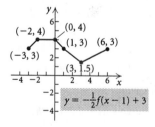

77. $g(x) = f(-x) + 3$

The graph of $g(x)$ is the graph of $f(x)$ reflected across the y-axis and shifted up 3 units. This is graph f.

79. $g(x) = -f(x) + 3$

The graph of $g(x)$ is the graph of $f(x)$ reflected across the x-axis and shifted up 3 units. This is graph f.

81. $g(x) = \dfrac{1}{3}f(x - 2)$

The graph of $g(x)$ is the graph of $f(x)$ shrunk vertically by a factor of 3 $\left(\text{that is, each } y\text{-coordinate is multiplied by } \dfrac{1}{3}\right)$ and then shifted right 2 units. This is graph d.

83. $g(x) = \dfrac{1}{3}f(x + 2)$

The graph of $g(x)$ is the graph of $f(x)$ shrunk vertically by a factor of 3 $\left(\text{that is, each } y\text{-coordinate is multiplied by } \dfrac{1}{3}\right)$ and then shifted left 2 units. This is graph c.

85. $f(-x) = 2(-x)^4 - 35(-x)^3 + 3(-x) - 5 = 2x^4 + 35x^3 - 3x - 5 = g(x)$

87. The graph of $f(x) = x^3 - 3x^2$ is shifted up 2 units. A formula for the transformed function is $g(x) = f(x) + 2$, or $g(x) = x^3 - 3x^2 + 2$.

89. The graph of $f(x) = x^3 - 3x^2$ is shifted left 1 unit. A formula for the transformed function is $k(x) = f(x + 1)$, or $k(x) = (x + 1)^3 - 3(x + 1)^2$.

91. Discussion and Writing

93. Discussion and Writing

95. $f(x) = 5x^2 - 7$

   a) $f(-3) = 5(-3)^2 - 7 = 5 \cdot 9 - 7 = 45 - 7 = 38$

   b) $f(3) = 5 \cdot 3^2 - 7 = 5 \cdot 9 - 7 = 45 - 7 = 38$

   c) $f(a) = 5a^2 - 7$

   d) $f(-a) = 5(-a)^2 - 7 = 5a^2 - 7$

97. First find the slope of the given line.

$$8x - y = 10$$
$$8x = y + 10$$
$$8x - 10 = y$$

The slope of the given line is 8. The slope of a line perpendicular to this line is the opposite of the reciprocal of 8, or $-\dfrac{1}{8}$.

$$y - y_1 = m(x - x_1)$$
$$y - 1 = -\frac{1}{8}[x - (-1)]$$
$$y - 1 = -\frac{1}{8}(x + 1)$$
$$y - 1 = -\frac{1}{8}x - \frac{1}{8}$$
$$y = -\frac{1}{8}x + \frac{7}{8}$$

99. $f(x) = x\sqrt{10 - x^2}$

$f(-x) = -x\sqrt{10 - (-x)^2} = -x\sqrt{10 - x^2}$

$-f(x) = -x\sqrt{10 - x^2}$

Since $f(-x) = -f(x)$, f is odd.

101. If the graph were folded on the x-axis, the parts above and below the x-axis would coincide, so the graph is symmetric with respect to the x-axis.

If the graph were folded on the y-axis, the parts to the left and right of the y-axis would coincide, so the graph is symmetric with respect to the y-axis.

If the graph were rotated 180°, the resulting graph would coincide with the original graph, so it is symmetric with respect to the origin.

103. If the graph were folded on the x-axis, the parts above and below the x-axis would coincide, so the graph is symmetric with respect to the x-axis.

If the graph were folded on the y-axis, the parts to the left and right of the y-axis would not coincide, so the graph is not symmetric with respect to the y-axis.

If the graph were rotated 180°, the resulting graph would not coincide with the original graph, so it is not symmetric with respect to the origin.

105. Think of the graph of $g(x) = \text{int}(x)$. Since $f(x) = g\left(x - \frac{1}{2}\right)$, the graph of $f(x) = \text{int}\left(x - \frac{1}{2}\right)$ is the graph of $g(x) = \text{int}(x)$ shifted right $\frac{1}{2}$ unit. The domain is the set of all real numbers; the range is the set of all integers.

107. On the graph of $y = 2f(x)$ each y-coordinate of $y = f(x)$ is multiplied by 2, so $(3, 4 \cdot 2)$, or $(3, 8)$ is on the transformed graph.

On the graph of $y = 2 + f(x)$, each y-coordinate of $y = f(x)$ is increased by 2 (shifted up 2 units), so $(3, 4 + 2)$, or $(3, 6)$ is on the transformed graph.

On the graph of $y = f(2x)$, each x-coordinate of $y = f(x)$ is multiplied by $\frac{1}{2}$ (or divided by 2), so $\left(\frac{1}{2} \cdot 3, 4\right)$, or $\left(\frac{3}{2}, 4\right)$ is on the transformed graph.

109. $f(2 - 3) = f(-1) = 5$, so $b = 5$.

(The graph of $y = f(x - 3)$ is the graph of $y = f(x)$ shifted right 3 units, so the point $(-1, 5)$ on $y = f(x)$ is transformed to the point $(-1 + 3, 5)$, or $(2, 5)$ on $y = f(x - 3)$.)

111. Let $f(x)$ and $g(x)$ be even functions. Then by definition, $f(x) = f(-x)$ and $g(x) = g(-x)$. Thus, $(f + g)(x) = f(x) + g(x) = f(-x) + g(-x) = (f + g)(-x)$ and $f + g$ is even. The statement is true.

113. See the answer section in the text.

115. a), b) See the answer section in the text.

Exercise Set 1.6

---

1. $y = kx$

$54 = k \cdot 12$

$\frac{54}{12} = k$, or $k = \frac{9}{2}$

The variation constant is $\frac{9}{2}$, or 4.5. The equation of variation is $y = \frac{9}{2}x$, or $y = 4.5x$.

3. $y = \frac{k}{x}$

$3 = \frac{k}{12}$

$36 = k$

The variation constant is 36. The equation of variation is $y = \frac{36}{x}$.

5. $y = kx$

$1 = k \cdot \frac{1}{4}$

$4 = k$

The variation constant is 4. The equation of variation is $y = 4x$.

7. $y = \frac{k}{x}$

$32 = \frac{k}{\frac{1}{8}}$

$\frac{1}{8} \cdot 32 = k$

$4 = k$

The variation constant is 4. The equation of variation is $y = \frac{4}{x}$.

9.
$$y = kx$$
$$\frac{3}{4} = k \cdot 2$$
$$\frac{1}{2} \cdot \frac{3}{4} = k$$
$$\frac{3}{8} = k$$

The variation constant is $\frac{3}{8}$. The equation of variation is $y = \frac{3}{8}x$.

11.
$$y = \frac{k}{x}$$
$$1.8 = \frac{k}{0.3}$$
$$0.54 = k$$

The variation constant is 0.54. The equation of variation is $y = \frac{0.54}{x}$.

13. $T = \frac{k}{P}$ T varies inversely as P.
$$5 = \frac{k}{7} \quad \text{Substituting}$$
$$35 = k \quad \text{Variation constant}$$

$$T = \frac{35}{P} \quad \text{Equation of variation}$$
$$T = \frac{35}{10} \quad \text{Substituting}$$
$$T = 3.5$$

It will take 10 bricklayers 3.5 hr to complete the job.

15. Let F = the number of grams of fat and w = the weight.

$F = kw$ F varies directly as w.
$60 = k \cdot 120$ Substituting
$\frac{60}{120} = k$, or   Solving for k
$\frac{1}{2} = k$ Variation constant

$F = \frac{1}{2}w$ Equation of variation
$F = \frac{1}{2} \cdot 180$ Substituting
$F = 90$

The maximum daily fat intake for a person weighing 180 lb is 90 g.

17. $W = \frac{k}{L}$ W varies inversely as L.
$$1200 = \frac{k}{8} \quad \text{Substituting}$$
$$9600 = k \quad \text{Variation constant}$$

$$W = \frac{9600}{L} \quad \text{Equation of variation}$$
$$W = \frac{9600}{14} \quad \text{Substituting}$$
$$W = 685\frac{5}{7}$$

A 14-m beam can support $685\frac{5}{7}$ kg, or about 686 kg.

19. $M = kE$ M varies directly as E.
$38 = k \cdot 95$ Substituting
$\frac{2}{5} = k$ Variation constant
$M = \frac{2}{5}E$ Equation of variation
$M = \frac{2}{5} \cdot 100$ Substituting
$M = 40$

A 100-lb person would weigh 40 lb on Mars.

21. $d = km$ d varies directly as m.
$40 = k \cdot 3$ Substituting
$\frac{40}{3} = k$ Variation constant

$d = \frac{40}{3}m$ Equation of variation
$d = \frac{40}{3} \cdot 5 = \frac{200}{3}$ Substituting
$d = 66\frac{2}{3}$

A 5-kg mass will stretch the spring $66\frac{2}{3}$ cm.

23. $P = \frac{k}{W}$ P varies inversely as W.
$$330 = \frac{k}{3.2} \quad \text{Substituting}$$
$$1056 = k \quad \text{Variation constant}$$

$$P = \frac{1056}{W} \quad \text{Equation of variation}$$

$$550 = \frac{1056}{W} \quad \text{Substituting}$$

$$550W = 1056 \quad \text{Multiplying by } W$$

$$W = \frac{1056}{550} \quad \text{Dividing by 550}$$

$$W = 1.92 \quad \text{Simplifying}$$

A tone with a pitch of 550 vibrations per second has a wavelength of 1.92 ft.

25.

$$y = \frac{k}{x^2}$$

$$0.15 = \frac{k}{(0.1)^2} \quad \text{Substituting}$$

$$0.15 = \frac{k}{0.01}$$

$$0.15(0.01) = k$$

$$0.0015 = k$$

The equation of variation is $y = \dfrac{0.0015}{x^2}$.

27.

$$y = kx^2$$

$$0.15 = k(0.1)^2 \quad \text{Substituting}$$

$$0.15 = 0.01k$$

$$\frac{0.15}{0.01} = k$$

$$15 = k$$

The equation of variation is $y = 15x^2$.

29.

$$y = kxz$$

$$56 = k \cdot 7 \cdot 8 \quad \text{Substituting}$$

$$56 = 56k$$

$$1 = k$$

The equation of variation is $y = xz$.

31.

$$y = kxz^2$$

$$105 = k \cdot 14 \cdot 5^2 \quad \text{Substituting}$$

$$105 = 350k$$

$$\frac{105}{350} = k$$

$$\frac{3}{10} = k$$

The equation of variation is $y = \dfrac{3}{10}xz^2$.

33.

$$y = k\frac{xz}{wp}$$

$$\frac{3}{28} = k\frac{3 \cdot 10}{7 \cdot 8} \quad \text{Substituting}$$

$$\frac{3}{28} = k \cdot \frac{30}{56}$$

$$\frac{3}{28} \cdot \frac{56}{30} = k$$

$$\frac{1}{5} = k$$

The equation of variation is $y = \dfrac{xz}{5wp}$.

35.

$$I = \frac{k}{d^2}$$

$$90 = \frac{k}{5^2} \quad \text{Substituting}$$

$$90 = \frac{k}{25}$$

$$2250 = k$$

The equation of variation is $I = \dfrac{2250}{d^2}$.

Substitute 40 for I and find d.

$$40 = \frac{2250}{d^2}$$

$$40d^2 = 2250$$

$$d^2 = 56.25$$

$$d = 7.5$$

The distance from 5 m to 7.5 m is $7.5 - 5$, or 2.5 m, so it is 2.5 m further to a point where the intensity is 40 W/m^2.

37.

$$d = kr^2$$

$$200 = k \cdot 60^2 \quad \text{Substituting}$$

$$200 = 3600k$$

$$\frac{200}{3600} = k$$

$$\frac{1}{18} = k$$

The equation of variation is $d = \dfrac{1}{18}r^2$.

Substitute 72 for d and find r.

$$72 = \frac{1}{18}r^2$$

$$1296 = r^2$$

$$36 = r$$

A car can travel 36 mph and still stop in 72 ft.

39. $E = \dfrac{kR}{I}$

We first find k.

$$3.18 = \frac{k \cdot 71}{201} \quad \text{Substituting}$$

$$3.18\left(\frac{201}{71}\right) = k \qquad \text{Multiplying by } \frac{201}{71}$$

$$9 \approx k$$

The equation of variation is $E = \frac{9R}{I}$.

Substitute 3.18 for E and 300 for I and solve R.

$$3.18 = \frac{9R}{300}$$

$$3.18\left(\frac{300}{9}\right) = R \qquad \text{Multiplying by } \frac{300}{9}$$

$$106 = R$$

Shawn Estes would have given up 106 earned runs if he had pitched 300 innings.

41. Discussion and Writing

43.

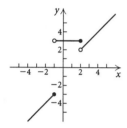

45. Test for symmetry with respect to the x-axis.

$$y^2 = x \quad \text{Original equation}$$

$$(-y)^2 = x \quad \text{Replacing } y \text{ by } -y$$

$$y^2 = x \quad \text{Simplifying}$$

The last equation is equivalent to t he original equation, so the graph is symmetric with respect to the x-axis.

Test for symmetry with respect to the y-axis:

$$y^2 = x \qquad \text{Original equation}$$

$$y^2 = -x \quad \text{Replacing } x \text{ by } -x$$

The last equation is not equivalent to the original equation, so the graph is not symmetric with respect to the y-axis.

Test for symmetry with respect to the origin:

$$y^2 = x \qquad \text{Original equation}$$

$$(-y)^2 = -x \quad \text{Replacing } x \text{ by } -x \text{ and}$$
$$\qquad\qquad\qquad y \text{ by } -y$$

$$y^2 = -x \quad \text{Simplifying}$$

The last equation is not equivalent to the original equation, so the graph is not symmetric with respect to the origin.

47. Let V represent the volume and p represent the price of a jar of peanut butter.

$$V = kp \qquad V \text{ varies directly as } p.$$

$$\pi\left(\frac{3}{2}\right)^2 (4) = k(1.2) \quad \text{Substituting}$$

$$7.5\pi = k \qquad \text{Variation constant}$$

$$V = 7.5\pi p \qquad \text{Equation of variation}$$

$$\pi(3)^2(6) = 7.5\pi p \quad \text{Substituting}$$

$$7.2 = p$$

The bigger jar should cost \$7.20.

49. We are told $A = kd^2$, and we know $A = \pi r^2$ so we have:

$$kd^2 = \pi r^2$$

$$kd^2 = \pi\left(\frac{d}{2}\right)^2 \qquad r = \frac{d}{2}$$

$$kd^2 = \frac{\pi d^2}{4}$$

$$k = \frac{\pi}{4} \qquad \text{Variation constant}$$

Exercise Set 1.7

1. Either point can be considered as (x_1, y_1).

$$d = \sqrt{(4-5)^2 + (6-9)^2}$$
$$= \sqrt{(-1)^2 + (-3)^2} = \sqrt{10} \approx 3.162$$

3. Either point can be considered as (x_1, y_1).

$$d = \sqrt{(6-9)^2 + (-1-5)^2}$$
$$= \sqrt{(-3)^2 + (-6)^2} = \sqrt{45} \approx 6.708$$

5. Either point can be considered as (x_1, y_1).

$$d = \sqrt{(-\sqrt{6}-\sqrt{3})^2 + (0-(-\sqrt{5}))^2}$$
$$= \sqrt{6 + 2\sqrt{18} + 3 + 5} = \sqrt{14 + 2\sqrt{9 \cdot 2}}$$
$$= \sqrt{14 + 2 \cdot 3\sqrt{2}} = \sqrt{14 + 6\sqrt{2}} \approx 4.742$$

7. First we find the length of the diameter:

$$d = \sqrt{(-3-9)^2 + (-1-4)^2}$$
$$= \sqrt{(-12)^2 + (-5)^2} = \sqrt{169} = 13$$

The length of the radius is one-half the length of the diameter, or $\frac{1}{2}(13)$, or 6.5.

9. First we find the distance between each pair of points.

For $(-4, 5)$ and $(6, 1)$:

$$d = \sqrt{(-4-6)^2 + (5-1)^2}$$
$$= \sqrt{(-10)^2 + 4^2} = \sqrt{116}$$

For $(-4, 5)$ and $(-8, -5)$:
$$d = \sqrt{(-4 - (-8))^2 + (5 - (-5))^2}$$
$$= \sqrt{4^2 + 10^2} = \sqrt{116}$$
For $(6, 1)$ and $(-8, -5)$:
$$d = \sqrt{(6 - (-8))^2 + (1 - (-5))^2}$$
$$= \sqrt{14^2 + 6^2} = \sqrt{232}$$
Since $(\sqrt{116})^2 + (\sqrt{116})^2 = (\sqrt{232})^2$, the points could be the vertices of a right triangle.

11. First we find the distance between each pair of points.
For $(-3, 4)$ and $(0, 5)$:
$$d = \sqrt{(-3 - 0)^2 + (4 - 5)^2}$$
$$= \sqrt{(-3)^2 + (-1)^2} = \sqrt{10}$$
For $(-3, 4)$ and $(3, -4)$:
$$d = \sqrt{(-3 - 3)^2 + (4 - (-4))^2}$$
$$= \sqrt{(-6)^2 + 8^2} = \sqrt{100} = 10$$
For $(0, 5)$ and $(3, -4)$:
$$d = \sqrt{(0 - 3)^2 + (5 - (-4))^2}$$
$$= \sqrt{(-3)^2 + 9^2} = \sqrt{90}$$
Since $(\sqrt{10})^2 + (\sqrt{90})^2 = 10^2$, the points are vertices of a right triangle.

13. We use the midpoint formula.
$$\left(\frac{4+5}{2}, \frac{6+9}{2} \right) = \left(\frac{9}{2}, \frac{15}{2} \right)$$

15. We use the midpoint formula.
$$\left(\frac{6+9}{2}, \frac{-1+5}{2} \right) = \left(\frac{15}{2}, 2 \right)$$

17. We use the midpoint formula.
$$\left(\frac{\sqrt{3} + (-\sqrt{6})}{2}, \frac{-\sqrt{5} + 0}{2} \right) = \left(\frac{\sqrt{3} - \sqrt{6}}{2}, -\frac{\sqrt{5}}{2} \right)$$

19.

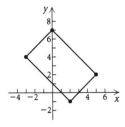

For the side with vertices $(-3, 4)$ and $(2, -1)$:
$$\left(\frac{-3+2}{2}, \frac{4 + (-1)}{2} \right) = \left(-\frac{1}{2}, \frac{3}{2} \right)$$
For the sides with vertices $(2, -1)$ and $(5, 2)$:
$$\left(\frac{2+5}{2}, \frac{-1+2}{2} \right) = \left(\frac{7}{2}, \frac{1}{2} \right)$$

For the sides with vertices $(5, 2)$ and $(0, 7)$:
$$\left(\frac{5+0}{2}, \frac{2+7}{2} \right) = \left(\frac{5}{2}, \frac{9}{2} \right)$$
For the sides with vertices $(0, 7)$ and $(-3, 4)$:
$$\left(\frac{0 + (-3)}{2}, \frac{7+4}{2} \right) = \left(-\frac{3}{2}, \frac{11}{2} \right)$$
For the quadrilateral whose vertices are the points found above, the diagonals have endpoints
$$\left(-\frac{1}{2}, \frac{3}{2} \right), \left(\frac{5}{2}, \frac{9}{2} \right) \text{ and } \left(\frac{7}{2}, \frac{1}{2} \right), \left(-\frac{3}{2}, \frac{11}{2} \right).$$
We find the length of each of these diagonals.
For $\left(-\frac{1}{2}, \frac{3}{2} \right), \left(\frac{5}{2}, \frac{9}{2} \right)$:
$$d = \sqrt{\left(-\frac{1}{2} - \frac{5}{2} \right)^2 + \left(\frac{3}{2} - \frac{9}{2} \right)^2}$$
$$= \sqrt{(-3)^2 + (-3)^2} = \sqrt{18}$$
For $\left(\frac{7}{2}, \frac{1}{2} \right), \left(-\frac{3}{2}, \frac{11}{2} \right)$:
$$d = \sqrt{\left(\frac{7}{2} - \left(-\frac{3}{2} \right) \right)^2 + \left(\frac{1}{2} - \frac{11}{2} \right)^2}$$
$$= \sqrt{5^2 + (-5)^2} = \sqrt{50}$$
Since the diagonals do not have the same lengths, the midpoints are not vertices of a rectangle.

21. We use the midpoint formula.
$$\left(\frac{\sqrt{7} + \sqrt{2}}{2}, \frac{-4+3}{2} \right) = \left(\frac{\sqrt{7} + \sqrt{2}}{2}, -\frac{1}{2} \right)$$

23. Square the viewing window. For the graph shown, one possibility is $[-12, 9, -4, 10]$.

25. $(x - h)^2 + (y - k)^2 = r^2$
$$(x - 2)^2 + (y - 3)^2 = \left(\frac{5}{3} \right)^2 \quad \text{Substituting}$$
$$(x - 2)^2 + (y - 3)^2 = \frac{25}{9}$$

27. The length of a radius is the distance between $(-1, 4)$ and $(3, 7)$:
$$r = \sqrt{(-1 - 3)^2 + (4 - 7)^2}$$
$$= \sqrt{(-4)^2 + (-3)^2} = \sqrt{25} = 5$$

$$(x - h)^2 + (y - k)^2 = r^2$$
$$[x - (-1)]^2 + (y - 4)^2 = 5^2$$
$$(x + 1)^2 + (y - 4)^2 = 25$$

29. The center is the midpoint of the diameter:
$$\left(\frac{7 + (-3)}{2}, \frac{13 + (-11)}{2} \right) = (2, 1)$$

Use the center and either endpoint of the diameter to find the length of a radius. We use the point $(7, 13)$:

$$r = \sqrt{(7-2)^2 + (13-1)^2}$$
$$= \sqrt{5^2 + 12^2} = \sqrt{169} = 13$$

$$(x-h)^2 + (y-k)^2 = r^2$$
$$(x-2)^2 + (y-1)^2 = 13^2$$
$$(x-2)^2 + (y-1)^2 = 169$$

31. Since the center is 2 units to the left of the y-axis and the circle is tangent to the y-axis, the length of a radius is 2.

$$(x-h)^2 + (y-k)^2 = r^2$$
$$[x-(-2)]^2 + (y-3)^2 = 2^2$$
$$(x+2)^2 + (y-3)^2 = 4$$

33.
$$x^2 + y^2 = 4$$
$$(x-0)^2 + (y-0)^2 = 2^2$$
Center: $(0,0)$; radius: 2

$$x^2 + y^2 = 4$$
$$y_1 = \sqrt{4-x^2}, \quad y_2 = -\sqrt{4-x^2}$$

35.
$$x^2 + (y-3)^2 = 16$$
$$(x-0)^2 + (y-3)^2 = 4^2$$
Center: $(0,3)$; radius: 4

$$x^2 + (y-3)^2 = 16$$
$$y_1 = 3 + \sqrt{16-x^2}, \quad y_2 = 3 - \sqrt{16-x^2}$$

37. $(x-1)^2 + (y-5)^2 = 36$
$$(x-1)^2 + (y-5)^2 = 6^2$$
Center: $(1,5)$; radius: 6

$$(x-1)^2 + (y-5)^2 = 36$$
$$y_1 = 5 + \sqrt{36-(x-1)^2}, \quad y_2 = 5 - \sqrt{36-(x-1)^2}$$

39.
$$(x+4)^2 + (y+5)^2 = 9$$
$$[x-(-4)]^2 + [y-(-5)]^2 = 3^2$$
Center: $(-4,-5)$; radius: 3

$$(x+4)^2 + (y+5)^2 = 9$$
$$y_1 = -5 + \sqrt{9-(x+4)^2}, \quad y_2 = -5 - \sqrt{9-(x+4)^2}$$

41. Discussion and Writing

43. The equation $y = -\dfrac{3}{5}x + 4$ is in the form $y = mx + b$. The slope is $-\dfrac{3}{5}$ and the y-intercept is $(0, 4)$.

45. Substitute 3 for m and -1 for b in the equation $y = mx + b$. We have $y = 3x - 1$.

47. Use the distance formula. Either point can be considered as (x_1, y_1).

$$d = \sqrt{(a+h-a)^2 + (\sqrt{a+h} - \sqrt{a})^2}$$
$$= \sqrt{h^2 + a + h - 2\sqrt{a^2 + ah} + a}$$
$$= \sqrt{h^2 + 2a + h - 2\sqrt{a^2 + ah}}$$

Next we use the midpoint formula.

$$\left(\frac{a+a+h}{2}, \frac{\sqrt{a}+\sqrt{a+h}}{2} \right) = \left(\frac{2a+h}{2}, \frac{\sqrt{a}+\sqrt{a+h}}{2} \right)$$

49. First use the formula for the area of a circle to find r^2:

$$A = \pi r^2$$
$$36\pi = \pi r^2$$
$$36 = r^2$$

Then we have:

$$(x-h)^2 + (y-k)^2 = r^2$$
$$(x-2)^2 + [y-(-7)]^2 = 36$$
$$(x-2)^2 + (y+7)^2 = 36$$

51. Label the drawing with additional information and lettering.

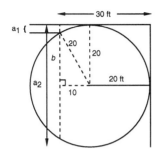

Find b using the Pythagorean theorem.

$$b^2 + 10^2 = 20^2$$
$$b^2 + 100 = 400$$
$$b^2 = 300$$
$$b = 10\sqrt{3}$$
$$b \approx 17.3$$

Find a_1:

$$a_1 = 20 - b \approx 20 - 17.3 \approx 2.7 \text{ ft}$$

Find a_2:

$$a_2 = 2b + a_1 \approx 2(17.3) + 2.7 \approx 37.3 \text{ ft}$$

53.

$$\frac{x^2 + y^2 = 1}{\left(\frac{\sqrt{3}}{2}\right)^2 + \left(-\frac{1}{2}\right)^2 \ ? \ 1}$$

$$\frac{3}{4} + \frac{1}{4}$$

$$1 \ \bigg| \ 1 \quad \text{TRUE}$$

$\left(\dfrac{\sqrt{3}}{2}, -\dfrac{1}{2}\right)$ lies on the unit circle.

55.

$$\frac{x^2 + y^2 = 1}{\left(\frac{\sqrt{2}}{2}\right)^2 + \left(\frac{\sqrt{2}}{2}\right)^2 \ ? \ 1}$$

$$\frac{2}{4} + \frac{2}{4}$$

$$1 \ \bigg| \ 1 \quad \text{TRUE}$$

$\left(\dfrac{\sqrt{2}}{2}, \dfrac{\sqrt{2}}{2}\right)$ lies on the unit circle.

57. Let $(0, y)$ be the required point. We set the distance between $(-2, 0)$ and $(0, y)$ equal to the distance between $(4, 6)$ and $(0, y)$ and solve for y.

$$\sqrt{[0 - (-2)]^2 + (y - 0)^2} = \sqrt{(0 - 4)^2 + (y - 6)^2}$$
$$\sqrt{4 + y^2} = \sqrt{16 + y^2 - 12y + 36}$$
$$4 + y^2 = 16 + y^2 - 12y + 36$$
$$\text{Squaring both sides}$$
$$-48 = -12y$$
$$4 = y$$

The point is $(0, 4)$.

59. a), b) See the answer section in the text.

Chapter 2

Functions and Equations: Zeros and Solutions

Exercise Set 2.1

1. a) The graph crosses the x-axis at $(4, 0)$. This is the x-intercept.

   b) The zero of the function is the first coordinate of the x-intercept. It is 4.

3.
$$x + 5 = 0 \qquad \text{Setting } f(x) = 0$$
$$x + 5 - 5 = 0 - 5 \quad \text{Subtracting 5 on both sides}$$
$$x = -5$$
The zero of the function is -5.

5.
$$-x + 18 = 0 \qquad \text{Setting } f(x) = 0$$
$$-x + 18 + x = 0 + x \quad \text{Adding } x \text{ on both sides}$$
$$18 = x$$
The zero of the function is 18.

7.
$$16 - x = 0 \qquad \text{Setting } f(x) = 0$$
$$16 - x + x = 0 + x \quad \text{Adding } x \text{ on both sides}$$
$$16 = x$$
The zero of the function is 16.

9. $f(x) = x + 12$

Graph the function in a window that displays the x-intercept. We use $[-15, 5, -10, 10]$. Then use the Zero feature.

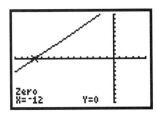

The zero of the function is -12.

11. $f(x) = -x + 6$

Graph the function in a window that displays the x-intercept. We use the standard window. Then use the Zero feature.

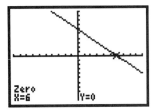

The zero of the function is 6.

13. $f(x) = 20 - x$

Graph the function in a window that displays the x-intercept. We use $[-5, 25, -10, 10]$, $\text{Xscl} = 5$. Then use the Zero feature.

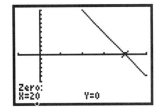

The zero of the function is 20.

15. $f(x) = x - 6$

Algebraic solution:
$$x - 6 = 0 \quad \text{Setting } f(x) = 0$$
$$x = 6 \quad \text{Adding 6 on both sides}$$

Graphical solution:

Graph the function in a window that displays the x-intercept. We use the standard window. Then use the Zero feature.

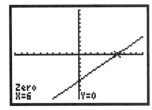

The zero of the function is 6.

17. $f(x) = x + 15$

Algebraic solution:
$$-x + 15 = 0 \quad \text{Setting } f(x) = 0$$
$$15 = x \quad \text{Adding } x \text{ on both sides}$$

Graphical solution:

Graph the function in a window that displays the x-intercept. We use $[-5, 20, -10, 10]$. Then use the Zero feature.

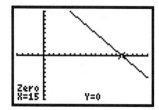

The zero of the function is 15.

19. $4x + 3 = 0$

$$4x = -3 \qquad \text{Subtracting 3}$$

$$x = -\frac{3}{4} \qquad \text{Dividing by 4}$$

The solution is $-\frac{3}{4}$.

21. $2x + 7 = x + 3$

$$x + 7 = 3 \qquad \text{Subtracting } x$$

$$x = -4 \qquad \text{Subtracting 7}$$

The solution is -4.

23. $3(x + 1) = 5 - 2(3x + 4)$

$$3x + 3 = 5 - 6x - 8 \qquad \text{Removing parentheses}$$

$$3x + 3 = -6x - 3 \qquad \text{Collecting like terms}$$

$$9x + 3 = -3 \qquad \text{Adding } 6x$$

$$9x = -6 \qquad \text{Subtracting 3}$$

$$x = -\frac{2}{3} \qquad \text{Dividing by 9}$$

25. $5x - 8 = 0$

The solution of this equation is the zero of the function $f(x) = 5x - 8$. Graph the function in a window that displays the x-intercept. We use the standard window. Then use the Zero feature.

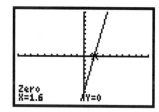

The solution is 1.6.

27. $3x - 5 = 2x + 1$

Graph $y_1 = 3x - 5$ and $y_2 = 2x + 1$ in a window that displays the point of intersection of the graphs. We use $[-10, 10, -5, 20]$, $\text{Yscl} = 5$. Then use the

Intersect feature to find the first coordinate of the point of intersection.

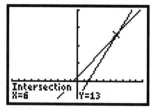

The solution is 6.

29. $2(x - 4) = 3 - 5(2x + 1)$

Graph $y_1 = 2(x - 4)$ and $y_2 = 3 - 5(2x + 1)$ in a window that displays the point of intersection of the graphs. We use the standard window. Then use the Intersect feature to find the first coordinate of the point of intersection.

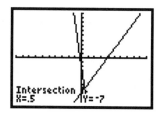

The solution is 0.5.

31. $8x + 25 = 0$

Algebraic solution:

$$8x + 25 = 0$$

$$8x = -25 \qquad \text{Subtracting 25}$$

$$x = -\frac{25}{8}, \text{ or } -3.125 \quad \text{Dividing by 8}$$

Graphical solution:

The solution of this equation is the zero of the function $f(x) = 8x + 25$. Graph the function in a window that displays the x-intercept. We use the standard window. Then use the Zero feature.

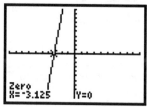

The solution is $-\dfrac{25}{8}$, or -3.125.

33. $6x - 5 = 3x + 10$

Algebraic solution:

$$6x - 5 = 3x + 10$$
$$3x - 5 = 10 \qquad \text{Subtracting } 3x$$
$$3x = 15 \qquad \text{Adding } 5$$
$$x = 5 \qquad \text{Dividing by } 3$$

Graphical solution:

Graph $y_1 = 6x - 5$ and $y_2 = 3x + 10$ in a window that displays the point of intersection of the graphs. We use $[-10, 10, -5, 35]$, Yscl $= 5$. Then use the Intersect feature to find the first coordinate of the point of intersection.

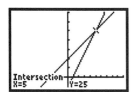

The solution is 5.

35. *Familiarize*. Let $x =$ the number of foreign tourists who visited the United States in 1998, in millions.

Translate. We reword the problem and translate to an equation.

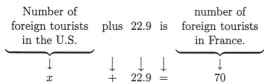

Carry out. We will solve algebraically.

$$x + 22.9 = 70$$
$$x = 47.1 \qquad \text{Subtracting } 22.9$$

Check. $47.1 + 22.9 = 70$, so the answer checks.

State. In 1998, 47.1 million foreign tourists visited the United States.

37. *Familiarize*. Let $c =$ the number of catalogs each household received.

Translate.

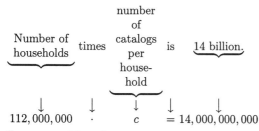

Carry out. We will solve algebraically.

$$112,000,000c = 14,000,000,000$$
$$c = 125$$

Check. If each of 112 million households receives 125 catalogs, then $112,000,000 \cdot 125$, or 14,000,000,000 catalogs are received. The answer checks.

State. Each household received 125 catalogs.

39. *Familiarize*. Let $p =$ the number of Braille pages published in 1977, in millions.

Translate.

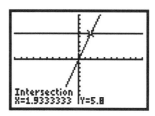

Carry out. We will solve graphically. Graph $y_1 = 3x$ and $y_2 = 5.8$ and find the first coordinate of the point of intersection of the graphs.

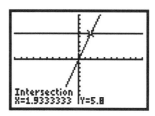

The possible solution is approximately 1.93 million.

Check. $3(1.93$ million$) = 5.79$ million $\approx$ 5.8 million, so the answer checks. (Recall that we rounded the possible solution to 2 decimal places.)

State. Approximately 1.93 million Braille pages were published in 1977.

41. *Familiarize*. Using the labels on the drawing in the text, we let $w =$ the width of the test plot and $w + 25 =$ the length. Recall that for a rectangle, Perimeter $= 2 \cdot$ length $+ 2 \cdot$ width.

Translate.

$$\underbrace{\text{Perimeter}}_{322} = \underbrace{2 \cdot \text{ length}}_{= 2(w + 25)} + \underbrace{2 \cdot \text{ width}}_{+ \quad 2 \cdot w}$$

Carry out. We solve the equation.

$$322 = 2(w + 25) + 2 \cdot w$$
$$322 = 2w + 50 + 2w$$
$$322 = 4w + 50$$
$$272 = 4w$$
$$68 = w$$

We could also find the first coordinate of the point of intersection of $y_1 = 322$ and $y_2 = 2(x + 25) + 2x$ using the Intersect feature.

When $w = 68$, then $w + 25 = 68 + 25 = 93$.

Check. The length is 25 m more than the width: $93 = 68 + 25$. The perimeter is $2 \cdot 93 + 2 \cdot 68$, or $186 + 136$, or 322 m. The answer checks.

State. The length is 93 m; the width is 68 m.

43. Familiarize. We make a drawing.

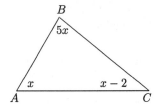

We let $x = $ the measure of angle A. Then $5x = $ the measure of angle B, and $x - 2 = $ the measure of angle C. The sum of the angle measures is $180°$.

Translate.

Measure of angle A	+	Measure of angle B	+	Measure of angle C	= 180.
x	+	$5x$	+	$x - 2$	$= 180$

Carry out. We solve the equation.

$$x + 5x + x - 2 = 180$$
$$7x - 2 = 180$$
$$7x = 182$$
$$x = 26$$

We could also find the first coordinate of the point of intersection of $y_1 = x + 5x + x - 2$ and $y_2 = 180$ using the Intersect feature.

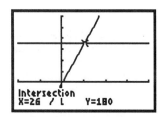

If $x = 26$, then $5x = 5 \cdot 26$, or 130, and $x - 2 = 26 - 2$, or 24.

Check. The measure of angle B, $130°$, is five times the measure of angle A, $26°$. The measure of angle C, $24°$, is $2°$ less than the measure of angle A, $26°$. The sum of the angle measures is $26° + 130° + 24°$, or $180°$. The answer checks.

State. The measure of angles A, B, and C are $26°$, $130°$, and $24°$, respectively.

45. Familiarize. Let $n = $ the former number of enrollees.

Translate. We will express 48.4% as 0.484.

Former number of enrollees	+	48.4% of		Former number of enrollees	is	416,300.
n	+	0.484	$\cdot$	n	=	416,300

Carry out. We solve the equation.

$$n + 0.484n = 416,300$$
$$1.484n = 416,300 \quad \text{Collecting like terms}$$
$$n \approx 280,526$$

We could also find the first coordinate of the point of intersection of $y_1 = x + 0.484x$ and $y_2 = 416,300$ using the Intersect feature.

Check. 48.4% of $280,526 \approx 135,775$ and $280,526 + 135,775 = 416,301 \approx 416,300$. The answer checks.

State. The former number of enrollees was about 280,526.

47. Familiarize. We make a drawing. Let $r = $ the speed of the Central Railway freight train. Then $r + 14 = $ the speed of the Amtrak passenger train. Also let $t = $ the time the train travels.

Passenger train
r mph t hr   330 mi

Passenger train
$r + 14$ mph t hr   400 mi

We can also organize the information in a table.

d	$=$	r	$\cdot$	t
	Distance	Rate	Time
Freight train	330	r	t
Passenger train	400	$r + 14$	t

Translate. Using the formula $d = rt$ in each row of the table, we get two equations.

$$330 = rt \quad \text{and} \quad 400 = (r + 14)t$$

Solve each equation for t.

$$\frac{330}{r} = t \quad \text{and} \quad \frac{400}{r + 14} = t$$

Thus, we have the equation

$$\frac{330}{r} = \frac{400}{r + 14}.$$

Carry out. We solve the equation.

$$\frac{330}{r} = \frac{400}{r+14}, \text{ LCD is } r(r+14)$$

$$r(r+14) \cdot \frac{330}{r} = r(r+14) \cdot \frac{400}{r+14}$$

$$330(r+14) = 400r$$

$$330r + 4620 = 400r$$

$$4620 = 70r$$

$$66 = r$$

We could also find the first coordinate of the point of intersection of $y_1 = \frac{330}{x}$ and $y_2 = \frac{400}{x+14}$ using the Intersect feature.

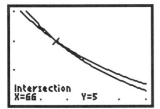

When $r = 66$, $r + 14 = 66 + 14 = 80$.

Check. The freight train travels 330 mi in 330/66, or 5 hr and the passenger train travels 400 mi in 400/80, or 5 hr. Since the times are the same the answer checks.

State. The speed of the Central Railway freight train is 66 mph, and the speed of the Amtrak passenger train is 80 mph.

49. *Familiarize*. Let w = the number of pounds of Jocelyn's body weight that is water.

Translate.

50% of body weight is water.

$$\downarrow \quad \downarrow \qquad \downarrow \qquad \downarrow \quad \downarrow$$
$$0.5 \quad \times \qquad 135 \qquad = \quad w$$

Carry out.

$$0.5 \times 135 = w$$

$$67.5 = w$$

Check. Since 50% of 138 is 67.5, the answer checks.

State. 67.5 lb of Jocelyn's boyd weight is water.

51. *Familiarize*. Let n = the number of Americans who had cardiovascular procedures in 1996, in millions. Then 9% more than this number is $n + 9\%n$, or $n + 0.09n$, or $1.09n$. This is the number of procedures performed in 1999.

Translate.

The number of procedures in 1999 is 6 million.

$$\downarrow \qquad\qquad \downarrow \quad \downarrow$$
$$1.09n \qquad\quad = \qquad 6$$

Carry out. We will solve graphically. Graph $y_1 = 1.09x$ and $y_2 = 6$ and find the first coordinate of the point of intersection of the graphs.

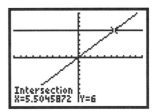

The possible solution is approximately 5.5 million.

Check. 9% of 5.5 is 0.495 and $5.5 + 0.495 = 5.995 \approx 6$. The answer checks.

State. Approximately 5.5 million Americans had cardiovascular procedures in 1996.

53. *Familiarize*. Let c = the number of 30-sec commercials NBC must sell on *ER* in order to break even on a episode of the show.

Translate.

Cost per commercial times number of commercials is $13 million.

$$\downarrow \qquad\qquad \downarrow \qquad \downarrow \qquad \downarrow \qquad \downarrow$$
$$565,000 \qquad \times \qquad c \qquad = \quad 13,000,000$$

Carry out. We will solve algebraically.

$$565,000c = 13,000,000$$

$$c \approx 23$$

Check. If 23 commercials are sold at $565,000 each, then $23(\$565,000) = \$12,995,000 \approx \$13,000,000$ is raised. The answer checks.

State. NBC must sell 23 30-sec commercials on *ER* in order to break even on the episode.

55. *Familiarize*. Let x = the number of years it will take Horseshoe Falls to migrate one-fourth mile upstream. We will express one-fourth mile in feet:

$$\frac{1}{4} \text{ mi} \times \frac{5280 \text{ ft}}{1 \text{ mi}} = 1320 \text{ ft}$$

Translate.

Rate per year times number of years is 1320 ft.

$$\downarrow \qquad\qquad \downarrow \qquad \downarrow \qquad \downarrow \qquad \downarrow$$
$$2 \qquad\qquad \cdot \qquad x \qquad = \quad 1320$$

Carry out. We will solve algebraically.

$$2x = 1320$$
$$x = 660$$

Check. At a rate of 2 ft per year, in 660 yr the falls will migrate $2 \cdot 660$, or 1320 ft. The answer checks.

State. It will take Horseshoe Falls 660 yr to migrate one-fourth mile upstream.

57. Discussion and Writing

59. First find the slope of the given line.

$$3x + 4y = 7$$
$$4y = -3x + 7$$
$$y = -\frac{3}{4}x + \frac{7}{4}$$

The slope is $-\frac{3}{4}$. Now write a slope-intersect equation of the line containing $(-1, 4)$ with slope $-\frac{3}{4}$.

$$y - 4 = -\frac{3}{4}[x - (-1)]$$
$$y - 4 = -\frac{3}{4}(x + 1)$$
$$y - 4 = -\frac{3}{4}x - \frac{3}{4}$$
$$y = -\frac{3}{4}x + \frac{13}{4}$$

61. $f(x) = 2x - 1$, $g(x) = x^2 - 5$

$(f + g)(x) = f(x) + g(x) = (2x - 1) + (x^2 - 5) = x^2 + 2x - 6$

63. $f(x) = 7 - \frac{3}{2}x = -\frac{3}{2}x + 7$

The function can be written in the form $y = mx + b$, so it is a linear function.

65. $f(x) = x^2 + 1$ cannot be written in the form $f(x) = mx + b$, so it is not a linear function.

Exercise Set 2.2

1. $(-5 + 3i) + (7 + 8i)$

$= (-5 + 7) + (3i + 8i)$ Collecting the real parts
                              and the imaginary parts

$= 2 + (3 + 8)i$

$= 2 + 11i$

3. $(4 - 9i) + (1 - 3i)$

$= (4 + 1) + (-9i - 3i)$ Collecting the real parts
                              and the imaginary parts

$= 5 + (-9 - 3)i$

$= 5 - 12i$

5. $(3 + \sqrt{-16}) + (2 + \sqrt{-25}) = (3 + 4i) + (2 + 5i)$

$\qquad\qquad\qquad\qquad = (3 + 2) + (4i + 5i)$

$\qquad\qquad\qquad\qquad = 5 + 9i$

7. $(10 + 7i) - (5 + 3i)$

$= (10 - 5) + (7i - 3i)$ The 5 and the $3i$ are
                              both being subtracted.

$= 5 + 4i$

9. $(13 + 9i) - (8 + 2i)$

$= (13 - 8) + (9i - 2i)$ The 8 and the $2i$ are
                              both being subtracted.

$= 5 + 7i$

11. $(1 + 3i)(1 - 4i)$

$= 1 - 4i + 3i - 12i^2$ Using FOIL

$= 1 - 4i + 3i - 12(-1)$ $i^2 = -1$

$= 1 - i + 12$

$= 13 - i$

13. $(2 + 3i)(2 + 5i)$

$= 4 + 10i + 6i + 15i^2$ Using FOIL

$= 4 + 10i + 6i - 15$ $i^2 = -1$

$= -11 + 16i$

15. $7i(2 - 5i)$

$= 14i - 35i^2$ Using the distributive law

$= 14i + 35$ $i^2 = -1$

$= 35 + 14i$ Writing in the form $a + bi$

17. $(3 + \sqrt{-16})(2 + \sqrt{-25})$

$= (3 + 4i)(2 + 5i)$

$= 6 + 15i + 8i + 20i^2$

$= 6 + 15i + 8i - 20$ $i^2 = -1$

$= -14 + 23i$

19. $(5 - 4i)(5 + 4i) = 5^2 - (4i)^2$

$\qquad\qquad\qquad\quad = 25 - 16i^2$

$\qquad\qquad\qquad\quad = 25 + 16$ $i^2 = -1$

$\qquad\qquad\qquad\quad = 41$

21. $(4 + 2i)^2$

$= 16 + 2 \cdot 4 \cdot 2i + (2i)^2$ Recall $(A + B)^2 = A^2 + 2AB + B^2$

$= 16 + 16i + 4i^2$

$= 16 + 16i - 4$ $i^2 = -1$

$= 12 + 16i$

23.

$(-2 + 7i)^2$

$= (-2)^2 + 2(-2)(7i) + (7i)^2$ Recall $(A + B)^2 =$
$A^2 + 2AB + B^2$

$= 4 - 28i + 49i^2$

$= 4 - 28i - 49$ $i^2 = -1$

$= -45 - 28i$

25.

$\dfrac{2 + \sqrt{3}i}{5 - 4i}$

$= \dfrac{2 + \sqrt{3}i}{5 - 4i} \cdot \dfrac{5 + 4i}{5 + 4i}$ $5 + 4i$ is the conjugate of the divisor.

$= \dfrac{(2 + \sqrt{3}i)(5 + 4i)}{(5 - 4i)(5 + 4i)}$

$= \dfrac{10 + 8i + 5\sqrt{3}i + 4\sqrt{3}i^2}{25 - 16i^2}$

$= \dfrac{10 + 8i + 5\sqrt{3}i - 4\sqrt{3}}{25 + 16}$ $i^2 = -1$

$= \dfrac{10 - 4\sqrt{3} + (8 + 5\sqrt{3})i}{41}$

$= \dfrac{10 - 4\sqrt{3}}{41} + \dfrac{8 + 5\sqrt{3}}{41}i$ Writing in the form $a + bi$

27.

$\dfrac{4 + i}{-3 - 2i}$

$= \dfrac{4 + i}{-3 - 2i} \cdot \dfrac{-3 + 2i}{-3 + 2i}$ $-3 + 2i$ is the conjugate of the divisor.

$= \dfrac{(4 + i)(-3 + 2i)}{(-3 - 2i)(-3 + 2i)}$

$= \dfrac{-12 + 5i + 2i^2}{9 - 4i^2}$

$= \dfrac{-12 + 5i - 2}{9 + 4}$ $i^2 = -1$

$= \dfrac{-14 + 5i}{13}$

$= -\dfrac{14}{13} + \dfrac{5}{13}i$ Writing in the form $a + bi$

29.

$\dfrac{3}{5 - 11i}$

$= \dfrac{3}{5 - 11i} \cdot \dfrac{5 + 11i}{5 + 11i}$ $5 - 11i$ is the conjugate of $5 + 11i$.

$= \dfrac{3(5 + 11i)}{(5 - 11i)(5 + 11i)}$

$= \dfrac{15 + 33i}{25 - 121i^2}$

$= \dfrac{15 + 33i}{25 + 121}$ $i^2 = -1$

$= \dfrac{15 + 33i}{146}$

$= \dfrac{15}{146} + \dfrac{33}{146}i$ Writing in the form $a + bi$

31.

$\dfrac{1 + i}{(1 - i)^2}$

$= \dfrac{1 + i}{1 - 2i + i^2}$

$= \dfrac{1 + i}{1 - 2i - 1}$ $i^2 = -1$

$= \dfrac{1 + i}{-2i}$

$= \dfrac{1 + i}{-2i} \cdot \dfrac{2i}{2i}$ $2i$ is the conjugate of $-2i$.

$= \dfrac{(1 + i)(2i)}{(-2i)(2i)}$

$= \dfrac{2i + 2i^2}{-4i^2}$

$= \dfrac{2i - 2}{4}$ $i^2 = -1$

$= -\dfrac{2}{4} + \dfrac{2}{4}i$

$= -\dfrac{1}{2} + \dfrac{1}{2}i$

33.
$$\frac{4-2i}{1+i}+\frac{2-5i}{1+i}$$

$$=\frac{6-7i}{1+i} \qquad \text{Adding}$$

$$=\frac{6-7i}{1+i}\cdot\frac{1-i}{1-i} \qquad \begin{array}{l}1-i \text{ is the conjugate}\\ \text{of } 1+i.\end{array}$$

$$=\frac{(6-7i)(1-i)}{(1+i)(1-i)}$$

$$=\frac{6-13i+7i^2}{1-i^2}$$

$$=\frac{6-13i-7}{1+1} \qquad i^2=-1$$

$$=\frac{-1-13i}{2}$$

$$=-\frac{1}{2}-\frac{13}{2}i$$

35. $i^{11}=i^{10}\cdot i=(i^2)^5\cdot i=(-1)^5\cdot i=-1\cdot i=-i$

37. $i^{35}=i^{34}\cdot i=(i^2)^{17}\cdot i=(-1)^{17}\cdot i=-1\cdot i=-i$

39. $i^{64}=(i^2)^{32}=(-1)^{32}=1$

41. $(-i)^{71}=(-1\cdot i)^{71}=(-1)^{71}\cdot i^{71}=-i^{70}\cdot i=$
$-(i^2)^{35}\cdot i=-(-1)^{35}\cdot i=-(-1)i=i$

43. $(5i)^4=5^4\cdot i^4=625(i^2)^2=625(-1)^2=625\cdot1=625$

45. Discussion and Writing

47. First find the slope of the given line.
$$3x-6y=7$$
$$-6y=-3x+7$$
$$y=\frac{1}{2}x-\frac{7}{6}$$

The slope is $\frac{1}{2}$. The slope of the desired line is the opposite of the reciprocal of $\frac{1}{2}$, or -2. Write a slope-intercept equation of the line containing $(3,-5)$ with slope -2.
$$y-(-5)=-2(x-3)$$
$$y+5=-2x+6$$
$$y=-2x+1$$

49. $(f/g)(2)=\dfrac{f(2)}{g(2)}=\dfrac{2^2+4}{3\cdot2+5}=\dfrac{4+4}{6+5}=\dfrac{8}{11}$

51. $(a+bi)+(a-bi)=2a$, a real number. Thus, the statement is true.

53. $(a+bi)(c+di)=(ac-bd)+(ad+bc)i$. The conjugate of the product is $(ac-bd)-(ad+bc)i=(a-bi)(c-di)$, the product of the conjugates of the individual complex numbers. Thus, the statement is true.

55. $z\bar{z}=(a+bi)(a-bi)=a^2-b^2i^2=a^2+b^2$

Exercise Set 2.3

1. a) The graph crosses the x-axis at $(-4,0)$ and at $(2,0)$. These are the x-intercepts.

b) The zeros of the function are the first coordinates of the x-intercepts of the graph. They are -4 and 2.

3.
$$x^2+6x=7$$
$$x^2+6x+9=7+9 \qquad \begin{array}{l}\text{Completing the square:}\\ \frac{1}{2}\cdot6=3 \text{ and } 3^2=9\end{array}$$
$$(x+3)^2=16 \qquad \text{Factoring}$$
$$x+3=\pm4 \qquad \begin{array}{l}\text{Using the principle}\\ \text{of square roots}\end{array}$$
$$x=-3\pm4$$
$$x=-3-4 \quad or \quad x=-3+4$$
$$x=-7 \qquad or \quad x=1$$
The solutions are -7 and 1.

5.
$$x^2=8x-9$$
$$x^2-8x=-9 \qquad \text{Subtracting } 8x$$
$$x^2-8x+16=-9+16 \qquad \begin{array}{l}\text{Completing the square:}\\ \frac{1}{2}(-8)=-4 \text{ and } (-4)^2=16\end{array}$$
$$(x-4)^2=7 \qquad \text{Factoring}$$
$$x-4=\pm\sqrt{7} \qquad \begin{array}{l}\text{Using the principle}\\ \text{of square roots}\end{array}$$
$$x=4\pm\sqrt{7}$$
The solutions are $4-\sqrt{7}$ and $4+\sqrt{7}$, or $4\pm\sqrt{7}$.

7.
$$x^2+8x+25=0$$
$$x^2+8x=-25 \qquad \text{Subtracting } 25$$
$$x^2+8x+16=-25+16 \qquad \begin{array}{l}\text{Completing the}\\ \text{square:}\end{array}$$
$$\frac{1}{2}\cdot8=4 \text{ and } 4^2=16$$
$$(x+4)^2=-9 \qquad \text{Factoring}$$
$$x+4=\pm3i \qquad \begin{array}{l}\text{Using the principle}\\ \text{of square roots}\end{array}$$
$$x=-4\pm3i$$
The solutions are $-4-3i$ and $-4+3i$, or $-4\pm3i$.

9. $3x^2 + 5x - 2 = 0$

$\qquad 3x^2 + 5x = 2 \qquad$ Adding 2

$\qquad x^2 + \dfrac{5}{3}x = \dfrac{2}{3} \qquad$ Dividing by 3

$\qquad x^2 + \dfrac{5}{3}x + \dfrac{25}{36} = \dfrac{2}{3} + \dfrac{25}{36} \qquad$ Completing the
$\qquad\qquad\qquad\qquad\qquad\qquad$ square:

$\qquad\qquad \frac{1}{2} \cdot \frac{5}{3} = \frac{5}{6}$ and $(\frac{5}{6})^2 = \frac{25}{36}$

$\qquad \left(x + \dfrac{5}{6} \right)^2 = \dfrac{49}{36} \qquad$ Factoring and
$\qquad\qquad\qquad\qquad\qquad$ simplifying

$\qquad x + \dfrac{5}{6} = \pm\dfrac{7}{6} \qquad$ Using the principle
$\qquad\qquad\qquad\qquad\qquad$ of square roots

$\qquad\qquad x = -\dfrac{5}{6} \pm \dfrac{7}{6}$

$x = -\dfrac{5}{6} - \dfrac{7}{6}$ $\quad or \quad$ $x = -\dfrac{5}{6} + \dfrac{7}{6}$

$x = -\dfrac{12}{6}$ $\quad or \quad$ $x = \dfrac{2}{6}$

$x = -2$ $\quad or \quad$ $x = \dfrac{1}{3}$

The solutions are -2 and $\dfrac{1}{3}$.

11. $\qquad x^2 - 2x = 15$

$\qquad x^2 - 2x - 15 = 0$

$\qquad (x-5)(x+3) = 0 \quad$ Factoring

$\qquad x - 5 = 0 \ \ or \ \ x + 3 = 0$

$\qquad\quad x = 5 \ \ or \ \qquad x = -3$

The solutions are 5 and -3.

13. $\qquad 5m^2 + 3m = 2$

$\qquad 5m^2 + 3m - 2 = 0$

$\qquad (5m - 2)(m + 1) = 0 \quad$ Factoring

$\qquad 5m - 2 = 0 \ \ or \ \ m + 1 = 0$

$\qquad\quad m = \dfrac{2}{5} \ \ or \ \qquad m = -1$

The solutions are $\dfrac{2}{5}$ and -1.

15. $\qquad 3x^2 + 6 = 10x$

$\qquad 3x^2 - 10x + 6 = 0$

We use the quadratic formula. Here $a = 3$, $b = -10$, and $c = 6$.

$x = \dfrac{-b \pm \sqrt{b^2 - 4ac}}{2a}$

$\quad = \dfrac{-(-10) \pm \sqrt{(-10)^2 - 4 \cdot 3 \cdot 6}}{2 \cdot 3} \quad$ Substituting

$\quad = \dfrac{10 \pm \sqrt{28}}{6} = \dfrac{10 \pm 2\sqrt{7}}{6}$

$\quad = \dfrac{2(5 \pm \sqrt{7})}{2 \cdot 3} = \dfrac{5 \pm \sqrt{7}}{3}$

The solutions are $\dfrac{5 - \sqrt{7}}{3}$ and $\dfrac{5 + \sqrt{7}}{3}$, or $\dfrac{5 \pm \sqrt{7}}{3}$.

17. $x^2 + x + 2 = 0$

We use the quadratic formula. Here $a = 1$, $b = 1$, and $c = 2$.

$x = \dfrac{-b \pm \sqrt{b^2 - 4ac}}{2a}$

$\quad = \dfrac{-1 \pm \sqrt{1^2 - 4 \cdot 1 \cdot 2}}{2 \cdot 1} \quad$ Substituting

$\quad = \dfrac{-1 \pm \sqrt{-7}}{2}$

$\quad = \dfrac{-1 \pm \sqrt{7}i}{2} = -\dfrac{1}{2} \pm \dfrac{\sqrt{7}}{2}i$

The solutions are $-\dfrac{1}{2} - \dfrac{\sqrt{7}}{2}i$ and $-\dfrac{1}{2} + \dfrac{\sqrt{7}}{2}i$, or $-\dfrac{1}{2} \pm \dfrac{\sqrt{7}}{2}i$.

19. $\qquad 5t^2 - 8t = 3$

$\qquad 5t^2 - 8t - 3 = 0$

We use the quadratic formula. Here $a = 5$, $b = -8$, and $c = -3$.

$t = \dfrac{-b \pm \sqrt{b^2 - 4ac}}{2a}$

$\quad = \dfrac{-(-8) \pm \sqrt{(-8)^2 - 4 \cdot 5(-3)}}{2 \cdot 5}$

$\quad = \dfrac{8 \pm \sqrt{124}}{10} = \dfrac{8 \pm 2\sqrt{31}}{10}$

$\quad = \dfrac{2(4 \pm \sqrt{31})}{2 \cdot 5} = \dfrac{4 \pm \sqrt{31}}{5}$

The solutions are $\dfrac{4 - \sqrt{31}}{5}$ and $\dfrac{4 + \sqrt{31}}{5}$, or $\dfrac{4 \pm \sqrt{31}}{5}$.

21. $\qquad 3x^2 + 4 = 5x$

$\qquad 3x^2 - 5x + 4 = 0$

We use the quadratic formula. Here $a = 3$, $b = -5$, and $c = 4$.

$$x = \frac{-b \pm \sqrt{b^2 - 4ac}}{2a}$$

$$= \frac{-(-5) \pm \sqrt{(-5)^2 - 4 \cdot 3 \cdot 4}}{2 \cdot 3}$$

$$= \frac{5 \pm \sqrt{-23}}{6} = \frac{5 \pm \sqrt{23}i}{6}$$

$$= \frac{5}{6} \pm \frac{\sqrt{23}}{6}i$$

The solutions are $\frac{5}{6} - \frac{\sqrt{23}}{6}i$ and $\frac{5}{6} + \frac{\sqrt{23}}{6}i$, or $\frac{5}{6} \pm \frac{\sqrt{23}}{6}i$.

23. $4x^2 = 8x + 5$

$4x^2 - 8x - 5 = 0$

$a = 4,\ b = -8,\ c = -5$

$b^2 - 4ac = (-8)^2 - 4 \cdot 4(-5) = 144$

Since $b^2 - 4ac > 0$, there are no imaginary solutions.

25. $x^2 + 3x + 4 = 0$

$a = 1,\ b = 3,\ c = 4$

$b^2 - 4ac = 3^2 - 4 \cdot 1 \cdot 4 = -7$

Since $b^2 - 4ac < 0$, there are imaginary solutions.

27. $5t^2 - 7t = 0$

$a = 5,\ b = -7,\ c = 0$

$b^2 - 4ac = (-7)^2 - 4 \cdot 5 \cdot 0 = 49$

Since $b^2 - 4ac > 0$, there are no imaginary solutions.

29. Graph $y = x^2 - 8x + 12$ and use the Zero feature twice.

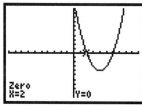

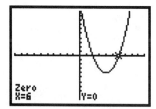

The solutions are 2 and 6.

31. Graph $y = 7x^2 - 43x + 6$ and use the Zero feature twice.

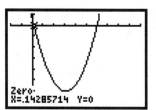

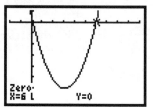

One solution is approximately 0.143 and the other is 6.

33. Graph $y_1 = 6x + 1$ and $y_2 = 4x^2$ and use the Intersect feature twice.

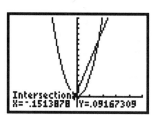

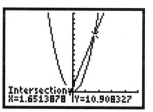

The solutions are approximately -0.151 and 1.651.

35. Graph $y_1 = 2x^2 - 4$ and $y_2 = 5x$ and use the Intersect feature twice.

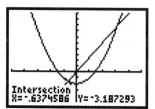

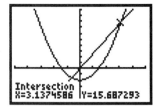

The solutions are approximately -0.637 and 3.137.

37. $x^2 + 6x + 5 = 0$ Setting $f(x) = 0$

$(x+5)(x+1) = 0$ Factoring

$x + 5 = 0$ or $x + 1 = 0$

$x = -5$ or $x = -1$

The solutions are -5 and -1.

39. $x^2 - 3x - 3 = 0$

$a = 1,\ b = -3,\ c = -3$

$x = \dfrac{-b \pm \sqrt{b^2 - 4ac}}{2a}$

$= \dfrac{-(-3) \pm \sqrt{(-3)^2 - 4 \cdot 1 \cdot (-3)}}{2 \cdot 1}$

$= \dfrac{3 \pm \sqrt{9 + 12}}{2}$

$= \dfrac{3 \pm \sqrt{21}}{2}$

The solutions are $\dfrac{3 - \sqrt{21}}{2}$ and $\dfrac{3 + \sqrt{21}}{2}$, or

$\dfrac{3 \pm \sqrt{21}}{2}$.

41. $x^2 - 5x + 1 = 0$

$a = 1,\ b = -5,\ c = 1$

$x = \dfrac{-b \pm \sqrt{b^2 - 4ac}}{2a}$

$= \dfrac{-(-5) \pm \sqrt{(-5)^2 - 4 \cdot 1 \cdot 1}}{2 \cdot 1}$

$= \dfrac{5 \pm \sqrt{25 - 4}}{2}$

$= \dfrac{5 \pm \sqrt{21}}{2}$

The solutions are $\dfrac{5 - \sqrt{21}}{2}$ and $\dfrac{5 + \sqrt{21}}{2}$, or

$\dfrac{5 \pm \sqrt{21}}{2}$.

43. Graph $y = x^2 + 2x - 5$ and use the Zero feature twice.

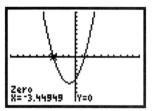

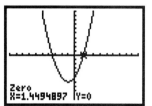

The zeros are approximately -3.449 and 1.449.

45. Graph $y = 3x^2 + 2x - 4$ and use the Zero feature twice.

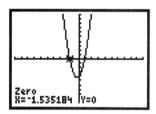

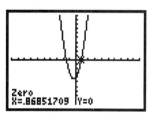

The zeros are approximately -1.535 and 0.869.

47. Graph $y = 5.02x^2 - 4.19x - 2.057$ and use the Zero feature twice.

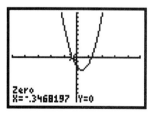

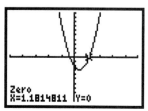

The zeros are approximately -0.347 and 1.181.

49. $x^4 - 3x^2 + 2 = 0$

Let $u = x^2$.

$\quad u^2 - 3u + 2 = 0 \quad$ Substituting u for x^2

$\quad (u - 1)(u - 2) = 0$

$\quad u - 1 = 0 \quad or \quad u - 2 = 0$

$\qquad u = 1 \quad or \qquad u = 2$

Now substitute x^2 for u and solve for x.

$\quad x^2 = 1 \quad or \quad x^2 = 2$

$\quad x = \pm 1 \quad or \quad x = \pm\sqrt{2}$

The solutions are -1, 1, $-\sqrt{2}$, and $\sqrt{2}$.

51. $x - 3\sqrt{x} - 4 = 0$

Let $u = \sqrt{x}$.

$\quad u^2 - 3u - 4 = 0 \quad$ Substituting u for $\sqrt{x}$

$\quad (u + 1)(u - 4) = 0$

$\quad u + 1 = 0 \quad or \quad u - 4 = 0$

$\qquad u = -1 \quad or \qquad u = 4$

Now substitute $\sqrt{x}$ for u and solve for x.

$\quad \sqrt{x} = -1 \quad or \quad \sqrt{x} = 4$

$\quad$ No solution $\qquad x = 16$

Note that $\sqrt{x}$ must be nonnegative, so $\sqrt{x} = -1$ has no solution. The number 16 checks and is the solution. The solution is 16.

53. $m^{2/3} - 2m^{1/3} - 8 = 0$

Let $u = m^{1/3}$.

$\quad u^2 - 2u - 8 = 0 \quad$ Substituting u for $m^{1/3}$

$\quad (u + 2)(u - 4) = 0$

$\quad u + 2 = 0 \quad or \quad u - 4 = 0$

$\qquad u = -2 \quad or \qquad u = 4$

Now substitute $m^{1/3}$ for u and solve for m.

$\quad m^{1/3} = -2 \qquad or \qquad m^{1/3} = 4$

$\quad (m^{1/3})^3 = (-2)^3 \quad or \quad (m^{1/3})^3 = 4^3 \quad$ Using the

$\qquad\qquad\qquad\qquad\qquad\qquad\qquad$ principle of powers

$\qquad m = -8 \qquad or \qquad\quad m = 64$

The solutions are -8 and 64.

55. $(2x - 3)^2 - 5(2x - 3) + 6 = 0$

Let $u = 2x - 3$.

$\quad u^2 - 5u + 6 = 0 \quad$ Substituting u for $2x - 3$

$\quad (u - 2)(u - 3) = 0$

$\quad u - 2 = 0 \quad or \quad u - 3 = 0$

$\qquad u = 2 \quad or \qquad u = 3$

Now substitute $2x - 3$ for u and solve for x.

$\quad 2x - 3 = 2 \quad or \quad 2x - 3 = 3$

$\qquad 2x = 5 \quad or \qquad 2x = 6$

$\qquad x = \dfrac{5}{2} \quad or \qquad x = 3$

The solutions are $\dfrac{5}{2}$ and 3.

57. $(2t^2 + t)^2 - 4(2t^2 + t) + 3 = 0$

Let $u = 2t^2 + t$.

$\quad u^2 - 4u + 3 = 0 \quad$ Substituting u for $2t^2 + t$

$\quad (u - 1)(u - 3) = 0$

$\quad u - 1 = 0 \quad or \quad u - 3 = 0$

$\qquad u = 1 \quad or \qquad u = 3$

Now substitute $2t^2 + t$ for u and solve for t.

$\quad 2t^2 + t = 1 \quad or \qquad\quad 2t^2 + t = 3$

$\quad 2t^2 + t - 1 = 0 \quad or \quad 2t^2 + t - 3 = 0$

$\quad (2t - 1)(t + 1) = 0 \quad or \quad (2t + 3)(t - 1) = 0$

$\quad 2t-1=0 \ or \ t+1=0 \quad or \ 2t+3=0 \quad or \ t-1 = 0$

$\quad t = \dfrac{1}{2} \ or \quad t=-1 \ or \qquad t=-\dfrac{3}{2} or \quad t = 1$

The solutions are $\dfrac{1}{2}$, -1, $-\dfrac{3}{2}$ and 1.

59. *Familiarize and Translate*. We will use the formula $s = 16t^2$, substituting 2120 for s.

$\quad 2120 = 16t^2$

Carry out. We solve the equation.

$\quad 2120 = 16t^2$

$\quad 132.5 = t^2 \qquad$ Dividing by 16 on both sides

$\quad 11.5 \approx t \qquad$ Taking the square root on both sides

We could also find the first coordinates of the points of intersection of the graphs of $y_1 = 2120$ and $y_2 = 16x^2$. Note that since a negative result has no meaning in this application, we only need to consider the point of intersection in the first quadrant.

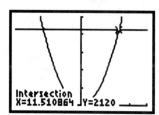

The possible solution is approximately 11.5.

Check. When $t = 11.5$, $s = 16(11.5)^2 = 2116 \approx 2120$. The answer checks.

State. It would take an object about 11.5 sec to reach the ground.

61. *Familiarize.* Let $w =$ the width of the rug. Then $w + 1 =$ the length.

Translate. We use the Pythagorean equation.
$$w^2 + (w+1)^2 = 5^2$$

Carry out. We solve the equation.
$$w^2 + (w+1)^2 = 5^2$$
$$w^2 + w^2 + 2w + 1 = 25$$
$$2w^2 + 2w + 1 = 25$$
$$2w^2 + 2w - 24 = 0$$
$$2(w+4)(w-3) = 0$$
$$w + 4 = 0 \quad or \quad w - 3 = 0$$
$$w = -4 \quad or \qquad w = 3$$

Since the width cannot be negative, we consider only 3. When $w = 3$, $w + 1 = 3 + 1 = 4$.

Check. The length, 4 ft, is 1 ft more than the width, 3 ft. The length of a diagonal of a rectangle with width 3 ft and length 4 ft is $\sqrt{3^2 + 4^2} = \sqrt{9 + 16} = \sqrt{25} = 5$. The answer checks.

State. The length is 4 ft, and the width is 3 ft.

63. *Familiarize.* Let $n =$ the smaller number. Then $n + 5 =$ the larger number.

Translate.

$$\underbrace{\text{The product of the numbers}}_{n(n+5)} \underset{=}{\underbrace{\text{is}}} \underset{36}{\underbrace{36}}$$

Carry out.
$$n(n+5) = 36$$
$$n^2 + 5n = 36$$
$$n^2 + 5n - 36 = 0$$
$$(n+9)(n-4) = 0$$
$$n + 9 = 0 \quad or \quad n - 4 = 0$$
$$n = -9 \quad or \qquad n = 4$$

If $n = -9$, then $n + 5 = -9 + 5 = -4$. If $n = 4$, then $n + 5 = 4 + 5 = 9$.

Check. The number -4 is 5 more than -9 and $(-4)(-9) = 36$, so the pair -9 and -4 check. The number 9 is 5 more than 4 and $9 \cdot 4 = 36$, so the pair 4 and 9 also check.

State. The numbers are -9 and -4 or 4 and 9.

65 *Familiarize.* We add labels to the drawing in the text.

We let x represent the length of a side of the square in each corner. Then the length and width of the resulting base are represented by $20 - 2x$ and $10 - 2x$, respectively. Recall that for a rectangle, Area = length × width.

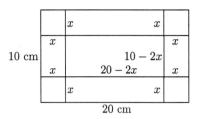

Translate.

$$\underbrace{\text{The area of the base}}_{(20-2x)(10-2x)} \text{ is } \underbrace{96 \text{ cm}^2}_{96}.$$

Carry out. We solve the equation.
$$200 - 60x + 4x^2 = 96$$
$$4x^2 - 60x + 104 = 0$$
$$x^2 - 15x + 26 = 0$$
$$(x-13)(x-2) = 0$$
$$x - 13 = 0 \quad or \quad x - 2 = 0$$
$$x = 13 \quad or \qquad x = 2$$

We could also find the first coordinates of the points of intersection of $y_1 = (20-2x)(10-2x)$ and $y_2 = 96$ using the Intersect feature.

Check. When $x = 13$, both $20 - 2x$ and $10 - 2x$ are negative numbers, so we only consider $x = 2$. When $x = 2$, then $20 - 2x = 20 - 2 \cdot 2 = 16$ and $10 - 2x = 10 - 2 \cdot 2 = 6$, and the area of the base is $16 \cdot 6$, or 96 cm^2. The answer checks.

State. The length of the sides of the squares is 2 cm.

67. $f(x) = 4 - 5x = -5x + 4$

The function can be written in the form $y = mx + b$, so it is a linear function.

69. $f(x) = 7x^2$

The function is in the form $f(x) = ax^2 + bx + c$, $a \neq 0$, so it is a quadratic function.

71. $f(x) = 1.2x - (3.6)^2$

The function is in the form $f(x) = mx + b$, so it is a linear function.

73. Discussion and Writing

75. Test for symmetry with respect to the x-axis:
$$3x^2 + 4y^2 = 5 \quad \text{Original equation}$$
$$3x^2 + 4(-y)^2 = 5 \quad \text{Replacing } y \text{ by } -y$$
$$3x^2 + 4y^2 = 5 \quad \text{Simplifying}$$

The last equation is equivalent to the original equation, so the graph is symmetric with respect to the x-axis.

Test for symmetry with respect to the y-axis:

$$3x^2 + 4y^2 = 5 \quad \text{Original equation}$$
$$3(-x)^2 + 4y^2 = 5 \quad \text{Replacing } x \text{ by } -x$$
$$3x^2 + 4y^2 = 5 \quad \text{Simplifying}$$

The last equation is equivalent to the original equation, so the equation is symmetric with respect to the y-axis.

Test for symmetry with respect to the origin:
$$3x^2 + 4y^2 = 5 \quad \text{Original equation}$$
$$3(-x)^2 + 4(-y)^2 = 5 \quad \text{Replacing } x \text{ by } -x$$
$$\text{and } y \text{ by } -y$$
$$3x^2 + 4y^2 = 5 \quad \text{Simplifying}$$

The last equation is equivalent to the original equation, so the equation is symmetric with respect to the origin.

77. $f(x) = 2x^3 - x$

$f(-x) = 2(-x)^3 - (-x) = -2x^3 + x$

$-f(x) = -2x^3 + x$

$f(x) \neq f(-x)$ so f is not even

$f(-x) = -f(x)$, so f is odd.

79. $x^2 + x - \sqrt{2} = 0$

$$x = \frac{-b \pm \sqrt{b^2 - 4ac}}{2a}$$

$$= \frac{-1 \pm \sqrt{1^2 - 4 \cdot 1(-\sqrt{2})}}{2 \cdot 1} = \frac{-1 \pm \sqrt{1 + 4\sqrt{2}}}{2}$$

The solutions are $\dfrac{-1 \pm \sqrt{1 + 4\sqrt{2}}}{2}$.

81.
$$2t^2 + (t - 4)^2 = 5t(t - 4) + 24$$
$$2t^2 + t^2 - 8t + 16 = 5t^2 - 20t + 24$$
$$0 = 2t^2 - 12t + 8$$
$$0 = t^2 - 6t + 4 \quad \text{Dividing by 2}$$

Use the quadratic formula.

$$t = \frac{-b \pm \sqrt{b^2 - 4ac}}{2a}$$

$$= \frac{-(-6) \pm \sqrt{(-6)^2 - 4 \cdot 1 \cdot 4}}{2 \cdot 1}$$

$$= \frac{6 \pm \sqrt{20}}{2} = \frac{6 \pm 2\sqrt{5}}{2}$$

$$= \frac{2(3 \pm \sqrt{5})}{2} = 3 \pm \sqrt{5}$$

The solutions are $3 \pm \sqrt{5}$.

83. $\sqrt{x - 3} - \sqrt[4]{x - 3} = 2$

Substitute u for $\sqrt[4]{x - 3}$.

$$u^2 - u - 2 = 0$$
$$(u - 2)(u + 1) = 0$$

$$u - 2 = 0 \quad or \quad u + 1 = 0$$
$$u = 2 \quad or \quad\quad u = -1$$

Substitute $\sqrt[4]{x - 3}$ for u and solve for x.

$$\sqrt[4]{x - 3} = 2 \quad or \quad \sqrt[4]{x - 3} = 1$$
$$x - 3 = 16 \quad\quad \text{No solution}$$
$$x = 19$$

The value checks. The solution is 19.

85.
$$\left(y + \frac{2}{y}\right)^2 + 3y + \frac{6}{y} = 4$$

$$\left(y + \frac{2}{y}\right)^2 + 3\left(y + \frac{2}{y}\right) - 4 = 0$$

Substitute u for $y + \dfrac{2}{y}$.

$$u^2 + 3u - 4 = 0$$
$$(u + 4)(u - 1) = 0$$
$$u = -4 \quad or \quad u = 1$$

Substitute $y + \dfrac{2}{y}$ for u and solve for y.

$$y + \frac{2}{y} = -4 \quad or \quad\quad y + \frac{2}{y} = 1$$
$$y^2 + 2 = -4y \quad or \quad\quad y^2 + 2 = y$$
$$y^2 + 4y + 2 = 0 \quad or \quad y^2 - y + 2 = 0$$

$$y = \frac{-4 \pm \sqrt{4^2 - 4 \cdot 1 \cdot 2}}{2 \cdot 1} \quad or$$

$$\quad\quad\quad\quad y = \frac{-(-1) \pm \sqrt{(-1)^2 - 4 \cdot 1 \cdot 2}}{2 \cdot 1}$$

$$y = \frac{-4 \pm \sqrt{8}}{2} \quad or \quad y = \frac{1 \pm \sqrt{-7}}{2}$$

$$y = \frac{-4 \pm 2\sqrt{2}}{2} \quad or \quad y = \frac{1 \pm \sqrt{7}i}{2}$$

$$y = -2 \pm \sqrt{2} \quad or \quad y = \frac{1}{2} \pm \frac{\sqrt{7}}{2}i$$

The solutions are $-2 \pm \sqrt{2}$ and $\dfrac{1}{2} \pm \dfrac{\sqrt{7}}{2}i$.

87. $\dfrac{1}{2}at + v_0 t + x_0 = 0$

Use the quadratic formula. Here $a = \dfrac{1}{2}a$, $b = v_0$, and $c = x_0$.

$$t = \frac{-v_0 \pm \sqrt{(v_0)^2 - 4 \cdot \frac{1}{2}a \cdot x_0}}{2 \cdot \frac{1}{2}a}$$

$$t = \frac{-v_0 \pm \sqrt{v_0^2 - 2ax_0}}{a}$$

Exercise Set 2.4

1. a) The minimum function value occurs at the vertex, so the vertex is $(-0.4999992, -2.25)$, or about $(-0.5, -2.25)$.

    b) The line of symmetry is a vertical line through the vertex. It is $x = -0.4999992$, or about $x = -0.5$.

    c) The minimum value of the function is -2.25.

3. $f(x) = x^2 - 8x + 12$ 16 completes the square for $x^2 - 8x$

$= x^2 - 8x + 16 - 16 + 12$ Adding $16-16$ on the right side

$= (x^2 - 8x + 16) - 16 + 12$

$= (x-4)^2 - 4$ Factoring and simplifying

$= (x-4)^2 + (-4)$ Writing in the form $f(x) = a(x-h)^2 + k$

    a) Vertex: $(4, -4)$

    b) Line of symmetry: $x = 4$

    c) Minimum value: -4

5. $f(x) = x^2 - 7x + 12$ $\dfrac{49}{4}$ completes the square for $x^2 - 7x$

$= x^2 - 7x + \dfrac{49}{4} - \dfrac{49}{4} + 12$ Adding

$\dfrac{49}{4} - \dfrac{49}{4}$ on the right side

$= \left(x^2 - 7x + \dfrac{49}{4}\right) - \dfrac{49}{4} + 12$

$= \left(x - \dfrac{7}{2}\right)^2 - \dfrac{1}{4}$ Factoring and simplifying

$= \left(x - \dfrac{7}{2}\right)^2 + \left(-\dfrac{1}{4}\right)$ Writing in the form $f(x) = a(x-h)^2 + k$

    a) Vertex: $\left(\dfrac{7}{2}, -\dfrac{1}{4}\right)$

    b) Line of symmetry: $x = \dfrac{7}{2}$

    c) Minimum value: $-\dfrac{1}{4}$

7. $g(x) = 2x^2 + 6x + 8$

$= 2(x^2 + 3x) + 8$ Factoring 2 out of the first two terms

$= 2\left(x^2 + 3x + \dfrac{9}{4} - \dfrac{9}{4}\right) + 8$ Adding

$\dfrac{9}{4} - \dfrac{9}{4}$ inside the parentheses

$= 2\left(x^2 + 3x + \dfrac{9}{4}\right) - 2 \cdot \dfrac{9}{4} + 8$ Removing $-\dfrac{9}{4}$ from within the parentheses

$= 2\left(x + \dfrac{3}{2}\right)^2 + \dfrac{7}{2}$ Factoring and simplifying

$= 2\left[x - \left(-\dfrac{3}{2}\right)\right]^2 + \dfrac{7}{2}$

    a) Vertex: $\left(-\dfrac{3}{2}, \dfrac{7}{2}\right)$

    b) Line of symmetry: $x = -\dfrac{3}{2}$

    c) Minimum value: $\dfrac{7}{2}$

9. $g(x) = -2x^2 + 2x + 1$

$= -2(x^2 - x) + 1$ Factoring -2 out of the first two terms

$= -2\left(x^2 - x + \dfrac{1}{4} - \dfrac{1}{4}\right) + 1$ Adding $\dfrac{1}{4} - \dfrac{1}{4}$ inside the parentheses

$= -2\left(x^2 - x + \dfrac{1}{4}\right) - 2\left(-\dfrac{1}{4}\right) + 1$

Removing $-\dfrac{1}{4}$ from within the parentheses

$= -2\left(x - \dfrac{1}{2}\right)^2 + \dfrac{3}{2}$

    a) Vertex: $\left(\dfrac{1}{2}, \dfrac{3}{2}\right)$

    b) Line of symmetry: $x = \dfrac{1}{2}$

    c) Maximum value: $\dfrac{3}{2}$

11. The graph of $y = (x+3)^2$ has vertex $(-3, 0)$ and opens up. It is graph (f).

13. The graph of $y = 2(x-4)^2 - 1$ has vertex $(4, -1)$ and opens up. It is graph (b).

15. The graph of $y = -\dfrac{1}{2}(x+3)^2 + 4$ has vertex $(-3, 4)$ and opens down. It is graph (h).

17. The graph of $y = -(x+3)^2 + 4$ has vertex $(-3, 4)$ and opens down. It is graph (c).

19. $f(x) = x^2 - 6x + 5$

a) The x-coordinate of the vertex is
$$-\frac{b}{2a} = -\frac{-6}{2 \cdot 1} = 3.$$
Since $f(3) = 3^2 - 6 \cdot 3 + 5 = -4$, the vertex is $(3, -4)$.

b) Since $a = 1 > 0$, the graph opens up so the second coordinate of the vertex, -4, is the minimum value of the function.

c) The range is $[-4, \infty)$.

d) Since the graph opens up, function values decrease to the left of the vertex and increase to the right of the vertex. Thus, $f(x)$ is increasing on $(3, \infty)$ and decreasing on $(-\infty, 3)$.

21. $f(x) = 2x^2 + 4x - 16$

a) The x-coordinate of the vertex is
$$-\frac{b}{2a} = -\frac{4}{2 \cdot 2} = -1.$$
Since $f(-1) = 2(-1)^2 + 4(-1) - 16 = -18$, the vertex is $(-1, -18)$.

b) Since $a = 2 > 0$, the graph opens up so the second coordinate of the vertex, -18, is the minimum value of the function.

c) The range is $[-18, \infty)$.

d) Since the graph opens up, function values decrease to the left of the vertex and increase to the right of the vertex. Thus, $f(x)$ is increasing on $(-1, \infty)$ and decreasing on $(-\infty, -1)$.

23. $f(x) = -\frac{1}{2}x^2 + 5x - 8$

a) The x-coordinate of the vertex is
$$-\frac{b}{2a} = -\frac{5}{2\left(-\frac{1}{2}\right)} = 5.$$
Since $f(5) = -\frac{1}{2} \cdot 5^2 + 5 \cdot 5 - 8 = \frac{9}{2}$, the vertex is $\left(5, \frac{9}{2}\right)$.

b) Since $a = -\frac{1}{2} < 0$, the graph opens down so the second coordinate of the vertex, $\frac{9}{2}$, is the maximum value of the function.

c) The range is $\left(-\infty, \frac{9}{2}\right]$.

d) Since the graph opens down, function values increase to the left of the vertex and decreases to the right of the vertex. Thus, $f(x)$ is increasing on $(-\infty, 5)$ and decreasing on $(5, \infty,)$.

25. $f(x) = 3x^2 + 6x + 5$

a) The x-coordinate of the vertex is
$$-\frac{b}{2a} = -\frac{6}{2 \cdot 3} = -1.$$
Since $f(-1) = 3(-1)^2 + 6(-1) + 5 = 2$, the vertex is $(-1, 2)$.

b) Since $a = 3 > 0$, the graph opens up so the second coordinate of the vertex, 2, is the minimum value of the function.

c) The range is $[2, \infty)$.

d) Since the graph opens up, function values decrease to the left of the vertex and increase to the right of the vertex. Thus, $f(x)$ is increasing on $(-1, \infty)$ and decreasing on $(-\infty, -1)$.

27. *Familiarize.* Using the label in the text, we let $x =$ the height of the file. Then the length $= 10$ and the width $= 18 - 2x$.

Translate. Since the volume of a rectangular solid is length × width × height we have
$$V(x) = 10(18 - 2x)x, \text{ or } -20x^2 + 180x.$$

Carry out. Since $V(x)$ is a quadratic function with $a = -20 < 0$, the maximum function value occurs at the vertex of the graph of the function. The first coordinate of the vertex is
$$-\frac{b}{2a} = -\frac{180}{2(-20)} = 4.5.$$

Check. When $x = 4.5$, then $18 - 2x = 9$ and $V(x) = 10 \cdot 9(4.5)$, or 405. As a partial check, we can find $V(x)$ for a value of x less than 4.5 and for a value of x greater than 4.5. For instance, $V(4.4) = 404.8$ and $V(4.6) = 404.8$. Since both of these values are less than 405, our result appears to be correct. We could also examine a table of values for $V(x)$ and/or examine its graph.

State. The file should be 4.5 in. tall in order to maximize the volume.

29. *Familiarize.* Let $b =$ the length of the base of the triangle. Then the height $= 20 - b$.

Translate. Since the area of a triangle is $\frac{1}{2} \times$ base × height, we have
$$A(b) = \frac{1}{2}b(20 - b), \text{ or } -\frac{1}{2}b^2 + 10b.$$

Carry out. Since $A(b)$ is a quadratic function with $a = -\frac{1}{2} < 0$, the maximum function value occurs at the vertex of the graph of the function. The first coordinate of the vertex is
$$-\frac{b}{2a} = -\frac{10}{2\left(-\frac{1}{2}\right)} = 10.$$

When $b = 10$, then $20 - b = 20 - 10 = 10$, and the area is $\frac{1}{2} \cdot 10 \cdot 10 = 50$ cm^2.

Check. As a partial check, we can find $A(b)$ for a value of b less than 10 and for a value of b greater than 10. For instance, $V(9.9) = 49.995$ and $V(10.1) = 49.995$. Since both of these values are less than 50, our result appears to be correct. We could also examine a table of values for $A(b)$ and/or examine its graph.

State. The area is a maximum when the base and the height are both 10 cm.

31. Familiarize. We let $s =$ the height of the elevator shaft, $t_1 =$ the time it takes the screwdriver to reach the bottom of the shaft, and $t_2 =$ the time it takes the sound to reach the top of the shaft.

Translate. We know that $t_1 + t_2 = 5$. Using the information in Example 4 we also know that

$$s = 16t_1^2, \quad or \quad t_1 = \frac{\sqrt{x}}{4} \quad and$$

$$s = 1100t_2, \quad or \quad t_2 = \frac{s}{1100}.$$

Then $\frac{\sqrt{s}}{4} + \frac{s}{1100} = 5$.

Carry out. We solve the last equation above.

$$\frac{\sqrt{s}}{4} + \frac{s}{1100} = 5$$

$$275\sqrt{s} + s = 5500 \quad \text{Multiplying by 1100}$$

$$2 + 275\sqrt{s} - 5500 = 0$$

Let $u = \sqrt{s}$ and substitute.

$$u^2 + 275u - 5500 = 0$$

$$u = \frac{-b + \sqrt{b^2 - 4ac}}{2a} \qquad \begin{array}{l}\text{We only want the} \\ \text{positive solution.}\end{array}$$

$$= \frac{-275 + \sqrt{275^2 - 4 \cdot 1(-5500)}}{2 \cdot 1}$$

$$= \frac{-275 + \sqrt{97,625}}{2} \approx 18.725$$

Since $u \approx 18.725$, we have $\sqrt{s} = 18.725$, so $s \approx 350.6$.

Check. If $s \approx 350.6$, then $t_1 = \frac{\sqrt{s}}{4} = \frac{\sqrt{350.6}}{4} \approx$ 4.68 and $t_2 = \frac{s}{1100} = \frac{350.6}{1100} \approx 0.32$, so $t_1 + t_2 = 4.68 + 0.32 = 5$.

The result checks.

State. The elevator shaft is about 350.6 ft tall.

33. $C(x) = 0.1x^2 - 0.7x + 2.425$

Since $C(x)$ is a quadratic function with $a = 0.1 > 0$, a minimum function value occurs at the vertex of the

graph of $C(x)$. The first coordinate of the vertex is

$$-\frac{b}{2a} = -\frac{-0.7}{2(0.1)} = 3.5.$$

Thus, 3.5 hundred, or 350 bicycles should be built to minimize the average cost per bicycle.

35. $P(x) = R(x) - C(x)$

$P(x) = (50x - 0.5x^2) - (10x + 3)$

$P(x) = -0.5x^2 + 40x - 3$

Since $P(x)$ is a quadratic function with $a = -0.5 < 0$, a maximum function value occurs at the vertex of the graph of the function. The first coordinate of the vertex is

$$-\frac{b}{2a} = -\frac{40}{2(-0.5)} = 40.$$

$P(40) = -0.5(40)^2 + 40 \cdot 40 - 3 = 797$

Thus, the maximum profit is \$797. It occurs when 40 units are sold.

37. Familiarize. Using the labels on the drawing in the text, we let $x =$ the width of each corral and $240 - 3x =$ the total length of the corrals.

Translate. Since the area of a rectangle is length × width, we have

$$A(x) = (240 - 3x)x = -3x^2 + 240x.$$

Carry out. Since $A(x)$ is a quadratic function with $a = -3 < 0$, the maximum function value occurs at the vertex of the graph of $A(x)$. The first coordinate of the vertex is

$$-\frac{b}{2a} = -\frac{240}{2(-3)} = 40.$$

$A(40) = -3(40)^2 + 240(40) = 4800$

Check. As a partial check we can find $A(x)$ for a value of x less than 40 and for a value of x greater than 40. For instance, $A(39.9) = 4799.97$ and $A(40.1) = 4799.97$. Since both of these values are less than 4800, our result appears to be correct. We could also examine a table of values for $A(x)$ and/or examine its graph.

State. The largest total area that can be enclosed is 4800 yd^2.

39. Discussion and Writing

41. Discussion and Writing

43. $f(x) = 2x^2 - x + 4$

$$f(x + h) = 2(x + h)^2 - (x + h) + 4$$
$$= 2(x^2 + 2xh + h^2) - (x + h) + 4$$
$$= 2x^2 + 4xh + 2h^2 - x - h - 4$$

$$\frac{f(x+h)-f(x)}{h}$$

$$=\frac{2x^2+4xh+2h^2-x-h-4-(2x^2-x+4)}{h}$$

$$=\frac{2x^2+4xh+2h^2-x-h-4-2x^2+x-4}{h}$$

$$=\frac{4xh+2h^2-h}{h}=\frac{h(4x+2h-1)}{h}$$

$$=4x+2h-1$$

45. The graph of $f(x)$ is stretched vertically and reflected across the x-axis.

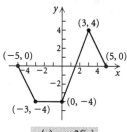

$$g(x)=-2f(x)$$

47. a)
$$kx^2-17x+33=0$$
$$k(3)^2-17(3)+33=0 \quad \text{Substituting 3 for } x$$
$$9k-51+33=0$$
$$9k=18$$
$$k=2$$

b)
$$2x^2-17x+33=0 \quad \text{Substituting 2 for } k$$
$$(2x-11)(x-3)=0$$
$$2x-11=0 \quad or \quad x-3=0$$
$$x=\frac{11}{2} \quad or \quad x=3$$

The other solution is $\frac{11}{2}$.

49. a)
$$(1+i)^2-k(1+i)+2=0 \quad \text{Substituting}$$
$$\qquad\qquad\qquad\qquad\qquad\quad 1+i \text{ for } x$$
$$1+2i-1-k-ki+2=0$$
$$2+2i=k+ki$$
$$2(1+i)=k(1+i)$$
$$2=k$$

b)
$$x^2-2x+2=0 \qquad \text{Substituting 2 for } k$$
$$x=\frac{-(-2)\pm\sqrt{(-2)^2-4\cdot 1\cdot 2}}{2\cdot 1}$$
$$=\frac{2\pm\sqrt{-4}}{2}$$
$$=\frac{2\pm 2i}{2}=1\pm i$$

The other solution is $1-i$.

51. $f(x)=-0.2x^2-3x+c$

The x-coordinate of the vertex of $f(x)$ is $-\dfrac{b}{2a}=$
$-\dfrac{-3}{2(-0.2)}=-7.5$. Now we find c such that
$f(-7.5)=$
-225.
$$-0.2(-7.5)^2-3(-7.5)+c=-225$$
$$-11.25+22.5+c=-225$$
$$c=-236.25$$

53. $y=(|x|-5)^2-3$

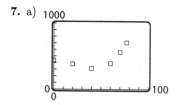

Exercise Set 2.5

1. The data points rise and then fall, suggesting that a quadratic function $f(x)=ax^2+bx+c$, $a<0$, might fit the data. The answer is (c).

3. The data points fall, then rise, and then fall again. Thus, neither a linear nor a quadratic function fits the data. The answer is (d).

5. The data points fall at a steady rate, so a linear function $f(x)=mx+b$ might fit the data. The answer is (a).

7. a)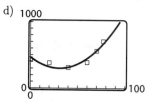

b) The data points fall and then rise, suggesting that a quadratic function might fit the data.

c) Using the quadratic regression feature on a grapher, we have $f(x)=0.1729980538x^2-10.70013387x+465.7917864$.

d)

e) In 2004, $x = 2004 - 1920$, or 84; $f(84) \approx 788$, so we predict that there will be about 788 morning newspapers in 2004.

In 2007, $x = 2007 - 1920$, or 87; $f(87) \approx 844$, so we predict that there will be about 844 morning newspapers in 2007.

9. a)

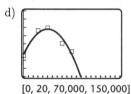

[0, 20, 70,000, 150,000]

b) The data points rise and then fall, suggesting that a quadratic function might fit the data.

c) Using the quadratic regression feature of a grapher, we have $f(x) = -1226.47072x^2 + 12,279.05135x + 97,260.77349$.

d)

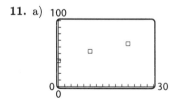

[0, 20, 70,000, 150,000]

e) In 2003, $x = 2003 - 1990$, or 13; $f(13) \approx 49,615$, so we predict that the assets of the Hospital Insurance program will be about \$49,615 million in 2003.

In 2007, $x = 2007 - 1990$, or 17; $f(17) \approx 5491$, so we predict that the assets of the Hospital Insurance program will be about \$5491 million in 2007.

11. a) 100

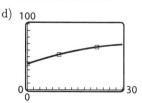

b) It appears that either a linear or a quadratic function could be fit to these data. We will use a quadratic function.

c) Using the quadratic regression feature on a grapher, we have $f(x) = -0.0223484848x^2 + 1.673484848x + 39$.

d) 100

e) In 2003, $x = 2003 - 1975$, or 28; $f(28) \approx 68.3$, so we predict that 68.3% of mothers in the labor force in 2003 will have children under 6 years old.

In 2010, $x = 2010 - 1975$, or 35; $f(35) \approx 70.2$, so we predict that 70.2% of mothers in the labor force in 2010 will have children under 6 years old.

13. a) 1500

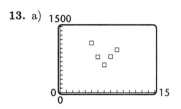

b) The data points fall and then rise, suggesting that a quadratic function might fit the data.

c) Using the quadratic regression feature on a grapher, we have $f(x) = 93.28571429x^2 - 1336x + 5460.828571$.

d) 1500

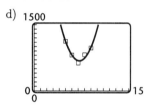

e) $f(7.5) \approx 688$, so we predict that the death rate of males who slept for an average of 7.5 hr per day was about 688 per 100,000.

$f(10) \approx 1429$, so we predict that the death rate of males who slept for an average of 10 hr per day was about 1429 per 100,000.

15. a) Using the quadratic regression feature of a grapher, we have $f(x) = -0.8630952381x^2 + 12.125x + 36.76785714$.

b) $f(5.5) \approx 77$ yr; $f(8.5) \approx 77$ yr

17. Discussion and Writing

19. $g(x) = \dfrac{x+5}{x-1}$

$g(-2) = \dfrac{-2+5}{-2-1} = \dfrac{3}{-3} = -1$

21. $g(x) = \dfrac{x+5}{x-1}$

$g(-5) = \dfrac{-5+5}{-5-1} = \dfrac{0}{-6} = 0$

23. a)

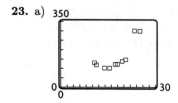

b) The data points fall and then rise, suggesting that a quadratic function might fit the data.

c) Using the quadratic regression feature on a grapher, we have $f(x) = 2.031904548x^2 - 59.04179598x + 527.2818092$.

d)

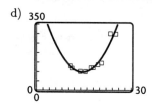

e) Graph $y_1 = f(x)$ and $y_2 = 180$ and find the first coordinate of the point of intersection of the graphs. We find that about \$20.9 million should be allotted for the advertising budget.

Exercise Set 2.6

1.
$$\frac{1}{4} + \frac{1}{5} = \frac{1}{t}, \text{ LCD is } 20t$$

$$20t\left(\frac{1}{4} + \frac{1}{5}\right) = 20t \cdot \frac{1}{t}$$

$$20t \cdot \frac{1}{4} + 20t \cdot \frac{1}{5} = 20t \cdot \frac{1}{t}$$

$$5t + 4t = 20$$

$$9t = 20$$

$$t = \frac{20}{9}$$

Check:

$$\frac{1}{4} + \frac{1}{5} = \frac{1}{t}$$

$$\frac{1}{4} + \frac{1}{5} \; ? \; \frac{1}{\frac{20}{9}}$$

$$\frac{5}{20} + \frac{4}{20} \quad \bigg| \quad 1 \cdot \frac{9}{20}$$

$$\frac{9}{20} \quad \bigg| \quad \frac{9}{20} \qquad \text{TRUE}$$

The solution is $\frac{20}{9}$.

3.
$$\frac{x+2}{4} - \frac{x-1}{5} = 15, \text{ LCD is } 20$$

$$20\left(\frac{x+2}{4} - \frac{x-1}{5}\right) = 20 \cdot 15$$

$$5(x+2) - 4(x-1) = 300$$

$$5x + 10 - 4x + 4 = 300$$

$$x + 14 = 300$$

$$x = 286$$

The solution is 286.

5.
$$\frac{1}{2} + \frac{2}{x} = \frac{1}{3} + \frac{3}{x}, \text{ LCD is } 6x$$

$$6x\left(\frac{1}{2} + \frac{2}{x}\right) = 6x\left(\frac{1}{3} + \frac{3}{x}\right)$$

$$3x + 12 = 2x + 18$$

$$3x - 2x = 18 - 12$$

$$x = 6$$

Check:

$$\frac{1}{2} + \frac{2}{x} = \frac{1}{3} + \frac{3}{x}$$

$$\frac{1}{2} + \frac{2}{6} \; ? \; \frac{1}{3} + \frac{3}{6}$$

$$\frac{1}{2} + \frac{1}{3} \quad \bigg| \quad \frac{1}{3} + \frac{1}{2} \qquad \text{TRUE}$$

The solution is 6.

7.
$$\frac{3x}{x+2} + \frac{6}{x} = \frac{12}{x^2 + 2x}$$

$$\frac{3x}{x+2} + \frac{6}{x} = \frac{12}{x(x+2)}, \text{ LCD is } x(x+2)$$

$$x(x+2)\left(\frac{3x}{x+2} + \frac{6}{x}\right) = x(x+2) \cdot \frac{12}{x(x+2)}$$

$$3x \cdot x + 6(x+2) = 12$$

$$3x^2 + 6x + 12 = 12$$

$$3x^2 + 6x = 0$$

$$3x(x+2) = 0$$

$$3x = 0 \quad or \quad x + 2 = 0$$

$$x = 0 \quad or \qquad x = -2$$

Neither 0 nor -2 checks, so the equation has no solution.

9.
$$\frac{4}{x^2 - 1} - \frac{2}{x - 1} = \frac{3}{x + 1},$$
LCD is $(x + 1)(x - 1)$

$$(x+1)(x-1)\left(\frac{4}{(x+1)(x-1)} - \frac{2}{x-1}\right) =$$
$$(x+1)(x-1) \cdot \frac{3}{x+1}$$
$$4 - 2(x + 1) = 3(x - 1)$$
$$4 - 2x - 2 = 3x - 3$$
$$2 - 2x = 3x - 3$$
$$2 + 3 = 3x + 2x$$
$$5 = 5x$$
$$1 = x$$

Check:
$$\frac{4}{x^2 - 1} - \frac{2}{x - 1} = \frac{3}{x + 1}$$
$$\frac{4}{1^2 - 1} - \frac{2}{1 - 1} \ ? \ \frac{3}{1 + 1}$$
$$\frac{4}{0} - \frac{2}{0} \ \Big| \ \frac{3}{2}$$

Division by zero is undefined.

There is no solution.

11.
$$\frac{490}{x^2 - 49} = \frac{5x}{x - 7} - \frac{35}{x + 7}$$
$$\frac{490}{(x + 7)(x - 7)} = \frac{5x}{x - 7} - \frac{35}{x + 7},$$
LCD is $(x + 7)(x - 7)$

$$(x+7)(x-7)\left(\frac{490}{(x+7)(x-7)}\right) =$$
$$(x+7)(x-7)\left(\frac{5x}{x-7} - \frac{35}{x+7}\right)$$
$$490 = 5x(x+7) - 35(x-7)$$
$$490 = 5x^2 + 35x - 35x + 245$$
$$0 = 5x^2 - 245$$
$$0 = 5(x + 7)(x - 7)$$

$$x + 7 = 0 \quad \text{or} \quad x - 7 = 0$$
$$x = -7 \quad \text{or} \quad x = 7$$

Neither -7 nor 7 checks, so the equation has no solution.

13.
$$\frac{1}{x - 6} - \frac{1}{x} = \frac{6}{x^2 - 6x}$$
$$\frac{1}{x - 6} - \frac{1}{x} = \frac{6}{x(x-6)}, \text{ LCD is } x(x-6)$$

$$x(x-6)\left(\frac{1}{x-6} - \frac{1}{x}\right) = x(x-6) \cdot \frac{6}{x(x-6)}$$
$$x - (x - 6) = 6$$
$$x - x + 6 = 6$$
$$6 = 6$$

We get an equation that is true for all real numbers. Note, however, that when $x = 6$ or $x = 0$, division by 0 occurs in the original equation. Thus, the solution set is $\{x|x \text{ is a real number } and \ x \neq 6 \ and \ x \neq 0\}$, or $(-\infty, 0) \cup (0, 6) \cup (6, \infty)$.

15.
$$\frac{8}{x^2 - 2x + 4} = \frac{x}{x + 2} + \frac{24}{x^3 + 8},$$
LCD is $(x + 2)(x^2 - 2x + 4)$

$$(x+2)(x^2-2x+4) \cdot \frac{8}{x^2-2x+4} =$$
$$(x+2)(x^2-2x+4)\left(\frac{x}{x+2} + \frac{24}{(x+2)(x^2-2x+4)}\right)$$
$$8(x + 2) = x(x^2 - 2x + 4) + 24$$
$$8x + 16 = x^3 - 2x^2 + 4x + 24$$
$$0 = x^3 - 2x^2 - 4x + 8$$
$$0 = x^2(x - 2) - 4(x - 2)$$
$$0 = (x - 2)(x^2 - 4)$$
$$0 = (x - 2)(x + 2)(x - 2)$$

$$x - 2 = 0 \quad \text{or} \quad x + 2 = 0 \quad \text{or} \quad x - 2 = 0$$
$$x = 2 \quad \text{or} \quad x = -2 \quad \text{or} \quad x = 2$$

Only 2 checks. The solution is 2.

17.
$$\sqrt{3x - 4} = 1$$
$$(\sqrt{3x - 4})^2 = 1^2$$
$$3x - 4 = 1$$
$$3x = 5$$
$$x = \frac{5}{3}$$

Check:
$$\sqrt{3x - 4} = 1$$
$$\sqrt{3 \cdot \frac{5}{3} - 4} \ ? \ 1$$
$$\sqrt{5 - 4}$$
$$\sqrt{1}$$
$$1 \ \Big| \ 1 \quad \text{TRUE}$$

The solution is $\frac{5}{3}$.

19. $\sqrt[4]{x^2 - 1} = 1$

$(\sqrt[4]{x^2 - 1})^4 = 1^4$

$x^2 - 1 = 1$

$x^2 = 2$

$x = \pm\sqrt{2}$

Check:

$$\frac{\sqrt[4]{x^2 - 1} = 1}{\sqrt[4]{(\pm\sqrt{2})^2 - 1} \ ? \ 1}$$

$\sqrt[4]{2 - 1}$

$\sqrt[4]{1}$

$1 \ | \ 1$ TRUE

The solutions are $\pm\sqrt{2}$.

21. $\sqrt{y - 1} + 4 = 0$

$\sqrt{y - 1} = -4$

The principal square root is never negative. Thus, there is no solution.

If we do not observe the above fact, we can continue and reach the same answer.

$(\sqrt{y - 1})^2 = (-4)^2$

$y - 1 = 16$

$y = 17$

Check:

$$\frac{\sqrt{y - 1} + 4 = 0}{\sqrt{17 - 1} + 4 \ ? \ 0}$$

$\sqrt{16} + 4$

$4 + 4$

$8 \ | \ 0$ FALSE

Since 17 does not check, there is no solution.

23. $\sqrt[3]{6x + 9} + 8 = 5$

$\sqrt[3]{6x + 9} = -3$

$(\sqrt[3]{6x + 9})^3 = (-3)^3$

$6x + 9 = -27$

$6x = -36$

$x = -6$

Check:

$$\frac{\sqrt[3]{6x + 9} + 8 = 5}{\sqrt[3]{6(-6) + 9} + 8 \ ? \ 5}$$

$\sqrt[3]{-27} + 8$

$-3 + 8$

$5 \ | \ 5$ TRUE

The solution is -6.

25. $\sqrt{x - 3} + \sqrt{x + 2} = 5$

$\sqrt{x + 2} = 5 - \sqrt{x - 3}$

$(\sqrt{x + 2})^2 = (5 - \sqrt{x - 3})^2$

$x + 2 = 25 - 10\sqrt{x - 3} + (x - 3)$

$x + 2 = 22 - 10\sqrt{x - 3} + x$

$10\sqrt{x - 3} = 20$

$\sqrt{x - 3} = 2$

$(\sqrt{x - 3})^2 = 2^2$

$x - 3 = 4$

$x = 7$

Check:

$$\frac{\sqrt{x - 3} + \sqrt{x + 2} = 5}{\sqrt{7 - 3} + \sqrt{7 + 2} \ ? \ 5}$$

$\sqrt{4} + \sqrt{9}$

$2 + 3$

$5 \ | \ 5$ TRUE

The solution is 7.

27. $\sqrt{3x - 5} + \sqrt{2x + 3} + 1 = 0$

$\sqrt{3x - 5} + \sqrt{2x + 3} = -1$

The principal square root is never negative. Thus the sum of two principal square roots cannot equal -1. There is no solution.

29. $\sqrt{x} - \sqrt{3x - 3} = 1$

$\sqrt{x} = \sqrt{3x - 3} + 1$

$(\sqrt{x})^2 = (\sqrt{3x - 3} + 1)^2$

$x = (3x - 3) + 2\sqrt{3x - 3} + 1$

$2 - 2x = 2\sqrt{3x - 3}$

$1 - x = \sqrt{3x - 3}$

$(1 - x)^2 = (\sqrt{3x - 3})^2$

$1 - 2x + x^2 = 3x - 3$

$x^2 - 5x + 4 = 0$

$(x - 4)(x - 1) = 0$

$x = 4$ or $x = 1$

The number 4 does not check, but 1 does. The solution is 1.

31. $\sqrt{2y-5} - \sqrt{y-3} = 1$

$$\sqrt{2y-5} = \sqrt{y-3} + 1$$
$$(\sqrt{2y-5})^2 = (\sqrt{y-3}+1)^2$$
$$2y - 5 = (y-3) + 2\sqrt{y-3} + 1$$
$$y - 3 = 2\sqrt{y-3}$$
$$(y-3)^2 = (2\sqrt{y-3})^2$$
$$y^2 - 6y + 9 = 4(y-3)$$
$$y^2 - 6y + 9 = 4y - 12$$
$$y^2 - 10y + 21 = 0$$
$$(y-7)(y-3) = 0$$
$$y = 7 \text{ or } y = 3$$

Both numbers check. The solutions are 7 and 3.

33. $x^{1/3} = -2$

$$(x^{1/3})^3 = (-2)^3 \qquad (x^{1/3} = \sqrt[3]{x})$$
$$x = -8$$

The value checks. The solution is -8.

35. $t^{1/4} = 3$

$$(t^{1/4})^4 = 3^4 \qquad (t^{1/4} = \sqrt[4]{t})$$
$$t = 81$$

The value checks. The solution is 81.

37. $|x| = 7$

The solutions are those numbers whose distance from 0 on a number line is 7. They are -7 and 7. That is,

$$x = -7 \text{ or } x = 7.$$

The solutions are -7 and 7.

39. $|x| = -10.7$

The absolute value of a number is nonnegative. Thus, the equation has no solution.

41. $|x - 1| = 4$

$$x - 1 = -4 \text{ or } x - 1 = 4$$
$$x = -3 \text{ or } \qquad x = 5$$

The solutions are -3 and 5.

43. $|3x| = 1$

$$3x = -1 \text{ or } 3x = 1$$
$$x = -\frac{1}{3} \text{ or } \quad x = \frac{1}{3}$$

The solutions are $-\frac{1}{3}$ and $\frac{1}{3}$.

45. $|x| = 0$

The distance of 0 from 0 on a number line is 0. That is,

$$x = 0.$$

The solution is 0.

47. $|3x + 2| = 1$

$$3x + 2 = -1 \text{ or } 3x + 2 = 1$$
$$3x = -3 \text{ or } \qquad 3x = -1$$
$$x = -1 \text{ or } \qquad x = -\frac{1}{3}$$

The solutions are -1 and $-\frac{1}{3}$.

49. $\left|\frac{1}{2}x - 5\right| = 17$

$$\frac{1}{2}x - 5 = -17 \text{ or } \frac{1}{2}x - 5 = 17$$
$$\frac{1}{2}x = -12 \text{ or } \qquad \frac{1}{2}x = 22$$
$$x = -24 \text{ or } \qquad x = 44$$

The solutions are -24 and 44.

51. $|x - 1| + 3 = 6$

$$|x - 1| = 3$$
$$x - 1 = -3 \text{ or } x - 1 = 3$$
$$x = -2 \text{ or } \qquad x = 4$$

The solutions are -2 and 4.

53. $\dfrac{P_1 V_1}{T_1} = \dfrac{P_2 V_2}{T_2}$

$$P_1 V_1 T_2 = P_2 V_2 T_1 \qquad \text{Multiplying by } T_1 T_2 \text{ on both sides}$$

$$\frac{P_1 V_1 T_2}{P_2 V_2} = T_1 \qquad \text{Dividing by } P_2 V_2 \text{ on both sides}$$

55.
$$\frac{1}{R} = \frac{1}{R_1} + \frac{1}{R_2}$$

$$RR_1 R_2 \cdot \frac{1}{R} = RR_1 R_2 \left(\frac{1}{R_1} + \frac{1}{R_2}\right)$$

Multiplying by $RR_1 R_2$ on both sides

$$R_1 R_2 = RR_2 + RR_1$$

$$R_1 R_2 - RR_2 = RR_1 \quad \text{Subtracting } RR_2 \text{ on both sides}$$

$$R_2(R_1 - R) = RR_1 \qquad \text{Factoring}$$

$$R_2 = \frac{RR_1}{R_1 - R} \qquad \text{Dividing by } R_1 - R \text{ on both sides}$$

57.
$$\frac{1}{F} = \frac{1}{m} + \frac{1}{p}$$

$$Fmp \cdot \frac{1}{F} = Fmp\left(\frac{1}{m} + \frac{1}{p}\right) \quad \text{Multiplying by } Fmp \text{ on both sides}$$

$$mp = Fp + Fm$$

$$mp - Fp = Fm \qquad \text{Subtracting } Fp \text{ on both sides}$$

$$p(m - F) = Fm \qquad \text{Factoring}$$

$$p = \frac{Fm}{m - F} \qquad \text{Dividing by } m - F \text{ on both sides}$$

59. Discussion and Writing

61. Graph $y = 15 - 2x$ and use the Zero feature.

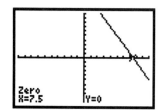

The zero is 7.5.

63. Familiarize. Let a = the number of adults who passed the GED test in 1997.

Translate.

Number of adults who passed the test in 1998	was	25,000	more than	number who passed the test in 1997.
↓	↓	↓	↓	↓
506,000	=	25,000	+	a

Carry out. We will solve algebraically.

$$506,000 = 25,000 + a$$

$$481,000 = a \qquad \text{Subtracting 25,000}$$

Check. 25,000 more than 481,000 is 25,000 + 481,000, or 506,000. The answer checks.

State. In 1997, 481,000 adults passed the GED test.

65.
$$(x - 3)^{2/3} = 2$$

$$[(x - 3)^{2/3}]^3 = 2^3$$

$$(x - 3)^2 = 8$$

$$x^2 - 6x + 9 = 8$$

$$x^2 - 6x + 1 = 0$$

$$a = 1, b = -6, c = 1$$

$$x = \frac{-b \pm \sqrt{b^2 - 4ac}}{2a}$$

$$= \frac{-(-6) \pm \sqrt{(-6)^2 - 4 \cdot 1 \cdot 1}}{2 \cdot 1}$$

$$= \frac{6 \pm \sqrt{32}}{2} = \frac{6 \pm 4\sqrt{2}}{2}$$

$$= \frac{2(3 \pm 2\sqrt{2})}{2} = 3 \pm 2\sqrt{2}$$

Both values check. The solutions are $3 \pm 2\sqrt{2}$.

67.
$$\sqrt{x + 5} + 1 = \frac{6}{\sqrt{x + 5}}, \quad \text{LCD is } \sqrt{x + 5}$$

$$x + 5 + \sqrt{x + 5} = 6 \qquad \text{Multiplying by } \sqrt{x + 5}$$

$$\sqrt{x + 5} = 1 - x$$

$$x + 5 = 1 - 2x + x^2$$

$$0 = x^2 - 3x - 4$$

$$0 = (x - 4)(x + 1)$$

$$x = 4 \quad \text{or} \quad x = -1$$

Only -1 checks. The solution set is -1.

Exercise Set 2.7

1. $x + 6 < 5x - 6$

$$6 + 6 < 5x - x \qquad \text{Subtracting } x \text{ and adding 6 on both sides}$$

$$12 < 4x$$

$$\frac{12}{4} < x \qquad \text{Dividing by 4 on both sides}$$

$$3 < x$$

This inequality could also be solved as follows:

$$x + 6 < 5x - 6$$

$$x - 5x < -6 - 6 \qquad \text{Subtracting } 5x \text{ and 6 on both sides}$$

$$-4x < -12$$

$$x > \frac{-12}{-4} \qquad \text{Dividing by } -4 \text{ on both sides and reversing the inequality symbol}$$

$$x > 3$$

The solution set is $\{x | x > 3\}$, or $(3, \infty)$.

3. $3x - 3 + 2x \geq 1 - 7x - 9$

$$5x - 3 \geq -7x - 8 \qquad \text{Collecting like terms}$$

$$5x + 7x \geq -8 + 3 \qquad \text{Adding } 7x \text{ and 3 on both sides}$$

$$12x \geq -5$$

$$x \geq -\frac{5}{12} \qquad \text{Dividing by 12 on both sides}$$

The solution set is $\left\{x \middle| x \geq -\frac{5}{12}\right\}$, or $\left[-\frac{5}{12}, \infty\right)$.

5. $14 - 5y \le 8y - 8$

$14 + 8 \le 8y + 5y$

$22 \le 13y$

$\dfrac{22}{13} \le y$

This inequality could also be solved as follows:

$14 - 5y \le 8y - 8$

$-5y - 8y \le -8 - 14$

$-13y \le -22$

$y \ge \dfrac{22}{13}$ Dividing by -13 on both sides and reversing the inequality symbol

The solution set is $\left\{ y \middle| y \ge \dfrac{22}{13} \right\}$, or $\left[\dfrac{22}{13}, \infty \right)$.

7. $-\dfrac{3}{4}x \ge -\dfrac{5}{8} + \dfrac{2}{3}x$

$\dfrac{5}{8} \ge \dfrac{3}{4}x + \dfrac{2}{3}x$

$\dfrac{5}{8} \ge \dfrac{9}{12}x + \dfrac{8}{12}x$

$\dfrac{5}{8} \ge \dfrac{17}{12}x$

$\dfrac{12}{17} \cdot \dfrac{5}{8} \ge \dfrac{12}{17} \cdot \dfrac{17}{12}x$

$\dfrac{15}{34} \ge x$

The solution set is $\left\{ x \middle| x \le \dfrac{15}{34} \right\}$, or $\left(-\infty, \dfrac{15}{34} \right]$.

9. $4x(x-2) < 2(2x-1)(x-3)$

$4x(x-2) < 2(2x^2 - 7x + 3)$

$4x^2 - 8x < 4x^2 - 14x + 6$

$-8x < -14x + 6$

$-8x + 14x < 6$

$6x < 6$

$x < \dfrac{6}{6}$

$x < 1$

The solution set is $\{x | x < 1\}$, or $(-\infty, 1)$.

11. The graph of $y_1 = 4x(x-2)$ lies below the graph of $y_2 = 2(2x-1)(x-3)$ for $x < 1$.

13. $-2 \le x + 1 < 4$

$-3 \le x < 3$ Subtracting 1

The solution set is $[-3, 3)$.

15. $5 \le x - 3 \le 7$

$8 \le x \le 10$ Adding 3

The solution set is $[8, 10]$.

17. $-3 \le x + 4 \le 3$

$-7 \le x \le -1$ Subtracting 4

The solution set is $[-7, -1]$.

19. $-2 < 2x + 1 < 5$

$-3 < 2x < 4$ Adding -1

$-\dfrac{3}{2} < x < 2$ Multiplying by $\dfrac{1}{2}$

The solution set is $\left(-\dfrac{3}{2}, 2 \right)$.

21. $-4 \le 6 - 2x < 4$

$-10 \le -2x < -2$ Adding -6

$5 \ge x > 1$ Multiplying by $-\dfrac{1}{2}$

or $1 < x \le 5$

The solution set is $(1, 5]$.

23. $-5 < \dfrac{1}{2}(3x + 1) \le 7$

$-10 < 3x + 1 \le 14$ Multiplying by 2

$-11 < 3x \le 13$ Adding -1

$-\dfrac{11}{3} < x \le \dfrac{13}{3}$ Multiplying by $\dfrac{1}{3}$

The solution set is $\left(-\dfrac{11}{3}, \dfrac{13}{3} \right]$.

25. $3x \le -6$ or $x - 1 > 0$

$x \le -2$ or $x > 1$

The solution set is $(-\infty, -2] \cup (1, \infty)$.

27. $2x + 3 \le -4$ or $2x + 3 \ge 4$

$2x \le -7$ or $2x \ge 1$

$x \le -\dfrac{7}{2}$ or $x \ge \dfrac{1}{2}$

The solution set is $\left(-\infty, -\dfrac{7}{2} \right] \cup \left[\dfrac{1}{2}, \infty \right)$.

29. $2x - 20 < -0.8$ or $2x - 20 > 0.8$

$2x < 19.2$ or $2x > 20.8$

$x < 9.6$ or $x > 10.4$

The solution set is $(-\infty, 9.6) \cup (10.4, \infty)$.

31. $x + 14 \leq -\dfrac{1}{4}$ *or* $x + 14 \geq \dfrac{1}{4}$

$\qquad x \leq -\dfrac{57}{4}$ *or* $\quad x \geq -\dfrac{55}{4}$

The solution set is $\left(-\infty, -\dfrac{57}{4}\right] \cup \left[-\dfrac{55}{4}, \infty\right)$.

33. The graph of $y_1 = 6 - 2x$ lies both on or above the graph of $y_2 = -4$ and below the graph of $y_3 = 4$ for $1 < x \leq 5$.

35. $|x| < 7$

To solve we look for all numbers x whose distance from 0 is less than 7. These are the numbers between -7 and 7. That is, $-7 < x < 7$. The solution set and its graph are as follows:

$(-7, 7)$

37. $|x| \geq 4.5$

To solve we look for all numbers x whose distance from 0 is greater than or equal to 4.5. That is, $x \leq -4.5$ or $x \geq 4.5$. The solution set and its graph are as follows.

$\{x | x \leq -4.5 \text{ or } x \geq 4.5\}$, or $(-\infty, -4.5] \cup [4.5, \infty)$

39. $|x + 8| < 9$

$\qquad -9 < x + 8 < 9$

$\qquad -17 < x < 1 \qquad$ Subtracting 8

The solution set is $(-17, 1)$.

41. $|x + 8| \geq 9$

$\qquad x + 8 \leq -9 \quad$ *or* $x + 8 \geq 9$

$\qquad \quad x \leq -17 \text{ or} \qquad x \geq 1 \quad$ Subtracting 8

The solution set is $(-\infty, -17] \cup [1, \infty)$.

43. $\left|x - \dfrac{1}{4}\right| < \dfrac{1}{2}$

$\qquad -\dfrac{1}{2} < x - \dfrac{1}{4} < \dfrac{1}{2}$

$\qquad -\dfrac{1}{4} < x < \dfrac{3}{4} \qquad$ Adding $\dfrac{1}{4}$

The solution set is $\left(-\dfrac{1}{4}, \dfrac{3}{4}\right)$.

45. $|3x| < 1$

$\qquad -1 < 3x < 1$

$\qquad -\dfrac{1}{3} < x < \dfrac{1}{3} \quad$ Dividing by 3

The solution set is $\left(-\dfrac{1}{3}, \dfrac{1}{3}\right)$.

47. $|2x + 3| \leq 9$

$\qquad -9 \leq 2x + 3 \leq 9$

$\qquad -12 \leq 2x \leq 6 \qquad$ Subtracting 3

$\qquad -6 \leq x \leq 3 \qquad$ Dividing by 2

The solution set is $[-6, 3]$.

49. $|x - 5| > 0.1$

$\qquad x - 5 < -0.1 \quad$ *or* $x - 5 > 0.1$

$\qquad \quad x < 4.9 \quad$ *or* $\qquad x > 5.1 \quad$ Adding 5

The solution set is $(-\infty, 4.9) \cup (5.1, \infty)$.

51. $|6 - 4x| \leq 8$

$\qquad -8 \leq 6 - 4x \leq 8$

$\qquad -14 \leq -4x \leq 2 \quad$ Subtracting 6

$\qquad \dfrac{14}{4} \geq x \geq -\dfrac{2}{4} \quad$ Dividing by -4 and revers-

$\qquad \qquad \qquad \qquad \qquad$ ing the inequality symbols

$\qquad \dfrac{7}{2} \geq x \geq -\dfrac{1}{2} \quad$ Simplifying

The solution set is $\left[-\dfrac{1}{2}, \dfrac{7}{2}\right]$.

53. $\left|x + \dfrac{2}{3}\right| \leq \dfrac{5}{3}$

$\qquad -\dfrac{5}{3} \leq x + \dfrac{2}{3} \leq \dfrac{5}{3}$

$\qquad -\dfrac{7}{3} \leq x \leq 1 \qquad$ Subtracting $\dfrac{2}{3}$

The solution set is $\left[-\dfrac{7}{3}, 1\right]$.

55. $\left|\dfrac{2x + 1}{3}\right| > 5$

$\qquad \dfrac{2x + 1}{3} < -5 \quad$ *or* $\dfrac{2x + 1}{3} > 5$

$\qquad 2x + 1 < -15 \text{ or } 2x + 1 > 15 \text{ Multiplying by 3}$

$\qquad \quad 2x < -16 \text{ or} \qquad 2x > 14 \text{ Subtracting 1}$

$\qquad \quad \ x < -8 \quad \text{ or} \qquad x > 7 \text{ Dividing by 2}$

The solution set is $\{x | x < -8 \text{ or } x > 7\}$, or $(-\infty, -8) \cup (7, \infty)$.

57. $|2x - 4| < -5$

Since $|2x - 4| \geq 0$ for all x, there is no x such that $|2x - 4|$ would be less than -5. There is no solution.

59. The graph of $y_1 = |x + 8|$ lies on or above the graph of $y_2 = 9$ for $x \leq -17$ or $x \geq 1$.

61. Discussion and Writing

63. Discussion and Writing

65. $(6 - 4i) - (-4 + 3i) = 6 - 4i + 4 - 3i$
$$= (6 + 4) + (-4 - 3)i$$
$$= 10 - 7i$$

67. $\dfrac{5 + 2i}{3 - 4i} = \dfrac{5 + 2i}{3 - 4i} \cdot \dfrac{3 + 4i}{3 + 4i}$

$$= \dfrac{(5 + 2i)(3 + 4i)}{(3 - 4i)(3 + 4i)}$$

$$= \dfrac{15 + 20i + 6i + 8i^2}{9 - 16i^2}$$

$$= \dfrac{15 + 20i + 6i - 8}{9 + 16}$$

$$= \dfrac{7 + 26i}{25}$$

$$= \dfrac{7}{25} + \dfrac{26}{25}i$$

69. $2x \leq 5 - 7x < 7 + x$

$2x \leq 5 - 7x$ and $5 - 7x < 7 + x$

$9x \leq 5$ $\quad$ and $\quad$ $-8x < 2$

$x \leq \dfrac{5}{9}$ $\quad$ and $\quad$ $x > -\dfrac{1}{4}$

The solution set is $\left(-\dfrac{1}{4}, \dfrac{5}{9} \right]$.

71. $|3x - 1| > 5x - 2$

$3x - 1 < -(5x - 2)$ $\quad$ or $\quad$ $3x - 1 > 5x - 2$

$3x - 1 < -5x + 2$ $\quad$ or $\quad$ $1 > 2x$

$8x < 3$ $\quad$ or $\quad$ $\dfrac{1}{2} > x$

$x < \dfrac{3}{8}$ $\quad$ or $\quad$ $\dfrac{1}{2} > x$

The solution set is $\left(-\infty, \dfrac{3}{8} \right) \cup \left(-\infty, \dfrac{1}{2} \right)$. This is equivalent to $\left(-\infty, \dfrac{1}{2} \right)$.

73. $|p - 4| + |p + 4| < 8$

If $p < -4$, then $|p - 4| = -(p - 4)$ and $|p + 4| = -(p + 4)$.

Solve: $-(p - 4) + [-(p + 4)] < 8$
$$-p + 4 - p - 4 < 8$$
$$-2p < 8$$
$$p > -4$$

Since this is false for all values of p in the interval $(-\infty, -4)$ there is no solution in this interval.

If $p \geq -4$, then $|p + 4| = p + 4$.

Solve: $|p - 4| + p + 4 < 8$
$$|p - 4| < 4 - p$$

$p - 4 > -(4 - p)$ $\quad$ and $\quad$ $p - 4 < 4 - p$

$p - 4 > p - 4$ $\quad$ and $\quad$ $2p < 8$

$-4 > -4$ $\quad$ and $\quad$ $p < 4$

Since $-4 > -4$ is false for all values of p, there is no solution in the interval $[-4, \infty)$.

Thus, $|p - 4| + |p + 4| < 8$ has no solution.

Chapter 3

Polynomial and Rational Functions

1. $g(x) = \frac{1}{2}x^3 - 10x + 8$

The degree of the polynomial is 3, so the polynomial is cubic. The leading term is $\frac{1}{2}x^3$ and the leading coefficient is $\frac{1}{2}$.

3. $h(x) = 0.9x - 0.13$

The degree of the polynomial is 1, so the polynomial is linear. The leading term is $0.9x$ and the leading coefficient is 0.9.

5. $g(x) = 305x^4 + 4021$

The degree of the polynomial is 4, so the polynomial is quartic. The leading term is $305x^4$ and the leading coefficient is 305.

7. $f(x) = \frac{1}{4}x^2 - 5$

The leading term is $\frac{1}{4}x^2$. The sign of the leading coefficient, $\frac{1}{4}$, is positive and the dgree, 2, is even, so we would choose either graph (b) or graph (d). Note also that $f(0) = -5$, so the y-intercept is $(0, -5)$. Thus, graph (d) is the graph of this function.

9. $f(x) = x^5 - x^4 + x^2 + 4$

The leading term is x^5. The sign of the leading coefficient, 1, is positive and the degree, 5, is odd. Thus, graph (f) is the graph of this function.

11. $f(x) = x^4 - 2x^3 + 12x^2 + x - 20$

The leading term is x^4. The sign of the leading coefficient, 1, is positive and the degree, 4, is even, so we would choose either graph (b) or graph (d). Note also that $f(0) = -20$, so the y-intercept is $(0, -20)$. Thus, graph (b) is the graph of this function.

13. Graph $y = x^3 - 3x - 1$ and use the Zero feature three times. The real-number zeros are about -1.532, -0.347, and 1.879.

15. Graph $y = x^4 - 2x^2$ and use the Zero feature three times. The real-number zeros are about -1.414, 0, and 1.414.

17. Graph $y = x^3 - x$ and use the Zero feature three times. The real-number zeros are -1, 0, and 1.

19. Graph $y = x^8 + 8x^7 - 28x^6 - 56x^5 + 70x^4 + 56x^3 - 28x^2 - 8x + 1$ and use the Zero feature eight times. The real-number zeros are about -10.153, -1.871, -0.821, -0.303, 0.098, 0.535, 1.219, and 3.297.

21. $g(x) = x^3 - 1.2x + 1$

Graph the function and use the Zero, Maximum, and Minimum features.

Zero: -1.368

Relative maximum: 1.506 when $x \approx -0.632$

Relative minimum: 0.494 when $x \approx 0.632$

Range: $(-\infty, \infty)$

23. $h(x) = -\frac{1}{2}x^4 + 3x^3 - 5x^2 + 3x + 6$

Graph the function and use the Zero, Maximum, and Minimum features.

Zeros: -0.720, 4.089

Relative maxima: 6.59375 when $x = 0.5$, 10.5 when $x = 3$

Relative minimum: 6.5 when $x = 1$

Range: $(-\infty, 10.5]$

25. $f(x) = x^6 - 3.8$

Graph the function and use the Zero and Minimum features.

Zeros: -1.249, 1.249

There is no relative maximum.

Relative minimum: -3.8 when $x = 0$

Range: $[-3.8, \infty)$

27. $g(x) = 12 - 3.14x$

Graph the function and use the Zero feature.

Zero: 3.822

There are no relative maxima or minima.

Range: $(-\infty, \infty)$

29. $f(x) = x^2 + 10x - x^5$

Graph the function and use the Zero, Maximum, and Minimum features.

Zeros: -1.697, 0, 1.856

Relative maximum: 11.012 when $x \approx 1.258$

Relative minimum: -8.183 when $x \approx -1.116$

Range: $(-\infty, \infty)$

31. $h(x) = -x^5 + 4x^3 - x$

Graph the function and use the Zero, Maximum, and Minimum features.

Zeros: $-1.932, -0.518, 0, 0.518, 1.932$

Relative maxima: 0.195 when $x \approx -0.294$, 4.414 when $x \approx 1.521$

Relative minima: -4.414 when $x \approx -1.521$, -0.195 when $x \approx 0.294$

Range: $(-\infty, \infty)$

33. $f(-5) = (-5)^3 + 3(-5)^2 - 9(-5) - 13 = -18$

$f(-4) = (-4)^3 + 3(-4)^2 - 9(-4) - 13 = 7$

By the intermediate value theorem, since $f(-5)$ and $f(-4)$ have opposite signs then $f(x)$ has a zero between -5 and -4.

35.
$$V = 48T^2$$
$$36 = 48T^2 \quad \text{Substituting}$$
$$0.75 = T^2$$
$$0.866 \approx T \quad \text{We only want the positive solution.}$$

Anfernee Hardaway's hang time is 0.866 sec.

37. First find the number of games played.

$N(x) = x^2 - x$

$N(9) = 9^2 - 9 = 72$

Now multiply the number of games by the cost per game to find the total cost.

$72 \cdot 45 = 3240$

It will cost $3240 to play the entire schedule.

39. $A = P(1 + i)^t$

a)
$$3610 = 2560(1 + i)^2 \quad \text{Substituting}$$
$$\frac{3610}{2560} = (1 + i)^2$$
$$\pm 1.1875 = 1 + i \quad \text{Taking the square root on both sides}$$
$$-1 \pm 1.1875 = i$$
$$-1 - 1.1875 = i \quad or \quad -1 + 1.1875 = i$$
$$-2.1875 = i \quad or \quad 0.1875 = i$$

Only the positive result has meaning in this application. The interest rate is 0.1875, or 18.75%.

b)
$$13,310 = 10,000(1 + i)^3 \quad \text{Substituting}$$
$$\frac{13,310}{10,000} = (1 + i)^3$$
$$1.10 \approx 1 + i \quad \text{Taking the cube root on both sides}$$
$$0.10 \approx i$$

The interest rate is 0.1, or 10%.

41. The sales drop and then rise. This suggests that a quadratic function that opens up might fit the data. Thus, we choose (b).

43. The sales rise and then drop. This suggests that a quadratic function that opens down might fit the data. Thus, we choose (c).

45. The sales fall steadily. Thus, we choose (a), a linear function.

47. The rate falls and then rises. This suggests that a quadratic function that opens up might fit the data. Thus, we choose (b).

49. a)

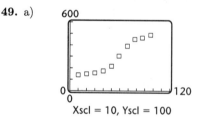

b) Linear: $y = 4.717948718x + 37.93076923$; $r^2 \approx 0.9306$

Quadratic: $y = 0.0281175273x^2 + 1.711977591x + 96.76946268$; $R^2 \approx 0.9504$

Cubic: $y = -0.0016285158x^3 + 0.2897410836x^2 - 10.07476028x + 227.1101167$; $R^2 \approx 0.9872$

Power: $y = 22.9451257x^{0.6396947491}$; $r^2 \approx 0.8142$

The value of R^2 is highest for the cubic function so we determine that this function fits the data best.

c)

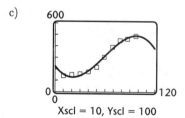

d) In 2000, $x = 2000 - 1900 = 100$; in 2010, $x = 2010 - 1900 - 110$.

For the linear function: $f(100) \approx 510$ and $f(110) \approx 557$.

For the quadratic function: $f(100) \approx 549$ and $f(110) \approx 625$.

For the cubic function: $f(100) \approx 489$ and $f(110) \approx 457$.

For the power function: $f(100) \approx 437$ and $f(110) \approx 464$.

It is reasonable to assume that the average acreage will continue to increase. For this reason we rule out the predictions made using the cubic and power functions. The rate of increase appears to be more like that predicted by the linear function than the quadratic function. Thus, we say that the prediction of 510 acres and 557 acres given by the linear function are the most realistic. Answers may vary.

51. a)
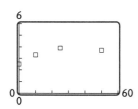

b) Linear: $y = 0.0216740088x + 2.889427313$; $r^2 \approx 0.5795$

Quadratic: $y = -0.001301354x^2 + 0.0887195358x + 2.513926499$; $R^2 \approx 0.9988$

Cubic: $y = 0.000008x^3 - 0.00188x^2 + 0.98x + 2.5$; $R^2 = 1$

The value of R^2 is highest for the cubic function. In fact, $R^2 = 1$ for this function so it fits the data perfectly.

c)
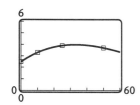

d) In 2060, $x = 2060 - 1985 = 75$.

For the linear function: $f(75) \approx 4.5$.

For the quadratic function: $f(75) \approx 1.8$.

For the cubic function: $f(75) \approx 2.65$.

Since the estimated concentration in 2035 is lower than the concentration in 2010, it would seem reasonable to assume that the level would continue to drop so we would not use the linear prediction. The rate of decrease appears to be more like that estimated by the cubic function than the quadratic function. Thus, we say that the estimate of 2.65 ppb given by the cubic function is most realistic. Answers may vary.

53. Discussion and Writing

55. Discussion and Writing

57.
$$d = \sqrt{(x_2 - x_1)^2 + (y_2 - y_1)^2}$$
$$= \sqrt{[-1 - (-5)]^2 + (0 - 3)^2}$$
$$= \sqrt{4^2 + (-3)^2} = \sqrt{16 + 9}$$
$$= \sqrt{25} = 5$$

59.
$$(x - 3)^2 + (y + 5)^2 = 49$$
$$(x - 3)^2 + [y - (-5)]^2 = 7^2$$

Center: $(3, -5)$; radius: 7

Exercise Set 3.2

1. $f(x) = x^3 - 9x^2 + 14x + 24$

$f(4) = 4^3 - 9 \cdot 4^2 + 14 \cdot 4 + 24 = 0$

Since $f(4) = 0$, 4 is a zero of $f(x)$.

$f(5) = 5^3 - 9 \cdot 5^2 + 14 \cdot 5 + 24 = -6$

Since $f(5) \neq 0$, 5 is not a zero of $f(x)$.

$f(-2) = (-2)^3 - 9(-2)^2 + 14(-2) + 24 = -48$

Since $f(-2) \neq 0$, -2 is not a zero of $f(x)$.

3. We divide to determine whether each binomial is a factor of $f(x)$.

a)
$$\begin{array}{r} x^2 - 5x - 6 \\ x - 4 \overline{\smash{\big)}\ x^3 - 9x^2 + 14x + 24} \\ \underline{x^3 - 4x^2} \\ -5x^2 + 14x \\ \underline{-5x^2 + 20x} \\ -6x + 24 \\ \underline{-6x + 24} \\ 0 \end{array}$$

Since the remainder is 0, $x - 4$ is a factor of $f(x)$.

b)

$$
\begin{array}{r}
x^2 - 4x - 6 \\
x - 5 \overline{\smash{\big)}\ x^3 - 9x^2 + 14x + 24} \\
\underline{x^3 - 5x^2} \\
-4x^2 + 14x \\
\underline{-4x^2 + 20x} \\
-6x + 24 \\
\underline{-6x + 30} \\
-6
\end{array}
$$

Since the remainder is not 0, $x - 5$ is not a factor of $f(x)$.

c)

$$
\begin{array}{r}
x^2 - 11x + 36 \\
x + 2 \overline{\smash{\big)}\ x^3 - 9x^2 + 14x + 24} \\
\underline{x^3 + 2x^2} \\
-11x^2 + 14x \\
\underline{-11x^2 - 22x} \\
36x + 24 \\
\underline{36x + 72} \\
-48
\end{array}
$$

Since the remainder is not 0, $x + 2$ is not a factor of $f(x)$.

5.

$$
\begin{array}{r}
x^2 - 2x + 4 \\
x + 2 \overline{\smash{\big)}\ x^3 + 0x^2 + 0x - 8} \\
\underline{x^3 + 2x^2} \\
-2x^2 + 0x \\
\underline{-2x^2 - 4x} \\
4x - 8 \\
\underline{4x + 8} \\
-16
\end{array}
$$

$x^3 - 8 = (x + 2)(x^2 - 2x + 4) - 16$

7.

$$
\begin{array}{r}
x^2 + 5 \\
x^2 + 4 \overline{\smash{\big)}\ x^4 + 9x^2 + 20} \\
\underline{x^4 + 4x^2} \\
5x^2 + 20 \\
\underline{5x^2 + 20} \\
0
\end{array}
$$

$x^4 + 9x^2 + 20 = (x^2 + 4)(x^2 + 5) + 0$

9. $(2x^4 + 7x^3 + x - 12) \div (x + 3)$

$= (2x^4 + 7x^3 + 0x^2 + x - 12) \div [x - (-3)]$

$$
\begin{array}{r|rrrrr}
-3 & 2 & 7 & 0 & 1 & -12 \\
   &   & -6 & -3 & 9 & -30 \\
\hline
   & 2 & 1 & -3 & 10 & -42
\end{array}
$$

The quotient is $2x^3 + x^2 - 3x + 10$. The remainder is -42.

11. $(x^3 - 2x^2 - 8) \div (x + 2)$

$= (x^3 - 2x^2 + 0x - 8) \div [x - (-2)]$

$$
\begin{array}{r|rrrr}
-2 & 1 & -2 & 0 & -8 \\
   &   & -2 & 8 & -16 \\
\hline
   & 1 & -4 & 8 & -24
\end{array}
$$

The quotient is $x^2 - 4x + 8$. The remainder is -24.

13. $(x^4 - 1) \div (x - 1)$

$= (x^4 + 0x^3 + 0x^2 + 0x - 1) \div (x - 1)$

$$
\begin{array}{r|rrrrr}
1 & 1 & 0 & 0 & 0 & -1 \\
  &   & 1 & 1 & 1 & 1 \\
\hline
  & 1 & 1 & 1 & 1 & 0
\end{array}
$$

The quotient is $x^3 + x^2 + x + 1$. The remainder is 0.

15. $(2x^4 + 3x^2 - 1) \div \left(x - \dfrac{1}{2}\right)$

$(2x^4 + 0x^3 + 3x^2 + 0x - 1) \div \left(x - \dfrac{1}{2}\right)$

$$
\begin{array}{r|rrrrr}
\frac{1}{2} & 2 & 0 & 3 & 0 & -1 \\
            &   & 1 & \frac{1}{2} & \frac{7}{4} & \frac{7}{8} \\
\hline
            & 2 & 1 & \frac{7}{2} & \frac{7}{4} & -\frac{1}{8}
\end{array}
$$

The quotient is $2x^3 + x^2 + \dfrac{7}{2}x + \dfrac{7}{4}$. The remainder is $-\dfrac{1}{8}$.

17. $(x^4 - y^4) \div (x - y)$

$= (x^4 + 0x^3 + 0x^2 + 0x - y^4) \div (x - y)$

$$
\begin{array}{r|rrrrr}
y & 1 & 0 & 0 & 0 & -y^4 \\
  &   & y & y^2 & y^3 & y^4 \\
\hline
  & 1 & y & y^2 & y^3 & 0
\end{array}
$$

The quotient is $x^3 + x^2y + xy^2 + y^3$. The remainder is 0.

19. $f(x) = x^3 - 6x^2 + 11x - 6$

Find $f(1)$.

$$
\begin{array}{r|rrrr}
1 & 1 & -6 & 11 & -6 \\
  &   & 1 & -5 & 6 \\
\hline
  & 1 & -5 & 6 & 0
\end{array}
$$

$f(1) = 0$

Find $f(-2)$.

$$
\begin{array}{r|rrrr}
-2 & 1 & -6 & 11 & -6 \\
   &   & -2 & 16 & -54 \\
\hline
   & 1 & -8 & 27 & -60
\end{array}
$$

$f(-2) = -60$

Find $f(3)$.

$$
\begin{array}{r|rrrr}
3 & 1 & -6 & 11 & -6 \\
  &   & 3 & -9 & 6 \\
\hline
  & 1 & -3 & 2 & 0
\end{array}
$$

$f(3) = 0$

21. $f(x) = 2x^5 - 3x^4 + 2x^3 - x + 8$

Find $f(20)$.

$$
\begin{array}{r|rrrrrr}
20 & 2 & -3 & 2 & 0 & -1 & 8 \\
   &   & 40 & 740 & 14{,}840 & 296{,}800 & 5{,}935{,}980 \\
\hline
   & 2 & 37 & 742 & 14{,}840 & 296{,}799 & 5{,}935{,}988
\end{array}
$$

$f(20) = 5{,}935{,}988$

Find $f(-3)$.

$$
\begin{array}{r|rrrrrr}
-3 & 2 & -3 & 2 & 0 & -1 & 8 \\
 & & -6 & 27 & -87 & 261 & -780 \\
\hline
 & 2 & -9 & 29 & -87 & 260 & \,|-772
\end{array}
$$

$f(-3) = -772$

23. $f(x) = x^4 - 16$

Find $f(2)$.

$$
\begin{array}{r|rrrrr}
2 & 1 & 0 & 0 & 0 & -16 \\
 & & 2 & 4 & 8 & 16 \\
\hline
 & 1 & 2 & 4 & 8 & \,|\;0
\end{array}
$$

$f(2) = 0$

Find $f(-2)$.

$$
\begin{array}{r|rrrrr}
-2 & 1 & 0 & 0 & 0 & -16 \\
 & & -2 & 4 & -8 & 16 \\
\hline
 & 1 & -2 & 4 & -8 & \,|\;0
\end{array}
$$

$f(-2) = 0$

Find $f(3)$.

$$
\begin{array}{r|rrrrr}
3 & 1 & 0 & 0 & 0 & -16 \\
 & & 3 & 9 & 27 & 81 \\
\hline
 & 1 & 3 & 9 & 27 & \,|\;65
\end{array}
$$

$f(3) = 65$

Find $f(1 - \sqrt{2})$.

$$
\begin{array}{r|rrrrr}
1-\sqrt{2} & 1 & 0 & 0 & 0 & -16 \\
 & & 1-\sqrt{2} & 3-2\sqrt{2} & 7-5\sqrt{2} & 17-12\sqrt{2} \\
\hline
 & 1 & 1-\sqrt{2} & 3-2\sqrt{2} & 7-5\sqrt{2} & \,|\;1-12\sqrt{2}
\end{array}
$$

$f(1 - \sqrt{2}) = 1 - 12\sqrt{2}$

25. $f(x) = 3x^3 + 5x^2 - 6x + 18$

If -3 is a zero of $f(x)$, then $f(-3) = 0$. Find $f(-3)$ using synthetic division.

$$
\begin{array}{r|rrrr}
-3 & 3 & 5 & -6 & 18 \\
 & & -9 & 12 & -18 \\
\hline
 & 3 & -4 & 6 & \,|\;0
\end{array}
$$

Since $f(-3) = 0$, -3 is a zero of $f(x)$.

If 2 is a zero of $f(x)$, then $f(2) = 0$. Find $f(2)$ using synthetic division.

$$
\begin{array}{r|rrrr}
2 & 3 & 5 & -6 & 18 \\
 & & 6 & 22 & 32 \\
\hline
 & 3 & 11 & 16 & \,|\;50
\end{array}
$$

Since $f(2) \neq 0$, 2 is not a zero of $f(x)$.

27. $f(x) = x^3 - \dfrac{7}{2}x^2 + x - \dfrac{3}{2}$

If -3 is a zero of $f(x)$, then $f(-3) = 0$. Find $f(-3)$ using synthetic division.

$$
\begin{array}{r|rrrr}
-3 & 1 & -\frac{7}{2} & 1 & -\frac{3}{2} \\
 & & -3 & \frac{39}{2} & -\frac{123}{2} \\
\hline
 & 1 & -\frac{13}{2} & \frac{41}{2} & \,|\;-63
\end{array}
$$

Since $f(-3) \neq 0$, -3 is not a zero of $f(x)$.

If $\dfrac{1}{2}$ is a zero of $f(x)$, then $f\left(\dfrac{1}{2}\right) = 0$.

Find $f\left(\dfrac{1}{2}\right)$ using synthetic division.

$$
\begin{array}{r|rrrr}
\frac{1}{2} & 1 & -\frac{7}{2} & 1 & -\frac{3}{2} \\
 & & \frac{1}{2} & -\frac{3}{2} & -\frac{1}{4} \\
\hline
 & 1 & -3 & -\frac{1}{2} & \,|\;-\frac{7}{4}
\end{array}
$$

Since $f\left(\dfrac{1}{2}\right) \neq 0$, $\dfrac{1}{2}$ is not a zero of $f(x)$.

29. $f(x) = x^3 + 4x^2 + x - 6$

Try $x - 1$. Use synthetic division to see whether $f(1) = 0$.

$$
\begin{array}{r|rrrr}
1 & 1 & 4 & 1 & -6 \\
 & & 1 & 5 & 6 \\
\hline
 & 1 & 5 & 6 & \,|\;0
\end{array}
$$

Since $f(1) = 0$, $x - 1$ is a factor of $f(x)$. Thus $f(x) = (x - 1)(x^2 + 5x + 6)$.

Factoring the trinomial we get

$f(x) = (x - 1)(x + 2)(x + 3)$.

To solve the equation $f(x) = 0$, use the principle of zero products.

$(x - 1)(x + 2)(x + 3) = 0$

$x - 1 = 0 \quad or \quad x + 2 = 0 \quad or \quad x + 3 = 0$

$x = 1 \quad or \qquad x = -2 \quad or \qquad x = -3$

The solutions are 1, -2, and -3.

31. $f(x) = x^3 - 6x^2 + 3x + 10$

Try $x - 1$. Use synthetic division to see whether $f(1) = 0$.

$$
\begin{array}{r|rrrr}
1 & 1 & -6 & 3 & 10 \\
 & & 1 & -5 & -2 \\
\hline
 & 1 & -5 & -2 & \,|\;8
\end{array}
$$

Since $f(1) \neq 0$, $x - 1$ is not a factor of $P(x)$.

Try $x + 1$. Use synthetic division to see whether $f(-1) = 0$.

$$
\begin{array}{r|rrrr}
-1 & 1 & -6 & 3 & 10 \\
 & & -1 & 7 & -10 \\
\hline
 & 1 & -7 & 10 & \,|\;0
\end{array}
$$

Since $f(-1) = 0$, $x + 1$ is a factor of $f(x)$.

Thus $f(x) = (x + 1)(x^2 - 7x + 10)$.

Factoring the trinomial we get

$f(x) = (x+1)(x-2)(x-5)$.

To solve the equation $f(x) = 0$, use the principle of zero products.

$(x+1)(x-2)(x-5) = 0$

$x+1 = 0 \quad or \quad x-2 = 0 \quad or \quad x-5 = 0$

$x = -1 \quad or \qquad x = 2 \quad or \qquad x = 5$

The solutions are -1, 2, and 5.

33. $f(x) = x^3 - x^2 - 14x + 24$

Try $x+1$, $x-1$, and $x+2$. Using synthetic division we find that $f(-1) \neq 0$, $f(1) \neq 0$ and $f(-2) \neq 0$. Thus $x+1$, $x-1$, and $x+2$, are not factors of $f(x)$.

Try $x-2$. Use synthetic division to see whether $P(2) = 0$.

$$\begin{array}{r|rrrr} 2 & 1 & -1 & -14 & 24 \\ & & 2 & 2 & -24 \\ \hline & 1 & 1 & -12 & 0 \end{array}$$

Since $f(2) = 0$, $x-2$ is a factor of $f(x)$. Thus $f(x) = (x-2)(x^2 + x - 12)$.

Factoring the trinomial we get

$f(x) = (x-2)(x+4)(x-3)$

To solve the equation $f(x) = 0$, use the principle of zero products.

$(x-2)(x+4)(x-3) = 0$

$x-2 = 0 \quad or \quad x+4 = 0 \quad or \quad x-3 = 0$

$x = 2 \quad or \qquad x = -4 \quad or \qquad x = 3$

The solutions are 2, -4, and 3.

35. $f(x) = x^4 - x^3 - 19x^2 + 49x - 30$

Try $x-1$. Use synthetic division to see whether $f(1) = 0$.

$$\begin{array}{r|rrrrr} 1 & 1 & -1 & -19 & 49 & -30 \\ & & 1 & 0 & -19 & 30 \\ \hline & 1 & 0 & -19 & 30 & 0 \end{array}$$

Since $f(1) = 0$, $x-1$ is a factor of $f(x)$. Thus $f(x) = (x-1)(x^3 - 19x + 30)$.

We continue to use synthetic division to factor $g(x) = x^3 - 19x + 30$. Trying $x-1$, $x+1$, and $x+2$ we find that $g(1) \neq 0$, $g(-1) \neq 0$, and $g(-2) \neq 0$. Thus $x-1$, $x+1$, and $x+2$ are not factors of $x^3 - 19x + 30$. Try $x-2$.

$$\begin{array}{r|rrrr} 2 & 1 & 0 & -19 & 30 \\ & & 2 & 4 & -30 \\ \hline & 1 & 2 & -15 & 0 \end{array}$$

Since $g(2) = 0$, $x-2$ is a factor of $x^3 - 19x + 30$.

Thus $f(x) = (x-1)(x-2)(x^2 + 2x - 15)$.

Factoring the trinomial we get

$f(x) = (x-1)(x-2)(x-3)(x+5)$.

To solve the equation $f(x) = 0$, use the principle of zero products.

$(x-1)(x-2)(x-3)(x+5) = 0$

$x-1 = 0 \ or \ x-2 = 0 \ or \ x-3 = 0 \ or \ x+5 = 0$

$x = 1 \ or \quad x = 2 \ or \quad x = 3 \ or \quad x = -5$

The solutions are 1, 2, 3, and -5.

37. Discussion and Writing

39. $2x - 7 = 5x + 8$

$\quad -7 = 3x + 8 \quad$ Subtracting $2x$ on both sides

$\quad -15 = 3x \qquad$ Subtracting 8 on both sides

$\quad -5 = x \qquad\quad$ Dividing by 3 on both sides

We could also graph $y_1 = 2x - 7$ and $y_2 = 5x + 8$ and use the Intersect feature to find the first coordinate of the point of intersection of the graphs.

The solution is -5.

41. $\qquad 7x^2 + 4x = 3$

$\qquad 7x^2 + 4x - 3 = 0$

$\qquad (7x - 3)(x + 1) = 0$

$\quad 7x - 3 = 0 \quad or \quad x + 1 = 0$

$\qquad 7x = 3 \quad or \qquad x = -1$

$\qquad x = \dfrac{3}{7} \quad or \qquad x = -1$

We could also graph $y_1 = 7x^2 + 4x$ and $y_2 = 3$ and use the Intersect feature to find the first coordinate of the point of intersection of the graphs.

The solutions are $\dfrac{3}{7}$ and -1.

43. a) -4, -3, 2, and 5 are zeros of the function, so $x + 4$, $x + 3$, $x - 2$, and $x - 5$ are factors.

b) We first write the product of the factors:

$P(x) = (x+4)(x+3)(x-2)(x-5)$

Note that $P(0) = 4 \cdot 3(-2)(-5) > 0$ and the graph shows a positive y-intercept, so this function is a correct one.

c) Yes; two examples are $f(x) = c \cdot P(x)$ for any non-zero constant c and $g(x) = (x-a)P(x)$.

d) No; only the function in part (b) has the given graph.

45. Divide $x^3 - kx^2 + 3x + 7k$ by $x + 2$.

$$\begin{array}{r|rrrr} -2 & 1 & -k & 3 & 7k \\ & & -2 & 2k+4 & -4k-14 \\ \hline & 1 & -k-2 & 2k+7 & 3k-14 \end{array}$$

Thus $P(-2) = 3k - 14$.

We know that if $x + 2$ is a factor of $f(x)$, then $f(-2) = 0$.

We solve $0 = 3k - 14$ for k.

$$0 = 3k - 14$$

$$\frac{14}{3} = k$$

47. $y = \frac{1}{13}x^3 - \frac{1}{14}x$

$$y = x\left(\frac{1}{13}x^2 - \frac{1}{14}\right)$$

We use the principle of zero products to find the zeros of the polynomial.

$$x = 0 \quad or \quad \frac{1}{13}x^2 - \frac{1}{14} = 0$$

$$x = 0 \quad or \quad \frac{1}{13}x^2 = \frac{1}{14}$$

$$x = 0 \quad or \quad x^2 = \frac{13}{14}$$

$$x = 0 \quad or \quad x = \pm\sqrt{\frac{13}{14}}$$

$$x = 0 \quad or \quad x \approx \pm 0.9636$$

Only 0 and 0.9636 are in the interval $[0, 2]$.

49. $\dfrac{2x^2}{x^2 - 1} + \dfrac{4}{x + 3} = \dfrac{32}{3x^2 - x - 3}$,

$$\text{LCM is } (x + 1)(x - 1)(x + 3)$$

$$(x + 1)(x - 1)(x + 3)\left[\frac{2x^2}{(x + 1)(x - 1)} + \frac{4}{x + 3}\right] =$$

$$(x + 1)(x - 1)(x + 3) \cdot \frac{32}{(x + 1)(x - 1)(x + 3)}$$

$$2x^2(x + 3) + 4(x + 1)(x - 1) = 32$$

$$2x^3 + 6x^2 + 4x^2 - 4 = 32$$

$$2x^3 + 10x^2 - 36 = 0$$

$$x^3 + 5x^2 - 18 = 0$$

Using synthetic division and several trials, we find that -3 is a factor of $f(x) = x^3 + 5x^2 - 18$:

$$\begin{array}{r|rrrr} -3 & 1 & 5 & 0 & -18 \\ & & -3 & -6 & 18 \\ \hline & 1 & 2 & -6 & 0 \end{array}$$

Then we have:

$$(x + 3)(x^2 + 2x - 6) = 0$$

$$x + 3 = 0 \quad or \quad x^2 + 2x - 6 = 0$$

$$x = -3 \quad or \quad x = -1 \pm \sqrt{7}$$

Only $-1 \pm \sqrt{7}$ check.

51. Answers may vary. One possibility is $P(x) = x^{15} - x^{14}$.

53. $\begin{array}{r|rrr} i & 1 & -3 & 7 \\ & & i & -3i - 1 \\ \hline & 1 & -3 + i & 6 - 3i \end{array}$ $\qquad (i^2 = -1)$

The answer is $x - 3 + i$, R $6 - 3i$.

Exercise Set 3.3

1. $f(x) = (x + 3)^2(x - 1) = (x + 3)(x + 3)(x - 1)$

The factor $x + 3$ occurs twice. Thus the zero -3 has a multiplicity of two.

The factor $x - 1$ occurs only one time. Thus the zero 1 has a multiplicity of one.

3. $f(x) = x^3(x - 1)^2(x + 4)$

$x \cdot x \cdot x(x - 1)(x - 1)(x + 4) = 0$

The factor x occurs three times. Thus the zero 0 has a multiplicity of three.

The factor $x - 1$ occurs twice. Thus the zero 1 has a multiplicity of two.

The factor $x + 4$ occurs only one time. Thus the zero -4 has a multiplicity of one.

5. $f(x) = x^4 - 4x^2 + 3$

We factor as follows:

$$P(x) = (x^2 - 3)(x^2 - 1)$$

$$= (x - \sqrt{3})(x + \sqrt{3})(x - 1)(x + 1)$$

The zeros of the polynomial are $\sqrt{3}$, $-\sqrt{3}$, 1, and -1. Each has multiplicity of one.

7. $f(x) = x^3 + 3x^2 - x - 3$

We factor by grouping:

$$P(x) = x^2(x + 3) - (x + 3)$$

$$= (x^2 - 1)(x + 3)$$

$$= (x - 1)(x + 1)(x + 3)$$

The zeros of the polynomial are 1, -1, and -3. Each has multiplicity of one.

9. Find a polynomial function of degree 3 with -2, 3, and 5 as zeros.

Such a polynomial has factors $x + 2$, $x - 3$, and $x - 5$, so we have

$$f(x) = a_n(x + 2)(x - 3)(x - 5).$$

The number a_n can be any nonzero number. The simplest polynomial will be obtained if we let it be 1. Multiplying the factors, we obtain

$$f(x) = (x + 2)(x - 3)(x - 5)$$

$$= (x^2 - x - 6)(x - 5)$$

$$= x^3 - 6x^2 - x + 30$$

11. Find a polynomial function of degree 3 with -3, $2i$, and $-2i$ as zeros.

Such a polynomial has factors $x+3$, $x-2i$, and $x+2i$, so we have

$f(x) = a_n(x+3)(x-2i)(x+2i).$

The number a_n can be any nonzero number. The simplest polynomial will be obtained if we let it be 1. Multiplying the factors, we obtain

$f(x) = (x+3)(x-2i)(x+2i)$
$\quad = (x+3)(x^2+4)$
$\quad = x^3 + 3x^2 + 4x + 12$

13. Find a polynomial function of degree 3 with $\sqrt{2}$, $-\sqrt{2}$, and $\sqrt{3}$ as zeros.

Such a polynomial has factors $x-\sqrt{2}$, $x+\sqrt{2}$, and $x-\sqrt{3}$, so we have

$f(x) = a_n(x-\sqrt{2})(x+\sqrt{2})(x-\sqrt{3}).$

The number a_n can be any nonzero number. The simplest polynomial will be obtained if we let it be 1. Multiplying the factors, we obtain

$f(x) = (x-\sqrt{2})(x+\sqrt{2})(x-\sqrt{3})$
$\quad = (x^2-2)(x-\sqrt{3})$
$\quad = x^3 - \sqrt{3}x^2 - 2x + 2\sqrt{3}$

15. A polynomial function of degree 5 has at most 5 zeros. Since 5 zeros are given, these are all of the zeros of the desired function.
$f(x) = (x+1)^3(x-0)(x-1)$
$\quad = (x^3+3x^2+3x+1)(x^2-x)$
$\quad = x^5 + 2x^4 - 2x^2 - x$

17. A polynomial function $f(x)$ of degree 5 has at most 5 zeros. Three of the zeros are 6, $-3+4i$, and $4-\sqrt{5}$. Since $f(x)$ has rational coefficients we know that the conjugates of $-3+4i$ and $4-\sqrt{5}$, or $-3-4i$ and $4+\sqrt{5}$, are also zeros.

19. Find a polynomial function of lowest degree with rational coefficients that has $1+i$ and 2 as some of its zeros.

$1-i$ is also a zero.

Thus the polynomial function is

$f(x) = a_n(x-2)[x-(1+i)][x-(1-i)].$

If we let $a_n = 1$, we obtain

$f(x) = (x-2)[(x-1)-i][(x-1)+i]$
$\quad = (x-2)[(x-1)^2 - i^2]$
$\quad = (x-2)(x^2-2x+1+1)$
$\quad = (x-2)(x^2-2x+2)$
$\quad = x^3 - 4x^2 + 6x - 4$

21. Find a polynomial function of lowest degree with rational coefficients that has $-4i$ and 5 as some of its zeros.

$4i$ is also a zero.

Thus the polynomial function is

$f(x) = a_n(x-5)(x+4i)(x-4i).$

If we let $a_n = 1$, we obtain

$f(x) = (x-5)[x^2 - (4i)^2]$
$\quad = (x-5)(x^2+16)$
$\quad = x^3 - 5x^2 + 16x - 80$

23. Find a polynomial function of lowest degree with rational coefficients that has $\sqrt{5}$ and $-3i$ as some of its zeros.

$-\sqrt{5}$ and $3i$ are also zeros.

Thus the polynomial function is

$f(x) = a_n(x-\sqrt{5})(x+\sqrt{5})(x+3i)(x-3i).$

If we let $a_n = 1$, we obtain

$f(x) = (x^2-5)(x^2+9)$
$\quad = x^4 + 4x^2 - 45$

25. If $-i$ is a zero of $f(x) = x^4 - 5x^3 + 7x^2 - 5x + 6$, i is also a zero. Thus $x+i$ and $x-i$ are factors of the polynomial. Since $(x+i)(x-i) = x^2+1$, we know that $f(x) = (x^2+1) \cdot Q(x)$. Divide $x^4 - 5x^3 + 7x^2 - 5x + 6$ by x^2+1.

$$
\begin{array}{r}
x^2 - 5x + 6 \\
x^2+1\overline{\smash{\big)}\ x^4 - 5x^3 + 7x^2 - 5x + 6} \\
\underline{x^4 \qquad + x^2} \\
-5x^3 + 6x^2 - 5x \\
\underline{-5x^3 \qquad - 5x} \\
6x^2 + 6 \\
\underline{6x^2 + 6} \\
0
\end{array}
$$

Thus
$x^4-5x^3+7x^2-5x+6 = (x+i)(x-i)(x^2-5x+6)$
$\qquad = (x+i)(x-i)(x-2)(x-3)$

Using the principle of zero products we find the other zeros to be i, 2, and 3.

27. $x^3 - 6x^2 + 13x - 20 = 0$

If 4 is a zero, then $x-4$ is a factor. Use synthetic division to find another factor.

$$
\begin{array}{r|rrrr}
4 & 1 & -6 & 13 & -20 \\
  &   & 4 & -8 & 20 \\
\hline
  & 1 & -2 & 5 & 0
\end{array}
$$

$(x-4)(x^2-2x+5) = 0$

$x-4=0$ or $x^2-2x+5=0$ Principle of
                                                          zero products

$x = 4$ or $x = \dfrac{2\pm\sqrt{4-20}}{2}$

                                          Quadratic formula

$x = 4$ or $x = \dfrac{2\pm4i}{2} = 1 \pm 2i$

The other zeros are $1 + 2i$ and $1 - 2i$.

29. $f(x) = x^5 - 3x^2 + 1$

According to the rational zeros theorem, any rational zero of f must be of the form p/q, where p is a factor of the constant term, 1, and q is a factor of the coefficient of x^5, 1.

$\dfrac{\text{Possibilities for } p}{\text{Possibilities for } q} : \dfrac{\pm 1}{\pm 1}$

Possibilities for p/q: $1, -1$

31. $f(x) = 15x^6 + 47x^2 + 2$

According to the rational zeros theorem, any rational zero of f must be of the form p/q, where p is a factor of 2 and q is a factor of 15.

$\dfrac{\text{Possibilities for } p}{\text{Possibilities for } q} : \dfrac{\pm 1, \pm 2}{\pm 1, \pm 3, \pm 5, \pm 15}$

Possibilities for p/q: $1, -1, 2, -2, \dfrac{1}{3}, -\dfrac{1}{3}, \dfrac{2}{3}, -\dfrac{2}{3}, \dfrac{1}{5},$
$\dfrac{1}{5}, \dfrac{2}{5}, -\dfrac{2}{5}, \dfrac{1}{15}, -\dfrac{1}{15}, \dfrac{2}{15}, -\dfrac{2}{15}$

33. $f(x) = x^3 + 3x^2 - 2x - 6$

a) $\dfrac{\text{Possibilities for } p}{\text{Possibilities for } q} : \dfrac{\pm 1, \pm 2, \pm 3, \pm 6}{\pm 1}$

Possibilities for p/q: $1, -1, 2, -2, 3, -3, 6, -6$

From the graph of $y = x^3 + 3x^2 - 2x - 6$, we see that, of the possibilities above, only -3 might be a zero. We use synthetic division to determine whether -3 is indeed a zero.

$$\begin{array}{r|rrrr} -3 & 1 & 3 & -2 & -6 \\ & & -3 & 0 & 6 \\ \hline & 1 & 0 & -2 & 0 \end{array}$$

Then we have $f(x) = (x + 3)(x^2 - 2)$.

We find the other zeros:
$$x^2 - 2 = 0$$
$$x^2 = 2$$
$$x = \pm\sqrt{2}.$$

There is only one rational zero, -3. The other zeros are $\pm\sqrt{2}$. (Note that we could have used factoring by grouping to find this result.)

b) $f(x) = (x + 3)(x - \sqrt{2})(x + \sqrt{2})$

35. $f(x) = x^3 - 3x + 2$

a) $\dfrac{\text{Possibilities for } p}{\text{Possibilities for } q} : \dfrac{\pm 1, \pm 2}{\pm 1}$

Possibilities for p/q: $1, -1, 2, -2$

From the graph of $y = x^3 - 3x + 2$, we see that, of the possibilities above, -2 and 1 might be a zeros. We use synthetic division to determine whether -2 is a zero.

$$\begin{array}{r|rrrr} -2 & 1 & 0 & -3 & 2 \\ & & -2 & 4 & -2 \\ \hline & 1 & -2 & 1 & 0 \end{array}$$

Then we have $f(x) = (x + 2)(x^2 - 2x + 1) = (x + 2)(x - 1)^2$.

Now $(x - 1)^2 = 0$ for $x = 1$. Thus, the rational zeros are -2 and 1. (The zero 1 has a multiplicity of 2.) These are the only zeros.

b) $f(x) = (x + 2)(x - 1)^2$

37. $f(x) = x^3 - 5x^2 + 11x + 17$

a) $\dfrac{\text{Possibilities for } p}{\text{Possibilities for } q} : \dfrac{\pm 1, \pm 17}{\pm 1}$

Possibilities for p/q: $1, -1, 17, -17$

From the graph of $y = x^3 - 5x^2 + 11x + 17$, we see that, of the possibilities above, we see that only -1 might be a zero. We use synthetic division to determine whether -1 is indeed a zero.

$$\begin{array}{r|rrrr} -1 & 1 & -5 & 11 & 17 \\ & & -1 & 6 & -17 \\ \hline & 1 & -6 & 17 & 0 \end{array}$$

Then we have $f(x) = (x + 1)(x^2 - 6x + 17)$. We use the quadratic formula to find the other zeros.

$$x^2 - 6x + 17 = 0$$

$$x = \frac{-(-6) \pm \sqrt{(-6)^2 - 4 \cdot 1 \cdot 17}}{2 \cdot 1}$$

$$= \frac{6 \pm \sqrt{-32}}{2} = \frac{6 \pm 4\sqrt{2}i}{2}$$

$$= 3 \pm 2\sqrt{2}i$$

The only rational zero is -1. The other zeros are $3 \pm 2\sqrt{2}i$.

b) $f(x) = (x + 1)[x - (3 + 2\sqrt{2}i)][x - (3 - 2\sqrt{2}i)]$
$= (x + 1)(x - 3 - 2\sqrt{2}i)(x - 3 + 2\sqrt{2}i)$

39. $f(x) = 5x^4 - 4x^3 + 19x^2 - 16x - 4$

a) $\dfrac{\text{Possibilities for } p}{\text{Possibilities for } q} : \dfrac{\pm 1, \pm 2, \pm 4}{\pm 1, \pm 5}$

Possibilities for p/q: $1, -1, 2, -2, 4, -4, \dfrac{1}{5}, -\dfrac{1}{5},$
$\dfrac{2}{5}, -\dfrac{2}{5}, \dfrac{4}{5}, -\dfrac{4}{5}$

From the graph of $y = 5x^4 - 4x^3 + 19x^2 - 16x - 4$, we see that, of the possibilities above, only $-\frac{2}{5}$, $-\frac{1}{5}$ and 1 might be zeros. We use synthetic division to determine whether 1 is a zero.

$$\begin{array}{r|rrrrr} 1 & 5 & -4 & 19 & -16 & -4 \\ & & 5 & 1 & 20 & 4 \\ \hline & 5 & 1 & 20 & 4 & 0 \end{array}$$

Then we have
$$\begin{aligned} f(x) &= (x-1)(5x^3 + x^2 + 20x + 4) \\ &= (x-1)[x^2(5x+1) + 4(5x+1)] \\ &= (x-1)(5x+1)(x^2+4). \end{aligned}$$

We find the other zeros:
$$5x + 1 = 0 \quad or \quad x^2 + 4 = 0$$
$$5x = -1 \quad or \quad x^2 = -4$$
$$x = -\frac{1}{5} \quad or \quad x = \pm 2i$$

The rational zeros are $-\frac{1}{5}$ and 1. The other zeros are $\pm 2i$.

b) From part (a) we see that
$$f(x) = (5x+1)(x-1)(x+2i)(x-2i).$$

41. $f(x) = x^4 - 3x^3 - 20x^2 - 24x - 8$

a) $\dfrac{\text{Possibilities for } p}{\text{Possibilities for } q} : \dfrac{\pm 1, \pm 2, \pm 4, \pm 8}{\pm 1}$

Possibilities for p/q: $1, -1, 2, -2, 4, -4, 8, -8$

From the graph of $y = x^4 - 3x^3 - 20x^2 - 24x - 8$, we see that, of the possibilities above, only -2 and -1 might be zeros. We use synthetic division to determine if -2 is a zero.

$$\begin{array}{r|rrrrr} -2 & 1 & -3 & -20 & -24 & -8 \\ & & -2 & 10 & 20 & 8 \\ \hline & 1 & -5 & -10 & -4 & 0 \end{array}$$

We see that -2 is a zero. Now we determine whether -1 is a zero.

$$\begin{array}{r|rrrr} -1 & 1 & -5 & -10 & -4 \\ & & -1 & 6 & 4 \\ \hline & 1 & -6 & -4 & 0 \end{array}$$

Then we have $f(x) = (x+2)(x+1)(x^2 - 6x - 4)$. Use the quadratic formula to find the other zeros.
$$x^2 - 6x - 4 = 0$$
$$x = \frac{-(-6) \pm \sqrt{(-6)^2 - 4\cdot 1\cdot(-4)}}{2\cdot 1}$$
$$= \frac{6 \pm \sqrt{52}}{2} = \frac{6 \pm 2\sqrt{13}}{2}$$
$$= 3 \pm \sqrt{13}$$

The rational zeros are -2 and -1. The other zeros are $3 \pm \sqrt{13}$.

b) $f(x) = (x+2)(x+1)[x-(3+\sqrt{13})][x-(3-\sqrt{13})]$
$= (x+2)(x+1)(x-3-\sqrt{13})(x-3+\sqrt{13})$

43. $f(x) = x^3 - 4x^2 + 2x + 4$

a) $\dfrac{\text{Possibilities for } p}{\text{Possibilities for } q} : \dfrac{\pm 1, \pm 2, \pm 4}{\pm 1}$

Possibilities for p/q: $1, -1, 2, -2, 4, -4$

From the graph of $y = x^3 - 4x^2 + 2x + 4$, we see that, of the possibilities above, only -1, 1, and 2 might be zeros. Synthetic division shows that neither -1 nor 1 is a zero. Try 2.

$$\begin{array}{r|rrrr} 2 & 1 & -4 & 2 & 4 \\ & & 2 & -4 & -4 \\ \hline & 1 & -2 & -2 & 0 \end{array}$$

Then we have $f(x) = (x-2)(x^2 - 2x - 2)$. Use the quadratic formula to find the other zeros.
$$x^2 - 2x - 2 = 0$$
$$x = \frac{-(-2) \pm \sqrt{(-2)^2 - 4\cdot 1\cdot(-2)}}{2\cdot 1}$$
$$= \frac{2 \pm \sqrt{12}}{2} = \frac{2 \pm 2\sqrt{3}}{2}$$
$$= 1 \pm \sqrt{3}$$

The only rational zero is 2. The other zeros are $1 \pm \sqrt{3}$.

b) $f(x) = (x-2)[x-(1+\sqrt{3})][x-(1-\sqrt{3})]$
$= (x-2)(x-1-\sqrt{3})(x-1+\sqrt{3})$

45. $f(x) = x^3 + 8$

a) $\dfrac{\text{Possibilities for } p}{\text{Possibilities for } q} : \dfrac{\pm 1, \pm 2, \pm 4, \pm 8}{\pm 1}$

Possibilities for p/q: $1, -1, 2, -2, 4, -4, 8, -8$

From the graph of $y = x^3 + 8$, we see that, of the possibilities above, only -2 might be a zero. We use synthetic division to see if it is.

$$\begin{array}{r|rrrr} -2 & 1 & 0 & 0 & 8 \\ & & -2 & 4 & -8 \\ \hline & 1 & -2 & 4 & 0 \end{array}$$

We have $f(x) = (x+2)(x^2 - 2x + 4)$. Use the quadratic formula to find the other zeros.
$$x^2 - 2x + 4 = 0$$
$$x = \frac{-(-2) \pm \sqrt{(-2)^2 - 4\cdot 1\cdot 4}}{2\cdot 1}$$
$$= \frac{2 \pm \sqrt{-12}}{2} = \frac{2 \pm 2\sqrt{3}i}{2}$$
$$= 1 \pm \sqrt{3}i$$

The only rational zero is -2. The other zeros are $1 \pm \sqrt{3}i$.

b) $f(x) = (x+2)[x - (1 + \sqrt{3}i)][x - (1 - \sqrt{3}i)]$
$= (x+2)(x - 1 - \sqrt{3}i)(x - 1 + \sqrt{3}i)$

47. $f(x) = \frac{1}{3}x^3 - \frac{1}{2}x^2 - \frac{1}{6}x + \frac{1}{6}$

$= \frac{1}{6}(2x^3 - 3x^2 - x + 1)$

a) The second form of the equation is equivalent to the first and has the advantage of having integer coefficients. Thus, we can use the rational zeros theorem for $g(x) = 2x^3 - 3x^2 - x + 1$. The zeros of $g(x)$ are the same as the zeros of $f(x)$. We find the zeros of $g(x)$.

$\dfrac{\text{Possibilities for } p}{\text{Possibilities for } q} : \dfrac{\pm 1}{\pm 1, \pm 2}$

Possibilities for p/q: $1, -1, \dfrac{1}{2}, -\dfrac{1}{2}$

From the graph of $y = 2x^3 - 3x^2 - x + 1$, we see that, of the possibilities above, only $-\dfrac{1}{2}$ and $\dfrac{1}{2}$ might be zeros. Synthetic division shows that $-\dfrac{1}{2}$ is not a zero. Try $\dfrac{1}{2}$.

$$\begin{array}{r|rrrr} \frac{1}{2} & 2 & -3 & -1 & 1 \\ & & 1 & -1 & -1 \\ \hline & 2 & -2 & -2 & 0 \end{array}$$

We have $g(x) = \left(x - \dfrac{1}{2}\right)(2x^2 - 2x - 2) =$ $\left(x - \dfrac{1}{2}\right)(2)(x^2 - x - 1)$. Use the quadratic formula to find the other zeros.
$x^2 - x - 1 = 0$

$$x = \frac{-(-1) \pm \sqrt{(-1)^2 - 4 \cdot 1 \cdot (-1)}}{2 \cdot 1}$$

$$= \frac{1 \pm \sqrt{5}}{2}$$

The only rational zero is $\dfrac{1}{2}$. The other zeros are $\dfrac{1 \pm \sqrt{5}}{2}$.

b) $f(x) = \dfrac{1}{6}g(x)$

$= \dfrac{1}{6}\left(x - \dfrac{1}{2}\right)(2)\left[x - \dfrac{1 + \sqrt{5}}{2}\right]\left[x - \dfrac{1 - \sqrt{5}}{2}\right]$

$= \dfrac{1}{3}\left(x - \dfrac{1}{2}\right)\left(x - \dfrac{1 + \sqrt{5}}{2}\right)\left(x - \dfrac{1 - \sqrt{5}}{2}\right)$

49. $f(x) = x^4 + 32$

According to the rational zeros theorem, the possible rational zeros are $\pm 1, \pm 2, \pm 4, \pm 8, \pm 16$, and ± 32.

The graph of $y = x^4 + 32$ has no x-intercepts, so $f(x)$ has no real-number zeros and hence no rational zeros.

51. $f(x) = x^3 - x^2 - 4x + 3$

According to the rational zeros theorem, the possible rational zeros are ± 1 and ± 3. The graph of $y = x^3 - x^2 - 4x + 3$ shows that none of these is a zero. Thus, there are no rational zeros.

53. $f(x) = x^4 + 2x^3 + 2x^2 - 4x - 8$

According to the rational zeros theorem, the possible rational zeros are $\pm 1, \pm 2, \pm 4$, and ± 8. The graph of $y = x^4 + 2x^3 + 2x^2 - 4x - 8$ shows that none of the possibilities is a zero. Thus, there are no rational zeros.

55. $f(x) = x^5 - 5x^4 + 5x^3 + 15x^2 - 36x + 20$

According to the rational zeros theorem, the possible rational zeros are $\pm 1, \pm 2, \pm 4, \pm 5, \pm 10$, and ± 20. The graph of $y = x^5 - 5x^4 + 5x^3 + 15x^2 - 36x + 20$ shows that, of these possibilities, only -2, 1 and 2 might be zeros. We try -2.

$$\begin{array}{r|rrrrrr} -2 & 1 & -5 & 5 & 15 & -36 & 20 \\ & & -2 & 14 & -38 & 46 & -20 \\ \hline & 1 & -7 & 19 & -23 & 10 & 0 \end{array}$$

Thus, -2 is a zero. Now try 1.

$$\begin{array}{r|rrrrr} 1 & 1 & -7 & 19 & -23 & 10 \\ & & 1 & -6 & 13 & 10 \\ \hline & 1 & -6 & 13 & -10 & 0 \end{array}$$

1 is also a zero. Try 2.

$$\begin{array}{r|rrrr} 2 & 1 & -6 & 13 & -10 \\ & & 2 & -8 & 10 \\ \hline & 1 & -4 & 5 & 0 \end{array}$$

2 is also a zero.

We have $f(x) = (x+2)(x-1)(x-2)(x^2 - 4x + 5)$. The discriminant of $x^2 - 4x + 5$ is $(-4)^2 - 4 \cdot 1 \cdot 5$, or $4 < 0$, so $x^2 - 4x + 5$ has two nonreal zeros. Thus, the rational zeros are -2, 1, and 2.

57. Discussion and Writing

59. $f(x) = x^2 - 8x + 10$

a) $-\dfrac{b}{2a} = -\dfrac{-8}{2 \cdot 1} = -(-4) = 4$
$f(4) = 4^2 - 8 \cdot 4 + 10 = -6$
The vertex is $(4, -6)$.

b) The line of symmetry is $x = 4$.

c) Since the coefficient of x^2 is positive, there is a minimum function value. It is the second coordinate of the vertex, -6. It occurs when $x = 4$.

61.　　$-\dfrac{4}{5}x + 8 = 0$

$$-\dfrac{4}{5}x = -8 \qquad \text{Subtracting 8}$$

$$-\dfrac{5}{4}\left(-\dfrac{4}{5}x\right) = -\dfrac{5}{4}(-8) \quad \text{Multiplying by } -\dfrac{5}{4}$$

$$x = 10$$

We can also graph $y = -\dfrac{4}{5}x + 8$ and use the Zero feature.

The solution is 10.

63.　$f(x) = 2x^3 - 5x^2 - 4x + 3$

a) $2x^3 - 5x^2 - 4x + 3 = 0$

$\dfrac{\text{Possibilities for } p}{\text{Possibilities for } q} : \dfrac{\pm 1, \pm 3}{\pm 1, \pm 2}$

Possibilities for p/q: $1, -1, 3, -3, \dfrac{1}{2}, -\dfrac{1}{2}, \dfrac{3}{2}, -\dfrac{3}{2}$

The first possibility that is a solution of $f(x) = 0$ is -1:

$$
\begin{array}{r|rrrr}
-1 & 2 & -5 & -4 & 3 \\
   &   & -2 & 7 & -3 \\
\hline
   & 2 & -7 & 3 & 0 \\
\end{array}
$$

Thus, -1 is a solution.

Then we have:

$(x + 1)(2x^2 - 7x + 3) = 0$

$(x + 1)(2x - 1)(x - 3) = 0$

The other solutions are $\dfrac{1}{2}$ and 3.

b) The graph of $y = f(x - 1)$ is the graph of $y = f(x)$ shifted 1 unit right. Thus, we add 1 to each solution of $f(x) = 0$ to find the solutions of $f(x - 1) = 0$. The solutions are $-1 + 1$, or 0; $\dfrac{1}{2} + 1$, or $\dfrac{3}{2}$; and $3 + 1$, or 4.

c) The graph of $y = f(x + 2)$ is the graph of $y = f(x)$ shifted 2 units left. Thus, we subtract 2 from each solution of $f(x) = 0$ to find the solutions of $f(x + 2) = 0$. The solutions are $-1 - 2$, or -3; $\dfrac{1}{2} - 2$, or $-\dfrac{3}{2}$; and $3 - 2$, or 1.

d) The graph of $y = f(2x)$ is a horizontal shrinking of the graph of $y = f(x)$ by a factor of 2. We divide each solution of $f(x) = 0$ by 2 to find the solutions of $f(2x) = 0$. The solutions are $\dfrac{-1}{2}$ or $-\dfrac{1}{2}$; $\dfrac{1/2}{2}$, or $\dfrac{1}{4}$; and $\dfrac{3}{2}$.

65. See the answer section in the text.

67. $P(x) = 2x^5 - 33x^4 - 84x^3 + 2203x^2 - 3348x - 10{,}080$

a) $2x^5 - 33x^4 - 84x^3 + 2203x^2 - 3348x - 10{,}080 = 0$

Trying some of the many possibilities for p/q, we find that 4 is a zero.

$$
\begin{array}{r|rrrrrr}
4 & 2 & -33 & -84 & 2203 & -3348 & -10{,}080 \\
  &   & 8 & -100 & -736 & 5868 & 10{,}080 \\
\hline
  & 2 & -25 & -184 & 1467 & 2520 & 0 \\
\end{array}
$$

Then we have:

$(x - 4)(2x^4 - 25x^3 - 184x^2 + 1467x + 2520) = 0$

We now use the fourth degree polynomial above to find another zero. Synthetic division shows that 4 is not a double zero, but 7 is a zero.

$$
\begin{array}{r|rrrrr}
7 & 2 & -25 & -184 & 1467 & 2520 \\
  &   & 14 & -77 & -1827 & -2520 \\
\hline
  & 2 & -11 & -261 & -360 & 0 \\
\end{array}
$$

Now we have:

$(x - 4)(x - 7)(2x^3 - 11x^2 - 261x - 360) = 0$

Use the third degree polynomial above to find a third zero. Synthetic division shows that 7 is not a double zero, but 15 is a zero.

$$
\begin{array}{r|rrrr}
15 & 2 & -11 & -261 & -360 \\
   &   & 30 & 285 & 360 \\
\hline
   & 2 & 19 & 24 & 0 \\
\end{array}
$$

We have:

$P(x) = (x - 4)(x - 7)(x - 15)(2x^2 + 19x + 24)$

$ = (x - 4)(x - 7)(x - 15)(2x + 3)(x + 8)$

The rational zeros are 4, 7, 15, $-\dfrac{3}{2}$, and -8.

Exercise Set 3.4

1. Graph (d) is the graph of $f(x) = \dfrac{8}{x^2 - 4}$.

$x^2 - 4 = 0$ when $x = \pm 2$, so $x = -2$ and $x = 2$ are vertical asymptotes.

The x-axis, $y = 0$, is the horizontal asymptote because the degree of the numerator is less than the degree of the denominator.

There is no oblique asymptote.

3. Graph (e) is the graph of $f(x) = \dfrac{8x}{x^2 - 4}$.

As in Exercise 1, $x = -2$ and $x = 2$ are vertical asymptotes.

The x-axis, $y = 0$, is the horizontal asymptote because the degree of the numerator is less than the degree of the denominator.

There is no oblique asymptote.

5. Graph (c) is the graph of $f(x) = \dfrac{8x^3}{x^2 - 4}$.

As in Exercise 1, $x = -2$ and $x = 2$ are vertical asymptotes.

The degree of the numerator is greater than the degree of the denominator, so there is no horizontal asymptote but there is a vertical asymptote. To find it we first divide to find an equivalent expression.

$$
\begin{array}{r}
8x \\
x^2 - 4 \overline{\smash{\big)}\, 8x^3} \\
\underline{8x^3 - 32x} \\
32x
\end{array}
$$

$$\frac{8x^3}{x^2 - 4} = 8x + \frac{32x}{x^2 - 4}$$

Now we multiply by 1, using $(1/x^2)/(1/x^2)$.

$$\frac{32x}{x^2 - 4} \cdot \frac{\dfrac{1}{x^2}}{\dfrac{1}{x^2}} = \frac{\dfrac{32}{x}}{1 - \dfrac{4}{x^2}}$$

As $|x|$ becomes very large, each expression with x in the denominator tends toward zero.

Then, as $|x| \to \infty$, we have

$$\frac{\dfrac{32}{x}}{1 - \dfrac{4}{x^2}} \to \frac{0}{1 - 0}, \text{ or } 0.$$

Thus, as $|x|$ becomes very large, the graph of $f(x)$ gets very close to the graph of $y = 8x$, so $y = 8x$ is the oblique asymptote.

7. $f(x) = \dfrac{1}{x + 3}$

1. -3 is the zero of the denominator, so the domain excludes -3. It is $(-\infty, -3) \cup (-3, \infty)$. The line $x = -3$ is the vertical asymptote.

2. Because the degree of the numerator is less than the degree of the denominator, the x-axis is the horizontal asymptote. There is no oblique asymptote.

3. The numerator has no zeros, so there is no x-intercept.

4. $f(0) = \dfrac{1}{0 + 3} = \dfrac{1}{3}$, so $\left(0, \dfrac{1}{3}\right)$ is the y-intercept.

5. Find other function values to determine the shape of the graph and then draw it.

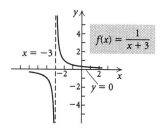

9. $f(x) = \dfrac{-2}{x - 5}$

1. 5 is the zero of the denominator, so the domain excludes 5. It is $(-\infty, 5) \cup (5, \infty)$. The line $x = 5$ is the vertical asymptote.

2. Because the degree of the numerator is less than the degree of the denominator, the x-axis is the horizontal asymptote. There is no oblique asymptote.

3. The numerator has no zeros, so there is no x-intercept.

4. $f(0) = \dfrac{-2}{0 - 5} = \dfrac{2}{5}$, so $\left(0, \dfrac{2}{5}\right)$ is the y-intercept.

5. Find other function values to determine the shape of the graph and then draw it.

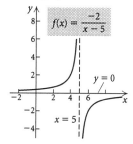

11. $f(x) = \dfrac{2x + 1}{x}$

1. 0 is the zero of the denominator, so the domain excludes 0. It is $(-\infty, 0) \cup (0, \infty)$. The line $x = 0$, or the y-axis, is the vertical asymptote.

2. The numerator and denominator have the same degree, so the horizontal asymptote is determined by the ratio of the leading coefficients, $2/1$, or 2. Thus, $y = 2$ is the horizontal asymptote. There is no oblique asymptote.

3. The zero of the numerator is the solution of $2x + 1 = 0$, or $-\dfrac{1}{2}$. The x-intercept is $\left(-\dfrac{1}{2}, 0\right)$.

4. Since 0 is not in the domain of the function, there is no y-intercept.

5. Find other function values to determine the shape of the graph and then draw it.

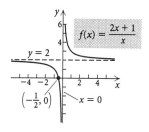

13. $f(x) = \dfrac{1}{(x-2)^2}$

1. 2 is the zero of the denominator, so the domain excludes 2. It is $(-\infty, 2) \cup (2, \infty)$. The line $x = 2$ is the vertical asymptote.

2. Because the degree of the numerator is less than the degree of the denominator, the x-axis is the horizontal asymptote. There is no oblique asymptote.

3. The numerator has no zeros, so there is no x-intercept.

4. $f(0) = \dfrac{1}{(0-2)^2} = \dfrac{1}{4}$, so $\left(0, \dfrac{1}{4}\right)$ is the y-intercept.

5. Find other function values to determine the shape of the graph and then draw it.

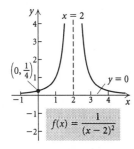

15. $f(x) = \dfrac{1}{x^2}$

1. 0 is the zero of the denominator, so the domain excludes 0. It is $(-\infty, 0) \cup (0, \infty)$. The line $x = 0$, or the y-axis, is the vertical asymptote.

2. Because the degree of the numerator is less than the degree of the denominator, the x-axis is the horizontal asymptote. There is no oblique asymptote.

3. The numerator has no zeros, so there is no x-intercept.

4. Since 0 is not in the domain of the function, there is no y-intercept.

5. Find other function values to determine the shape of the graph and then draw it.

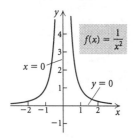

17. $f(x) = \dfrac{1}{x^2 + 3}$

1. The denominator has no real-number zeros, so the domain is the set of all real numbers and there is no vertical asymptote.

2. Because the degree of the numerator is less than the degree of the denominator, the x-axis is the horizontal asymptote. There is no oblique asymptote.

3. The numerator has no zeros, so there is no x-intercept.

4. $f(0) = \dfrac{1}{0^2 + 3} = \dfrac{1}{3}$, so $\left(0, \dfrac{1}{3}\right)$ is the y-intercept.

5. Find other function values to determine the shape of the graph and then draw it.

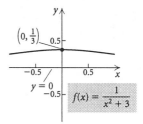

19. $f(x) = \dfrac{x^2 - 4}{x - 2} = \dfrac{(x+2)(x-2)}{x-2} = x + 2, \ x \neq 2$

The graph is the same as the graph of $f(x) = x + 2$ except at $x = 2$, where there is a hole. The zero of $f(x) = x + 2$ is -2, so the x-intercept is $(-2, 0)$; $f(0) = 2$, so the y-intercept is $(0, 2)$.

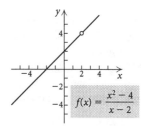

21. $f(x) = \dfrac{x-1}{x+2}$

1. -2 is the zero of the denominator, so the domain excludes -2. It is $(-\infty, -2) \cup (-2, \infty)$. The line $x = -2$ is the vertical asymptote.

2. The numerator and denominator have the same degree, so the horizontal asymptote is determined by the ratio of the leading coefficients, $1/1$, or 1. Thus, $y = 1$ is the horizontal asymptote. There is no oblique asymptote.

3. The zero of the numerator is 1, so the x-intercept is $(1, 0)$.

4. $f(0) = \dfrac{0-1}{0+2} = -\dfrac{1}{2}$, so $\left(0, -\dfrac{1}{2}\right)$ is the y-intercept.

5. Find other function values to determine the shape of the graph and then draw it.

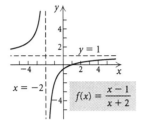

23. $f(x) = \dfrac{x+3}{2x^2 - 5x - 3}$

1. The zeros of the denominator are the solutions of $2x^2 - 5x - 3 = 0$. Since $2x^2 - 5x - 3 = (2x+1)(x-3)$, the zeros are $-\dfrac{1}{2}$ and 3. Thus, the domain is $\left(-\infty, -\dfrac{1}{2}\right) \cup \left(-\dfrac{1}{2}, 3\right) \cup (3, \infty)$ and the lines $x = -\dfrac{1}{2}$ and $x = 3$ are vertical asymptotes.

2. Because the degree of the numerator is less than the degree of the denominator, the x-axis is the horizontal asymptote. There is no oblique asymptote.

3. -3 is the zero of the numerator, so $(-3, 0)$ is the x-intercept.

4. $f(0) = \dfrac{0+3}{2 \cdot 0^2 - 5 \cdot 0 - 3} = -1$, so $(0, -1)$ is the y-intercept.

5. Find other function values to determine the shape of the graph and then draw it.

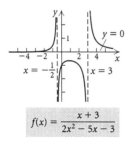

25. $f(x) = \dfrac{x^2 - 9}{x+1}$

1. -1 is the zero of the denominator, so the domain excludes -1. It is $(-\infty, -1) \cup (-1, \infty)$. The line $x = -1$ is the vertical asymptote.

2. Because the degree of the numerator is one greater than the degree of the denominator, there is an oblique asymptote. Using division, we find that $\dfrac{x^2 - 9}{x+1} = x - 1 + \dfrac{-8}{x+1}$. As $|x|$ becomes very large, the graph of $f(x)$ gets close to the graph of $y = x - 1$. Thus, the line $y = x - 1$ is the oblique asymptote.

3. Since $x^2 - 9 = (x+3)(x-3)$, the zeros of the numerator are -3 and 3. Thus, the x-intercepts are $(-3, 0)$ and $(3, 0)$.

4. $f(0) = \dfrac{0^2 - 9}{0+1} = -9$, so $(0, -9)$ is the y-intercept.

5. Find other function values to determine the shape of the graph and then draw it.

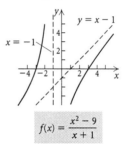

27. $f(x) = \dfrac{x^2 + x - 2}{2x^2 + 1}$

1. The denominator has no real-number zeros, so the domain is the set of all real numbers and there is no vertical asymptote.

2. The numerator and denominator have the same degree, so the horizontal asymptote is determined by the ratio of the leading coefficients, $1/2$. Thus, $y = 1/2$ is the horizontal asymptote. There is no oblique asymptote.

3. Since $x^2 + x - 2 = (x+2)(x-1)$, the zeros of the numerator are -2 and 1. Thus, the x-intercepts are $(-2, 0)$ and $(1, 0)$.

4. $f(0) = \dfrac{0^2 + 0 - 2}{2 \cdot 0^2 + 1} = -2$, so $(0, -2)$ is the y-intercept.

5. Find other function values to determine the shape of the graph and then draw it.

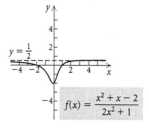

29. $g(x) = \dfrac{3x^2 - x - 2}{x - 1} = \dfrac{(3x+2)(x-1)}{x-1} = 3x + 2$, $x \neq 1$

The graph is the same as the graph of $g(x) = 3x + 2$ except at $x = 1$, where there is a hole.

The zero of $g(x) = 3x + 2$ is $-\dfrac{2}{3}$, so the x-intercept is $\left(-\dfrac{2}{3}, 0\right)$; $g(0) = 2$, so the y-intercept is $(0, 2)$.

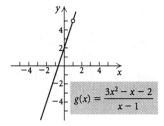

31. $f(x) = \dfrac{x - 1}{x^2 - 2x - 3}$

1. The zeros of the denominator are the solutions of
   $x^2 - 2x - 3 = 0$. Since $x^2 - 2x - 3 = (x+1)(x-3)$, the zeros are -1 and 3. Thus, the domain is $(-\infty, -1) \cup (-1, 3) \cup (3, \infty)$ and the lines $x = -1$ and $y = 3$ are the vertical asymptotes.

2. Because the degree of the numerator is less than the degree of the denominator, the x-axis is the horizontal asymptote. There is no oblique asymptote.

3. 1 is the zero of the numerator, so $(1, 0)$ is the x-intercept.

4. $f(0) = \dfrac{0 - 1}{0^2 - 2 \cdot 0 - 3} = \dfrac{1}{3}$, so $\left(0, \dfrac{1}{3}\right)$ is the y-intercept.

5. Find other function values to determine the shape of the graph and then draw it.

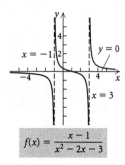

33. $f(x) = \dfrac{x - 3}{(x + 1)^3}$

1. -1 is the zero of the denominator, so the domain excludes -1. It is $(-\infty, -1) \cup (-1, \infty)$. The line $x = 1$ is the vertical asymptote.

2. Because the degree of the numerator is less than the degree of the denominator, the x-axis is the horizontal asymptote. There is no oblique asymptote.

3. 3 is the zero of the numerator, so $(3, 0)$ is the x-intercept.

4. $f(0) = \dfrac{0 - 3}{(0 + 1)^3} = -3$, so $(0, -3)$ is the y-intercept.

5. Find other function values to determine the shape of the graph and then draw it.

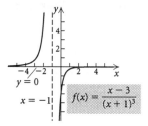

35. $f(x) = \dfrac{x^3 + 1}{x}$

1. 0 is the zero of the denominator, so the domain excludes 0. It is $(-\infty, 0) \cup (0, \infty)$. The line $x = 0$, or the y-axis, is the vertical asymptote.

2. Because the degree of the numerator is more than one greater than the degree of the denominator, there is no horizontal or oblique asymptote.

3. The real-number zero of the numerator is -1, so the x-intercept is $(-1, 0)$.

4. Since 0 is not in the domain of the function, there is no y-intercept.

5. Find other function values to determine the shape of the graph and then draw it.

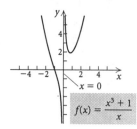

37. $f(x) = \dfrac{x^3 + 2x^2 - 15x}{x^2 - 5x - 14}$

1. The zeros of the denominator are the solutions of
   $x^2 - 5x - 14 = 0$. Since $x^2 - 5x - 14 = (x+2)(x-7)$, the zeros are -2 and 7. Thus, the domain is $(-\infty, -2) \cup (-2, 7) \cup (7, \infty)$ and the lines $x = -2$ and $x = 7$ are the vertical asymptotes.

2. Because the degree of the numerator is one greater than the degree of the denominator, there is an oblique asymptote. Using division, we find that $\dfrac{x^3 + 2x^2 - 15x}{x^2 - 5x - 14} = x + 7 + \dfrac{34x + 98}{x^2 - 5x - 14}$. As $|x|$ becomes very large, the graph of $f(x)$ gets close to the graph of $y = x + 7$. Thus, the line $y = x + 7$ is the oblique asymptote.

3. The zeros of the numerator are the solutions of $x^3 + 2x^2 - 15x = 0$. Since $x^3 + 2x^2 - 15x = x(x+5)(x-3)$, the zeros are 0, -5, and 3. Thus, the x-intercepts are $(-5, 0)$, $(0, 0)$, and $(3, 0)$.

4. From part (3) we see that $(0, 0)$ is the y-intercept.

5. Find other function values to determine the shape of the graph and then draw it.

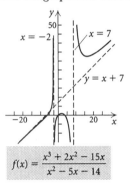

39. $f(x) = \dfrac{5x^4}{x^4 + 1}$

1. The denominator has no real-number zeros, so the domain is the set of all real numbers and there is no vertical asymptote.

2. The numerator and denominator have the same degree, so the horizontal asymptote is determined by the ratio of the leading coefficients, 5/1, or 5. Thus, $y = 5$ is the horizontal asymptote. There is no oblique asymptote.

3. The zero of the numerator is 0, so $(0, 0)$ is the x-intercept.

4. From part (3) we see that $(0, 0)$ is the y-intercept.

5. Find other function values to determine the shape of the graph and then draw it.

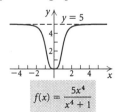

41. $f(x) = \dfrac{x^2}{x^2 - x - 2}$

1. The zeros of the denominator are the solutions of $x^2 - x - 2 = 0$. Since $x^2 - x - 2 = (x+1)(x-2)$, the zeros are -1 and 2. Thus, the domain is $(-\infty, -1) \cup (-1, 2) \cup (2, \infty)$ and the lines $x = -1$ and $x = 2$ are the vertical asymptotes.

2. The numerator and denominator have the same degree, so the horizontal asymptote is determined by the ratio of the leading coefficients, 1/1, or 1. Thus, $y = 1$ is the horizontal asymptote. There is no oblique asymptote.

3. The zero of the numerator is 0, so the x-intercept is $(0, 0)$.

4. From part (3) we see that $(0, 0)$ is the y-intercept.

5. Find other function values to determine the shape of the graph and then draw it.

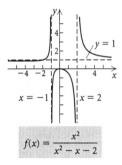

43. Answers may vary. The numbers -4 and 5 must be zeros of the denominator. A function that satisfies these conditions is
$$f(x) = \frac{1}{(x+4)(x-5)}, \text{ or } f(x) = \frac{1}{x^2 - x - 20}.$$

45. Answers may vary. The numbers -4 and 5 must be zeros of the denominator and -2 must be a zero of the numerator. In addition, the numerator and denominator must have the same degree and the ratio of their leading coefficients must be 3/2. A function that satisfies these conditions is
$$f(x) = \frac{3x(x+2)}{2(x+4)(x-5)}, \text{ or } f(x) = \frac{3x^2 + 6x}{2x^2 - 2x - 40}.$$
Another function that satisfies these conditions is
$$g(x) = \frac{3(x+2)^2}{2(x+4)(x-5)}, \text{ or } g(x) = \frac{3x^2 + 12x + 12}{2x^2 - 2x - 40}.$$

47. a)

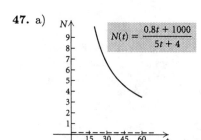

The horizontal asymptote of $N(t)$ is the ratio of the leading coefficients of the numerator and denominator, 0.8/5, or 0.16. Thus, $N(t) \to 0.16$ as $t \to \infty$.

b) The medication never completely disappears from the body; a trace amount remains.

49. a)

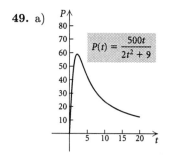

b) Use the table feature of a grapher (in ASK mode) or substitution.

$P(0) = 0$; $P(1) = 45.455$ thousand, or $45,455$;

$P(3) = 55.556$ thousand, or $55,556$;

$P(8) = 29.197$ thousand, or $29,197$

c) The degree of the numerator is less than the degree of the denominator, so the x-axis is the horizontal asymptote. Thus, $P(t) \to 0$ as $t \to \infty$.

d) Eventually, no one will live in Lordsburg.

e) Using the Maximum feature of a grapher, we find that the maximum value is $f(2.1213) = 58.926$ thousand, or $58,926$.

51. Discussion and Writing

53. Discussion and Writing

55. $y = \dfrac{k}{x}$

$\dfrac{2}{5} = \dfrac{k}{10}$ Substituting

$4 = k$ Variation constant

$y = \dfrac{4}{x}$ Equation of variation

57. $T = \dfrac{k}{P}$

$6 = \dfrac{k}{8}$ Substituting

$48 = k$ Variation constant

$T = \dfrac{48}{P}$ Equation of variation

$T = \dfrac{48}{12}$ Substituting

$T = 4$

It will take 4 hr for 12 employees to make the calls.

59. $f(x) = \dfrac{x^5 + 2x^3 + 4x^2}{x^2 + 2} = x^3 + 4 + \dfrac{-8}{x^2 + 2}$

As $|x| \to \infty$, $\dfrac{-8}{x^2 + 2} \to 0$ and the value of $f(x) \to x^3 + 4$. Thus, the nonlinear asymptote is $y = x^3 + 4$.

61.

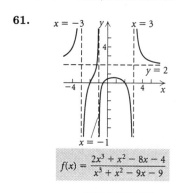

$$f(x) = \frac{2x^3 + x^2 - 8x - 4}{x^3 + x^2 - 9x - 9}$$

63. $f(x) = \sqrt{\dfrac{72}{x^2 - 4x - 21}}$

The radicand must be nonnegative and the denominator must be nonzero. Thus, the values of x for which $x^2 - 4x - 21 > 0$ comprise the domain. By inspecting the graph of $y = x^2 - 4x - 21$ we see that the domain is $\{x | x < -3 \text{ or } x > 7\}$, or $(-\infty, -3) \cup (7, \infty)$.

Exercise Set 3.5

1. First we find an equivalent inequality with 0 on one side.

$$x^3 + 6x^2 < x + 30$$

$$x^3 + 6x^2 - x - 30 < 30$$

From the graph we see that the x-intercepts of the related function occur at $x = -5$, $x = -3$, and $x = 2$. They divide the x-axis into the intervals $(-\infty, -5)$, $(-5, -3)$, $(-3, 2)$, and $(2, \infty)$. From the graph we see that the function has negative values only on $(-\infty, -5)$ and $(-3, 2)$. Thus, the solution set is $(-\infty, -5) \cup (-3, 2)$.

3. By observing the graph or the denomination of the function, we see that the function is not defined for $x = -2$ or $x = 2$. We also see that 0 is a zero of the function. These numbers divide the x-axis into the intervals $(-\infty, -2)$, $(-2, 0)$, $(0, 2)$, and $(2, \infty)$. From the graph we see that the function has positive values only on $(-2, 0)$ and $(2, \infty)$. Since the inequality symbol is $\geq$, 0 must be included in the solution set. It is $(-2, 0] \cup (2, \infty)$.

5. $(x - 1)(x + 4) < 0$

The related equation is $(x - 1)(x + 4) = 0$. Using the principle of zero products or by observing the graph of $y = (x - 1)(x + 4)$, we see that the solutions of the related equation are 1 and -4. These numbers divide the x-axis into the intervals $(-\infty, -4)$, $(-4, 1)$, and $(1, \infty)$. We let $f(x) = (x - 1)(x + 4)$ and test a value in each interval.

$(-\infty, -4)$: $f(-5) = 6 > 0$

$(-4, 1)$: $f(0) = -4 < 0$

$(1, \infty)$: $f(2) = 6 > 0$

Function values are negative only in the interval $(-4, 1)$. The graph of $y = (x - 1)(x + 4)$ can also be used to determine this. The solution set is $(-4, 1)$.

7. $(x - 4)(x + 2) \geq 0$

The related equation is $(x - 4)(x + 2) = 0$. Using the principle of zero products or by observing the graph of $y = (x - 4)(x + 2)$, we see that the solutions of the related equation are 4 and -2. These numbers divide the x-axis into the intervals $(-\infty, -2)$, $(-2, 4)$, and $(4, \infty)$. We let $f(x) = (x - 4)(x + 2)$ and test a value in each interval.

$(-\infty, -2)$: $f(-3) = 7 > 0$

$(-2, 4)$: $f(0) = -8 < 0$

$(4, \infty)$: $f(5) = 7 > 0$

Function values are positive on $(-\infty, -2)$ and $(4, \infty)$. The graph of $y = (x - 4)(x + 2)$ can also be used to determine this. Since the inequality symbol is $\geq$, the endpoints of the intervals must be included in the solution set. It is $(-\infty, -2] \cup [4, \infty)$.

9. $x^2 + x - 2 > 0$ Polynomial inequality

$x^2 + x - 2 = 0$ Related equation

$(x + 2)(x - 1) = 0$ Factoring

Using the principle of zero products or by observing the graph of $y = x^2 + x - 2$, we see that the solutions of the related equation are -2 and 1. These numbers divide the x-axis into the intervals $(-\infty, -2)$, $(-2, 1)$, and $(1, \infty)$. We let $f(x) = x^2 + x - 2$ and test a value in each interval.

$(-\infty, -2)$: $f(-3) = 4 > 0$

$(-2, 1)$: $f(0) = -2 < 0$

$(1, \infty)$: $f(2) = 4 > 0$

Function values are positive on $(-\infty, -2)$ and $(1, \infty)$. The graph of $y = x^2 + x - 2$ can also be used to determine this. The solution set is $(-\infty, -2) \cup (1, \infty)$.

11. $x^2 > 25$ Polynomial inequality

$x^2 - 25 > 0$ Equivalent inequality with 0 on one side

$x^2 - 25 = 0$ Related equation

$(x + 5)(x - 5) = 0$ Factoring

Using the principle of zero products or by observing the graph of $y = x^2 - 25$, we see that the solutions of the related equation are -5 and 5. These numbers divide the x-axis into the intervals $(-\infty, -5)$, $(-5, 5)$, and $(5, \infty)$. We let $f(x) = x^2 - 25$ and test a value in each interval.

$(-\infty, -5)$: $f(-6) = 11 > 0$

$(-5, 5)$: $f(0) = -25 < 0$

$(5, \infty)$: $f(6) = 11 > 0$

Function values are positive on $(-\infty, -5)$ and $(5, \infty)$. The graph of $y = x^2 - 25$ can also be used to determine this. The solution set is $(-\infty, -5) \cup (5, \infty)$.

13. $4 - x^2 \leq 0$ Polynomial inequality

$4 - x^2 = 0$ Related equation

$(2 + x)(2 - x) = 0$ Factoring

Using the principle of zero products or by observing the graph of $y = 4 - x^2$, we see that the solutions of the related equation are -2 and 2. These numbers divide the x-axis into the intervals $(-\infty, -2)$, $(-2, 2)$, and $(2, \infty)$. We let $f(x) = 4 - x^2$ and test a value in each interval.

$(-\infty, -2)$: $f(-3) = -5 < 0$

$(-2, 2)$: $f(0) = 4 > 0$

$(2, \infty)$: $f(3) = -5 < 0$

Function values are negative on $(-\infty, -2)$ and $(2, \infty)$. The graph of $y = 4 - x^2$ can also be used to determine this. Since the inequality symbol is $\leq$, the endpoints of the intervals must be included in the solution set. It is $(-\infty, -2] \cup [2, \infty)$.

15. $6x - 9 - x^2 < 0$ Polynomial inequality

$6x - 9 - x^2 = 0$ Related equation

$-(x^2 - 6x + 9) = 0$ Factoring out -1 and rearranging

$-(x - 3)(x - 3) = 0$ Factoring

Using the principle of zero products or by observing the graph of $y = 6x - 9 - x^2$, we see that the solution of the related equation is 3. This number divides the

x-axis into the intervals $(-\infty, 3)$ and $(3, \infty)$. We let $f(x) = 6x - 9 - x^2$ and test a value in each interval.

$(-\infty, 3)$: $f(-4) = -49 < 0$

$(3, \infty)$: $f(4) = -49 < 0$

Function values are negative on both intervals. The solution set is $(-\infty, 3) \cup (3, \infty)$.

17.
$$x^2 + 12 < 4x \quad \text{Polynomial inequality}$$
$$x^2 - 4x + 12 < 0 \quad \begin{array}{l}\text{Equivalent inequality with 0}\\\text{on one side}\end{array}$$
$$x^2 - 4x + 12 = 0 \quad \text{Related equation}$$

Using the quadratic formula or by observing the graph of $y = x^2 - 4x + 12$ we see that the related equation has no real-number solutions. The graph lies entirely above the x-axis, so the inequality has no solution. We could determine this algebraically by letting $f(x) = x^2 - 4x + 12$ and testing any real number (since there are no real-number solutions of $f(x) = 0$ to divide the x-axis into intervals). For example, $f(0) = 12 > 0$, so we see algebraically that the inequality has no solution. The solution set is $\emptyset$.

19.
$$4x^3 - 7x^2 \leq 15x \quad \begin{array}{l}\text{Polynomial}\\\text{inequality}\end{array}$$
$$4x^3 - 7x^2 - 15x \leq 0 \quad \begin{array}{l}\text{Equivalent}\\\text{inequality with 0}\\\text{on one side}\end{array}$$
$$4x^3 - 7x^2 - 15x = 0 \quad \text{Related equation}$$
$$x(4x + 5)(x - 3) = 0 \quad \text{Factoring}$$

Using the principle of zero products or by observing the graph of $y = 4x^3 - 7x^2 - 15x$, we see that the solutions of the related equation are 0, $-\dfrac{5}{4}$, and 3. These numbers divide the x-axis into the intervals $\left(-\infty, -\dfrac{5}{4}\right)$, $\left(-\dfrac{5}{4}, 0\right)$, $(0, 3)$, and $(3, \infty)$. We let $f(x) = 4x^3 - 7x^2 - 15x$ and test a value in each interval.

$\left(-\infty, -\dfrac{5}{4}\right)$: $f(-2) = -30 < 0$

$\left(-\dfrac{5}{4}, 0\right)$: $f(-1) = 4 > 0$

$(0, 3)$　$f(1) = -18 < 0$

$(3, \infty)$: $f(4) = 84 > 0$

Function values are negative on $\left(-\infty, -\dfrac{5}{4}\right)$ and $(0, 3)$. The graph of $y = 4x^3 - 7x^2 - 15x$ can also be used to determine this. Since the inequality symbol is $\leq$, the endpoints of the intervals must be included in the solution set. It is $\left(-\infty, -\dfrac{5}{4}\right] \cup [0, 3]$.

21.
$$x^3 + 3x^2 - x - 3 \geq 0 \quad \begin{array}{l}\text{Polynomial}\\\text{inequality}\end{array}$$
$$x^3 + 3x^2 - x - 3 = 0 \quad \text{Related equation}$$
$$x^2(x + 3) - (x + 3) = 0 \quad \text{Factoring}$$
$$(x^2 - 1)(x + 3) = 0$$
$$(x + 1)(x - 1)(x + 3) = 0$$

Using the principle of zero products or by observing the graph of $y = x^3 + 3x^2 - x - 3$, we see that the solutions of the related equation are -1, 1, and -3. These numbers divide the x-axis into the intervals $(-\infty, -3)$, $(-3, -1)$, $(-1, 1)$, and $(1, \infty)$. We let $f(x) = x^3 + 3x^2 - x - 3$ and test a value in each interval.

$(-\infty, -3)$: $f(-4) = -15 < 0$

$(-3, -1)$: $f(-2) = 3 > 0$

$(-1, 1)$: $f(0) = -3 < 0$

$(1, \infty)$: $f(2) = 15 > 0$

Function values are positive on $(-3, -1)$ and $(1, \infty)$. The graph of $y = x^3 + 3x^2 - x - 3$ can also be used to determine this. Since the inequality symbol is $\geq$, the endpoints of the intervals must be included in the solution set. It is $[-3, -1] \cup [1, \infty)$.

23.
$$x^3 - 2x^2 < 5x - 6 \quad \begin{array}{l}\text{Polynomial}\\\text{inequality}\end{array}$$
$$x^3 - 2x^2 - 5x + 6 < 0 \quad \begin{array}{l}\text{Equivalent}\\\text{inequality with 0}\\\text{on one side}\end{array}$$
$$x^3 - 2x^2 - 5x + 6 = 0 \quad \text{Related equation}$$

Using the techniques of Section 2.5, we find that the solutions of the related equation are -2, 1, and 3. We can also use the graph of $y = x^3 - 2x^2 - 5x + 6$ to find these solutions. They divide the x-axis into the intervals $(-\infty, -2)$, $(-2, 1)$, $(1, 3)$, and $(3, \infty)$. Let $f(x) = x^3 - 2x^2 - 5x + 6$ and test a value in each interval.

$(-\infty, -2)$: $f(-3) = -24 < 0$

$(-2, 1)$: $f(0) = 6 > 0$

$(1, 3)$: $f(2) = -4 < 0$

$(3, \infty)$: $f(4) = 18 > 0$

Function values are negative on $(-\infty, -2)$ and $(1, 3)$. This can also be determined graphically. The solution set is $(-\infty, -2) \cup (1, 3)$.

25.
$$x^5 + x^2 \geq 2x^3 + 2 \quad \text{Polynomial inequality}$$
$$x^5 - 2x^3 + x^2 - 2 \geq 0 \quad \begin{array}{l}\text{Related}\\\text{inequality with}\\\text{0 on one side}\end{array}$$
$$x^5 - 2x^3 + x^2 - 2 = 0 \quad \begin{array}{l}\text{Related}\\\text{equation}\end{array}$$
$$x^3(x^2 - 2) + x^2 - 2 = 0 \quad \text{Factoring}$$
$$(x^3 + 1)(x^2 - 2) = 0$$

Using the principle of zero products or by observing the graph of $y = x^5 - 2x^3 + x^2 - 2$, we see that the real-number solutions of the related equation are -1, $-\sqrt{2}$, and $\sqrt{2}$. These numbers divide the x-axis into the intervals $(-\infty, -\sqrt{2})$, $(-\sqrt{2}, -1)$, $(-1, \sqrt{2})$, and $(\sqrt{2}, \infty)$. We let $f(x) = x^5 - 2x^3 + x^2 - 2$ and test a value in each interval.

$(-\infty, -\sqrt{2})$: $f(-2) = -14 < 0$

$(-\sqrt{2}, -1)$: $f(-1.3) \approx 0.37107 > 0$

$(-1, \sqrt{2})$: $f(0) = -2 < 0$

$(\sqrt{2}, \infty)$: $f(2) = 18 > 0$

Function values are positive on $(-\sqrt{2}, -1)$ and $(\sqrt{2}, \infty)$. This can also be determined graphically. Since the inequality symbol is $\geq$, the endpoints of the intervals must be included in the solution set. It is $[-\sqrt{2}, -1] \cup [\sqrt{2}, \infty)$.

27.
$$2x^3 + 6 \leq 5x^2 + x \quad \text{Polynomial inequality}$$
$$2x^3 - 5x^2 - x + 6 \leq 0 \quad \begin{array}{l}\text{Equivalent}\\\text{inequality with 0}\\\text{on one side}\end{array}$$
$$2x^3 - 5x^2 - x + 6 = 0 \quad \text{Related equation}$$

Using the techniques of Section 2.5, we find that the solutions of the related equation are -1, $\frac{3}{2}$, and 2. We can also use the graph of $y = 2x^3 - 5x^2 - x + 6$ to find these solutions. They divide the x-axis into the intervals $(-\infty, -1)$, $\left(-1, \frac{3}{2}\right)$, $\left(\frac{3}{2}, 2\right)$, and $(2, \infty)$. Let $f(x) = 2x^3 - 5x^2 - x + 6$ and test a value in each interval.

$(-\infty, -1)$: $f(-2) = -28 < 0$

$\left(-1, \frac{3}{2}\right)$: $f(0) = 6 > 0$

$\left(\frac{3}{2}, 2\right)$: $f(1.6) = -0.208 < 0$

$(2, \infty)$: $f(3) = 12 > 0$

Function values are negative in $(-\infty, -1)$ and $\left(\frac{3}{2}, 2\right)$. This can also be determined graphically. Since the inequality symbol is $\leq$, the endpoints of

the intervals must be included in the solution set. The solution set is $\left(-\infty, -1\right] \cup \left[\frac{3}{2}, 2\right]$.

29.
$$x^3 + 5x^2 - 25x \leq 125 \quad \text{Polynomial inequality}$$
$$x^3 + 5x^2 - 25x - 125 \leq 0 \quad \begin{array}{l}\text{Equivalent}\\\text{inequality with 0}\\\text{on one side}\end{array}$$
$$x^3 + 5x^2 - 25x - 125 = 0 \quad \text{Related equation}$$
$$x^2(x + 5) - 25(x + 5) = 0 \quad \text{Factoring}$$
$$(x^2 - 25)(x + 5) = 0$$
$$(x + 5)(x - 5)(x + 5) = 0$$

Using the principle of zero products or by observing the graph of $y = x^3 + 5x^2 - 25x - 125$, we see that the solutions of the related equation are -5 and 5. These numbers divide the x-axis into the intervals $(-\infty, -5)$, $(-5, 5)$, and $(5, \infty)$. We let $f(x) = x^3 + 5x^2 - 25x - 125$ and test a value in each interval.

$(-\infty, -5)$: $f(-6) = -11 < 0$

$(-5, 5)$: $f(0) = -125 < 0$

$(5, \infty)$: $f(6) = 121 > 0$

Function values are negative on $(-\infty, -5)$ and $(-5, 5)$. This can also be determined graphically. Since the inequality symbol is $\leq$, the endpoints of the intervals must be included in the solution set. It is $(-\infty, -5] \cup [-5, 5]$ or $(-\infty, 5]$.

31. $0.1x^3 - 0.6x^2 - 0.1x + 2 < 0$ Polynomial inequality

$0.1x^3 - 0.6x^2 - 0.1x + 2 = 0$ Related equation

After trying all the possibilities, we find that the related equation has no rational zeros. Using the graph of $y = 0.1x^3 - 0.6x^2 - 0.1x + 2$, we find that the only real-number solutions of the related equation are approximately -1.680, 2.154, and 5.526. These numbers divide the x-axis into the intervals $(-\infty, -1.680)$, $(-1.680, 2.154)$, $(2.154, 5.526)$, and $(5.526, \infty)$. We let $f(x) = 0.1x^3 - 0.6x^2 - 0.1x + 2$ and test a value in each interval.

$(-\infty, -1.680)$: $f(-2) = -1 < 0$

$(-1.680, 2.154)$: $f(0) = 2 > 0$

$(2.154, 5.526)$: $f(3) = -1 < 0$

$(5.526, \infty)$: $f(6) = 1.4 > 0$

Function values are negative on $(-\infty, -1.680)$ and $(2.154, 5.526)$. The graph can also be used to determine this. The solution set is $(-\infty, -1.680) \cup (2.154, 5.526)$.

33. $\dfrac{1}{x+4} > 0$ Rational inequality

$\dfrac{1}{x+4} = 0$ Related equation

The denominator of $f(x) = \dfrac{1}{x+4}$ is 0 when $x = -4$, so the function is not defined for $x = -4$. The related equation has no solution. Thus, the only critical value is -4. It divides the x-axis into the intervals $(-\infty, -4)$ and $(-4, \infty)$. We test a value in each interval.

$(-\infty, -4)$: $f(-5) = -1 < 0$

$(-4, \infty)$: $f(0) = \dfrac{1}{4} > 0$

Function values are positive on $(-4, \infty)$. This can also be determined from the graph of $y = \dfrac{1}{x+4}$. The solution set is $(-4, \infty)$.

35. $\dfrac{-4}{2x+5} < 0$ Rational inequality

$\dfrac{-4}{2x+5} = 0$ Related equation

The denominator of $f(x) = \dfrac{-4}{2x+5}$ is 0 when $x = -\dfrac{5}{2}$, so the function is not defined for $x = -\dfrac{5}{2}$. The related equation has no solution. Thus, the only critical value is $-\dfrac{5}{2}$. It divides the x-axis into the intervals $\left(-\infty, -\dfrac{5}{2}\right)$ and $\left(-\dfrac{5}{2}, \infty\right)$. We test a value in each interval.

$\left(-\infty, -\dfrac{5}{2}\right)$: $f(-3) = 4 > 0$

$\left(-\dfrac{5}{2}, \infty\right)$: $f(0) = -\dfrac{4}{5} < 0$

Function values are negative on $\left(-\dfrac{5}{2}, \infty\right)$. This can also be determined from the graph of $y = \dfrac{-4}{2x+5}$. The solution set is $\left(-\dfrac{5}{2}, \infty\right)$.

37. $\dfrac{x-4}{x+3} - \dfrac{x+2}{x-1} \le 0$

The denominator of $f(x) = \dfrac{x-4}{x+3} - \dfrac{x+2}{x-1}$ is 0 when $x = -3$ or $x = 1$, so the function is not defined for these values of x. We solve the related equation $f(x) = 0$.

$\dfrac{x-4}{x+3} - \dfrac{x+2}{x-1} = 0$

$(x+3)(x-1)\left(\dfrac{x-4}{x+3} - \dfrac{x+2}{x-1}\right) = (x+3)(x-1)\cdot 0$

$(x-1)(x-4) - (x+3)(x+2) = 0$

$x^2 - 5x + 4 - (x^2 + 5x + 6) = 0$

$-10x - 2 = 0$

$-10x = 2$

$x = -\dfrac{1}{5}$

The critical values are -3, $-\dfrac{1}{5}$, and 1. They divide the x-axis into the intervals $(-\infty, -3)$, $\left(3, -\dfrac{1}{5}\right)$, $\left(-\dfrac{1}{5}, 1\right)$, and $(1, \infty)$. We test a value in each interval.

$(-\infty, -3)$: $f(-4) = 7.6 > 0$

$\left(-3, -\dfrac{1}{5}\right)$: $f(-1) = -2 < 0$

$\left(-\dfrac{1}{5}, 1\right)$: $f(0) = \dfrac{2}{3} > 0$

$(1, \infty)$: $f(2) = -4.4 < 0$

Function values are negative on $\left(-3, -\dfrac{1}{5}\right)$ and $(1, \infty)$. This can also be determined from the graph of $y = \dfrac{x-4}{x+3} - \dfrac{x+2}{x-1}$. Note that since the inequality symbol is $\le$ and $f\left(-\dfrac{1}{5}\right) = 0$, then $-\dfrac{1}{5}$ must be included in the solution set. Note also that since neither -3 nor 1 is in the domain of $f(x)$, they are not included in the solution set. It is $\left(-3, -\dfrac{1}{5}\right] \cup (1, \infty)$.

39. $\dfrac{2x-1}{x+3} \ge \dfrac{x+1}{3x+1}$ Rational inequality

$\dfrac{2x-1}{x+3} - \dfrac{x+1}{3x+1} \ge 0$ Equivalent inequality with 0 on one side

The denominator of $f(x) = \dfrac{2x-1}{x+3} - \dfrac{x+1}{3x+1}$ is 0 when $x = -3$ or $x = -\dfrac{1}{3}$, so the function is not defined for these values of x. We solve the related equation $f(x) = 0$.

$$\frac{2x-1}{x+3} - \frac{x+1}{3x+1} = 0$$

$$(x+3)(3x+1)\left(\frac{2x-1}{x+3} - \frac{x+1}{3x+1}\right) =$$
$$(x+3)(3x+1)\cdot 0$$
$$(3x+1)(2x-1) - (x+3)(x+1) = 0$$
$$6x^2 - x - 1 - (x^2 + 4x + 3) = 0$$
$$5x^2 - 5x - 4 = 0$$

Using the quadratic formula we find that $x = \dfrac{5 \pm \sqrt{105}}{10}$. Then the critical values are -3, $\dfrac{5 - \sqrt{105}}{10}$, $-\dfrac{1}{3}$, and $\dfrac{5 + \sqrt{105}}{10}$. They divide the x-axis into the intervals $(-\infty, -3)$, $\left(-3, \dfrac{5 - \sqrt{105}}{10}\right)$, $\left(\dfrac{5 - \sqrt{105}}{10}, -\dfrac{1}{3}\right)$, $\left(-\dfrac{1}{3}, \dfrac{5 + \sqrt{105}}{10}\right)$, and $\left(\dfrac{5 + \sqrt{105}}{10}, \infty\right)$. We test a value in each interval.

$(-\infty, -3)$: $f(-4) \approx 8.7273 > 0$

$\left(-3, \dfrac{5 - \sqrt{105}}{10}\right)$: $f(-2) = -5.2 < 0$

$\left(\dfrac{5 - \sqrt{105}}{10}, -\dfrac{1}{3}\right)$: $f(-0.4) \approx 2.3077 > 0$

$\left(-\dfrac{1}{3}, \dfrac{5 + \sqrt{105}}{10}\right)$: $f(0) \approx -1.333 < 0$

$\left(\dfrac{5 + \sqrt{105}}{10}, \infty\right)$: $f(2) \approx 0.1714 > 0$

Function values are positive on $(-\infty, -3)$, $\left(\dfrac{5 - \sqrt{105}}{10}, -\dfrac{1}{3}\right)$ and $\left(\dfrac{5 + \sqrt{105}}{10}, \infty\right)$. This can also be determined from the graph of $y = \dfrac{2x-1}{x+3} - \dfrac{x+1}{3x+1}$. Note that since the inequality symbol is $\geq$ and $f\left(\dfrac{5 \pm \sqrt{105}}{10}\right) = 0$, then $\dfrac{5 \pm \sqrt{105}}{10}$ must be included in the solution set. Note also that since neither -3 nor $-\dfrac{1}{3}$ is in the domain of $f(x)$, they are not included in the solution set. It is

$$(-\infty, -3) \cup \left[\frac{5 - \sqrt{105}}{10}, -\frac{1}{3}\right) \cup \left[\frac{5 + \sqrt{105}}{10}, \infty\right).$$

41. $\quad \dfrac{x+1}{x-2} \geq 3 \quad$ Rational inequality

$\dfrac{x+1}{x-2} - 3 \geq 0 \quad$ Equivalent inequality with 0 on one side

The denominator of $f(x) = \dfrac{x+1}{x-2} - 3$ is 0 when $x = 2$, so the function is not defined for this value of x. We solve the related equation $f(x) = 0$.

$$\frac{x+1}{x-2} - 3 = 0$$

$$(x-2)\left(\frac{x+1}{x-2} - 3\right) = (x-2)\cdot 0$$
$$x + 1 - 3(x-2) = 0$$
$$x + 1 - 3x + 6 = 0$$
$$-2x + 7 = 0$$
$$-2x = -7$$
$$x = \frac{7}{2}$$

The critical values are 2 and $\dfrac{7}{2}$. They divide the x-axis into the intervals $(-\infty, 2)$, $\left(2, \dfrac{7}{2}\right)$, and $\left(\dfrac{7}{2}, \infty\right)$. We test a value in each interval.

$(-\infty, 2)$: $f(0) = -3.5 < 0$

$\left(2, \dfrac{7}{2}\right)$: $f(3) = 1 > 0$

$\left(\dfrac{7}{2}, \infty\right)$: $f(4) = -0.5 < 0$

Function values are positive on $\left(2, \dfrac{7}{2}\right)$. This can also be determined from the graph of $y = \dfrac{x+1}{x-2} - 3$. Note that since the inequality symbol is $\geq$ and $f\left(\dfrac{7}{2}\right) = 0$, then $\dfrac{7}{2}$ must be included in the solution set. Note also that since 2 is not in the domain of $f(x)$, it is not included in the solution set. It is $\left(2, \dfrac{7}{2}\right]$.

43. $\quad x - 2 > \dfrac{1}{x} \quad$ Rational inequality

$x - 2 - \dfrac{1}{x} > 0 \quad$ Equivalent inequality with 0 on one side

The denominator of $f(x) = x - 2 - \dfrac{1}{x}$ is 0 when $x = 0$, so the function is not defined for this value of x. We solve the related equation $f(x) = 0$.

$$x - 2 - \frac{1}{x} = 0$$

$$x\left(x - 2 - \frac{1}{x}\right) = x \cdot 0$$

$$x^2 - 2x - x \cdot \frac{1}{x} = 0$$

$$x^2 - 2x - 1 = 0$$

Using the quadratic formula we find that $x = 1 \pm \sqrt{2}$. The critical values are $1 - \sqrt{2}$, 0, and $1 + \sqrt{2}$. They divide the x-axis into the intervals $(-\infty, 1 - \sqrt{2})$,

$(1 - \sqrt{2}, 0)$, $(0, 1 + \sqrt{2})$, and $(1 + \sqrt{2}, \infty)$. We test a value in each interval.

$(-\infty, 1 - \sqrt{2})$: $f(-1) = -2 < 0$

$(1 - \sqrt{2}, 0)$: $f(-0.1) = 7.9 > 0$

$(0, 1 + \sqrt{2})$: $f(1) = -2 < 0$

$(1 + \sqrt{2}, \infty)$: $f(3) = \dfrac{2}{3} > 0$

Function values are positive on $(1 - \sqrt{2}, 0)$ and $(1 + \sqrt{2}, \infty)$. This can also be determined from the graph of $y = x - 2 - \dfrac{1}{x}$. The solution set is $(1 - \sqrt{2}, 0) \cup (1 + \sqrt{2}, \infty)$.

45.
$$\frac{2}{x^2 - 4x + 3} \leq \frac{5}{x^2 - 9}$$

$$\frac{2}{x^2 - 4x + 3} - \frac{5}{x^2 - 9} \leq 0$$

$$\frac{2}{(x-1)(x-3)} - \frac{5}{(x+3)(x-3)} \leq 0$$

The denominator of $f(x) = \dfrac{2}{(x-1)(x-3)} - \dfrac{5}{(x+3)(x-3)}$ is 0 when $x = 1$, 3, or -3, so the function is not defined for these values of x. We solve the related equation $f(x) = 0$.

$$\frac{2}{(x-1)(x-3)} - \frac{5}{(x+3)(x-3)} = 0$$

$$(x-1)(x-3)(x+3)\left(\frac{2}{(x-1)(x-3)} - \frac{5}{(x+3)(x-3)}\right)$$
$$= (x-1)(x-3)(x+3) \cdot 0$$

$$2(x+3) - 5(x-1) = 0$$

$$2x + 6 - 5x + 5 = 0$$

$$-3x + 11 = 0$$

$$-3x = -11$$

$$x = \frac{11}{3}$$

The critical values are -3, 1, 3, and $\dfrac{11}{3}$. They divide the x-axis into the intervals $(-\infty, -3)$, $(-3, 1)$, $(1, 3)$, $\left(3, \dfrac{11}{3}\right)$, and $\left(\dfrac{11}{3}, \infty\right)$. We test a value in each interval.

$(-\infty, -3)$: $f(-4) \approx -0.6571 < 0$

$(-3, 1)$: $f(0) \approx 1.2222 > 0$

$(1, 3)$: $f(2) = -1 < 0$

$\left(3, \dfrac{11}{3}\right)$: $f(3.5) \approx 0.6154 > 0$

$\left(\dfrac{11}{3}, \infty\right)$: $f(4) \approx -0.0476 < 0$

Function values are negative on $(-\infty, -3)$, $(1, 3)$, and $\left(\dfrac{11}{3}, \infty\right)$. They can also be determined from

the graph of $y = \dfrac{2}{(x-1)(x-3)} - \dfrac{5}{(x+3)(x-3)}$. Note that since the inequality symbol is $\leq$ and $f\left(\dfrac{11}{3}\right) = 0$, then $\dfrac{11}{3}$ must be included in the solution set. Note also that since -3, 1, and 3 are not in the domain of $f(x)$, they are not included in the solution set. It is $(-\infty, -3) \cup (1, 3) \cup \left[\dfrac{11}{3}, \infty\right)$.

47.
$$\frac{3}{x^2 + 1} \geq \frac{6}{5x^2 + 2}$$

$$\frac{3}{x^2 + 1} - \frac{6}{5x^2 + 2} \geq 0$$

The denominator of $f(x) = \dfrac{3}{x^2 + 1} - \dfrac{6}{5x^2 + 2}$ has no real-number zeros. We solve the related equation $f(x) = 0$.

$$\frac{3}{x^2 + 1} - \frac{6}{5x^2 + 2} = 0$$

$$(x^2 + 1)(5x^2 + 2)\left(\frac{3}{x^2 + 1} - \frac{6}{5x^2 + 2}\right) =$$
$$(x^2 + 1)(5x^2 + 2) \cdot 0$$

$$3(5x^2 + 2) - 6(x^2 + 1) = 0$$

$$15x^2 + 6 - 6x^2 - 6 = 0$$

$$9x^2 = 0$$

$$x^2 = 0$$

$$x = 0$$

The only critical value is 0. It divides the x-axis into the intervals $(-\infty, 0)$ and $(0, \infty)$. We test a value in each interval.

$(-\infty, 0)$: $f(-1) \approx 0.64286 > 0$

$(0, 0)$: $f(1) \approx 0.64286 > 0$

Function values are positive on both intervals. Note that since the inequality symbol is $\geq$ and $f(0) = 0$, then 0 must be included in the solution set. It is $(-\infty, 0] \cup [0, \infty)$, or $(-\infty, \infty)$.

49.
$$\frac{5}{x^2 + 3x} < \frac{3}{2x + 1}$$

$$\frac{5}{x^2 + 3x} - \frac{3}{2x + 1} < 0$$

$$\frac{5}{x(x + 3)} - \frac{3}{2x + 1} < 0$$

The denominator of $f(x) = \dfrac{5}{x(x + 3)} - \dfrac{3}{2x + 1}$ is 0 when $x = 0$, -3, or $-\dfrac{1}{2}$, so the function is not defined for these values of x. We solve the related equation $f(x) = 0$.

$$\frac{5}{x(x+3)} - \frac{3}{2x+1} = 0$$

$$x(x+3)(2x+1)\left(\frac{5}{x(x+3)} - \frac{3}{2x+1}\right) =$$

$$x(x+3)(2x+1)\cdot 0$$

$$5(2x+1) - 3x(x+3) = 0$$

$$10x + 5 - 3x^2 - 9x = 0$$

$$-3x^2 + x + 5 = 0$$

Using the quadratic formula we find that

$x = \dfrac{1 \pm \sqrt{61}}{6}$. The critical values are -3, $\dfrac{1-\sqrt{61}}{6}$,

$-\dfrac{1}{2}$, 0, and $\dfrac{1+\sqrt{61}}{6}$. They divide the x-axis into

the intervals $(-\infty, -3)$, $\left(-3, \dfrac{1-\sqrt{61}}{6}\right)$,

$\left(\dfrac{1-\sqrt{61}}{6}, -\dfrac{1}{2}\right)$, $\left(-\dfrac{1}{2}, 0\right)$, $\left(0, \dfrac{1+\sqrt{61}}{6}\right)$, and

$\left(\dfrac{1+\sqrt{61}}{6}, \infty\right)$.

We test a value in each interval.

$(-\infty, -3)$: $f(-4) \approx 1.6786 > 0$

$\left(-3, \dfrac{1-\sqrt{61}}{6}\right)$: $f(-2) = -1.5 < 0$

$\left(\dfrac{1-\sqrt{61}}{6}, -\dfrac{1}{2}\right)$: $f(-1) = 0.5 > 0$

$\left(-\dfrac{1}{2}, 0\right)$: $f(-0.1) \approx -20.99 < 0$

$\left(0, \dfrac{1+\sqrt{61}}{6}\right)$: $f(1) = 0.25 > 0$

$\left(\dfrac{1+\sqrt{61}}{6}, \infty\right)$: $f(2) = -0.1 < 0$

Function values are negative on $\left(-3, \dfrac{1-\sqrt{61}}{6}\right)$,

$\left(-\dfrac{1}{2}, 0\right)$ and $\left(\dfrac{1+\sqrt{61}}{6}, \infty\right)$. This can also be

determined from the graph of

$y = \dfrac{5}{x(x+3)} - \dfrac{3}{2x+1}$.

The solution set is

$\left(-3, \dfrac{1-\sqrt{61}}{6}\right) \cup \left(-\dfrac{1}{2}, 0\right) \cup \left(\dfrac{1+\sqrt{61}}{6}, \infty\right)$.

51.
$$\frac{5x}{7x-2} > \frac{x}{x+1}$$

$$\frac{5x}{7x-2} - \frac{x}{x+1} > 0$$

The denominator of $f(x) = \dfrac{5x}{7x-2} - \dfrac{x}{x+1}$ is 0

when $x = \dfrac{2}{7}$ or $x = -1$, so the function is not defined

for these values of x. We solve the related equation $f(x) = 0$.

$$\frac{5x}{7x-2} - \frac{x}{x+1} = 0$$

$$(7x-2)(x+1)\left(\frac{5x}{7x-2} - \frac{x}{x+1}\right) = (7x-2)(x+1)\cdot 0$$

$$5x(x+1) - x(7x-2) = 0$$

$$5x^2 + 5x - 7x^2 + 2x = 0$$

$$-2x^2 + 7x = 0$$

$$-x(2x-7) = 0$$

$$x = 0 \quad or \quad x = \frac{7}{2}$$

The critical values are -1, 0, $\dfrac{2}{7}$, and $\dfrac{7}{2}$. They divide the x-axis into the intervals $(-\infty, -1)$, $(-1, 0)$,

$\left(0, \dfrac{2}{7}\right)$, $\left(\dfrac{2}{7}, \dfrac{7}{2}\right)$, and $\left(\dfrac{7}{2}, \infty\right)$. We test a value in each interval.

$(-\infty, -1)$: $f(-2) = -1.375 < 0$

$(-1, 0)$: $f(-0.5) \approx 1.4545 > 0$

$\left(0, \dfrac{2}{7}\right)$: $f(0.1) \approx -0.4755 < 0$

$\left(\dfrac{2}{7}, \dfrac{7}{2}\right)$: $f(1) = 0.5 > 0$

$\left(\dfrac{7}{2}, \infty\right)$: $f(4) \approx -0.0308 < 0$

Function values are positive on $(-1, 0)$ and $\left(\dfrac{2}{7}, \dfrac{7}{2}\right)$. This can also be determined from the graph of

$y = \dfrac{5x}{7x-2} - \dfrac{x}{x+1}$. The solution set is

$(-1, 0) \cup \left(\dfrac{2}{7}, \dfrac{7}{2}\right)$.

53.
$$\frac{x}{x^2+4x-5} + \frac{3}{x^2-25} \le \frac{2x}{x^2-6x+5}$$

$$\frac{x}{x^2+4x-5} + \frac{3}{x^2-25} - \frac{2x}{x^2-6x+5} \le 0$$

$$\frac{x}{(x+5)(x-1)} + \frac{3}{(x+5)(x-5)} - \frac{2x}{(x-5)(x-1)} \le 0$$

The denominator of

$$f(x) = \frac{x}{(x+5)(x-1)} + \frac{3}{(x+5)(x-5)} - \frac{2x}{(x-5)(x-1)}$$

is 0 when $x = -5$, 1, or 5, so the function is not defined for these values of x. We solve the related equation $f(x) = 0$.

$$\frac{x}{(x+5)(x-1)} + \frac{3}{(x+5)(x-5)} - \frac{2x}{(x-5)(x-1)} = 0$$

$$x(x-5) + 3(x-1) - 2x(x+5) = 0$$

Multiplying by $(x+5)(x-1)(x-5)$

$$x^2 - 5x + 3x - 3 - 2x^2 - 10x = 0$$

$$-x^2 - 12x - 3 = 0$$

$$x^2 + 12x + 3 = 0$$

Using the quadratic formula, we find that $x = -6 \pm \sqrt{33}$. The critical values are $-6-\sqrt{33}$, -5, $-6+\sqrt{33}$, 1, and 5. They divide the x-axis into the intervals $(-\infty, -6-\sqrt{33})$, $(-6-\sqrt{33}, -5)$, $(-5, -6+\sqrt{33})$, $(-6+\sqrt{33}, 1)$, $(1, 5)$, and $(5, \infty)$. We test a value in each interval.

$(-\infty, -6-\sqrt{33})$: $f(-12) \approx 0.00194 > 0$

$(-6-\sqrt{33}, -5)$: $f(-6) \approx -0.4286 < 0$

$(-5, -6+\sqrt{33})$: $f(-1) \approx 0.16667 > 0$

$(-6+\sqrt{33}, 1)$: $f(0) = -0.12 < 0$

$(1, 5)$: $f(2) \approx 1.4762 > 0$

$(5, \infty)$: $f(6) \approx -2.018 < 0$

Function values are negative on $(-6-\sqrt{33}, -5)$, $(-6+\sqrt{33}, 1)$, and $(5, \infty)$. They can also be determined from the graph of $y = \frac{x}{(x+5)(x-1)} + \frac{3}{(x+5)(x-5)} - \frac{2x}{(x-5)(x-1)}$. Note that since the inequality symbol is $\leq$ and $f(-6 \pm \sqrt{33}) = 0$, then $-6-\sqrt{33}$ and $-6+\sqrt{33}$ must be included in the solution set. Note also that since -5, 1, and 5 are not in the domain of $f(x)$, they are not included in the solution set. It is $[-6-\sqrt{33}, -5) \cup [-6+\sqrt{33}, 1) \cup (5, \infty)$.

55. We write and solve a rational inequality.

$$\frac{4t}{t^2+1} + 98.6 > 100$$

$$\frac{4t}{t^2+1} - 1.4 > 0$$

The denominator of $f(t) = \frac{4t}{t^2+1} - 1.4$ has no real-number zeros. We solve the related equation $f(t) = 0$.

$$\frac{4t}{t^2+1} - 1.4 = 0$$

$$4t - 1.4(t^2+1) = 0 \quad \text{Multiplying by } t^2+1$$

$$4t - 1.4t^2 - 1.4 = 0$$

Using the quadratic formula, we find that $t = \frac{4 \pm \sqrt{8.16}}{2.8}$; that is, $t \approx 0.408$ or $t \approx 2.449$. This can also be determined from the graph of $y = \frac{4x}{x^2+1} - 1.4$. These numbers divide the t-axis

into the intervals $(-\infty, 0.408)$, $(0.408, 2.449)$, and $(2.449, \infty)$. We test a value in each interval.

$(-\infty, 0.408)$: $f(0) = -1.4 < 0$

$(0.408, 2.449)$: $f(1) = 0.6 > 0$

$(2.449, \infty)$: $f(3) = -0.2 < 0$

Function values are positive on $(0.408, 2.449)$. This can also be determined from the graph. The solution set is $(0.408, 2.449)$.

57. a) We write and solve a polynomial inequality.

$$-3x^2 + 630x - 6000 > 0 \quad (x \geq 0)$$

We first solve the related equation.

$$-3x^2 + 630x - 6000 = 0$$

$$x^2 - 210x + 2000 = 0 \quad \text{Dividing by } -3$$

$$(x-10)(x-200) = 0 \quad \text{Factoring}$$

Using the principle of zero products or by observing the graph of $y = -3x^2 + 630 - 6000$, we see that the solutions of the related equation are 10 and 200. These numbers divide the x-axis into the intervals $(-\infty, 10)$, $(10, 200)$, and $(200, \infty)$. Since we are restricting our discussion to non-negative values of x, we consider the intervals $[0, 10)$, $(10, 200)$, and $(200, \infty)$.

We let $f(x) = -3x^2 + 630x - 6000$ and test a value in each interval.

$[0, 10)$: $f(0) = -6000 < 0$

$(10, 200)$: $f(11) = 567 > 0$

$(200, \infty)$: $f(201) = -573 < 0$

Function values are positive only on $(10, 200)$. This can also be determined graphically. The solution set is $\{x | 10 < x < 200\}$, or $(10, 200)$.

b) From part (a), we see that function values are negative on $[0, 10)$ and $(200, \infty)$. Thus, the solution set is $\{x | 0 < x < 10 \text{ or } x > 200\}$, or $(0, 10) \cup (200, \infty)$.

59. We write an inequality.

$$27 \leq \frac{n(n-3)}{2} \leq 230$$

$$54 \leq n(n-3) \leq 460 \quad \text{Multiplying by 2}$$

$$54 \leq n^2 - 3n \leq 460$$

We write this as two inequalities.

$$54 \leq n^2 - 3n \quad \text{and} \quad n^2 - 3n \leq 460$$

Solve each inequality.

$$n^2 - 3n \geq 54$$

$$n^2 - 3n - 54 \geq 0$$

$$n^2 - 3n - 54 = 0 \quad \text{Related equation}$$

$$(n+6)(n-9) = 0$$

$n = -6 \ or \ n = 9$

Since only positive values of n have meaning in this application, we consider the intervals $(0, 9)$ and $(9, \infty)$. Let $f(n) = n^2 - 3n - 54$ and test a value in each interval.

$(0, 9)$: $f(1) = -56 < 0$

$(9, \infty)$: $f(10) = 16 > 0$

Function values are positive on $(9, \infty)$. Since the inequality symbol is $\geq$, 9 must also be included in the solution set for this portion of the inequality. It is $\{n | n \geq 9\}$.

Now solve the second inequality.

$$n^2 - 3n \leq 460$$
$$n^2 - 3n - 460 \leq 0$$
$$n^2 - 3n - 460 = 0 \quad \text{Related equation}$$
$$(n + 20)(n - 23) = 0$$
$$n = -20 \ or \ n = 23$$

We consider only positive values of n as above. Thus, we consider the intervals $(0, 23)$ and $(23, \infty)$. Let $f(n) = n^2 - 3n - 460$ and test a value in each interval.

$(0, 23)$: $f(1) = -462 < 0$

$(23, \infty)$: $f(24) = 44 > 0$

Function values are negative on $(0, 23)$. Since the inequality symbol is $\leq$, 23 must also be included in the solution set for this portion of the inequality. It is $\{n | 0 < n \leq 23\}$.

The solution set of the original inequality is $\{n | n \geq 9 \ and \ 0 < n \leq 23\}$, or $\{n | 9 \leq n \leq 23\}$.

61. Discussion and Writing

63.
$$(x - h)^2 + (y - k)^2 = r^2$$
$$[x - (-2)]^2 + (y - 4)^2 = 3^2$$
$$(x + 2)^2 + (y - 4)^2 = 9$$

65. $h(x) = -2x^2 + 3x - 8$

a) $\quad -\dfrac{b}{2a} = -\dfrac{3}{2(-2)} = \dfrac{3}{4}$

$h\left(\dfrac{3}{4}\right) = -2\left(\dfrac{3}{4}\right)^2 + 3 \cdot \dfrac{3}{4} - 8 = -\dfrac{55}{8}$

The vertex is $\left(\dfrac{3}{4}, -\dfrac{55}{8}\right)$.

b) The coefficient of x^2 is negative, so there is a maximum value. It is the second coordinate of the vertex, $-\dfrac{55}{8}$. It occurs at $x = \dfrac{3}{4}$.

c) The range is $\left(-\infty, -\dfrac{55}{8}\right]$.

67.
$$x^2 + 9 \leq 6x$$
$$x^2 - 6x + 9 \leq 0$$
$$(x - 3)^2 \leq 0 \quad \text{Factoring}$$

Note that $(x - 3)^2 \geq 0$ for all values of x and that $(x - 3)^2 = 0$ when $x = 3$. Thus, the solution set is $\{3\}$.

69.
$$x^4 + 3x^2 > 4x - 15$$
$$x^4 + 3x^2 - 4x + 15 > 0$$

The graph of $y = x^4 + 3x^2 - 4x + 15$ lies entirely above the x-axis. Thus, the solution set is the set of all real numbers, or $(-\infty, \infty)$.

71. $|x^2 - 5| = |5 - x^2| = 5 - x^2$ when $5 - x^2 \geq 0$. Thus we solve $5 - x^2 \geq 0$.

$$5 - x^2 \geq 0$$
$$5 - x^2 = 0 \quad \text{Related equation}$$
$$5 = x^2$$
$$\pm\sqrt{5} = x$$

Let $f(x) = 5 - x^2$ and test a value in each of the intervals determined by the solutions of the related equation.

$(-\infty, -\sqrt{5})$: $f(-3) = -4 < 0$

$(-\sqrt{5}, \sqrt{5})$: $f(0) = 5 > 0$

$(\sqrt{5}, \infty)$: $f(3) = -4 < 0$

Function values are positive on $(-\sqrt{5}, \sqrt{5})$. Since the inequality symbol is $\geq$, the endpoints of the interval must be included in the solution set. It is $\left[-\sqrt{5}, \sqrt{5}\right]$.

73.
$$2|x|^2 - |x| + 2 \leq 5$$
$$2|x|^2 - |x| - 3 \leq 0$$
$$2|x|^2 - |x| - 3 = 0 \quad \text{Related equation}$$
$$(2|x| - 3)(|x| + 1) = 0 \quad \text{Factoring}$$
$$2|x| - 3 = 0 \ or \ |x| + 1 = 0$$
$$|x| = \dfrac{3}{2} \ or \qquad |x| = -1$$

The solution of the first equation is $x = -\dfrac{3}{2}$ or $x = \dfrac{3}{2}$. The second equation has no solution. Let $f(x) = 2|x|^2 - |x| - 3$ and test a value in each interval determined by the solutions of the related equation.

$\left(-\infty, -\dfrac{3}{2}\right)$: $f(-2) = 3 > 0$

$\left(-\dfrac{3}{2}, \dfrac{3}{2}\right)$: $f(0) = -3 < 0$

$\left(\dfrac{3}{2}, \infty\right)$: $f(2) = 3 > 0$

Function values are negative on $\left(-\frac{3}{2},\frac{3}{2}\right)$. Since the inequality symbol is $\leq$, the endpoints of the interval must also be included in the solution set. It is $\left[-\frac{3}{2},\frac{3}{2}\right]$.

75.
$$\left|1+\frac{1}{x}\right|<3$$
$$-3<1+\frac{1}{x}<3$$
$$-3<1+\frac{1}{x}\quad and\quad 1+\frac{1}{x}<3$$

First solve $-3<1+\frac{1}{x}$.
$$0<4+\frac{1}{x},\quad or\quad \frac{1}{x}+4>0$$

The denominator of $f(x)=\frac{1}{x}+4$ is 0 when $x=0$, so the

function is not defined for this value of x. Now solve the related equation.
$$\frac{1}{x}+4=0$$
$$1+4x=0\qquad \text{Multiplying by }x$$
$$x=-\frac{1}{4}$$

The critical values are $-\frac{1}{4}$ and 0. Test a value in each of the intervals determined by them.

$\left(\infty,-\frac{1}{4}\right)$: $f(-1)=3>0$

$\left(-\frac{1}{4},0\right)$: $f(-0.1)=-6<0$

$(0,\infty)$: $f(1)=5>0$

The solution set for this portion of the inequality is $\left(-\infty,-\frac{1}{4}\right)\cup(0,\infty)$.

Next solve $1+\frac{1}{x}<3$, or $\frac{1}{x}-2<0$. The denominator of $f(x)=\frac{1}{x}-2$ is 0 when $x=0$, so the function is not defined for this value of x. Now solve the related equation.
$$\frac{1}{x}-2=0$$
$$1-2x=0\qquad \text{Multiplying by }x$$
$$x=\frac{1}{2}$$

The critical values are 0 and $\frac{1}{2}$. Test a value in each of the intervals determined by them.

$(-\infty,0)$: $f(-1)=-3<0$

$\left(0,\frac{1}{2}\right)$: $f(0.1)=8>0$

$\left(\frac{1}{2},\infty\right)$: $f(1)=-1<0$

The solution set for this portion of the inequality is $(-\infty,0)\cup\left(\frac{1}{2},\infty\right)$.

The solution set of the original inequality is
$$\left(\left(-\infty,-\frac{1}{4}\right)\cup(0,\infty)\right)and\left((-\infty,0)\cup\left(\frac{1}{2},\infty\right)\right),$$
or $\left(-\infty,-\frac{1}{4}\right)\cup\left(\frac{1}{2},\infty\right)$.

77.
$$|x^2+3x-1|<3$$
$$-3<x^2+3x-1<3$$
$$-3<x^2+3x-1\quad and\quad x^2+3x-1<3$$

First solve $-3<x^2+3x-1$, or $x^2+3x-1>-3$.
$$x^2+3x-1>-3$$
$$x^2+3x+2>0$$
$$x^2+3x+2=0\qquad \text{Related equation}$$
$$(x+2)(x+1)=0$$
$$x=-2\quad or\quad x=-1$$

Let $f(x)=x^2+3x+2$ and test a value in each of the intervals determined by the solution of the related equation.

$(-\infty,-2)$: $f(-3)=2>0$

$(-2,-1)$: $f(-1.5)=-0.25<0$

$(-1,\infty)$: $f(0)=2>0$

The solution set for this portion of the inequality is $(-\infty,-2)\cup(-1,\infty)$.

Next solve $x^2+3x-1<3$.
$$x^2+3x-1<3$$
$$x^2+3x-4<0$$
$$x^2+3x-4=0\qquad \text{Related equation}$$
$$(x+4)(x-1)=0$$
$$x=-4\quad or\quad x=1$$

Let $f(x)=x^2+3x-4$ and test a value in each of the intervals determined by the solution of the related equation.

$(-\infty,-4)$: $f(-5)=6>0$

$(-4,1)$: $f(0)=-4<0$

$(1,\infty)$: $f(2)=6>0$

The solution set for this portion of the inequality is $(-4,1)$.

The solution set of the original inequality is
$$((-\infty,-2)\cup(-1,\infty))\ and\ (-4,1),\ or$$
$$(-4,-2)\cup(-1,1).$$

79. First find a quadratic equation with solutions -4 and 3.

$$(x + 4)(x - 3) = 0$$
$$x^2 + x - 12 = 0$$

Observe that the graph of $y = x^2 + x - 12$ lies below the x-axis on $(-4, 3)$. Thus, a quadratic inequality for which the solution set is $(-4, 3)$ is $x^2 + x - 12 < 0$. Answers may vary.

Chapter 4

Exponential and Logarithmic Functions

Exercise Set 4.1

1. $(f \circ g)(x) = f(g(x)) = f(x - 3) = x - 3 + 3 = x$
$(g \circ f)(x) = g(f(x)) = g(x + 3) = x + 3 - 3 = x$

3. $(f \circ g)(x) = f(g(x)) = f\left(\dfrac{x + 7}{3}\right) =$

$3\left(\dfrac{x + 7}{3}\right) - 7 = x + 7 - 7 = x$

$(g \circ f)(x) = g(f(x)) = g(3x - 7) = \dfrac{(3x - 7) + 7}{3} =$

$\dfrac{3x}{3} = x$

5. $(f \circ g)(x) = f(g(x)) = f(0.05) = 20$
$(g \circ f)(x) = g(f(x)) = g(20) = 0.05$

7. $(f \circ g)(x) = f(g(x)) = f(x^2 - 5) =$
$\sqrt{x^2 - 5 + 5} = \sqrt{x^2} = |x|$
$(g \circ f)(x) = g(f(x)) = g(\sqrt{x + 5}) =$
$(\sqrt{x + 5})^2 - 5 = x + 5 - 5 = x$

9. $(f \circ g)(x) = f(g(x)) = f\left(\dfrac{1}{1 + x}\right) =$

$\dfrac{1 - \left(\dfrac{1}{1 + x}\right)}{\dfrac{1}{1 + x}} \dfrac{\dfrac{1 + x - 1}{1 + x}}{\dfrac{1}{1 + x}} =$

$\dfrac{x}{1 + x} \cdot \dfrac{1 + x}{1} = x$

$(g \circ f)(x) = g(f(x)) = g\left(\dfrac{1 - x}{x}\right) =$

$\dfrac{1}{1 + \left(\dfrac{1 - x}{x}\right)} = \dfrac{1}{\dfrac{x + 1 - x}{x}} =$

$\dfrac{1}{\dfrac{1}{x}} = 1 \cdot \dfrac{x}{1} = x$

11. $(f \circ g)(x) = f(g(x)) = f(x + 1) =$
$(x + 1)^3 - 5(x + 1)^2 + 3(x + 1) + 7 =$
$x^3 + 3x^2 + 3x + 1 - 5x^2 - 10x - 5 + 3x + 3 + 7 =$
$x^3 - 2x^2 - 4x + 6$

$(g \circ f)(x) = g(f(x)) = g(x^3 - 5x^2 + 3x + 7) =$
$x^3 - 5x^2 + 3x + 7 + 1 = x^3 - 5x^2 + 3x + 8$

13. $h(x) = (4 + 3x)^5$

This is $4 + 3x$ to the 5th power. The most obvious answer is $f(x) = x^5$ and $g(x) = 4 + 3x$.

15. $h(x) = \dfrac{1}{(x - 2)^4}$

This is 1 divided by $(x - 2)$ to the 4th power. One obvious answer is $f(x) = \dfrac{1}{x^4}$ and $g(x) = (x - 2)$.
Another possibility is $f(x) = \dfrac{1}{x}$ and $g(x) = (x - 2)^4$.

17. $f(x) = \dfrac{x - 1}{x + 1}$, $g(x) = x^3$

19. $f(x) = x^6$, $g(x) = \dfrac{2 + x^3}{2 - x^3}$

21. $f(x) = \sqrt{x}$, $g(x) = \dfrac{x - 5}{x + 2}$

23. $f(x) = x^3 - 5x^2 + 3x - 1$, $g(x) = x + 2$

25. $f(x) = y(x) \circ s(x) = y(s(x))$
$f(x) = 2(x - 32 + 12) = 2(x - 20)$

27. We interchange the first and second coordinates of each ordered pair to find the inverse of the relation. It is

$\{(8, 7),\ (8, -2),\ (-4, 3),\ (-8, 8)\}.$

29. We interchange the first and second coordinates of each ordered pair to find the inverse of the relation. It is

$\{(-1, -1),\ (4, -3)\}.$

31. Interchange x and y.

$y = 4x - 5$
$\downarrow \quad \downarrow$
$x = 4y - 5$

33. Interchange x and y.

$$x^3 y = -5$$
$$\downarrow \; \downarrow$$
$$y^3 x = -5$$

35. Graph $x = y^2 - 3$. Some points on the graph are $(-3,0)$, $(-2,-1)$, $(-2,1)$, $(1,-2)$, and $(1,2)$. Plot these points and draw the curves. Then reflect the graph across the line $y = x$.

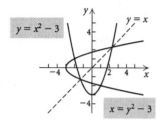

37. Graph $y = |x|$. Some points on the graph are $(0,0)$, $(-2,2)$, $(2,2)$, $(-5,5)$, and $(5,5)$. Plot these points and draw the graph. Then reflect the graph across the line $y = x$.

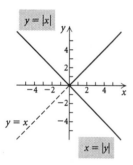

39. The function is one-to-one, because no horizontal line crosses the graph more than once.

41. The function is not one-to-one, because there are many horizontal lines that cross the graph more than once.

43. The function is not one-to-one, because there are many horizontal lines that cross the graph more than once.

45. The function is one-to-one, because no horizontal line crosses the graph more than once.

47. The graph of $f(x) = 5x - 8$ is shown below.

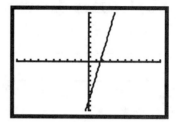

Since there is no horizontal line that crosses the graph more than once, the function is one-to-one.

49. The graph of $f(x) = 1 - x^2$ is shown below.

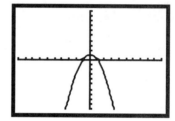

Since there are many horizontal lines that cross the graph more than once, the function is not one-to-one.

51. The graph of $f(x) = |x + 2|$ is shown below.

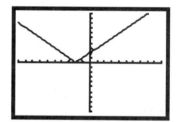

Since there are many horizontal lines that cross the graph more than once, the function is not one-to-one.

53. The graph of $f(x) = -\dfrac{4}{x}$ is shown below.

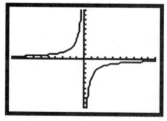

Since there is no horizontal line that crosses the graph more than once, the function is one-to-one.

55. $y_1 = 0.8x + 1.7,$
$$y_2 = \frac{x - 1.7}{0.8}$$

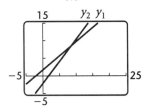

Both the domain and the range of f are the set of all real numbers. Then both the domain and the range of f^{-1} are also the set of all real numbers.

57. $y_1 = \frac{1}{2}x - 4,$
$y_2 = 2x + 8$

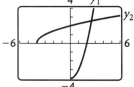

Both the domain and the range of f are the set of all real numbers. Then both the domain and the range of f^{-1} are also the set of all real numbers.

59. $y_1 = \sqrt{x - 3},$
$y_2 = x^2 + 3, x \geqslant 0$

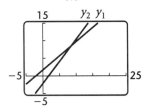

The domain of f is $[3, \infty)$ and the range of f is $[0, \infty)$. Then the domain of f^{-1} is $[0, \infty)$ and the range of f^{-1} is $[3, \infty)$.

61. $y_1 = x^2 - 4, x \geqslant 0; \ y_2 = \sqrt{4 + x}$

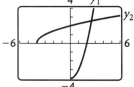

Since it is specified that $x \geq 0$, the domain of f is $[0, \infty)$. The range of f is $[-4, \infty)$. Then the domain of f^{-1} is $[-4, \infty)$ and the range of f^{-1} is $[0, \infty)$.

63. $y_1 = (3x - 9)^3, \ y_2 = \dfrac{\sqrt[3]{x} + 9}{3}$

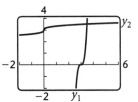

Both the domain and the range of f are the set of all real numbers. Then both the domain and the range of f^{-1} are also the set of all real numbers.

65. a) The graph of $f(x) = x + 4$ is shown below. It passes the horizontal line test, so it is one-to-one.

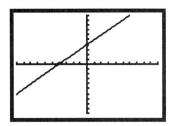

b) Replace $f(x)$ with y: $y = x + 4$

Interchange x and y: $x = y + 4$

Solve for y: $x - 4 = y$

Replace y with $f^{-1}(x)$: $f^{-1}(x) = x - 4$

67. a) The graph of $f(x) = 2x - 1$ is shown below. It passes the horizontal line test, so it is one-to-one.

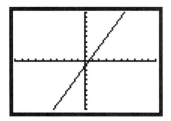

b) Replace $f(x)$ with y: $y = 2x - 1$

Interchange x and y: $x = 2y - 1$

Solve for y: $\dfrac{x + 1}{2} = y$

Replace y with $f^{-1}(x)$: $f^{-1}(x) = \dfrac{x + 1}{2}$

69. a) The graph of $f(x) = \dfrac{4}{x+7}$ is shown below. It passes the horizontal line test, so the function is one-to-one.

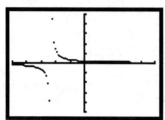

b) Replace $f(x)$ with y: $y = \dfrac{4}{x+7}$

Interchange x and y: $x = \dfrac{4}{y+7}$

Solve for y: $x(y+7) = 4$

$$y + 7 = \frac{4}{x}$$

$$y = \frac{4}{x} - 7$$

Replace y with $f^{-1}(x)$: $f^{-1}(x) = \dfrac{4}{x} - 7$

71. a) The graph of $f(x) = \dfrac{x+4}{x-3}$ is shown below. It passes the horizontal line test, so the function is one-to-one.

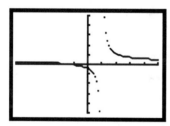

b) Replace $f(x)$ with y: $y = \dfrac{x+4}{x-3}$

Interchange x and y: $x = \dfrac{y+4}{y-3}$

Solve for y: $(y-3)x = y+4$

$$xy - 3x = y + 4$$

$$xy - y = 3x + 4$$

$$y(x-1) = 3x + 4$$

$$y = \frac{3x+4}{x-1}$$

Replace y with $f^{-1}(x)$: $f^{-1}(x) = \dfrac{3x+4}{x-1}$

73. a) The graph of $f(x) = x^3 - 1$ is shown below. It passes the horizontal line test, so the function is one-to-one.

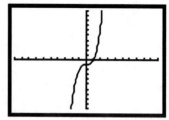

b) Replace $f(x)$ with y: $y = x^3 - 1$

Interchange x and y: $x = y^3 - 1$

Solve for y: $x + 1 = y^3$

$$\sqrt[3]{x+1} = y$$

Replace y with $f^{-1}(x)$: $f^{-1}(x) = \sqrt[3]{x+1}$

75. a) The graph of $f(x) = x\sqrt{4 - x^2}$ is shown below. Since there are many horizontal lines that cross the graph more than once, the function is not one-to-one and thus does not have an inverse that is a function.

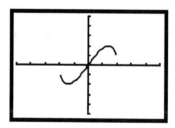

77. a) The graph of $f(x) = 5x^2 - 2$, $x \geq 0$ is shown below. It passes the horizontal line test, so it is one-to-one.

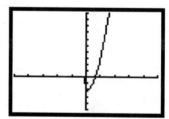

b) Replace $f(x)$ with y: $y = 5x^2 - 2$

Interchange x and y: $x = 5y^2 - 2$

Solve for y: $x + 2 = 5y^2$

$$\frac{x + 2}{5} = y^2$$

$$\sqrt{\frac{x + 2}{5}} = y$$

(We take the principal square root, because $x \geq 0$ in the original equation.)

Replace y with $f^{-1}(x)$: $f^{-1}(x) = \sqrt{\dfrac{x + 2}{5}}$ for all x in the range of $f(x)$, or $f^{-1}(x) = \sqrt{\dfrac{x + 2}{5}}$, $x \geq -2$

79. a) The graph of $f(x) = \sqrt{x + 1}$ is shown below. It passes the horizontal line test, so the function is one-to-one.

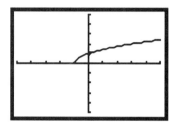

b) Replace $f(x)$ with y: $y = \sqrt{x + 1}$

Interchange x and y: $x = \sqrt{y + 1}$

Solve for y: $x^2 = y + 1$

$$x^2 - 1 = y$$

Replace y with $f^{-1}(x)$: $f^{-1}(x) = x^2 - 1$ for all x in the range of $f(x)$, or $f^{-1}(x) = x^2 - 1$, $x \geq 0$.

81. $f(x) = 3x$

The function f multiplies an input by 3. Then to reverse this procedure, f^{-1} would divide each of its inputs by 3. Thus, $f^{-1}(x) = \dfrac{x}{3}$, or $f^{-1}(x) = \dfrac{1}{3}x$.

83. $f(x) = -x$

The outputs of f are the opposites, or additive inverses, of the inputs. Then the outputs of f^{-1} are the opposites of its inputs. Thus, $f^{-1}(x) = -x$.

85. $f(x) = \sqrt[3]{x - 5}$

The function f subtracts 5 from each input and then takes the cube root of the result. To reverse this procedure, f^{-1} would raise each input to the third power and then add 5 to the result. Thus, $f^{-1}(x) = x^3 + 5$.

87. We find $(f^{-1} \circ f)(x)$ and $(f \circ f^{-1})(x)$ and check to see that each is x.

$$(f^{-1} \circ f)(x) = f^{-1}(f(x)) = f^{-1}\left(\frac{7}{8}x\right) =$$

$$\frac{8}{7}\left(\frac{7}{8}x\right) = x$$

$$(f \circ f^{-1})(x) = f(f^{-1}(x)) = f\left(\frac{8}{7}x\right) = \frac{7}{8}\left(\frac{8}{7}x\right) = x$$

89. We find $(f^{-1} \circ f)(x)$ and $(f \circ f^{-1})(x)$ and check to see that each is x.

$$(f^{-1} \circ f)(x) = f^{-1}(f(x)) = f^{-1}\left(\frac{1 - x}{x}\right) =$$

$$\frac{1}{\dfrac{1 - x}{x} + 1} = \frac{1}{\dfrac{1 - x + x}{x}} = \frac{1}{\dfrac{1}{x}} = x$$

$$(f \circ f^{-1})(x) = f(f^{-1}(x)) = f\left(\frac{1}{x + 1}\right) =$$

$$\frac{1 - \dfrac{1}{x + 1}}{\dfrac{1}{x + 1}} = \frac{\dfrac{x + 1 - 1}{x + 1}}{\dfrac{1}{x + 1}} = \frac{\dfrac{x}{x + 1}}{\dfrac{1}{x + 1}} = x$$

91. Since $f(f^{-1}(x)) = f^{-1}(f(x)) = x$, then $f(f^{-1}(5)) = 5$ and $f^{-1}(f(a)) = a$.

93. a) $g(x) = 2(x + 12)$

$g(6) = 2(6 + 12) = 2 \cdot 18 = 36$

$g(8) = 2(8 + 12) = 2 \cdot 20 = 40$

$g(10) = 2(10 + 12) = 2 \cdot 22 = 44$

$g(14) = 2(14 + 12) = 2 \cdot 26 = 52$

$g(18) = 2(18 + 12) = 2 \cdot 30 = 60$

b) The graph passes the horizontal line test and thus has an inverse that is a function.

Replace $g(x)$ with y: $y = 2(x + 12)$

Interchange x and y: $x = 2(y + 12)$

Solve for y: $\dfrac{x - 24}{2} = y$

Replace y with $g^{-1}(x)$: $g^{-1}(x) = \dfrac{x - 24}{2}$, or $\dfrac{x}{2} - 12$

c) $g^{-1}(36) = \dfrac{36 - 24}{2} = \dfrac{12}{2} = 6$

$g^{-1}(40) = \dfrac{40 - 24}{2} = \dfrac{16}{2} = 8$

$g^{-1}(44) = \dfrac{44 - 24}{2} = \dfrac{20}{2} = 10$

$g^{-1}(52) = \dfrac{52 - 24}{2} = \dfrac{28}{2} = 14$

$g^{-1}(60) = \dfrac{60 - 24}{2} = \dfrac{36}{2} = 18$

95. a) $D(r) = \dfrac{11r + 5}{10}$

$$D(0) = \frac{11 \cdot 0 + 5}{10} = \frac{5}{10} = 0.5$$

$$D(10) = \frac{11 \cdot 10 + 5}{10} = \frac{115}{10} = 11.5$$

$$D(20) = \frac{11 \cdot 20 + 5}{10} = \frac{225}{10} = 22.5$$

$$D(50) = \frac{11 \cdot 50 + 5}{10} = \frac{555}{10} = 55.5$$

$$D(65) = \frac{11 \cdot 65 + 5}{10} = \frac{720}{10} = 72$$

b), d) $\quad y_1 = \dfrac{11x + 5}{10}, \quad y_2 = \dfrac{10x - 5}{11}$

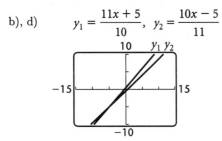

c) Replace $D(r)$ with y: $\quad y = \dfrac{11r + 5}{10}$

Interchange r and y: $\quad r = \dfrac{11y + 5}{10}$

Solve for y: $\qquad 10r = 11y + 5$

$$10r - 5 = 11y$$

$$\frac{10r - 5}{11} = y$$

Replace y with $D^{-1}(r)$: $\quad D^{-1}(r) = \dfrac{10r - 5}{11}$

$D^{-1}(r)$ represents the speed, in miles per hour, that the car is traveling when the reaction distance is r feet.

d) See part (b).

97. Discussion and Writing

99. Graph $y = \dfrac{3}{2}x - 4$.

When $x = 0$, $y = \dfrac{3}{2} \cdot 0 - 4 = 0 - 4 = -4$.

When $x = 2$, $y = \dfrac{3}{2} \cdot 2 - 4 = 3 - 4 = -1$.

When $x = 4$, $y = \dfrac{3}{2} \cdot 4 - 4 = 6 - 4 = 2$.

Plot the points $(0, -4)$, $(2, -1)$, and $(4, 2)$ and connect them with a straight line.

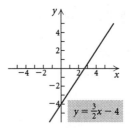

$y = \frac{3}{2}x - 4$

101. Graph $y = \dfrac{x + 3}{x^2 - x - 2}$.

1. The zeros of the denominator are -1 and 2, so the domain is $(-\infty, -1) \cup (-1, 2) \cup (2, \infty)$ and the lines $x = -1$ and $x = 2$ are vertical asymptotes.

2. Because the degree of the numerator is less than the degree of the denominator, the x-axis is the horizontal asymptote. There is no oblique asymptote.

3. The zero of the numerator is -3, so the x-intercept is $(-3, 0)$.

4. When $x = 0$, $y = \dfrac{0 + 3}{0^2 - 0 - 2} = -\dfrac{3}{2}$, so the y-intercept is $\left(0, -\dfrac{3}{2}\right)$.

5. Find other function values to determine the shape of the graph and then draw it.

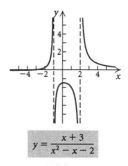

$y = \dfrac{x + 3}{x^2 - x - 2}$

103. Graph $y_1 = f(x)$ and $y_2 = g(x)$ and observe that the graphs are reflections of each other across the line $y = x$. Thus, the functions are inverses of each other.

105. Graph $y_1 = f(x)$ and $y_2 = g(x)$ and observe that the graphs are not reflections of each other across the line $y = x$. Thus, the functions are not inverses of each other.

107. The graph of $f(x) = x^2 - 3$ is a parabola with vertex $(0, -3)$. If we consider x-values such that $x \geq 0$, then the graph is the right-hand side of the parabola and it passes the horizontal line test. We find the inverse of $f(x) = x^2 - 3$, $x \geq 0$.

Replace $f(x)$ with y: $y = x^2 - 3$

Interchange x and y: $x = y^2 - 3$

Solve for y: $x + 3 = y^2$

$$\sqrt{x+3} = y$$

(We take the principal square root, because $x \geq 0$ in the original equation.)

Replace y with $f^{-1}(x)$: $f^{-1}(x) = \sqrt{x+3}$ for all x in the range of $f(x)$, or $f^{-1}(x) = \sqrt{x+3}$, $x \geq -3$.

Answers may vary. There are other restrictions that also make $f(x)$ one-to-one.

109. Answers may vary. $f(x) = \dfrac{3}{x}$, $f(x) = 1 - x$, $f(x) = x$.

Exercise Set 4.2

1. $e^4 \approx 54.5982$

3. $e^{-2.458} \approx 0.0856$

5. Graph $f(x) = 3^x$.

Compute some function values, plot the corresponding points, and connect them with a smooth curve.

x	$y = f(x)$	(x, y)
-3	$\dfrac{1}{27}$	$\left(-3, \dfrac{1}{27}\right)$
-2	$\dfrac{1}{9}$	$\left(-2, \dfrac{1}{9}\right)$
-1	$\dfrac{1}{3}$	$\left(-1, \dfrac{1}{3}\right)$
0	1	$(0, 1)$
1	3	$(1, 3)$
2	9	$(2, 9)$
3	27	$(3, 27)$

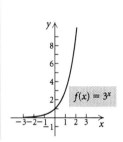

7. Graph $f(x) = 6^x$.

Compute some function values, plot the corresponding points, and connect them with a smooth curve.

x	$y = f(x)$	(x, y)
-3	$\dfrac{1}{216}$	$\left(-3, \dfrac{1}{216}\right)$
-2	$\dfrac{1}{36}$	$\left(-2, \dfrac{1}{36}\right)$
-1	$\dfrac{1}{6}$	$\left(-1, \dfrac{1}{6}\right)$
0	1	$(0, 1)$
1	6	$(1, 6)$
2	36	$(2, 36)$
3	216	$(3, 216)$

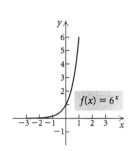

9. Graph $f(x) = \left(\dfrac{1}{4}\right)^x$.

Compute some function values, plot the corresponding points, and connect them with a smooth curve.

x	$y = f(x)$	(x, y)
-3	64	$(-3, 64)$
-2	16	$(-2, 16)$
-1	4	$(-1, 4)$
0	1	$(0, 1)$
1	$\dfrac{1}{4}$	$\left(1, \dfrac{1}{4}\right)$
2	$\dfrac{1}{16}$	$\left(2, \dfrac{1}{16}\right)$
3	$\dfrac{1}{64}$	$\left(3, \dfrac{1}{64}\right)$

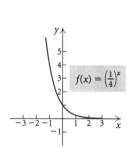

11. Graph $x = 3^y$.

Choose values for y and compute the corresponding x-values. Plot the points (x, y) and connect them with a smooth curve.

x	y	(x, y)
$\dfrac{1}{27}$	-3	$\left(\dfrac{1}{27}, -3\right)$
$\dfrac{1}{9}$	-2	$\left(\dfrac{1}{9}, -2\right)$
$\dfrac{1}{3}$	-1	$\left(\dfrac{1}{3}, -1\right)$
1	0	$(1, 0)$
3	1	$(3, 1)$
9	2	$(9, 2)$
27	3	$(27, 3)$

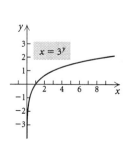

13. Graph $x = \left(\dfrac{1}{2}\right)^y$.

Choose values for y and compute the corresponding x-values. Plot the points (x,y) and connect them with a smooth curve.

x	y	(x,y)
8	-3	$(8,-3)$
4	-2	$(4,-2)$
2	-1	$(2,-1)$
1	0	$(1,0)$
$\dfrac{1}{2}$	1	$\left(\dfrac{1}{2},1\right)$
$\dfrac{1}{4}$	2	$\left(\dfrac{1}{4},2\right)$
$\dfrac{1}{8}$	3	$\left(\dfrac{1}{8},3\right)$

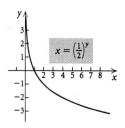

15. Graph $y = \dfrac{1}{4}e^x$.

Choose values for x and compute the corresponding y-values. Plot the points (x,y) and connect them with a smooth curve.

x	y	(x,y)
-3	0.0124	$(-3,0.0124)$
-2	0.0338	$(-2,0.0338)$
-1	0.0920	$(-1,0.0920)$
0	0.25	$(0,0.25)$
1	0.6796	$(1,0.6796)$
2	1.8473	$(2,1.8473)$
3	5.0214	$(3,5.0214)$

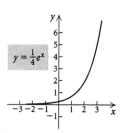

17. Graph $f(x) = 1 - e^{-x}$.

Compute some function values, plot the corresponding points, and connect them with a smooth curve.

x	y	(x,y)
-3	-19.0855	$(-3,-19.0855)$
-2	-6.3891	$(-2,-6.3891)$
-1	-1.7183	$(-1,-1.7183)$
0	0	$(0,0)$
1	0.6321	$(1,0.6321)$
2	0.8647	$(2,0.8647)$
3	0.9502	$(3,0.9502)$

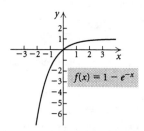

$f(x) = 1 - e^{-x}$

19. Shift the graph of $y = 2^x$ left 1 unit.

$y = 2^{x+1}$

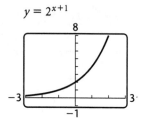

21. Shift the graph of $y = 2^x$ down 3 units.

$y = 2^x - 3$

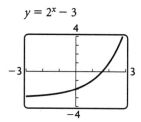

23. Reflect the graph of $y = 3^x$ across the y-axis, then across the x-axis, and then shift it up 4 units.

$y = 4 - 3^{-x}$

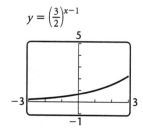

25. Shift the graph of $y = \left(\dfrac{3}{2}\right)^x$ right 1 unit.

$y = \left(\dfrac{3}{2}\right)^{x-1}$

27. Shift the graph of $y = 2^x$ left 3 units and down 5 units.

$$y = 2^{x+3} - 5$$

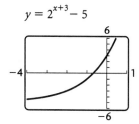

29. Shrink the graph of $y = e^x$ horizontally.

$$y = e^{2x}$$

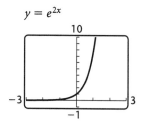

31. Shift the graph of $y = e^x$ left 1 unit and reflect it across the y-axis.

$$y = e^{-x+1}$$

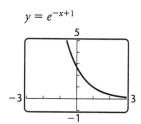

33. Reflect the graph of $y = e^x$ across the y-axis and then across the x-axis; shift it up 1 unit and then stretch it vertically.

$$y = 2(1 - e^{-x}), x > 0$$

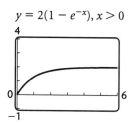

35. a) In 1998, $t = 1998 - 1992 = 6$.

   $N(6) = 274,390(1.195)^6 \approx 799,053$

   b) In 2001, $t = 2001 - 1992 = 9$.

   $N(9) = 274,390(1.195)^9 \approx 1,363,576$

c) $y = 274,390(1.195)^t$

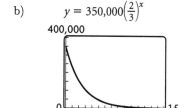

d) Using Intersect feature, we find that the first coordinate of the point of intersection of $y_1 = 274,390(1.195)^x$ and $y_2 = 3,000,000$ is about 13. Thus, it will take about 13 years for the number of AIDS cases reported in the United States to reach 3 million.

We could also graph $y = 274,390(1.195)^x - 3,000,000$ and use the Zero method to find the result.

37. a) $N(0) = 350,000\left(\dfrac{2}{3}\right)^0 = 350,000$

   $N(1) = 350,000\left(\dfrac{2}{3}\right)^1 \approx 233,333$

   $N(4) = 350,000\left(\dfrac{2}{3}\right)^4 \approx 69,136$

   $N(10) = 350,000\left(\dfrac{2}{3}\right)^{10} \approx 6070$

b) $y = 350,000\left(\dfrac{2}{3}\right)^x$

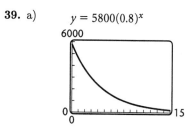

c) Using the Intersect feature, we find that the first coordinate of the point of intersection of $y_1 = 350,000\left(\dfrac{2}{3}\right)^x$ and $y_2 = 2000$ is about 13. Thus, 2000 cans will still be in use after about 13 years.

39. a) $y = 5800(0.8)^x$

6000

0 ——————— 15
0

b) $V(0) = 5800(0.8)^0 = \$5800$

$V(1) = 5800(0.8)^1 = \$4640$

$V(2) = 5800(0.8)^2 = \$3712$

$V(5) = 5800(0.8)^5 \approx \$1900.54$

$V(10) = 5800(0.8)^{10} \approx \$622.77$

c) Using the Intersect feature, we find that the first coordinate of the point of intersection of $y_1 = 5800(0.8)^x$ and $y_2 = 500$ is about 11. Thus, the machine will be replaced after about 11 years.

41. a) $y = 46.6(1.018)^x$

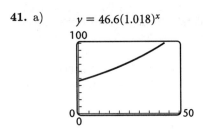

b) In 2000, $t = 2000 - 1981 = 19$.

$N(19) = 46.6(1.018)^{19} \approx 65.4$ billion cubic feet

In 2010, $t = 2010 - 1981 = 29$.

$N(29) = 46.6(1.018)^{29} \approx 78.2$ billion cubic feet

c) Using the Intersect feature, we find that the first coordinate of the point of intersection of $y_1 = 46.6(1.018)^x$ and $y_2 = 93.4$ is about 39. Then the demand for timber will be 93.4 billion cubic feet after about 39 years.

43. a) $y = 100(1 - e^{-0.04x})$

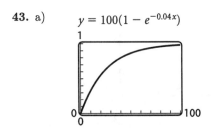

b) $f(25) = 100(1 - e^{-0.04(25)}) \approx 63\%$.

c) $90 = 100(1 - e^{-0.04t})$

$0.9 = 1 - e^{-0.04t}$

$-0.1 = -e^{-0.04t}$

$0.1 = e^{-0.04t}$

$\ln 0.1 = \ln e^{-0.04t}$

$\ln 0.1 = -0.04t$

$\dfrac{\ln 0.01}{-0.04} = t$

$58 \approx t$

After about 58 days, 90% of the target market will have bought the product.

45. Graph (c) is the graph of $y = 3^x - 3^{-x}$.

47. Graph (a) is the graph of $f(x) = -2.3^x$.

49. Graph (l) is the graph of $y = 2^{-|x|}$.

51. Graph (g) is the graph of $f(x) = (0.58)^x - 1$.

53. Graph (i) is the graph of $g(x) = e^{|x|}$.

55. Graph (k) is the graph of $y = 2^{-x^2}$.

57. Graph (m) is the graph of $g(x) = \dfrac{e^x - e^{-x}}{2}$.

59. Graph $y_1 = |1 - 3^x|$ and $y_2 = 4 + 3^{-x^2}$. Use the Intersect feature to find their point of intersection, $(1.481, 4.090)$.

61. Graph $y_1 = 2e^x - 3$ and $y_2 = \dfrac{e^x}{x}$. Use the Intersect feature to find their points of intersection, $(-0.402, -1.662)$ and $(1.051, 2.722)$.

63. Graph $y_1 = 5.3^x - 4.2^x$ and $y_2 = 1073$. Use the Intersect feature to find the first coordinate of their point of intersection. The solution is 4.448.

65. Graph $y_1 = 2^x$ and $y_2 = 1$. Use the Intersect feature to find their point of intersection, $(0, 1)$. Note that the graph of y_1 lies above the graph of y_2 for all points to the right of this point. Thus the solution set is $(0, \infty)$.

67. Graph $y_1 = 2^x + 3^x$ and $y_2 = x^2 + x^3$. Use the Intersect feature to find the first coordinates of their points of intersection. The solutions are 2.294 and 3.228.

69. Discussion and Writing

71. Discussion and Writing

73. Graph $y = x^4 - 5x + 3$ and use the Zero feature twice. The real-number zeros are approximately 0.63188461 and 1.4252375.

75. Graph $y = x^3 + 6x^2 - 16x$ and use the Zero feature three times. The solutions are -8, 0, and 2.

77. Scrolling through the table set in AUTO mode, we see that function values approach 1 as x increases. Similarly, we could set the table in ASK mode and enter increasing values of x. In either case, we see that the horizontal asymptote is $y = 1$.

79. a) $y = e^{-x^2}$

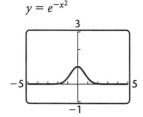

b) There are no x-intercepts, so the function has no zeros.

c) Relative minimum: none;

relative maximum: 1 at $x = 0$

Exercise Set 4.3

1. Graph $y = \log_3 x$.

The equation $y = \log_3 x$ is equivalent to $x = 3^y$. We can find ordered pairs that are solutions by choosing values for y and computing the corresponding x-values.

For $y = -2$, $x = 3^{-2} = \dfrac{1}{9}$.

For $y = -1$, $x = 3^{-1} = \dfrac{1}{3}$.

For $y = 0$, $x = 3^0 = 1$.

For $y = 1$, $x = 3^1 = 3$.

For $y = 2$, $x = 3^2 = 9$.

x, or 3^y	y
$\dfrac{1}{9}$	-2
$\dfrac{1}{3}$	-1
1	0
3	1
9	2

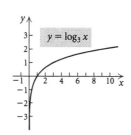

3. Graph $f(x) = \log x$.

Think of $f(x)$ as y. The equation $y = \log x$ is equivalent to $x = 10^y$. We can find ordered pairs that are solutions by choosing values for y and computing the corresponding x-values.

For $y = -2$, $x = 10^{-2} = 0.01$.

For $y = -1$, $x = 10^{-1} = 0.1$.

For $y = 0$, $x = 10^0 = 1$.

For $y = 1$, $x = 10^1 = 10$.

For $y = 2$, $x = 10^2 = 100$.

x, or 10^y	y
0.01	-2
0.1	-1
1	0
10	1
100	2

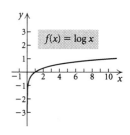

5. $\log_2 16 = 4$ because the exponent to which we raise 2 to get 16 is 4.

7. $\log_5 125 = 3$, because the exponent to which we raise 5 to get 125 is 3.

9. $\log 0.001 = -3$, because the exponent to which we raise 10 to get 0.001 is -3.

11. $\log_2 \dfrac{1}{4} = -2$, because the exponent to which we raise 2 to get $\dfrac{1}{4}$ is -2.

13. $\ln 1 = 0$, because the exponent to which we raise e to get 1 is 0.

15. $\log 10 = 1$, because the exponent to which we raise 10 to get 10 is 1.

17.

The exponent is the logarithm.

$10^3 = 1000 \Rightarrow 3 = \log_{10} 1000$

The base remains the same.

19.

The exponent is the logarithm.

$8^{1/3} = 2 \Rightarrow \log_8 2 = \dfrac{1}{3}$

The base remains the same.

21. $e^3 = t \Rightarrow \log_e t = 3$

23. $e^2 = 7.3891 \Rightarrow \log_e 7.3891 = 2$

25. $p^k = 3 \Rightarrow \log_p 3 = k$

27.

The logarithm is the exponent.

$\log_5 5 = 1 \Rightarrow 5^1 = 5$

The base remains the same.

29. $\log 0.01 = -2$ is equivalent to $\log_{10} 0.01 = -2$.

The logarithm is the exponent.

$$\log_{10} 0.01 = -2 \Rightarrow 10^{-2} = 0.01$$

The base remains the same.

31. $\ln 30 = 3.4012 \Rightarrow e^{3.4012} = 30$

33. $\log_a M = -x \Rightarrow a^{-x} = M$

35. $\log_a T^3 = x \Rightarrow a^x = T^3$

37. $\log 3 \approx 0.4771$

39. $\log 532 \approx 2.7259$

41. $\log 0.57 \approx -0.2441$

43. $\log(-2)$ does not exist. (The grapher gives an error message.)

45. $\ln 2 \approx 0.6931$

47. $\ln 809.3 \approx 6.6962$

49. $\ln(-1.32)$ does not exist. (The grapher gives an error message.)

51. Let $a = 10$, $b = 4$, and $M = 100$ and substitute in the change-of-base formula.

$$\log_4 100 = \frac{\log_{10} 100}{\log_{10} 4}$$
$$\approx \frac{2}{0.602060}$$
$$\approx 3.3219$$

53. Let $a = e$, $b = 100$, and $M = 0.3$ and substitute in the change-of-base formula.

$$\log_{100} 0.3 = \frac{\ln 0.3}{\ln 100}$$
$$\approx \frac{-1.203973}{4.605170}$$
$$\approx -0.2614$$

55. Let $a = 10$, $b = 200$, and $M = 50$ and substitute in the change-of-base formula.

$$\log_{200} 50 = \frac{\log_{10} 50}{\log_{10} 200}$$
$$\approx \frac{1.698970}{2.301030}$$
$$\approx 0.7384$$

57. Shift the graph of $y = \log_2 x$ left 3 units. To use the grapher, we must first change the base. Here we change from base 2 to base 10. We get $y = \dfrac{\log(x+3)}{\log 2}$, using $\log_b M = \dfrac{\log_a M}{\log_a b}$ with $b = 2$, $M = x + 3$, and $a = 10$.

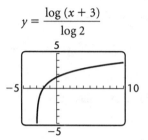

$$y = \frac{\log(x+3)}{\log 2}$$

Domain: $(-3, \infty)$

Vertical asymptote: $x = -3$

59. Shift the graph of $y = \log_3 x$ down 1 unit. To use the grapher, we must first change the base. Here we change from base 3 to base 10. We get $y = \dfrac{\log x}{\log 3} - 1$, using $\log_b M = \dfrac{\log_a M}{\log_a b}$ with $b = 3$, $M = x$, and $a = 10$.

$$y = \frac{\log x}{\log 3} - 1$$

Domain: $(0, \infty)$

Vertical asymptote: $x = 0$

61. Stretch the graph of $y = \ln x$ vertically.

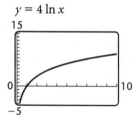

$$y = 4 \ln x$$

Domain: $(0, \infty)$

Vertical asymptote: $x = 0$

63. Reflect the graph of $y = \ln x$ across the x-axis and then shift it up 2 units.

$y = 2 - \ln x$

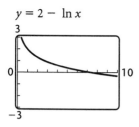

Domain: $(0, \infty)$

Vertical asymptote: $x = 0$

65. Graph $y_1 = 3^x$ and $y_2 = \log_3 x = \dfrac{\log x}{\log 3}$ in the same window, or graph $y = 3^x$ and use the grapher feature that automatically graphs inverses to graph $y = \log_3 x$. On a hand-drawn graph, we can graph $y = 3^x$ and then reflect this graph across the line $y = x$ to get the graph of $y = \log_3 x$.

$$y_1 = 3^x,$$
$$y_2 = \frac{\log x}{\log 3}$$

67. Graph $y_1 = \log x$ and $y_2 = 10^x$ in the same window, or graph $y = \log x$ and use the grapher feature that automatically graphs inverses to graph $y = 10^x$. On a hand-drawn graph, we can graph $y = \log x$ and then reflect this graph across the line $y = x$ to get the graph of $y = 10^x$.

$$y_1 = \log x,$$
$$y_2 = 10^x$$

69. a) We substitute 419.681 for P, since P is in thousands.

$$w(419.681) = 0.37 \ln 419.681 + 0.05$$
$$\approx 2.3 \text{ ft/sec}$$

b) We substitute 2721.547 for P, since P is in thousands.

$$w(2721.547) = 0.37 \ln 2721.547 + 0.05$$
$$\approx 3.0 \text{ ft/sec}$$

c) We substitute 350.363 for P, since P is in thousands.

$$w(350.363) = 0.37 \ln 350.363 + 0.05$$
$$\approx 2.2 \text{ ft/sec}$$

d) We substitute 149.799 for P, since P is in thousands.

$$w(149.799) = 0.37 \ln 149.799.4 + 0.05$$
$$\approx 1.9 \text{ ft/sec}$$

e) We substitute 94.466 for P, since P is in thousands.

$$w(94.466) = 0.37 \ln 94.466 + 0.05$$
$$\approx 1.7 \text{ ft/sec}$$

71. a) $S(0) = 78 - 15 \log(0 + 1)$
$$= 78 - 15 \log 1$$
$$= 78 - 15 \cdot 0$$
$$= 78\%$$

b) $S(4) = 78 - 15 \log(4 + 1)$
$$= 78 - 15 \log 5$$
$$\approx 78 - 15(0.698970)$$
$$\approx 67.5\%$$

$S(24) = 78 - 15 \log(24 + 1)$
$$= 78 - 15 \log 25$$
$$\approx 78 - 15(1.397940)$$
$$\approx 57\%$$

c) $y = 78 - 15 \log(x + 1), x \geq 0$

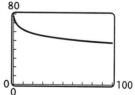

d) $50 = 78 - 15 \log(x + 1)$
$$-28 = -15 \log(x + 1) \quad \text{Subtracting 78}$$
$$\frac{28}{15} = \log(x + 1) \quad \text{Dividing by } -15$$
$$x + 1 = 10^{28/15} \quad \text{Using the definition of logarithm}$$
$$x = 10^{28/15} - 1$$
$$x \approx 73 \text{ months}$$

73. a) $7 = -\log[H^+]$

$-7 = \log[H^+]$

$H^+ = 10^{-7}$ Using the definition of logarithm

b) $5.4 = -\log[H^+]$

$-5.4 = \log[H^+]$

$H^+ = 10^{-5.4}$ Using the definition of logarithm

$H^+ \approx 4.0 \times 10^{-6}$

c) $3.2 = -\log[H^+]$

$-3.2 = \log[H^+]$

$H^+ = 10^{-3.2}$ Using the definition of logarithm

$H^+ \approx 6.3 \times 10^{-4}$

d) $4.8 = -\log[H^+]$

$-4.8 = \log[H^+]$

$H^+ = 10^{-4.8}$ Using the definition of logarithm

$H^+ \approx 1.6 \times 10^{-5}$

75. a) $L = 10 \log \dfrac{2510 \cdot I_0}{I_0}$

$= 10 \log 2510$

$\approx 34 \text{ decibels}$

b) $L = 10 \log \dfrac{2,500,000 \cdot I_0}{I_0}$

$= 10 \log 2,500,000$

$\approx 64 \text{ decibels}$

c) $L = 10 \log \dfrac{10^6 \cdot I_0}{I_0}$

$= 10 \log 10^6$

$= 60 \text{ decibels}$

d) $L = 10 \log \dfrac{10^9 \cdot I_0}{I_0}$

$= 10 \log 10^9$

$= 90 \text{ decibels}$

77. Discussion adn Writing

79.

$$\begin{array}{r|rrrr} -5 & 1 & -6 & 3 & 10 \\ & & -5 & 55 & -290 \\ \hline & 1 & -11 & 58 & -280 \end{array}$$

The remainder is -280, so $f(-5) = -280$.

81. $f(x) = (x - \sqrt{7})(x + \sqrt{7})(x - 0)$

$= (x^2 - 7)(x)$

$= x^3 - 7x$

83. Using the change-of-base formula, we get

$\dfrac{\log_5 8}{\log_2 8} = \log_2 8 = 3.$

85. $f(x) = \log_5 x^3$

x^3 must be positive. Since $x^3 > 0$ for $x > 0$, the domain is $(0, \infty)$.

87. $f(x) = \ln|x|$

$|x|$ must be positive. Since $|x| > 0$ for $x \neq 0$, the domain is $(-\infty, 0) \cup (0, \infty)$.

89. Graph $y = \log_2(2x + 5) = \dfrac{\log(2x + 5)}{\log 2}$. Observe that outputs are negative for inputs between $-\dfrac{5}{2}$ and -2. Thus, the solution set is $\left(-\dfrac{5}{2}, -2\right)$.

91. Graph (d) is the graph of $f(x) = \ln|x|$.

93. Graph (b) is the graph of $f(x) = \ln x^2$.

95. a) $y = x \ln x$

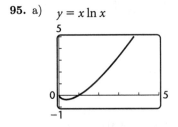

b) Use the Zero feature. The zero is 1.

c) Use the Minimum feature. The relative minimum is -0.368 at $x = 0.368$. There is no relative maximum.

97. a) $y = \dfrac{\ln x}{x^2}$

b) Use the Zero feature. The zero is 1.

c) Use the Maximum feature. There is no relative minimum. The relative maximum is 0.184 at $x = 1.649$.

Exercise Set 4.4

1. Use the product rule.

$\log_3(81 \cdot 27) = \log_3 81 + \log_3 27$

3. Use the product rule.

$\log_5(5 \cdot 125) = \log_5 5 + \log_5 125$

5. Use the product rule.

$\log_t 8Y = \log_t 8 + \log_t Y$

7. Use the power rule.

$\log_b t^3 = 3\log_b t$

9. Use the power rule.

$\log y^8 = 8\log y$

11. Use the power rule.

$\log_c K^{-6} = -6\log_c K$

13. Use the difference rule.

$\log_t \dfrac{M}{8} = \log_t M - \log_t 8$

15. Use the difference rule.

$\log_a \dfrac{x}{y} = \log_a x - \log_a y$

17. $\quad \log_a 6xy^5 z^4$

$= \log_a 6 + \log_a x + \log_a y^5 + \log_a z^4$

$\qquad\qquad\qquad\qquad\qquad$ Product rule

$= \log_a 6 + \log_a x + 5\log_a y + 4\log_a z$

$\qquad\qquad\qquad\qquad\qquad$ Power rule

19. $\quad \log_b \dfrac{p^2 q^5}{m^4 b^9}$

$= \log_b p^2 q^5 - \log_b m^4 b^9 \quad$ Quotient rule

$= \log_b p^2 + \log_b q^5 - (\log_b m^4 + \log_b b^9)$

$\qquad\qquad\qquad\qquad\qquad$ Product rule

$= \log_b p^2 + \log_b q^5 - \log_b m^4 - \log_b b^9$

$= \log_b p^2 + \log_b q^5 - \log_b m^4 - 9 \quad (\log_b b^9 = 9)$

$= 2\log_b p + 5\log_b q - 4\log_b m - 9 \quad$ Power rule

21. $\quad \dfrac{1}{2}\log_a \dfrac{x^6}{p^5 q^8}$

$= \dfrac{1}{2}[\log_a x^6 - \log_a(p^5 q^8)] \qquad$ Quotient rule

$= \dfrac{1}{2}[\log_a x^6 - (\log_a p^5 + \log_a q^8)] \quad$ Product rule

$= \dfrac{1}{2}(\log_a x^6 - \log_a p^5 - \log_a q^8)$

$= \dfrac{1}{2}(6\log_a x - 5\log_a p - 8\log_a q) \quad$ Power rule

$= 3\log_a x - \dfrac{5}{2}\log_a p - 4\log_a q$

23. $\quad \log_a \sqrt[4]{\dfrac{m^8 n^{12}}{a^3 b^5}}$

$= \dfrac{1}{4}\log_a \dfrac{m^8 n^{12}}{a^3 b^5} \qquad\qquad$ Power rule

$= \dfrac{1}{4}(\log_a m^8 n^{12} - \log_a a^3 b^5) \quad$ Quotient rule

$= \dfrac{1}{4}[\log_a m^8 + \log_a n^{12} - (\log_a a^3 + \log_a b^5)]$

$\qquad\qquad\qquad\qquad\qquad$ Product rule

$= \dfrac{1}{4}(\log_a m^8 + \log_a n^{12} - \log_a a^3 - \log_a b^5)$

$= \dfrac{1}{4}(\log_a m^8 + \log_a n^{12} - 3 - \log_a b^5)$

$\qquad\qquad\qquad\qquad\qquad (\log_a a^3 = 3)$

$= \dfrac{1}{4}(8\log_a m + 12\log_a n - 3 - 5\log_a b)$

$\qquad\qquad\qquad\qquad\qquad$ Power rule

$= 2\log_a m + 3\log_a n - \dfrac{3}{4} - \dfrac{5}{4}\log_a b$

25. $\quad \log_a 75 + \log_a 2$

$= \log_a(75 \cdot 2) \qquad$ Product rule

$= \log_a 150$

27. $\quad \log 10,000 - \log 100$

$= \log \dfrac{10,000}{100} \qquad$ Quotient rule

$= \log 100$

$= 2$

29. $\quad \dfrac{1}{2}\log_a x + 4\log_a y - 3\log_a x$

$= \log_a x^{1/2} + \log_a y^4 - \log_a x^3 \quad$ Power rule

$= \log_a x^{1/2} y^4 - \log_a x^3 \qquad$ Product rule

$= \log_a \dfrac{x^{1/2} y^4}{x^3} \qquad\qquad$ Quotient rule

$= \log_a x^{-5/2} y^4, \text{ or } \log_a \dfrac{y^4}{x^{5/2}} \quad$ Simplifying

31. $\ln x^2 - 2\ln\sqrt{x}$

$= \ln x^2 - \ln(\sqrt{x})^2$ Power rule

$= \ln x^2 - \ln x$ $[(\sqrt{x})^2 = x]$

$= \ln\dfrac{x^2}{x}$ Quotient rule

$= \ln x$

33. $\ln(x^2 - 4) - \ln(x + 2)$

$= \ln\dfrac{x^2 - 4}{x + 2}$ Quotient rule

$= \ln\dfrac{(x + 2)(x - 2)}{x + 2}$ Factoring

$= \ln(x - 2)$ Removing a factor of 1

35. $\ln x - 3[\ln(x - 5) + \ln(x + 5)]$

$= \ln x - 3\ln[(x - 5)(x + 5)]$ Product rule

$= \ln x - 3\ln(x^2 - 25)$

$= \ln x - \ln(x^2 - 25)^3$ Power rule

$= \ln\dfrac{x}{(x^2 - 25)^3}$ Quotient rule

37. $\dfrac{3}{2}\ln 4x^6 - \dfrac{4}{5}\ln 2y^{10}$

$= \dfrac{3}{2}\ln 2^2 x^6 - \dfrac{4}{5}\ln 2y^{10}$ Writing 4 as 2^2

$= \ln(2^2 x^6)^{3/2} - \ln(2y^{10})^{4/5}$ Power rule

$= \ln(2^3 x^9) - \ln(2^{4/5}y^8)$

$= \ln\dfrac{2^3 x^9}{2^{4/5}y^8}$ Quotient rule

$= \ln\dfrac{2^{11/5} x^9}{y^8}$

39. $\log_b\dfrac{3}{5} = \log_b 3 - \log_b 5$ Quotient rule

$= 1.0986 - 1.6094$

$= -0.5108$

41. $\log_b\dfrac{1}{5} = \log_b 1 - \log_b 5$ Quotient rule

$= 0 - 1.6094$ $(\log_b 1 = 0)$

$= -1.6094$

43. $\log_b\sqrt{b} = \log_b b^{1/2}$

$= \dfrac{1}{2}\log_b b$ Power rule

$= \dfrac{1}{2}\cdot 1$ $(\log_b b = 1)$

$= \dfrac{1}{2}$

45. $\log_b 5b = \log_b 5 + \log_b b$ Product rule

$= 1.6094 + 1$ $(\log_b b = 1)$

$= 2.6094$

47. $\log_b 75 = \log_b(3 \cdot 5^2)$

$= \log_b 3 + \log_b 5^2$ Product rule

$= \log_b 3 + 2\log_b 5$ Power rule

$= 1.0986 + 2(1.6094)$

$= 4.3174$

49. $\log_p p^3 = 3$ $(\log_a a^x = x)$

51. $\log_e e^{|x-4|} = |x - 4|$ $(\log_a a^x = x)$

53. $3^{\log_3 4x} = 4x$ $(a^{\log_a x} = x)$

55. $10^{\log w} = w$ $(a^{\log_a x} = x)$

57. $\ln e^{8t} = 8t$ $(\log_a a^x = x)$

59. Discussion and Writing

61. $(1 - 4i)(7 + 6i) = 7 + 6i - 28i - 24i^2$

$= 7 + 6i - 28i + 24$

$= 31 - 22i$

63. Graph $y = 2x^2 - 13x - 7$ and use the Zero feature twice. The x-intercepts are $(-0.5, 0)$ and $(7, 0)$. The zeros are -0.5 and 7.

65. $5^{\log_5 8} = 2x$

$8 = 2x$ $(a^{\log_a x} = x)$

$4 = x$

The solution is 4.

67. $\log_a(x^2 + xy + y^2) + \log_a(x - y)$

$= \log_a[(x^2 + xy + y^2)(x - y)]$ Product rule

$= \log_a(x^3 - y^3)$ Multiplying

69. $\log_a\dfrac{x - y}{\sqrt{x^2 - y^2}}$

$= \log_a\dfrac{x - y}{(x^2 - y^2)^{1/2}}$

$= \log_a(x - y) - \log_a(x^2 - y^2)^{1/2}$ Quotient rule

$= \log_a(x - y) - \dfrac{1}{2}\log_a(x^2 - y^2)$ Power rule

$= \log_a(x - y) - \dfrac{1}{2}\log_a[(x + y)(x - y)]$

$= \log_a(x - y) - \dfrac{1}{2}[\log_a(x + y) + \log_a(x - y)]$

Product rule

$= \log_a(x - y) - \dfrac{1}{2}\log_a(x + y) - \dfrac{1}{2}\log_a(x - y)$

$= \dfrac{1}{2}\log_a(x - y) - \dfrac{1}{2}\log_a(x + y)$

71.

$$\log_a \frac{\sqrt[4]{y^2 z^5}}{\sqrt[4]{x^3 z^{-2}}}$$

$$= \log_a \sqrt[4]{\frac{y^2 z^5}{x^3 z^{-2}}}$$

$$= \log_a \sqrt[4]{\frac{y^2 z^7}{x^3}}$$

$$= \log_a \left(\frac{y^2 z^7}{x^3}\right)^{1/4}$$

$$= \frac{1}{4} \log_a \left(\frac{y^2 z^7}{x^3}\right) \qquad \text{Power rule}$$

$$= \frac{1}{4}(\log_a y^2 z^7 - \log_a x^3) \qquad \text{Quotient rule}$$

$$= \frac{1}{4}(\log_a y^2 + \log_a z^7 - \log_a x^3) \quad \text{Product rule}$$

$$= \frac{1}{4}(2\log_a y + 7\log_a z - 3\log_a x) \quad \text{Power rule}$$

$$= \frac{1}{4}(2 \cdot 3 + 7 \cdot 4 - 3 \cdot 2)$$

$$= \frac{1}{4} \cdot 28$$

$$= 7$$

73. $\log_a M - \log_a N = \log_a \dfrac{M}{N}$

This is the quotient rule, so it is true.

75. $\dfrac{\log_a M}{x} = \dfrac{1}{x}\log_a M = \log_a M^{1/x}$. The statement is true by the power rule.

77. $\log_a 8x = \log_a 8 + \log_a x = \log_a x + \log_a 8$. The statement is true by the product rule and the commutative property of addition.

79. $\log_a \left(\dfrac{1}{x}\right) = \log_a x^{-1} = -1 \cdot \log_a x = -1 \cdot 2 = -2$

81. We use the change-of-base formula.

$$\log_{10} 11 \cdot \log_{11} 12 \cdot \log_{12} 13 \cdots$$
$$\log_{998} 999 \cdot \log_{999} 1000$$

$$= \log_{10} 11 \cdot \frac{\log_{10} 12}{\log_{10} 11} \cdot \frac{\log_{10} 13}{\log_{10} 12} \cdots$$
$$\frac{\log_{10} 999}{\log_{10} 998} \cdot \frac{\log_{10} 1000}{\log_{10} 999}$$

$$= \frac{\log_{10} 11}{\log_{10} 11} \cdot \frac{\log_{10} 12}{\log_{10} 12} \cdots \frac{\log_{10} 999}{\log_{10} 999} \cdot \log_{10} 1000$$

$$= \log_{10} 1000$$

$$= 3$$

83.

$$\log_a \left(\frac{x + \sqrt{x^2 - 5}}{5}\right)$$

$$= \log_a \left(\frac{x + \sqrt{x^2 - 5}}{5} \cdot \frac{x - \sqrt{x^2 - 5}}{x - \sqrt{x^2 - 5}}\right)$$

$$= \log_a \left(\frac{5}{5(x - \sqrt{x^2 - 5})}\right) = \log_a \left(\frac{1}{x - \sqrt{x^2 - 5}}\right)$$

$$= \log_a 1 - \log_a(x - \sqrt{x^2 - 5})$$

$$= -\log_a(x - \sqrt{x^2 - 5})$$

Exercise Set 4.5

1. $3^x = 81$

$\quad 3^x = 3^4$

$\quad\quad x = 4 \qquad$ The exponents are the same.

The solution is 4.

3. $2^{2x} = 8$

$\quad 2^{2x} = 2^3$

$\quad\quad 2x = 3 \qquad$ The exponents are the same.

$\quad\quad\quad x = \dfrac{3}{2}$

The solution is $\dfrac{3}{2}$.

5. $\quad 2^x = 33$

$\quad \log 2^x = \log 33 \qquad$ Taking the common logarithm on both sides

$\quad x \log 2 = \log 33 \qquad$ Power rule

$\quad\quad\quad x = \dfrac{\log 33}{\log 2}$

$\quad\quad\quad x \approx \dfrac{1.5185}{0.3010}$

$\quad\quad\quad x \approx 5.044$

The solution is 5.044.

7. $\quad 5^{4x-7} = 125$

$\quad\quad 5^{4x-7} = 5^3$

$\quad\quad 4x - 7 = 3$

$\quad\quad\quad 4x = 10$

$\quad\quad\quad\quad x = \dfrac{10}{4} = \dfrac{5}{2}$

The solution is $\dfrac{5}{2}$.

9. $27 = 3^{5x} \cdot 9^{x^2}$

$3^3 = 3^{5x} \cdot (3^2)^{x^2}$

$3^3 = 3^{5x} \cdot 3^{2x^2}$

$3^3 = 3^{5x+2x^2}$

$3 = 5x + 2x^2$

$0 = 2x^2 + 5x - 3$

$0 = (2x - 1)(x + 3)$

$x = \dfrac{1}{2} \quad or \quad x = -3$

The solutions are -3 and $\dfrac{1}{2}$.

11. $84^x = 70$

$\log 84^x = \log 70$

$x \log 84 = \log 70$

$x = \dfrac{\log 70}{\log 84}$

$x \approx \dfrac{1.8451}{1.9243}$

$x \approx 0.959$

The solution is 0.959.

13. $e^t = 1000$

$\ln e^t = \ln 1000$

$t = \ln 1000 \quad$ Using $\log_a a^x = x$

$t \approx 6.908$

The solution is 6.908.

15. $e^{-0.03t} = 0.08$

$\ln e^{-0.03t} = \ln 0.08$

$-0.03t = \ln 0.08$

$t = \dfrac{\ln 0.08}{-0.03}$

$t \approx \dfrac{-2.5257}{-0.03}$

$t \approx 84.191$

The solution is 84.191.

17. $3^x = 2^{x-1}$

$\ln 3^x = \ln 2^{x-1}$

$x \ln 3 = (x - 1) \ln 2$

$x \ln 3 = x \ln 2 - \ln 2$

$\ln 2 = x \ln 2 - x \ln 3$

$\ln 2 = x(\ln 2 - \ln 3)$

$\dfrac{\ln 2}{\ln 2 - \ln 3} = x$

$\dfrac{0.6931}{0.6931 - 1.0986} \approx x$

$-1.710 \approx x$

The solution is -1.710.

19. $(3.9)^x = 48$

$\log(3.9)^x = \log 48$

$x \log 3.9 = \log 48$

$x = \dfrac{\log 48}{\log 3.9}$

$x \approx \dfrac{1.6812}{0.5911}$

$x \approx 2.844$

The solution is 2.844.

21. $e^x + e^{-x} = 5$

$e^{2x} + 1 = 5e^x \quad$ Multiplying by e^x

$e^{2x} - 5e^x + 1 = 0 \quad$ This equation is quadratic in e^x.

$e^x = \dfrac{5 \pm \sqrt{21}}{2}$

$x = \ln\left(\dfrac{5 \pm \sqrt{21}}{2}\right) \approx \pm 1.567$

The solutions are -1.567 and 1.567.

23. $\dfrac{e^x + e^{-x}}{e^x - e^{-x}} = 3$

$e^x + e^{-x} = 3e^x - 3e^{-x} \quad$ Multiplying by $e^x - e^{-x}$

$4e^{-x} = 2e^x \quad$ Subtracting e^x and adding e^{-x}

$2e^{-x} = e^x$

$2 = e^{2x} \quad$ Multiplying by e^x

$\ln 2 = \ln e^{2x}$

$\ln 2 = 2x$

$\dfrac{\ln 2}{2} = x$

$0.347 \approx x$

The solution is 0.347.

25. $\log_5 x = 4$

$x = 5^4 \quad$ Writing an equivalent exponential equation

$x = 625$

The solution is 625.

27. $\log x = -4 \quad$ The base is 10.

$x = 10^{-4}$, or 0.0001

The solution is 0.0001.

29. $\ln x = 1 \quad$ The base is e.

$x = e^1 = e$

The solution is e.

31. $\log_2(10 + 3x) = 5$

$$2^5 = 10 + 3x$$
$$32 = 10 + 3x$$
$$22 = 3x$$
$$\frac{22}{3} = x$$

The answer checks. The solution is $\dfrac{22}{3}$.

33. $\log x + \log(x - 9) = 1$ The base is 10.

$$\log_{10}[x(x - 9)] = 1$$
$$x(x - 9) = 10^1$$
$$x^2 - 9x = 10$$
$$x^2 - 9x - 10 = 0$$
$$(x - 10)(x + 1) = 0$$
$$x = 10 \text{ or } x = -1$$

Check: For 10:

$$\frac{\log x + \log(x - 9) = 1}{\log 10 + \log(10 - 9) \;?\; 1}$$
$$\log 10 + \log 1$$
$$1 + 0$$
$$1 \;\Big|\; 1 \quad \text{TRUE}$$

For -1:

$$\frac{\log x + \log(x - 9) = 1}{\log(-1) + \log(-1 - 9) \;?\; 1}$$

The number -1 does not check, because negative numbers do not have logarithms. The solution is 10.

35. $\log_8(x + 1) - \log_8 x = 2$

$$\log_8\left(\frac{x + 1}{x}\right) = 2 \quad \text{Quotient rule}$$
$$\frac{x + 1}{x} = 8^2$$
$$\frac{x + 1}{x} = 64$$
$$x + 1 = 64x$$
$$1 = 63x$$
$$\frac{1}{63} = x$$

The answer checks. The solution is $\dfrac{1}{63}$.

37. $\log_4(x + 3) + \log_4(x - 3) = 2$

$$\log_4[(x + 3)(x - 3)] = 2 \quad \text{Product rule}$$
$$(x + 3)(x - 3) = 4^2$$
$$x^2 - 9 = 16$$
$$x^2 = 25$$
$$x = \pm 5$$

The number 5 checks, but -5 does not. The solution is 5.

39. $\log(2x + 1) - \log(x - 2) = 1$

$$\log\left(\frac{2x + 1}{x - 2}\right) = 1 \quad \text{Quotient rule}$$
$$\frac{2x + 1}{x - 2} = 10^1 = 10$$
$$2x + 1 = 10x - 20$$
$$\qquad\qquad \text{Multiplying by } x - 2$$
$$21 = 8x$$
$$\frac{21}{8} = x$$

The answer checks. The solution is $\dfrac{21}{8}$.

41. $e^{7.2x} = 14.009$

Graph $y_1 = e^{7.2x}$ and $y_2 = 14.009$ and find the first coordinate of the point of intersection using the Intersect feature. The solution is 0.367.

43. $xe^{3x} - 1 = 3$

Graph $y_1 = xe^{3x} - 1$ and $y_2 = 3$ and find the first coordinate of the point of intersection using the Intersect feature. The solution is 0.621.

45. $4\ln(x + 3.4) = 2.5$

Graph $y_1 = 4\ln(x + 3.4)$ and $y_2 = 2.5$ and find the first coordinate of the point of intersection using the Intersect feature. The solution is -1.532.

47. $\log_8 x + \log_8(x + 2) = 2$

Graph $y_1 = \dfrac{\log x}{\log 8} + \dfrac{\log(x + 2)}{\log 8}$ and $y_2 = 2$ and find the first coordinate of the point of intersection using the intersect feature. The solution is 7.062.

49. $\log_5(x + 7) - \log_5(2x - 3) = 1$

Graph $y_1 = \dfrac{\log(x + 7)}{\log 5} - \dfrac{\log(2x - 3)}{\log 5}$ and $y_2 = 1$ and find the first coordinate of the point of intersection using the Intersect feature. The solution is 2.444.

51. Solving the first equation for y, we get $y = \dfrac{12.4 - 2.3x}{3.8}$. Graph $y_1 = \dfrac{12.4 - 2.3x}{3.8}$ and $y_2 = 1.1\ln(x - 2.05)$ and use the Intersect feature to find the point of intersection. It is $(4.093, 0.786)$.

53. Graph $y_1 = 2.3\ln(x + 10.7)$ and $y_2 = 10e^{-0.007x^2}$ and use the Intersect feature to find the point of intersection. It is $(7.586, 6.684)$.

55. Discussion and Writing

57. $g(x) = x^2 - 6$

a) $-\dfrac{b}{2a} = -\dfrac{0}{2 \cdot 1} = 0$

$g(0) = 0^2 - 6 = -6$

The vertex is $(0, -6)$.

b) The line of symmetry is $x = 0$.

c) Since the coefficient of the x^2-term is positive, the function has a minimum value. It is the second coordinate of the vertex, -6, and it occurs when $x = 0$.

59. $G(x) = -2x^2 - 4x - 7$

a) $-\dfrac{b}{2a} = -\dfrac{-4}{2(-2)} = -1$

$G(-1) = -2(-1)^2 - 4(-1) - 7 = -5$

The vertex is $(-1, -5)$.

b) The line of symmetry is $x = -1$.

c) Since the coefficient of the x^2-term is negative, the function has a maximum value. It is the second coordinate of the vertex, -5, and it occurs when $x = -1$.

61. $\ln(\log x) = 0$

$\log x = e^0$

$\log x = 1$

$x = 10^1 = 10$

The answer checks. The solution is 10.

63. $\sqrt{\ln x} = \ln \sqrt{x}$

$\sqrt{\ln x} = \dfrac{1}{2}\ln x$ Power rule

$\ln x = \dfrac{1}{4}(\ln x)^2$ Squaring both sides

$0 = \dfrac{1}{4}(\ln x)^2 - \ln x$

Let $u = \ln x$ and substitute.

$\dfrac{1}{4}u^2 - u = 0$

$u\left(\dfrac{1}{4}u - 1\right) = 0$

$u = 0$ or $\dfrac{1}{4}u - 1 = 0$

$u = 0$ or $\dfrac{1}{4}u = 1$

$u = 0$ or $u = 4$

$\ln x = 0$ or $\ln x = 4$

$x = e^0 = 1$ or $x = e^4 \approx 54.598$

Both answers check. The solutions are 1 and e^4, or 1 and 54.598.

65. $(\log_3 x)^2 - \log_3 x^2 = 3$

$(\log_3 x)^2 - 2\log_3 x - 3 = 0$

Let $u = \log_3 x$, substitute:

$u^2 - 2u - 3 = 0$

$(u - 3)(u + 1) = 0$

$u = 3$ or $u = -1$

$\log_3 x = 3$ or $\log_3 x = -1$

$x = 3^3$ or $x = 3^{-1}$

$x = 27$ or $x = \dfrac{1}{3}$

Both answers check. The solutions are $\dfrac{1}{3}$ and 27.

67. $\ln x^2 = (\ln x)^2$

$2\ln x = (\ln x)^2$

$0 = (\ln x)^2 - 2\ln x$

Let $u = \ln x$ and substitute.

$0 = u^2 - 2u$

$0 = u(u - 2)$

$u = 0$ or $u = 2$

$\ln x = 0$ or $\ln x = 2$

$x = 1$ or $x = e^2 \approx 7.389$

Both answers check. The solutions are 1 and e^2, or 1 and 7.389.

69. $5^{2x} - 3 \cdot 5^x + 2 = 0$

$(5^x - 1)(5^x - 2) = 0$ This equation is quadratic in 5^x.

$5^x = 1$ or $5^x = 2$

$\log 5^x = \log 1$ or $\log 5^x = \log 2$

$x \log 5 = 0$ or $x \log 5 = \log 2$

$x = 0$ or $x = \dfrac{\log 2}{\log 5} \approx 0.431$

The solutions are 0 and 0.431.

71. $\log_3 |x| = 2$

$|x| = 3^2$

$|x| = 9$

$x = -9$ or $x = 9$

Both answers check. The solutions are -9 and 9.

73. $\ln x^{\ln x} = 4$

$\ln x \cdot \ln x = 4$

$(\ln x)^2 = 4$

$\ln x = \pm 2$

$$\ln x = -2 \quad or \quad \ln x = 2$$
$$x = e^{-2} \quad or \quad x = e^2$$
$$x \approx 0.135 \quad or \quad x \approx 7.389$$

Both answers check. The solutions are e^{-2} and e^2, or 0.135 and 7.389.

75.
$$\frac{\sqrt{(e^{2x} \cdot e^{-5x})^{-4}}}{e^x \div e^{-x}} = e^7$$

$$\frac{\sqrt{e^{12x}}}{e^{x-(-x)}} = e^7$$

$$\frac{e^{6x}}{e^{2x}} = e^7$$

$$e^{4x} = e^7$$

$$4x = 7$$

$$x = \frac{7}{4}$$

The solution is $\frac{7}{4}$.

77. $\ln(x - 2) > 4$

Graph $y_1 = \ln(x-2)$ and $y_2 = 4$. Using the Intersect feature, we find that the first coordinate of the point of intersection of the graphs is 56.598. Observe that the graph of y_1 lies above the graph of y_2 for all x-values greater than 56.598. Thus, the solution set is $(56.598, \infty)$.

79. $|\log_5 x| + 3\log_5 |x| = 4$

Note that we must have $x > 0$. First consider the case when $0 < x < 1$. When $0 < x < 1$, then $\log_5 x < 0$, so $|\log_5 x| = -\log_5 x$ and $|x| = x$. Thus we have:

$$-\log_5 x + 3\log_5 x = 4$$
$$2\log_5 x = 4$$
$$\log_5 x^2 = 4$$
$$x^2 = 5^4$$
$$x = 5^2$$
$$x = 25 \quad \text{(Recall that } x > 0.)$$

25 cannot be a solution since we assumed $0 < x < 1$. Now consider the case when $x > 1$. In this case $\log_5 x > 0$, so $|\log_5 x| = \log_5 x$ and $|x| = x$. Thus we have:

$$\log_5 x + 3\log_5 x = 4$$
$$4\log_5 x = 4$$
$$\log_5 x = 1$$
$$x = 5$$

This answer checks. The solution is 5.

81. $a = \log_8 225$, so $8^a = 225 = 15^2$.

$b = \log_2 15$, so $2^b = 15$.

Then
$$8^a = (2^b)^2$$
$$(2^3)^a = 2^{2b}$$
$$2^{3a} = 2^{2b}$$
$$3a = 2b$$
$$a = \frac{2}{3}b.$$

83. $\log_2[\log_3(\log_4 x)] = 0$ yields $x = 64$.

$\log_3[\log_2(\log_4 y)] = 0$ yields $y = 16$.

$\log_4[\log_3(\log_2 z)] = 0$ yields $z = 8$.

Then $x + y + z = 64 + 16 + 8 = 88$.

Exercise Set 4.6

1. a) Substitute 6.0 for P_0 and 0.013 for k in $P(t) = P_0 e^{kt}$. We have:

$P(t) = 6.0e^{0.013t}$, where $P(t)$ is in billions and t is the number of years after 1999.

b) In 2005, $t = 2005 - 1999 = 6$.

$P(6) = 6.0e^{0.013(6)} \approx 6.5$ billion

In 2010, $t = 2010 - 1999 = 11$.

$P(11) = 6.0e^{0.013(11)} \approx 6.9$ billion

c) Substitute 8 for $P(t)$ and solve for t.

$$8 = 6.0e^{0.013t}$$
$$\frac{4}{3} = e^{0.013t}$$
$$\ln\frac{4}{3} = \ln e^{0.013t}$$
$$\ln\frac{4}{3} = 0.013t$$
$$\frac{\ln\frac{4}{3}}{0.013} = t$$
$$22.1 \approx t$$

The world population will be 8 billion about 22.1 yr after 1999.

d) $T = \dfrac{\ln 2}{0.013} \approx 53.3$ yr

3. a) $T = \dfrac{\ln 2}{0.019} \approx 36.5$ yr

b) $k = \dfrac{\ln 2}{346} \approx 0.2\%$ per yr

c) $T = \dfrac{\ln 2}{0.033} \approx 21.0$ yr

d) $T = \dfrac{\ln 2}{0.005} \approx 138.6$ yr

e) $k = \dfrac{\ln 2}{20.4} \approx 3.4\%$ per yr

f) $k = \dfrac{\ln 2}{31.5} \approx 2.2\%$ per yr

5.
$$P(t) = P_0 e^{kt}$$
$$24{,}313{,}062{,}400 = 5{,}644{,}000 e^{0.026t}$$
$$\frac{24{,}313{,}062{,}400}{5{,}644{,}000} = e^{0.026t}$$
$$\ln\left(\frac{24{,}313{,}062{,}400}{5{,}644{,}000}\right) = \ln e^{0.026t}$$
$$\ln\left(\frac{24{,}313{,}062{,}400}{5{,}644{,}000}\right) = 0.026t$$
$$\frac{\ln\left(\dfrac{24{,}313{,}062{,}400}{5{,}644{,}000}\right)}{0.026} = t$$
$$322 \approx t$$

There will be one person for every square yard of land about 322 yr after 1998.

7. a) Substitute 10,000 for P_0 and 5.4%, or 0.054 for k.
$$P(t) = 10{,}000 e^{0.054t}$$

b) We can use the TABLE feature of a grapher, set in ASK mode, to evaluate $P(t) = 10{,}000 e^{0.054t}$ for the desired values of t. We enter $y = 10{,}000 e^{0.054x}$.
$$P(1) \approx \$10{,}555$$
$$P(2) \approx \$11{,}140$$
$$P(5) \approx \$13{,}100$$
$$P(10) \approx \$17{,}160$$

c) $T = \dfrac{\ln 2}{0.054} \approx 12.8$ yr

9. We use the function found in Example 5. If the mummy has lost 46% of its carbon-14 from an initial amount P_0, then $54\% P_0$, or $0.54 P_0$ remains. We substitute in the function.
$$0.54 P_0 = P_0 e^{-0.00012t}$$
$$0.54 = e^{-0.00012t}$$
$$\ln 0.54 = \ln e^{-0.00012t}$$
$$\ln 0.54 = -0.00012t$$
$$\frac{\ln 0.54}{-0.00012} = t$$
$$5135 \approx t$$

The mummy is about 5135 years old.

11. a) $K = \dfrac{\ln 2}{3} \approx 0.231$, or 23.1% per min

b) $k = \dfrac{\ln 2}{22} \approx 0.0315$, or 3.15% per yr

c) $T = \dfrac{\ln 2}{0.096} \approx 7.2$ days

d) $T = \dfrac{\ln 2}{0.063} \approx 11$ yr

e) $k = \dfrac{\ln 2}{25} \approx 0.028$, or 2.8% per yr

f) $k = \dfrac{\ln 2}{4560} \approx 0.00015$, or 0.015% per yr

g) $k = \dfrac{\ln 2}{23{,}105} \approx 0.00003$, or 0.003% per yr

13. a) Substitute $1996 - 1985$, or 11, for t; 80 for P_0; and 67 for $P(11)$ and solve for k.
$$67 = 80 e^{-11k}$$
$$0.8375 = e^{-11k}$$
$$\ln 0.8375 = \ln e^{-11k}$$
$$\ln 0.8375 = -11k$$
$$\frac{\ln 0.8375}{-11} = k$$
$$0.016 \approx k$$

The desired equation is $P(t) = 80 e^{-0.016t}$.

b) In 2002, $t = 2002 - 1985 = 17$.
$$P(17) = 80 e^{-0.016(17)} \approx 61 \text{ lb per person}$$

c)
$$20 = 80 e^{-0.016t}$$
$$0.25 = e^{-0.016t}$$
$$\ln 0.25 = \ln e^{-0.016t}$$
$$\ln 0.25 = -0.016t$$
$$\frac{\ln 0.25}{-0.016} = t$$
$$86.6 \approx t$$

The average annual consumption of beef will be 20 lb per person after about 86.6 years.

15. a) $y = \dfrac{2000}{1 + 19.9 e^{-0.6x}}$

b) $N(0) = \dfrac{2000}{1 + 19.9 e^{-0.6(0)}} \approx 96$

c) $N(2) = \dfrac{2000}{1 + 19.9 e^{-0.6(2)}} \approx 286$

$N(5) = \dfrac{2000}{1 + 19.9 e^{-0.6(5)}} \approx 1005$

$N(8) = \dfrac{2000}{1 + 19.9 e^{-0.6(8)}} \approx 1719$

$$N(12) = \frac{2000}{1 + 19.9e^{-0.6(12)}} \approx 1971$$

$$N(16) = \frac{2000}{1 + 19.9e^{-0.6(16)}} \approx 1997$$

17. a) $y = \dfrac{2500}{1 + 5.25e^{-0.32x}}$

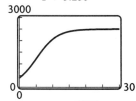

b) $P(0) = \dfrac{2500}{1 + 5.25e^{-0.32(0)}} = 400$

$P(1) = \dfrac{2500}{1 + 5.25e^{-0.32(1)}} \approx 520$

$P(5) = \dfrac{2500}{1 + 5.25e^{-0.32(5)}} \approx 1214$

$P(10) = \dfrac{2500}{1 + 5.25e^{-0.32(10)}} \approx 2059$

$P(15) = \dfrac{2500}{1 + 5.25e^{-0.32(15)}} \approx 2396$

$P(20) = \dfrac{2500}{1 + 5.25e^{-0.32(20)}} \approx 2478$

19. The data have an S-shaped pattern, so function (f) might be used as a model.

21. The data fit the pattern of a polynomial function with degree greater than two. Thus, function (b) might be used as a model.

23. The data have the pattern of an increasing exponential function, so function (c) might be used as a model.

25. a) $y = 4.195491964(1.025306189)^x$

We can convert this equation to an equation with base e, if desired.

$$y = 4.195491964e^{x(\ln 1.025306189)}$$

$$= 4.195491964e^{0.024991289x}$$

In each case x is the number of years after 2000 and y is in millions.

The coefficient of correlation r is approximately 0.9954. Since this is close to 1, the function is a good fit.

b)

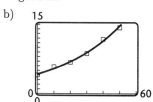

c) In 2005, $t = 2005 - 2000 = 5$.

$y = 4.195491964(1.025306189)^5 \approx 4.8$ million

In 2025, $t = 2025 - 2000 = 25$.

$y = 4.195491964(1.025306189)^{25} \approx 7.8$ million

In 2100, $t = 2100 - 2000 = 100$.

$y = 4.195491964(1.025306189)^{100} \approx 51.1$ million

27. a)

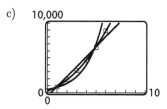

b) Linear: $y = 1429.214286x - 530.0714286$; $r^2 \approx 0.9641$

Quadratic: $y = 158.0952381x^2 + 480.6428571x + 260.4047619$; $R^2 \approx 0.9995$

Exponential: $y = 445.8787388(1.736315606)^x$; $r^2 \approx 0.9361$

The value of R^2 is highest for the quadratic function, os we determine that this function fits the data best.

c)   10,000

d) In 2010, $x = 2010 - 1996 = 14$.

For the linear function, when $x = 14$, $y \approx \$19,479$ million, or $\$19.479$ billion.

For the quadratic function, when $x = 14$, $y \approx \$37,976$ million, or $\$37.976$ billion.

For the exponential function, when $x = 14$, $y \approx \$1,009,295$ million, or $\$1009.295$ billion.

Given the rate at which the revenues in the table are increasing, it appears that the quadratic function provides the most realistic prediction. Answers may vary.

29. Discussion and Writing

31. $y = 6 = 0 \cdot x + 6$

Slope: 0; y-intercept $(0, 6)$

33. $y = 2x - \dfrac{3}{13}$

Slope: 2, y-intercept: $\left(0, -\dfrac{3}{13}\right)$

35.
$$P(t) = P_0 e^{kt}$$
$$50,000 = P_0 e^{0.07(18)}$$
$$\frac{50,000}{e^{0.07(18)}} = P_0$$
$$\$14,182.70 \approx P_0$$

37. $480e^{-0.003p} = 150e^{0.004p}$
$$\frac{480}{150} = \frac{e^{0.004p}}{e^{-0.003p}}$$
$$3.2 = e^{0.007p}$$
$$\ln 3.2 = \ln e^{0.007p}$$
$$\ln 3.2 = 0.007p$$
$$\frac{\ln 3.2}{0.007} = p$$
$$\$166.16 \approx p$$

39. To find k we substitute 105 for T_1, 0 for T_0, 5 for t, and 70 for $T(t)$ and solve for k.
$$70 = 0 + |105 - 0|e^{-5k}$$
$$70 = 105e^{-5k}$$
$$\frac{70}{105} = e^{-5k}$$
$$\ln \frac{70}{105} = \ln e^{-5k}$$
$$\ln \frac{70}{105} = -5k$$
$$\frac{\ln \frac{70}{105}}{-5} = k$$
$$0.081 \approx k$$

The function is $T(t) = 105e^{-0.081t}$.

Now we find $T(10)$.

$$T(10) = 105e^{-0.081(10)} \approx 46.7 \text{ °F}$$

41.
$$i = \frac{V}{R}\left[1 - e^{-(R/L)t}\right]$$
$$\frac{iR}{V} = 1 - e^{-(R/L)t}$$
$$e^{-(R/L)t} = 1 - \frac{iR}{V}$$
$$\ln e^{-(R/L)t} = \ln\left(1 - \frac{iR}{V}\right)$$
$$-\frac{R}{L}t = \ln\left(1 - \frac{iR}{V}\right)$$
$$t = -\frac{L}{R}\left[\ln\left(1 - \frac{iR}{V}\right)\right]$$

43.
$$y = ae^x$$
$$\ln y = \ln(ae^x)$$
$$\ln y = \ln a + \ln e^x$$
$$\ln y = \ln a + x$$
$$Y = x + \ln a$$

This function is of the form $y = mx + b$, so it is linear.

Chapter 5

The Trigonometric Functions

Exercise Set 5.1

1. See the answer section in the text.

3. See the answer section in the text.

5. Clockwise M is at $\frac{2}{3} \cdot \pi$, or $\frac{2\pi}{3}$. $\left(\text{We could also}\right.$

say that M is at $\frac{1}{3} \cdot 2\pi.\Big)$ Counterclockwise M is at

$\frac{2}{3} \cdot 2\pi$, or $\frac{4\pi}{3}$ so the number $-\frac{4\pi}{3}$ determines M.

$\Big($We could also say that M is at $\frac{8}{6} \cdot \pi$ moving coun-

terclockwise, so again the result is $-\frac{4\pi}{3}.\Big)$

Clockwise N is at $\frac{3}{4} \cdot 2\pi$, or $\frac{3\pi}{2}$.

Counterclockwise N is at $\frac{1}{4} \cdot 2\pi$, or $\frac{\pi}{2}$, so the number

$-\frac{\pi}{2}$ determines N.

Clockwise P is at $\frac{5}{8} \cdot 2\pi$, or $\frac{5\pi}{4}$. $\Big($We could also say

that P is at $\frac{5}{4} \cdot \pi.\Big)$

Counterclockwise P is at $\frac{3}{4} \cdot \pi$, or $\frac{3\pi}{4}$, so the number

$-\frac{3\pi}{4}$ determines P. $\Big($We could also say that P is at

$\frac{3}{8} \cdot 2\pi$ moving counterclockwise, so again the result

is $-\frac{3\pi}{4}.\Big)$

Clockwise Q is at $\frac{11}{6} \cdot \pi$, or $\frac{11\pi}{6}$. $\Big($We could also

say that Q is at $\frac{11}{12} \cdot 2\pi.\Big)$

Counterclockwise Q is at $\frac{1}{6} \cdot \pi$, or $\frac{\pi}{6}$ so the number

$-\frac{\pi}{6}$ determines Q. $\Big($We could also say that Q is at

$\frac{1}{12} \cdot 2\pi$ moving counterclockwise, so again the result

is $-\frac{\pi}{6}.\Big)$

7. a) $\frac{2.4}{2\pi} \approx 0.38$, or $2.4 \approx 0.38(2\pi)$, so we move counterclockwise about 0.38 of the way around the
circle.

b) $\frac{7.5}{2\pi} \approx 1.19$, or $7.5 \approx 1.19(2\pi)$, or $1(2\pi) +$
$0.19(2\pi)$, so we move completely around the
circle counterclockwise once and then continue
about another 0.19 of the way around.

c) $\frac{32}{2\pi} \approx 5.09$, or $32 \approx 5.09(2\pi)$, or $5(2\pi)+0.09(2\pi)$,
so we move completely around the circle coun-
terclockwise 5 times and then continue about another 0.09 of the way around.

d) $\frac{320}{2\pi} \approx 50.93$, or $320 \approx 50.93(2\pi)$, or $50(2\pi) +$
$0.93(2\pi)$, so we move completely around the cir-
cle counterclockwise 50 times and then continue
about another 0.93 of the way around.

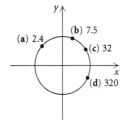

9.

a) $\left(-\dfrac{3}{4}, -\dfrac{\sqrt{7}}{4}\right)$, b) $\left(\dfrac{3}{4}, \dfrac{\sqrt{7}}{4}\right)$, c) $\left(\dfrac{3}{4}, -\dfrac{\sqrt{7}}{4}\right)$

11.

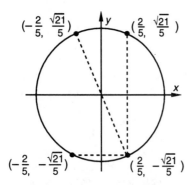

$\left(-\frac{2}{5}, \frac{\sqrt{21}}{5}\right)$ $\left(\frac{2}{5}, \frac{\sqrt{21}}{5}\right)$

$\left(-\frac{2}{5}, -\frac{\sqrt{21}}{5}\right)$ $\left(\frac{2}{5}, -\frac{\sqrt{21}}{5}\right)$

a) $\left(\frac{2}{5}, \frac{\sqrt{21}}{5}\right)$, b) $\left(-\frac{2}{5}, -\frac{\sqrt{21}}{5}\right)$, c) $\left(-\frac{2}{5}, \frac{\sqrt{21}}{5}\right)$

13. a) Reflecting a point across the x-axis changes the sign of the second coordinate. We have $\left(\frac{\sqrt{11}}{6}, -\frac{5}{6}\right)$.

b) Reflecting a point across the y-axis changes the sign of the first coordinate. We have $\left(-\frac{\sqrt{11}}{6}, \frac{5}{6}\right)$.

c) Reflecting a point across the origin changes the signs of both coordinates. We have $\left(-\frac{\sqrt{11}}{6}, -\frac{5}{6}\right)$.

15. The point determined by $-\pi/4$ is a reflection across the x-axis of the point determined by $\pi/4$, $\left(\frac{\sqrt{2}}{2}, \frac{\sqrt{2}}{2}\right)$. Thus, the coordinates of the point determined by $-\pi/4$ are $\left(\frac{\sqrt{2}}{2}, -\frac{\sqrt{2}}{2}\right)$.

17. Discussion and Writing

19. $(x-8)^2 + (y+13)^2 = 25$
$(x-8)^2 + [y-(-13)]^2 = 5^2$ Standard form
The center is $(8, -13)$, and the radius is 5.

21. Answers may vary. The simplest polynomial is obtained as follows.
$f(x) = (x-10)(x-3i)(x+3i)$
$f(x) = (x-10)(x^2+9)$
$f(x) = x^3 - 10x^2 + 9x - 90$

23. M corresponds to the real number $\frac{2\pi}{3}$. We add 2π to find a corresponding real number between 2π and 4π.
$$\frac{2\pi}{3} + 2\pi = \frac{2\pi}{3} + \frac{6\pi}{3} = \frac{8\pi}{3}$$

N corresponds to the real number $\frac{3\pi}{2}$. We add 2π.
$$\frac{3\pi}{2} + 2\pi = \frac{3\pi}{2} + \frac{4\pi}{2} = \frac{7\pi}{2}$$

P corresponds to the real number $\frac{5\pi}{4}$. We add 2π.
$$\frac{5\pi}{4} + 2\pi = \frac{5\pi}{4} + \frac{8\pi}{4} = \frac{13\pi}{4}$$

Q corresponds to the real number $\frac{11\pi}{6}$. We add 2π.
$$\frac{11\pi}{6} + 2\pi = \frac{11\pi}{6} + \frac{12\pi}{6} = \frac{23\pi}{6}$$

25. $x^2 + y^2 = 1$
$\left(-\frac{1}{3}\right)^2 + y^2 = 1$
$\frac{1}{9} + y^2 = 1$
$y^2 = \frac{8}{9}$
$y = \pm\frac{\sqrt{8}}{3} = \pm\frac{2\sqrt{2}}{3}$
The y-coordinate is $\frac{2\sqrt{2}}{3}$ or $-\frac{2\sqrt{2}}{3}$.

Exercise Set 5.2

1. The coordinates of the point determined by π are $(-1, 0)$. Thus,
$$\sin \pi = y = 0.$$

3. The point determined by $\frac{7\pi}{6}$ is a reflection across the origin of the point determined by $\pi/6$, $\left(\frac{\sqrt{3}}{2}, \frac{1}{2}\right)$. Its coordinates are $\left(-\frac{\sqrt{3}}{2}, -\frac{1}{2}\right)$. Thus,
$$\cot\frac{7\pi}{6} = \frac{x}{y} = \frac{-\frac{\sqrt{3}}{2}}{-\frac{1}{2}} = \sqrt{3}.$$

5. The coordinates of the point determined by -3π are $(-1, 0)$. Thus,
$$\sin(-3\pi) = y = 0.$$

7. The point determined by $5\pi/6$ is a reflection across the y-axis of the point determined by $\pi/6$, $\left(\frac{\sqrt{3}}{2}, \frac{1}{2}\right)$. Its coordinates are $\left(-\frac{\sqrt{3}}{2}, \frac{1}{2}\right)$. Thus,
$$\cos\frac{5\pi}{6} = x = -\frac{\sqrt{3}}{2}.$$

9. The coordinates of the point determined by 10π are $(1, 0)$. Thus,
$$\cos 10\pi = x = 1.$$

11. The coordinates of the point determined by $\pi/6$ are $\left(\dfrac{\sqrt{3}}{2}, \dfrac{1}{2}\right)$. Thus,
$$\cos\frac{\pi}{6} = x = \frac{\sqrt{3}}{2}.$$

13. The point determined by $5\pi/4$ is a reflection across the origin of the point determined by $\pi/4$, $\left(\dfrac{\sqrt{2}}{2}, \dfrac{\sqrt{2}}{2}\right)$. Its coordinates are $\left(-\dfrac{\sqrt{2}}{2}, -\dfrac{\sqrt{2}}{2}\right)$. Thus,
$$\sin\frac{5\pi}{4} = y = -\frac{\sqrt{2}}{2}.$$

15. The coordinates of the point determined by -5π are $(-1, 0)$. Thus,
$$\sin(-5\pi) = y = 0.$$

17. The coordinates of the point determined by $5\pi/2$ are $(0, 1)$. Thus,
$$\cot\frac{5\pi}{2} = \frac{x}{y} = \frac{0}{1} = 0.$$

19. Use a grapher set in radian mode.
$$\tan\frac{\pi}{7} \approx 0.4816$$

21. Use a grapher set in radian mode.
$$\sec 37 = \frac{1}{\cos 37} \approx 1.3065$$

23. Use a grapher set in radian mode.
$$\cot 342 = \frac{1}{\tan 342} \approx -2.1599$$

25. Use a grapher set in radian mode.
$$\cos 6\pi = 1$$

27. Use a grapher set in radian mode.
$$\csc 4.16 = \frac{1}{\sin 4.16} \approx -1.1747$$

29. Use a grapher set in radian mode.
$$\tan\frac{7\pi}{4} = -1$$

31. Use a grapher set in radian mode.
$$\sin\left(-\frac{\pi}{4}\right) \approx -0.7071$$

33. Use a grapher set in radian mode.
$$\sin 0 = 0$$

35. Use a grapher set in radian mode.
$$\tan\frac{2\pi}{9} \approx 0.8391$$

37. a)

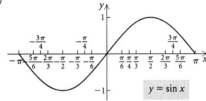

b) Reflect the graph of $y = \sin x$ across the y-axis.

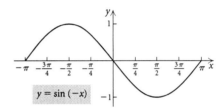

c) Reflect the graph of $y = \sin x$ across the x-axis. The graph is the same as the graph in part (b).

d) They are the same.

39. a) See Exercise 37(a).

b) Shift the graph of $y = \sin x$ left π units.

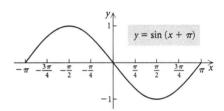

c) Reflect the graph of $y = \sin x$ across the x-axis. The graph is the same as the graph in part (b).

d) They are the same.

41. a)

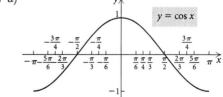

b) Shift the graph of $y = \cos x$ left π units.

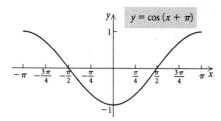

c) Reflect the graph of $y = \cos x$ across the x-axis. The graph is the same as the graph in part (b).

d) They are the same.

43. $\sin(-x) = -\sin x$

The sine function is an odd function.

$\cos(-x) = \cos x$

The cosine function is an even function.

$$\tan(-x) = \frac{\sin(-x)}{\cos(-x)} = \frac{-\sin x}{\cos x}$$

$$= -\frac{\sin x}{\cos x} = -\tan x$$

The tangent function is an odd function.

$$\csc(-x) = \frac{1}{\sin(-x)} = \frac{1}{-\sin x} = -\frac{1}{\sin x} = -\csc x$$

The cosecant function is an odd function.

$$\sec(-x) = \frac{1}{\cos(-x)} = \frac{1}{\cos x} = \sec x$$

The secant function is an even function.

$$\cot(-x) = \frac{\cos(-x)}{\sin(-x)} = \frac{\cos x}{-\sin x}$$

$$= -\frac{\cos x}{\sin x} = -\cot x$$

The cotangent function is an odd function.

Thus the cosine and secant functions are even; the sine, tangent, cosecant, and cotangent functions are odd.

45. Consider a point (x, y) on the unit circle determined by an angle s. Then $\tan s = \dfrac{y}{x}$. The coordinates x and y have the same sign in quadrants I and III, so the tangent function is positive in these quadrants; x and y have opposite signs in quadrants II and IV, so the tangent function is negative in these quadrants.

47. Consider a point (x, y) on the unit circle determined by an angle s. Then $\cos s = x$. Since x is positive in quadrants I and IV, the cosine function is positive in these quadrants; x is negative in quadrants II and III, so the cosine function is negative in these quadrants.

49. Discussion and Writing

51. See the answer section in the text.

53. See the answer section in the text.

55. Start with $y = x^3$.

Reflect it across the x-axis: $y = -x^3$

Shift it right 2 units: $y = -(x - 2)^3$

Shift it down 1 unit: $y = -(x - 2)^3 - 1$

57. Any real numbers x and $-x$ determine points on the unit circle that are symmetric with respect to the x-axis. Their first coordinates are the same, so $\cos(-x) = \cos x$.

59. Any real numbers x and $x + 2k\pi$ determine the same point on the unit circle, so $\sin(x + 2k\pi) = \sin x$.

61. Any real numbers x and $\pi - x$ determine points on the unit circle that are symmetric with respect to the y-axis. Their second coordinates are the same, so $\sin(\pi - x) = \sin x$.

63. Any real numbers x and $x - \pi$ determine points on the unit circle that are symmetric with respect to the origin. Their first coordinates are opposites, so $\cos(x - \pi) = -\cos x$.

65. Any real number x and $x + \pi$ determine points on the unit circle that are symmetric with respect to the origin. Their second coordinates are opposites, so $\sin(x + \pi) = -\sin x$.

67. a) $\sin \dfrac{\pi}{2} = 1$

$$\sin\left(\frac{\pi}{2} + 2\pi\right) = 1 \qquad \sin\left(\frac{\pi}{2} - 2\pi\right) = 1$$

$$\sin\left(\frac{\pi}{2} + 2 \cdot 2\pi\right) = 1 \qquad \sin\left(\frac{\pi}{2} - 2 \cdot 2\pi\right) = 1$$

$$\sin\left(\frac{\pi}{2} + 3 \cdot 2\pi\right) = 1 \qquad \sin\left(\frac{\pi}{2} - 3 \cdot 2\pi\right) = 1$$

$$\sin\left(\frac{\pi}{2} + k \cdot 2\pi\right) = 1, k \text{ an integer}$$

Thus $x = \dfrac{\pi}{2} + 2k\pi$, k an integer.

b) $\cos \pi = -1$

$$\cos(\pi + 2\pi) = -1 \qquad \cos(\pi - 2\pi) = -1$$

$$\cos(\pi + 2 \cdot 2\pi) = -1 \qquad \cos(\pi - 2 \cdot 2\pi) = -1$$

$$\cos(\pi + 3 \cdot 2\pi) = -1 \qquad \cos(\pi - 3 \cdot 2\pi) = -1$$

$$\cos(\pi + k \cdot 2\pi) = 1, \ k \text{ an integer}$$

Thus $x = \pi + 2k\pi$, or $x = (2k+1)\pi$, k an integer.

c) $\sin 0 = 0$

   $\sin \pi = 0 \qquad \sin(-\pi) = 0$

   $\sin 2\pi = 0 \qquad \sin(-2\pi) = 0$

   $\sin 3\pi = 0 \qquad \sin(-3\pi) = 0$

   $\sin k\pi = 0$, k an integer

   Thus $x = k\pi$, k an integer.

d) A grapher shows that intervals on which $\sin x < \cos x$ are $\left(-\dfrac{11\pi}{4}, -\dfrac{7\pi}{4}\right)$, $\left(-\dfrac{3\pi}{4}, \dfrac{\pi}{4}\right)$, and $\left(\dfrac{5\pi}{4}, \dfrac{9\pi}{4}\right)$.

   In general $\sin x < \cos x$ on the intervals $\left(-\dfrac{3\pi}{4} + 2k\pi, \ \dfrac{\pi}{4} + 2k\pi\right)$, k an integer.

69. Graph $y = (\sin x)^2$.

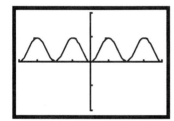

From the graph we find the following information.

Domain: $(-\infty, \infty)$

Range: $[0, 1]$

Period: π

Amplitude: $\dfrac{1}{2}(1 - 0)$, or $\dfrac{1}{2}$

71. $f(x) = \sqrt{\cos x}$

The domain consists of the values of x for which $\cos x \geq 0$. From the graph of the cosine function we see that the domain consists of the intervals $\left[-\dfrac{\pi}{2} + 2k\pi, \ \dfrac{\pi}{2} + 2k\pi\right]$, k an integer.

73. $f(x) = \dfrac{\sin x}{\cos x}$

The domain consists of the values of x for which $\cos x \neq 0$. From the graph of the cosine function we see that the domain is $\left\{x \middle| x \neq \dfrac{\pi}{2} + k\pi, \ k \text{ an integer}\right\}$.

75. $f(x) = \dfrac{\sin x}{x}$, when $0 < x < \dfrac{\pi}{2}$

$\dfrac{\sin \pi/2}{\pi/2} \approx 0.6369$

$\dfrac{\sin 3\pi/8}{3\pi/8} \approx 0.7846$

$\dfrac{\sin \pi/4}{\pi/4} \approx 0.9008$

$\dfrac{\sin \pi/8}{\pi/8} \approx 0.9750$

The limiting value of $\dfrac{\sin x}{x}$ as x approaches 0 is 1.

77. See the answer section in the text.

79. See the answer section in the text.

Exercise Set 5.3

1.

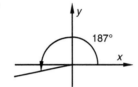

The terminal side lies in quadrant III.

3.

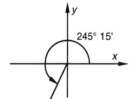

The terminal side lies in quadrant III.

5.

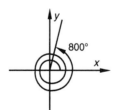

The terminal side lies in quadrant I.

7.

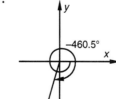

The terminal side lies in quadrant III.

9.

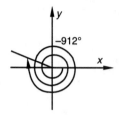

−912°

The terminal side lies in quadrant II.

11.

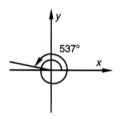

537°

The terminal side lies in quadrant II.

13. We add and subtract multiples of 360°. Many answers are possible.

$74° + 360° = 434°$;

$74° + 2(360°) = 794°$;

$74° − 360° = −286°$;

$74° − 2(360°) = −646°$

15. We add and subtract multiples of 360°. Many answers are possible.

$115.3° + 360° = 475.3°$;

$115.3° + 2(360°) = 835.3°$;

$115.3° − 360° = −244.7°$;

$115.3° − 2(360°) = −604.7°$

17. We add and subtract multiples of 360°. Many answers are possible.

$−180° + 360° = 180°$;

$−180° + 2(360°) = 540°$;

$−180° − 360° = −540°$;

$−180° − 2(360°) = −900°$

19. We add and subtract multiples of 2π. Answers may vary.

Positive angle: $\dfrac{\pi}{4} + 2\pi = \dfrac{\pi}{4} + \dfrac{8\pi}{4} = \dfrac{9\pi}{4}$

Negative angle: $\dfrac{\pi}{4} − 2\pi = \dfrac{\pi}{4} − \dfrac{8\pi}{4} = −\dfrac{7\pi}{4}$

21. We add and subtract multiples of 2π. Answers may vary.

Positive angle: $\dfrac{7\pi}{6} + 2\pi = \dfrac{7\pi}{6} + \dfrac{12\pi}{6} = \dfrac{19\pi}{6}$

Negative angle: $\dfrac{7\pi}{6} − 2\pi = \dfrac{7\pi}{6} − \dfrac{12\pi}{6} = −\dfrac{5\pi}{6}$

23. Using a grapher, enter $9° \, 43'$. The result is $9° \, 43' \approx 9.72°$.

25. Using a grapher, enter $35° \, 50''$. The result is $35° \, 50'' \approx 35.01°$.

27. Using a grapher, enter $3° \, 2'$. The result is $3° \, 2' \approx 3.03°$.

29. Using a grapher, enter $49° \, 38' \, 46''$. The result is $49° \, 38' \, 46'' \approx 49.65°$.

31. Enter 17.6° on a grapher and use the DMS feature: $17.6° = 17° \, 36'$

33. Enter 83.025° on a grapher and use the DMS feature: $83.025° = 83° \, 1' \, 30''$

35. Enter 11.75° on a grapher and use the DMS feature: $11.75° = 11° \, 45'$

37. Enter 47.8268° on a grapher and use the DMS feature: $47.8268° \approx 47° \, 49' \, 36''$

39. $90° − 17.11° = 72.89°$

$180° − 17.11° = 162.89°$

The complement of 17.11° is 72.89° and the supplement is 162.89°.

41.
$$90° = \begin{array}{r} 89°59'60'' \\ −12° \ 3'14'' \\ \hline 77°56'46'' \end{array}$$

$$180° = \begin{array}{r} 179°59'60'' \\ −12° \ 3'14'' \\ \hline 167°56'46'' \end{array}$$

The complement of $12°3'14''$ is $77°56'46''$ and the supplement is $167°56'46''$.

43. $90° − 45.2° − 44.8°$

$180° − 45.2° = 134.8°$

The complement of 45.2° is 44.8° and the supplement is 134.8°.

45. Complement: $\dfrac{\pi}{2} − \dfrac{\pi}{3} = \dfrac{3\pi}{6} − \dfrac{2\pi}{6} = \dfrac{\pi}{6}$

Supplement: $\pi − \dfrac{\pi}{3} = \dfrac{3\pi}{3} − \dfrac{\pi}{3} = \dfrac{2\pi}{3}$

47. Complement: $\dfrac{\pi}{2} − \dfrac{3\pi}{8} = \dfrac{4\pi}{8} − \dfrac{3\pi}{8} = \dfrac{\pi}{8}$

Supplement: $\pi − \dfrac{3\pi}{8} = \dfrac{8\pi}{8} − \dfrac{3\pi}{8} = \dfrac{5\pi}{8}$

49. $75° = 75° \cdot \dfrac{\pi \text{ radians}}{180°} = \dfrac{5\pi}{12}$ radians

51. $200° = 200° \cdot \dfrac{\pi \text{ radians}}{180°} = \dfrac{10\pi}{9}$ radians

53. $-214.6° = -214.6° \cdot \dfrac{\pi \text{ radians}}{180°} = -\dfrac{214.6\pi}{180}$ radians

$\left(\text{If we multiply by 5/5 to clear the decimal, we have}\right.$

$\left.-\dfrac{1073\pi}{900} \text{ radians.}\right)$

55. $-180° = -180° \cdot \dfrac{\pi \text{ radians}}{180°} = -\pi$ radians

57. $240° = 240° \cdot \dfrac{\pi \text{ radians}}{180°} = \dfrac{240\pi}{180}$ radians $\approx$ 4.19 radians

59. $-60° = -60° \cdot \dfrac{\pi \text{ radians}}{180°} = -\dfrac{60\pi}{180}$ radians $\approx$ -1.05 radians

61. $117.8° = 117.8° \cdot \dfrac{\pi \text{ radians}}{180°} = \dfrac{117.8\pi}{180}$ radians $\approx$ 2.06 radians

63. $1.354° = 1.354° \cdot \dfrac{\pi \text{ radians}}{180°} = \dfrac{1.354\pi}{180}$ radians $\approx$ 0.02 radians

65. $-\dfrac{3\pi}{4} = -\dfrac{3\pi}{4}$ radians $\cdot \dfrac{180°}{\pi \text{ radians}} = -\dfrac{3}{4} \cdot 180° =$ $-135°$

67. $8\pi = 8\pi$ radians $\cdot \dfrac{180°}{\pi \text{ radians}} = 8 \cdot 180° = 1440°$

69. $1 = 1$ radian $\cdot \dfrac{180°}{\pi \text{ radians}} \approx \dfrac{180°}{\pi} \approx 57.30°$

71. $2.347 = 2.347$ radians $\cdot \dfrac{180°}{\pi \text{ radians}} =$ $\dfrac{2.347(180°)}{\pi} \approx 134.47°$

73. $0° = 0° \cdot \dfrac{\pi \text{ radians}}{180°} = 0;$

$30° = 30° \cdot \dfrac{\pi \text{ radians}}{180°} = \dfrac{\pi}{6};$

$45° = 45° \cdot \dfrac{\pi \text{ radians}}{180°} = \dfrac{\pi}{4};$

$60° = 60° \cdot \dfrac{\pi \text{ radians}}{180°} = \dfrac{\pi}{3};$

$90° = 90° \cdot \dfrac{\pi \text{ radians}}{180°} = \dfrac{\pi}{2};$

$135° = 135° \cdot \dfrac{\pi \text{ radians}}{180°} = \dfrac{3\pi}{4};$

$180° = 180° \cdot \dfrac{\pi \text{ radians}}{180°} = \pi;$

$225° = 225° \cdot \dfrac{\pi \text{ radians}}{180°} = \dfrac{5\pi}{4};$

$270° = 270° \cdot \dfrac{\pi \text{ radians}}{180°} = \dfrac{3\pi}{2};$

$315° = 315° \cdot \dfrac{\pi \text{ radians}}{180°} = \dfrac{7\pi}{4};$

$360° = 360° \cdot \dfrac{\pi \text{ radians}}{180°} = 2\pi$

75. $\theta = \dfrac{s}{r}$

θ is the radian measure of the central angle, s is arc length, and r is radius length.

$\theta = \dfrac{132 \text{ cm}}{120 \text{ cm}}$ Substituting 132 cm for s and 120 cm for r

$\theta = \dfrac{11}{10}$, or 1.1   The unit is understood to be radians.

$1.1 = 1.1$ radians $\cdot \dfrac{180°}{\pi \text{ radians}} \approx 63°$

77. We use the formula $\theta = \dfrac{s}{r}$, or $s = r\theta$.

$s = r\theta = (2 \text{ yd})(1.6) = 3.2$ yd

79. Let $r =$ the length of the minute hand. In 60 minutes the minute hand travels the circumference of a circle with radius r, or $2\pi r$. Then in 50 minutes the minute hand travels $\dfrac{50}{60} \cdot 2\pi r$, or $\dfrac{5\pi r}{3}$.

Now find the angle through which the minute hand rotates in 50 minutes.

$\theta = \dfrac{s}{r}$

$\theta = \dfrac{\dfrac{5\pi r}{3}}{r}$

$\theta = \dfrac{5\pi r}{3} \cdot \dfrac{1}{r}$

$\theta = \dfrac{5\pi}{3} \approx 5.24$

The minute hand rotates through about 5.24 radians.

81. Since the linear speed must be in cm/min, the given angular speed, 7 radians/sec, must be changed to radians/min.

$\omega = \dfrac{7 \text{ radians}}{1 \text{ sec}} \cdot \dfrac{60 \text{ sec}}{1 \text{ min}} = \dfrac{420 \text{ radians}}{1 \text{ min}}$

$r = \dfrac{d}{2} = \dfrac{15 \text{ cm}}{2} = 7.5$ cm

Using $v = r\omega$, we have:

$v = 7.5 \text{ cm} \cdot \dfrac{420}{1 \text{ min}}$ Substituting and omitting the word radians

$= 3150 \dfrac{\text{cm}}{\text{min}}$

The linear speed of a point on the rim is 3150 cm/min.

83. First convert 18.33 ft/sec to in./hr.

$$18.33 \text{ ft/sec} = \frac{18.33 \text{ ft}}{1 \text{ sec}} \cdot \frac{12 \text{ in.}}{1 \text{ ft}} \cdot \frac{60 \text{ sec}}{1 \text{ min}} \cdot \frac{60 \text{ min}}{1 \text{ hr}} =$$

791,856 in./hr

$$r = \frac{13.37 \text{ in.}}{2} = 6.685 \text{ in.}$$

Using $v = r\omega$, we have

791, 856 in./hr = 6.685 in · ω

118, 453 radians/hr ≈ ω

Now convert 118,453 radians/hr to revolutions/hr.

118, 453 radians/hr =

$$\frac{118,453 \text{ radians}}{1 \text{ hr}} \cdot \frac{1 \text{ revolution}}{2\pi \text{ radians}} \approx$$

18,852 revolutions/hr

The angular speed of the cylinder is about 18,852 revolutions/hr (or about 18,852 IPH).

85. First find ω in radians per hour.

$$\omega = \frac{2\pi}{24 \text{ hr}} = \frac{\pi}{12 \text{ hr}}$$

Using $v = r\omega$, we have

$$v = 4000 \text{ mi} \cdot \frac{\pi}{12 \text{ hr}} \approx 1047 \text{ mph}.$$

The linear speed of a point on the equator is about 1047 mph.

87. First find ω in radians per hour.

$$\omega = \frac{14 \cdot 2\pi}{1 \text{ min}} \cdot \frac{60 \text{ min}}{1 \text{ hr}} = \frac{1680\pi}{\text{hr}}$$

Next find r in miles.

$$r = 10 \text{ ft} \cdot \frac{1 \text{ mi}}{5280 \text{ ft}} = \frac{1}{528} \text{ mi}$$

Using $v = r\omega$, we have

$$v = \frac{1}{528} \text{ mi} \cdot \frac{1680\pi}{1 \text{ hr}} \approx 10 \text{ mph}.$$

The speed of the river is about 10 mph.

89. First convert 22 mph to inches/second.

$$v = \frac{22 \text{ mi}}{1 \text{ hr}} \cdot \frac{5280 \text{ ft}}{1 \text{ mi}} \cdot \frac{12 \text{ in.}}{1 \text{ ft}} \cdot \frac{1 \text{ hr}}{60 \text{ min}} \cdot \frac{1 \text{ min}}{60 \text{ sec}} =$$

387.2 in./sec

Using $v = r\omega$, we have

$$387.2 \frac{\text{in.}}{\text{sec}} = 23 \text{ in.} \cdot \omega$$

so $\omega = \dfrac{387.2 \text{ in./sec}}{23 \text{ in.}} \approx \dfrac{16.835}{1 \text{ sec}}.$

Then in 12 sec,

$$\theta = \omega t = \frac{16.835}{1 \text{ sec}} \cdot 12 \text{ sec} \approx 202.$$

The wheel rotates through an angle of 202 radians.

91. Discussion and Writing

93. $5^x = 625$

$5^x = 5^4$

$x = 4$ 　　Equating exponents

The solution is 4.

95. $\log_7 x = 3$

$x = 7^3 = 343$

The solution is 343.

97. One degree of latitude is $\dfrac{1}{360}$ of the circumference of the earth.

$c = \pi d$, or $2\pi r$

When $r = 6400$ km, $C = 2\pi \cdot 6400 = 12,800\pi$ km.

Thus 1° of latitude is $\dfrac{1}{360} \cdot 12,800\pi$ km = 111.7 km.

When $r = 4000$ mi, $C = 2\pi \cdot 4000 = 8000\pi$ mi.

Thus 1° of latitude is $\dfrac{1}{360} \cdot 8000\pi$ mi ≈ 69.8 mi.

99. a)　　　100 mils

$= 100 \text{ mils} \cdot \dfrac{90°}{1600 \text{ mils}}$

$= 5.625°$

$= 5°37'30''$ 　Using the DMS feature of a grapher

b)　　　350 mils

$= 350 \text{ mils} \cdot \dfrac{90°}{1600 \text{ mils}}$

$= 19.6875°$

$= 19°41'15''$　Using the DMS feature of a grapher

b) $\dfrac{5\pi}{7} = \dfrac{5\pi}{7}$ radians $\cdot \dfrac{100 \text{ grads}}{\dfrac{\pi}{2} \text{ radians}} = \dfrac{1000}{7}$ grads ≈

142.86 grads

101. Let ω_1 = the angular speed of the smaller wheel and ω_2 = the angular speed of the larger wheel. The wheels have the same linear speed, so we have $v = 40\omega_1 = 50\omega_2$.

Convert the angular speed of the smaller wheel, ω_1, to radians per second.

$$\omega_1 = 20 \text{ rpm} = \frac{20 \cdot 2\pi}{1 \text{ min}} \cdot \frac{1 \text{ min}}{60 \text{ sec}} = \frac{2}{3}\pi/\text{sec}$$

Then

$$40\omega_1 = 50\omega_2$$

$$40 \cdot \frac{2}{3}\pi/\text{sec} = 50\omega_2$$

$$\frac{8\pi}{15}/\text{sec} = \omega_2$$

$$1.676/\text{sec} \approx \omega_2.$$

The angular speed of the larger wheel is about 1.676 radians/sec.

103.

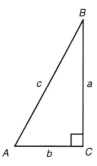

$a = 38°28'45'' - 38°27'30'' = 1'15'' = 1.25' =$
1.25 nautical miles.

$b = 82°57'15'' - 82°56'30'' = 45'' = 0.75' =$
0.75 nautical miles.

$c = \sqrt{a^2 + b^2} = \sqrt{(1.25)^2 + (0.75)^2} \approx$
1.46 nautical miles

Exercise Set 5.4

1. We first determine r.

$r = \sqrt{x^2 + y^2}$

$r = \sqrt{(-12)^2 + 5^2}$

$\quad = \sqrt{144 + 25} = \sqrt{169} = 13$

Substituting -12 for x, 5 for y, and 13 for r, the trigonometric function values of θ are

$\sin \beta = \dfrac{y}{r} = \dfrac{5}{13}$

$\cos \beta = \dfrac{x}{r} = \dfrac{-12}{13} = -\dfrac{12}{13}$

$\tan \beta = \dfrac{y}{x} = \dfrac{5}{-12} = -\dfrac{5}{12}$

$\csc \beta = \dfrac{r}{y} = \dfrac{13}{5}$

$\sec \beta = \dfrac{r}{x} = \dfrac{13}{-12} = -\dfrac{13}{12}$

$\cot \beta = \dfrac{x}{y} = \dfrac{-12}{5} = -\dfrac{12}{5}$

3. We first determine r.

$r = \sqrt{x^2 + y^2}$

$r = \sqrt{(-2\sqrt{3})^2 + (-4)^2} = \sqrt{4 \cdot 3 + 16}$

$\quad = \sqrt{12 + 16} = \sqrt{28} = 2\sqrt{7}$

Substituting $-2\sqrt{3}$ for x, -4 for y and $2\sqrt{7}$ for r, the trigonometric function values are

$\sin \phi = \dfrac{y}{r} = \dfrac{-4}{2\sqrt{7}} = -\dfrac{2}{\sqrt{7}}$, or $-\dfrac{2\sqrt{7}}{7}$

$\cos \phi = \dfrac{x}{r} = \dfrac{-2\sqrt{3}}{2\sqrt{7}} = -\dfrac{\sqrt{3}}{\sqrt{7}}$, or $-\dfrac{\sqrt{21}}{7}$

$\tan \phi = \dfrac{y}{x} = \dfrac{-4}{-2\sqrt{3}} = \dfrac{2}{\sqrt{3}}$, or $\dfrac{2\sqrt{3}}{3}$

$\csc \phi = \dfrac{r}{y} = \dfrac{2\sqrt{7}}{-4} = -\dfrac{\sqrt{7}}{2}$

$\sec \phi = \dfrac{r}{x} = \dfrac{2\sqrt{7}}{-2\sqrt{3}} = -\dfrac{\sqrt{7}}{\sqrt{3}}$, or $-\dfrac{\sqrt{21}}{3}$

$\cot \phi = \dfrac{x}{y} = \dfrac{-2\sqrt{3}}{-4} = \dfrac{\sqrt{3}}{2}$

5. $r = \sqrt{x^2 + y^2}$

$r = \sqrt{(-\sqrt{5})^2 + 2^2} = \sqrt{5 + 4}$

$\quad = \sqrt{9} = 3$

$\sin \phi = \dfrac{y}{r} = \dfrac{2}{3}$

$\cos \phi = \dfrac{x}{r} = \dfrac{-\sqrt{5}}{3} = -\dfrac{\sqrt{5}}{3}$

$\tan \phi = \dfrac{y}{x} = \dfrac{2}{-\sqrt{5}} = -\dfrac{2\sqrt{5}}{5}$

$\csc \phi = \dfrac{r}{y} = \dfrac{3}{2}$

$\sec\phi = \dfrac{r}{x} = \dfrac{3}{-\sqrt{5}} = -\dfrac{3\sqrt{5}}{5}$

$\cot \phi = \dfrac{x}{y} = \dfrac{-\sqrt{5}}{2} = -\dfrac{\sqrt{5}}{2}$

7. $r = \sqrt{x^2 + y^2}$

$r = \sqrt{(-1)^2 + (-6)^2} = \sqrt{37}$

$\sin \phi = \dfrac{y}{r} = \dfrac{-6}{\sqrt{37}} = -\dfrac{6\sqrt{37}}{37}$

$\cos \phi = \dfrac{x}{r} = \dfrac{-1}{\sqrt{37}} = -\dfrac{\sqrt{37}}{37}$

$\tan \phi = \dfrac{y}{x} = \dfrac{-6}{-1} = 6$

$\csc \phi = \dfrac{r}{y} = \dfrac{\sqrt{37}}{-6} = -\dfrac{\sqrt{37}}{6}$

$\sec \phi = \dfrac{r}{x} = \dfrac{\sqrt{37}}{-1} = -\sqrt{37}$

$\cot \phi = \dfrac{x}{y} = \dfrac{-1}{-6} = \dfrac{1}{6}$

9. $r = \sqrt{x^2 + y^2}$

$r = \sqrt{(\sqrt{2})^2 + (\sqrt{10})^2} = \sqrt{2 + 10}$

$r = \sqrt{12} = 2\sqrt{3}$

$$\sin\phi = \frac{y}{r} = \frac{\sqrt{10}}{2\sqrt{3}} = \frac{\sqrt{30}}{6}$$

$$\cos\phi = \frac{x}{r} = \frac{\sqrt{2}}{2\sqrt{3}} = \frac{\sqrt{6}}{6}$$

$$\tan\phi = \frac{y}{x} = \frac{\sqrt{10}}{\sqrt{2}} = \sqrt{5}$$

$$\csc\phi = \frac{r}{y} = \frac{2\sqrt{3}}{\sqrt{10}} = \frac{2\sqrt{30}}{10} = \frac{\sqrt{30}}{5}$$

$$\sec\phi = \frac{r}{x} = \frac{2\sqrt{3}}{\sqrt{2}} = \frac{2\sqrt{6}}{2} = \sqrt{6}$$

$$\cot\phi = \frac{x}{y} = \frac{\sqrt{2}}{\sqrt{10}} = \frac{\sqrt{5}}{5}$$

11. First we draw the graph of $2x+3y = 0$ and determine a quadrant IV solution of the equation.

We let $x = 3$ and find the corresponding y-value.

$$2x + 3y = 0$$
$$2\cdot 3 + 3y = 0 \qquad \text{Substituting}$$
$$3y = -6$$
$$y = -2$$

Thus, $(3, -2)$ is a point on the terminal side of the angle θ.

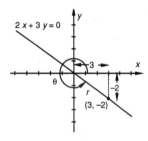

Using $(3, -2)$, we determine r:
$$r = \sqrt{3^2 + (-2)^2} = \sqrt{13}$$
Then using $x = 3$, $y = -2$, and $r = \sqrt{13}$, we find

$$\sin\theta = \frac{-2}{\sqrt{13}}, \text{ or } -\frac{2\sqrt{13}}{13},$$

$$\cos\theta = \frac{3}{\sqrt{13}}, \text{ or } \frac{3\sqrt{13}}{13},$$

$$\tan\theta = \frac{-2}{3} = -\frac{2}{3}.$$

13. First we draw the graph of $5x-4y = 0$ and determine a quadrant I solution of the equation.

We let $x = 4$ and find the corresponding y-value.

$$5x - 4y = 0$$
$$5\cdot 4 - 4y = 0 \qquad \text{Substituting}$$
$$-4y = -20$$
$$y = 5$$

Thus, $(4, 5)$ is a point on the terminal side of the angle θ.

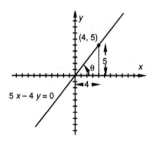

Using $(4, 5)$, we determine r:
$$r = \sqrt{4^2 + 5^2} = \sqrt{41}$$
Then using $x = 4$, $y = 5$, and $r = \sqrt{41}$, we find

$$\sin\theta = \frac{5}{\sqrt{41}}, \text{ or } \frac{5\sqrt{41}}{41},$$

$$\cos\theta = \frac{4}{\sqrt{41}}, \text{ or } \frac{4\sqrt{41}}{41},$$

$$\tan\theta = \frac{5}{4}.$$

15. First we find a quadrant II solution of the equation. Let $x = -3$ and find the corresponding y-value.

$$y = -\frac{2}{3}x$$
$$y = -\frac{2}{3}(-3)$$
$$y = 2$$

Thus, $(-3, 2)$ is a point on the terminal side of the angle. Using $(-3, 2)$, we find r.
$$r = \sqrt{(-3)^2 + 2^2} = \sqrt{13}$$
Then using $x = -3$, $y = 2$, and $r = \sqrt{13}$, we find

$$\sin\phi = \frac{2}{\sqrt{13}} = \frac{2\sqrt{13}}{13},$$

$$\cos\phi = \frac{-3}{\sqrt{13}} = -\frac{3\sqrt{13}}{13},$$

$$\tan\phi = \frac{2}{-3} = -\frac{2}{3}.$$

17. First we sketch a third-quadrant angle and a reference triangle. Since $\sin\theta = -\frac{1}{3} = \frac{-1}{3}$ the length of the vertical leg is 1 and the length of the hypotenuse is 3. The other leg must then have length $\sqrt{8}$, or $2\sqrt{2}$.

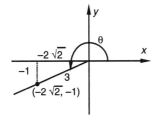

Now we can read off the appropriate ratios:

$$\cos\theta = \frac{-2\sqrt{2}}{3}, \text{ or } -\frac{2\sqrt{2}}{3}$$

$$\tan\theta = \frac{-1}{-2\sqrt{2}} = \frac{1}{2\sqrt{2}}, \text{ or } \frac{\sqrt{2}}{4}$$

$$\csc\theta = \frac{3}{-1} = -3$$

$$\sec\theta = \frac{3}{-2\sqrt{2}} = -\frac{3}{2\sqrt{2}}, \text{ or } -\frac{3\sqrt{2}}{4}$$

$$\cot\theta = \frac{-2\sqrt{2}}{-1} = 2\sqrt{2}$$

19. Since θ is in quadrant IV, we have $\cot\theta = -2 = \frac{2}{-1}$.

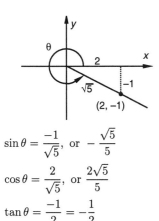

$$\sin\theta = \frac{-1}{\sqrt{5}}, \text{ or } -\frac{\sqrt{5}}{5}$$

$$\cos\theta = \frac{2}{\sqrt{5}}, \text{ or } \frac{2\sqrt{5}}{5}$$

$$\tan\theta = \frac{-1}{2} = -\frac{1}{2}$$

$$\csc\theta = \frac{\sqrt{5}}{-1} = -\sqrt{5}$$

$$\sec\theta = \frac{\sqrt{5}}{2}$$

21. $\cos\phi = \dfrac{3}{5}$

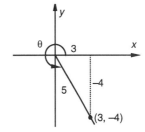

$$\sin\theta = -\frac{4}{5}$$

$$\tan\theta = -\frac{4}{3}$$

$$\csc\theta = -\frac{5}{4}$$

$$\sec\theta = \frac{5}{3}$$

$$\cot\theta = -\frac{3}{4}$$

23. $\tan\phi = \dfrac{y}{x} = \dfrac{1}{-2}$ for ϕ in quadrant II

$$r = \sqrt{(-2)^2 + 1^2} = \sqrt{5}$$

$$\sin\phi = \frac{1}{\sqrt{5}} = \frac{\sqrt{5}}{5}$$

$$\cos\phi = \frac{-2}{\sqrt{5}} = -\frac{2\sqrt{5}}{5}$$

$$\csc\phi = \frac{\sqrt{5}}{1} = \sqrt{5}$$

$$\sec\phi = \frac{\sqrt{5}}{-2} = -\frac{\sqrt{5}}{2}$$

$$\cot\phi = \frac{-2}{1} = -2$$

25. $\csc\beta = \dfrac{r}{y} - \sqrt{3} = \dfrac{\sqrt{3}}{1}$

$$\sqrt{3} = \sqrt{x^2 + 1^2}$$

$$(\sqrt{3})^2 = (\sqrt{x^2+1})^2$$

$$3 = x^2 + 1$$

$$2 = x^2$$

$$\sqrt{2} = x \quad (\beta \text{ is in quadrant I})$$

$$\sin\beta = \frac{1}{\sqrt{3}} = \frac{\sqrt{3}}{3}$$

$$\cos\beta = \frac{\sqrt{2}}{\sqrt{3}} = \frac{\sqrt{6}}{3}$$

$$\tan\beta = \frac{1}{\sqrt{2}} = \frac{\sqrt{2}}{2}$$

$$\sec\beta = \frac{\sqrt{3}}{\sqrt{2}} = \frac{\sqrt{6}}{2}$$

$$\cot\beta = \frac{\sqrt{2}}{1} = \sqrt{2}$$

27. A point on the terminal side of $-180°$ is $(-1,0)$. Then $\sin(-180°) = \frac{0}{1} = 0$.

29. A point on the terminal side of $-270°$ is $(0,1)$. Then $\csc(-270°) = \frac{1}{1} = 1$.

31. Note that $450°$ is coterminal with $90°$. Then $\tan 450° = \tan 90° = \frac{1}{0}$, so $\tan 450°$ is undefined.

33. Note that $540°$ is coterminal with $180°$. Then $\cos 540° = \cos 180° = \frac{-1}{1} = -1$.

35. Note that $-720°$ is coterminal with $0°$. Then $\sec(-720)° = \sec 0° = \frac{1}{1} = 1$.

37. Note that $1440°$ is coterminal with $0°$. Then $\cot 1440° = \cot 0° = \frac{1}{0}$, so $\cot 1440°$ is undefined.

39. Discussion and Writing

41. $f(x) = \frac{1}{x^2 - 25}$

1. The zeros of the denominator are -5 and 5, so $x = -5$ and $x = 5$ are vertical asymptotes.
2. Because the degree of the numerator is less than the degree of the denominator, the x-axis is the horizontal asymptote. There is no oblique asymptote.
3. The numerator has no zeros, so there is no x-intercept.
4. $f(0) = \frac{1}{0^2 - 25} = -\frac{1}{25}$, so $\left(0, -\frac{1}{25}\right)$ is the y-intercept.
5. Find other function values as needed to determine the general shape and then draw the graph.

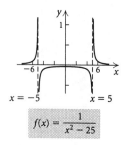

43. $f(x) = \frac{x-4}{x+2}$

The denominator is zero for $x = -2$, so the domain is $\{x | x \neq -2\}$.

Examining the graph of the function, we see that the range is $\{x | x \neq 1\}$.

45. Since $\sin\theta$ is negative, θ can be in quadrant III or IV. Since $\sin\theta = -\frac{\sqrt{2}}{2}$, we know that θ is coterminal with $\frac{5\pi}{4}$ or $\frac{7\pi}{4}$. We add 2π to each of these angles to find the possible values of θ in $(2\pi, 4\pi)$:

$$\frac{5\pi}{4} + 2\pi = \frac{13\pi}{4} \text{ and } \frac{7\pi}{4} + 2\pi = \frac{15\pi}{4}.$$

47. Since $\csc\theta$ is positive, θ can be in quadrant I or II. Since $\csc\theta = \frac{2\sqrt{3}}{3} = \frac{2}{\sqrt{3}}$, we know that θ is coterminal with $\frac{\pi}{3}$ or $\frac{2\pi}{3}$. We add 2π to each of these angles to find the possible values of θ in $(2\pi, 3\pi)$:

$$\frac{\pi}{3} + 2\pi = \frac{7\pi}{3} \text{ and } \frac{2\pi}{3} + 2\pi = \frac{8\pi}{3}.$$

Exercise Set 5.5

1. We use the definitions.

$$\sin\phi = \frac{\text{opp}}{\text{hyp}} = \frac{15}{17}$$
$$\cos\phi = \frac{\text{adj}}{\text{hyp}} = \frac{8}{17}$$
$$\tan\phi = \frac{\text{opp}}{\text{adj}} = \frac{15}{8}$$
$$\csc\phi = \frac{\text{hyp}}{\text{opp}} = \frac{17}{15}$$
$$\sec\phi = \frac{\text{hyp}}{\text{adj}} = \frac{17}{8}$$
$$\cot\phi = \frac{\text{adj}}{\text{opp}} = \frac{8}{15}$$

3. First we use the Pythagorean theorem to find the length of the hypotenuse, c.

$$a^2 + b^2 = c^2$$
$$4^2 + 7^2 = c^2$$
$$65 = c^2$$
$$\sqrt{65} = c$$

Then we use the definitions to find the trigonometric function values of ϕ.

$$\sin\phi = \frac{\text{opp}}{\text{hyp}} = \frac{7}{\sqrt{65}}, \text{ or } \frac{7\sqrt{65}}{65}$$
$$\cos\phi = \frac{\text{adj}}{\text{hyp}} = \frac{4}{\sqrt{65}}, \text{ or } \frac{4\sqrt{65}}{65}$$

$$\tan \phi = \frac{\text{opp}}{\text{adj}} = \frac{7}{4}$$

$$\csc \phi = \frac{\text{hyp}}{\text{opp}} = \frac{\sqrt{65}}{7}$$

$$\sec \phi = \frac{\text{hyp}}{\text{adj}} = \frac{\sqrt{65}}{4}$$

$$\cot \phi = \frac{\text{adj}}{\text{opp}} = \frac{4}{7}$$

5. We know from the definition of the sine function that the ratio $\frac{24}{25}$ is $\frac{\text{opp}}{\text{hyp}}$. Let's consider a right triangle in which the hypotenuse has length 25 and the side opposite θ has length 24.

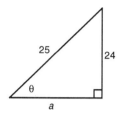

Use the Pythagorean theorem to find the length of the side adjacent to θ.

$$a^2 + b^2 = c^2$$
$$a^2 + 24^2 = 25^2$$
$$a^2 = 625 - 576 = 49$$
$$a = 7$$

Use the lengths of the three sides to find the other five ratios.
$$\cos \theta = \frac{7}{25}, \tan \theta = \frac{24}{7}, \csc \theta = \frac{25}{24}, \sec \theta = \frac{25}{7},$$
$$\cot \theta = \frac{7}{24}$$

7. We know from the definition of the tangent function that 2, or the ratio $\frac{2}{1}$ is $\frac{\text{opp}}{\text{adj}}$. Let's consider a right triangle in which the side opposite ϕ has length 2 and the side adjacent to ϕ has length 1.

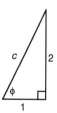

Use the Pythagorean theorem to find the length of the hypotenuse.

$$a^2 + b^2 = c^2$$
$$1^2 + 2^2 = c^2$$
$$1 + 4 = c^2$$
$$5 = c^2$$
$$\sqrt{5} = c$$

Use the lengths of the three sides to find the other five ratios.
$$\sin \phi = \frac{2}{\sqrt{5}}, \text{ or } \frac{2\sqrt{5}}{5}; \cos \phi = \frac{1}{\sqrt{5}}, \text{ or } \frac{\sqrt{5}}{5};$$
$$\csc \phi = \frac{\sqrt{5}}{2}; \sec \phi = \sqrt{5}; \cot \phi = \frac{1}{2}$$

9. $\csc \theta = 1.5 = \frac{1.5}{1} = \frac{15}{10} = \frac{3}{2}$

We know from the definition of the cosecant function that the ratio $\frac{3}{2}$ is $\frac{\text{hyp}}{\text{opp}}$. Let's consider a right triangle in which the hypotenuse has length 3 and side opposite θ has length 2.

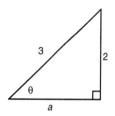

Use the Pythagorean theorem to find the length of the side adjacent to θ.

$$a^2 + b^2 = c^2$$
$$a^2 + 2^2 = 3^2$$
$$a^2 = 9 - 4 = 5$$
$$a = \sqrt{5}$$

Use the lengths of the three sides to find the other five ratios.
$$\sin \theta = \frac{2}{3}; \cos \theta = \frac{\sqrt{5}}{3}; \tan \theta = \frac{2}{\sqrt{5}}, \text{ or } \frac{2\sqrt{5}}{5};$$
$$\sec \theta = \frac{3}{\sqrt{5}}, \text{ or } \frac{3\sqrt{5}}{5}; \cot \theta = \frac{\sqrt{5}}{2}$$

11.

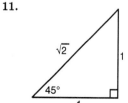

$$\cos 45° = \frac{1}{\sqrt{2}}, \text{ or } \frac{\sqrt{2}}{2}$$

13.

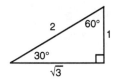

$$\sec 60° = \frac{2}{1} = 2$$

15. See the triangle in Exercise 13.

$$\cot 60° = \frac{1}{\sqrt{3}}, \text{ or } \frac{\sqrt{3}}{3}$$

17. See the triangle in Exercise 13.

$$\sin 30° = \frac{1}{2}$$

19. Since we know the measure of an acute angle of a right triangle and the lengths of the sides opposite and adjacent to the angle, we can use the sine, cosine, cosecant, or secant ratio to determine the length of the hypotenuse. We will use the cosecant function.

$$\csc 45° = \frac{\text{hyp}}{\text{opp}} = \frac{h}{90}$$
$$90 \csc 45° = h$$
$$90\sqrt{2} = h$$
$$127.3 \approx h$$

The distance from third base to first base is about 127.3 ft.

21. Use a grapher set in degree mode.

$$\cos 51° \approx 0.6293$$

23. Use a grapher set in degree mode.

$$\tan 4°13' \approx 0.0737$$

25. Use a grapher set in degree mode. We find the reciprocal of $\cos 38.43°$ by entering $\dfrac{1}{\cos 38.43}$ or $(\cos 38.43)^{-1}$.

$$\sec 38.43° \approx 1.2765$$

27. Use a grapher set in degree mode.

$$\cos 40.35° \approx 0.7621$$

29. Use a grapher set in degree mode.

$$\sin 69° \approx 0.9336$$

31. Use a grapher set in degree mode. We find the reciprocal of $\tan 30°25'6''$ by entering $\dfrac{1}{\tan 30'25'6'}$ or $(\tan 30'25'6')^{-1}$.

$$\cot 30°25'6'' \approx 1.7032$$

33. Press $\boxed{\text{2nd}}$ $\boxed{\text{SIN}}$ .5125 $\boxed{\text{ENTER}}$.

$\theta = 30.8°$

35. $\tan \theta = 0.2226$

Press $\boxed{\text{2nd}}$ $\boxed{\text{TAN}}$ .2226 $\boxed{\text{ENTER}}$.

$\theta = 12.5°$

37. $\cos \theta = 0.6879$

Press $\boxed{\text{2nd}}$ $\boxed{\text{COS}}$ .6879 $\boxed{\text{ENTER}}$.

$\theta = 46.5°$

39. $\cot \theta = \dfrac{1}{\tan \theta} = 2.127$

Thus, $\tan \theta = \dfrac{1}{2.127}$.

Press $\boxed{\text{2nd}}$ $\boxed{\text{TAN}}$ $(1 \div 2.127)$ $\boxed{\text{ENTER}}$.

$\theta \approx 25.2°$

41.

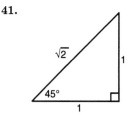

$\sin \theta = \dfrac{\sqrt{2}}{2}$, so $\theta = 45°$.

43.

$\cos \theta = \dfrac{1}{2}$, so $\theta = 60°$.

45. See the triangle in Exercise 41.

$\tan \theta = 1$, so $\theta = 45°$.

47. The cosine and sine functions are cofunctions. The cosine and secant functions are reciprocals.

$$\cos 20° = \sin 70° = \frac{1}{\sec 20°}$$

49. The tangent and cotangent functions are cofunctions and reciprocals.

$$\tan 52° = \cot 38° = \frac{1}{\cot 52°}$$

51. Since 25° and 65° are complementary angles, we have

$$\sin 25° = \cos 65° \approx 0.4226,$$
$$\cos 25° = \sin 65° \approx 0.9063,$$
$$\tan 25° = \cot 65° \approx 0.4663,$$
$$\csc 25° = \sec 65° \approx 2.366,$$
$$\sec 25° = \csc 65° \approx 1.103,$$
$$\cot 25° = \tan 65° \approx 2.145.$$

53. Since $180° - 150° = 30°$, the reference angle is $30°$. Note that $150°$ is a second-quadrant angle, so the cosine is negative. Recalling that $\cos 30° = \sqrt{3}/2$, we have
$$\cos 150° = -\frac{\sqrt{3}}{2}.$$

55. $-135° + 360° = 225°$, so $-135°$ and $225°$ are coterminal. Since $180° + 45° = 225°$, the reference angle is $45°$. Note that $-135°$ is a third-quadrant angle, so the tangent is positive. Recalling that $\tan 45° = 1$, we have
$$\tan(-135°) = 1.$$

57. $495° - 360° = 135°$, so $495°$ and $135°$ are coterminal. Since $180° - 135° = 45°$, the reference angle is $45°$. Note that $495°$ is a second-quadrant angle, so the cosine is negative. Recalling that $\cos 45° = \sqrt{2}/2$ we have
$$\cos 495° = -\frac{\sqrt{2}}{2}.$$

59. $-210° + 360° = 150°$, so $-210°$ and $150°$ are coterminal. Since $180° - 150° = 30°$, the reference angle is $30°$. Note that $-210°$ is a second-quadrant angle, so the cosecant is positive. Recalling that $\csc 30° = 2$, we have
$$\csc(-210°) = 2.$$

61. $570° - 360° = 210°$, so $570°$ and $210°$ are coterminal. Since $180° + 30° = 210°$, the reference angle is $30°$. Note that $570°$ is a third-quadrant angle, so the cotangent is positive. Recalling that $\cot 30° = \sqrt{3}$, we have
$$\cot 570° = \sqrt{3}.$$

63. $360° - 330° = 30°$, so the reference angle is $30°$. Note that $330°$ is a fourth-quadrant angle, so the tangent is negative. Recalling that $\tan 30° = \sqrt{3}/3$, we have
$$\tan 330° = -\frac{\sqrt{3}}{3}.$$

65. $240° - 180° = 60°$, so the reference angle is $60°$. Note that $240°$ is a third-quadrant angle, so the tangent is positive. Recalling that $\tan 60° = \sqrt{3}$, we have
$$\tan 240° = \sqrt{3}.$$

67. $225° - 180° = 45°$, so the reference angle is $45°$. Note that $225°$ is a third-quadrant angle, so the cosecant is negative. Recalling that $\csc 45° = \sqrt{2}$, we have
$$\csc 225° = -\sqrt{2}.$$

69. $480° - 360° = 120°$, so $480°$ and $120°$ are coterminal. Since $180° - 120° = 60°$, the reference angle is $60°$. Note that $480°$ is a second-quadrant angle, so the tangent is negative. Recalling that $\tan 60° = \sqrt{3}$, we have
$$\tan 480° = -\sqrt{3}.$$

71. $319°$ is a fourth-quadrant angle, so the cosine and secant function values are positive and the sine, cosecant, tangent, and cotangent are negative.

73. $194°$ is a third-quadrant angle, so the tangent and cotangent function values are positive and the sine, cosecant, cosine, and secant are negative.

75. $-215°$ is a second-quadrant angle, so the sine and cosecant function values are positive and the cosine, secant, tangent, and cotangent are negative.

77. $360° - 319° = 41°$, so $41°$ is the reference angle for $319°$. Note that $319°$ is a fourth-quadrant angle. Use the given trigonometric function values for $41°$ to find the trigonometric function values for $319°$. In the fourth quadrant, the cosine and secant functions are positive and the other four are negative.

$$\sin 319° = -\sin 41° = -0.6561$$
$$\cos 319° = \cos 41° = 0.7547$$
$$\tan 319° = -\tan 41° = -0.8693$$
$$\csc 319° = \frac{1}{\sin 319°} = \frac{1}{-0.6561} \approx -1.5242$$
$$\sec 319° = \frac{1}{\cos 319°} = \frac{1}{0.7547} \approx 1.3250$$
$$\cot 319° = \frac{1}{\tan 319°} = \frac{1}{-0.8693} \approx -1.1504$$

79.

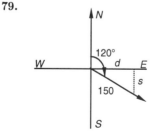

The reference angle is $30°$. First find d, the airplane's distance east of the airport.
$$\cos 30° = \frac{d}{150}$$
$$d = 150 \cos 30°$$
$$d \approx 130$$

The airplane is about 130 km east of the airport.

Now find s, the airplane's distance south of the airport.

$$\sin 30° = \frac{s}{150}$$

$$s = 150 \sin 30°$$

$$s = 75$$

The airplane is 75 km south of the airport.

81.

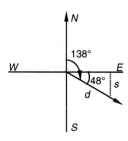

The reference angle is 48°. Use the formula $d = rt$ to find the airplane's distance d from Omaha at the end of 2 hr:

$$d = 150 \frac{\text{km}}{\text{h}} \cdot 2 \text{ hr} = 300 \text{ km}$$

Then find s, the airplane's distance south of Omaha.

$$\sin 48° = \frac{s}{300}$$

$$s = 300 \sin 48°$$

$$s \approx 223$$

The plane is about 223 km south of Omaha.

83. Use a grapher set in degree mode.

$$\tan 310.8° \approx -1.1585$$

85. Use a grapher set in degree mode.

$$\cot 146.15° = \frac{1}{\tan 146.15°} = -1.4910$$

87. Use a grapher set in degree mode.

$$\sin 118°42' \approx 0.8771$$

89. Use a grapher set in degree mode.

$$\cos(-295.8°) \approx 0.4352$$

91. Use a grapher set in degree mode.

$$\csc 520° = \frac{1}{\sin 520°} \approx 2.9238$$

93. $\sin \theta = -0.9956$, $270° < \theta < 360°$

We ignore the fact that $\sin \theta$ is negative and press
2nd SIN .9956 ENTER to find that the reference
angle is approximately 84.6°. Since θ is a fourth-
quadrant angle, we find θ by subtracting 84.6° from
360°:

$$360° - 84.6° = 275.4°.$$

Thus, $\theta \approx 275.4°$.

95. $\cos \theta = -0.9388$, $180° < \theta < 270°$

We ignore the fact that $\cos \theta$ is negative and press
2nd COS .9388 ENTER to find that the refer-
ence angle is approximately 20.1°. Since θ is a third-
quadrant angle, we find θ by adding 180° and 20.1°:

$$180° + 20.1° = 200.1°.$$

Thus, $\theta \approx 200.1°$.

97. Discussion and Writing

99.

x	$f(x)$
-4	0.1353
-2	0.3679
0	1
2	2.7183
4	7.3891

101.

x	$h(x)$
0.5	-0.6931
1	0
2	0.6932
4	1.3863
5	1.6094

103. Since 49.2° and 40.8° are complementary angles, we
have $\sin 40.8° = \cos 49.2° = \dfrac{1}{\sec 49.2°} = \dfrac{1}{1.5304} \approx$
0.6534.

105. Area $= \dfrac{1}{2} \times \text{base} \times \text{height}$

$$= \frac{1}{2}ab$$

$$= \frac{1}{2}c \sin A \cdot b$$

$$\left(\sin A = \frac{a}{c}, \text{ so } c \sin A = a \right)$$

$$= \frac{1}{2}bc \sin A$$

107.

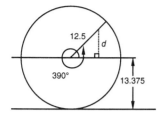

The reference angle is 30°.

Let d be the vertical distance of the valve cap above the center of the wheel.

$$\sin 30° = \frac{d}{12.5}$$
$$d = 12.5 \sin 30°$$
$$d = 6.25$$

The distance above the ground is 6.25 in.+ 13.375 in. = 19.625 in.

Exercise Set 5.6

1. $y = \sin x + 1$

$A = 1$, $B = 1$, $C = 1$, $D = 0$

Amplitude: $|A| = |1| = 1$

Period: $\left|\dfrac{2\pi}{C}\right| = \left|\dfrac{2\pi}{1}\right| = 2\pi$

Phase shift: $\dfrac{D}{C} = \dfrac{0}{1} = 0$

Translate the graph of $y = \sin x$ up 1 unit.

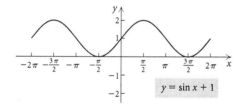

3. $y = -3\cos x$

$A = -3$, $B = 0$, $C = 1$, $D = 0$

Amplitude: $|A| = |-3| = 3$

Period: $\left|\dfrac{2\pi}{C}\right| = \left|\dfrac{2\pi}{1}\right| = 2\pi$

Phase shift: $\dfrac{D}{C} = \dfrac{0}{1} = 0$

Stretch the graph of $y = \cos x$ vertically by a factor of 3 and reflect the graph across the x-axis.

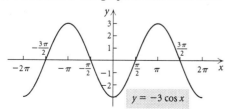

5. $y = 2\sin\left(\dfrac{1}{2}x\right)$

$A = 2$, $B = 0$, $C = \dfrac{1}{2}$, $D = 0$

Amplitude: $|A| = |2| = 2$

Period: $\left|\dfrac{2\pi}{C}\right| = \left|\dfrac{2\pi}{\frac{1}{2}}\right| = 4\pi$

Phase shift: $\dfrac{D}{C} = \dfrac{0}{\frac{1}{2}} = 0$

Stretch the graph of $y = \sin x$ horizontally by a factor of 2 and stretch it vertically, also by a factor of 2.

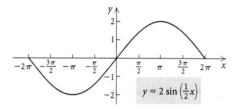

7. $y = \dfrac{1}{2}\sin\left(x + \dfrac{\pi}{2}\right) = \dfrac{1}{2}\sin\left[x - \left(-\dfrac{\pi}{2}\right)\right]$

$A = \dfrac{1}{2}$, $B = 0$, $C = 1$, $D = -\dfrac{\pi}{2}$

Amplitude: $|A| = \left|\dfrac{1}{2}\right| = \dfrac{1}{2}$

Period: $\left|\dfrac{2\pi}{C}\right| = \left|\dfrac{2\pi}{1}\right| = 2\pi$

Phase shift: $\dfrac{D}{C} = \dfrac{-\frac{\pi}{2}}{1} = -\dfrac{\pi}{2}$

Shrink the graph of $y = \sin x$ vertically by a factor of 2 and translate it to the left $\dfrac{\pi}{2}$ units.

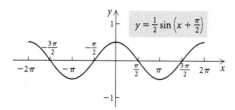

9. $y = 3\cos(x - \pi)$

$A = 3$, $B = 0$, $C = 1$, $D = \pi$

Amplitude: $|A| = |3| = 3$

Period: $\left|\dfrac{2\pi}{C}\right| = \left|\dfrac{2\pi}{1}\right| = 2\pi$

Phase shift: $\dfrac{D}{C} = \dfrac{\pi}{1} = \pi$

Stretch the graph of $y = \cos x$ vertically by a factor of 3 and shift it π units to the right.

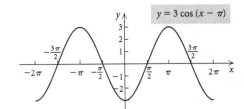

11. $y = \dfrac{1}{3}\sin x - 4$

$A = \dfrac{1}{3}$, $B = -4$, $C = 1$, $D = 0$

Amplitude: $|A| = \left|\dfrac{1}{3}\right| = \dfrac{1}{3}$

Period: $\left|\dfrac{2\pi}{C}\right| = \left|\dfrac{2\pi}{1}\right| = 2\pi$

Phase shift: $\dfrac{D}{C} = \dfrac{0}{1} = 0$

Shrink the graph of $y = \sin x$ vertically by a factor of 3 and shift it down 4 units.

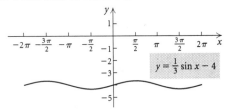

13. $y = -\cos(-x) + 2$

$A = -1$, $B = 2$, $C = -1$, $D = 0$

Amplitude: $|A| = |-1| = 1$

Period: $\left|\dfrac{2\pi}{C}\right| = \left|\dfrac{2\pi}{-1}\right| = 2\pi$

Phase shift: $\dfrac{D}{C} = \dfrac{0}{-1} = 0$

Reflect the graph of $y = \cos x$ across the y-axis and across the x-axis and translate it up 2 units.

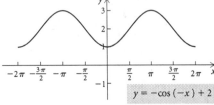

15. $y = 2\cos\left(\dfrac{1}{2}x - \dfrac{\pi}{2}\right) = 2\cos\left[\dfrac{1}{2}(x - \pi)\right]$

$A = 2$, $B = 0$, $C = \dfrac{1}{2}$, $D = \dfrac{\pi}{2}$

Amplitude: $|A| = |2| = 2$

Period: $\left|\dfrac{2\pi}{C}\right| = \left|\dfrac{2\pi}{\frac{1}{2}}\right| = 4\pi$

Phase shift: $\dfrac{D}{C} = \pi$

$y = 2\cos\left(\dfrac{1}{2}x - \dfrac{\pi}{2}\right)$

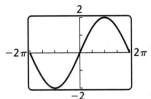

17. $y = -\dfrac{1}{2}\sin\left(2x + \dfrac{\pi}{2}\right) = -\dfrac{1}{2}\sin\left[2\left(x - \left(-\dfrac{\pi}{4}\right)\right)\right]$

$A = -\dfrac{1}{2}$, $B = 0$, $C = 2$, $D = -\dfrac{\pi}{2}$

Amplitude: $|A| = \left|-\dfrac{1}{2}\right| = \dfrac{1}{2}$

Period: $\left|\dfrac{2\pi}{C}\right| = \left|\dfrac{2\pi}{2}\right| = \pi$

Phase shift: $\dfrac{D}{C} = -\dfrac{\pi}{4}$

$y = -\dfrac{1}{2}\sin\left(2x + \dfrac{\pi}{2}\right)$

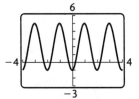

19. $y = 2 + 3\cos(\pi x - 3) = 3\cos\left[\pi\left(x - \dfrac{3}{\pi}\right)\right] + 2$

$A = 3$, $B = 2$, $C = \pi$, $D = 3$

Amplitude: $|A| = |3| = 3$

Period: $\left|\dfrac{2\pi}{C}\right| = \left|\dfrac{2\pi}{\pi}\right| = 2$

Phase shift: $\dfrac{D}{C} = \dfrac{3}{\pi}$

$y = 2 + 3\cos(\pi x - 3)$

21. $y = -\dfrac{1}{2}\cos(2\pi x) + 2$

$A = -\dfrac{1}{2}$, $B = 2$, $C = 2\pi$, $D = 0$

Amplitude: $|A| = \left| -\dfrac{1}{2} \right| = \dfrac{1}{2}$

Period: $\left| \dfrac{2\pi}{C} \right| = \left| \dfrac{2\pi}{2\pi} \right| = 1$

Phase shift: $\dfrac{D}{C} = \dfrac{0}{2\pi} = 0$

$y = -\dfrac{1}{2}\cos(2\pi x) + 2$

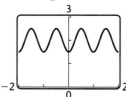

23. $y = -\sin\left(\dfrac{1}{2}x - \dfrac{\pi}{2} \right) + \dfrac{1}{2} = -\sin\left[\dfrac{1}{2}(x - \pi) \right] + \dfrac{1}{2}$

$A = -1,\ B = \dfrac{1}{2},\ C = \dfrac{1}{2},\ D = \dfrac{\pi}{2}$

Amplitude: $|A| = |-1| = 1$

Period: $\left| \dfrac{2\pi}{C} \right| = \left| \dfrac{2\pi}{\frac{1}{2}} \right| = 4\pi$

Phase shift: $\dfrac{D}{C} = \pi$

$y = -\sin\left(\dfrac{1}{2}x - \dfrac{\pi}{2} \right) + \dfrac{1}{2}$

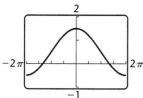

25. $y = \cos(-2\pi x) + 2$

$A = 1,\ B = 2,\ C = -2\pi,\ D = 0$

Amplitude: $|A| = |1| = 1$

Period: $\left| \dfrac{2\pi}{C} \right| = \left| \dfrac{2\pi}{-2\pi} \right| = 1$

Phase shift: $\dfrac{D}{C} = \dfrac{0}{-2\pi} = 0$

$y = \cos(-2\pi x) + 2$

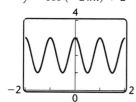

27. $y = -\dfrac{1}{4}\cos(\pi x - 4) = -\dfrac{1}{4}\cos\left[\pi\left(x - \dfrac{4}{\pi} \right) \right]$

$A = -\dfrac{1}{4},\ B = 0,\ C = \pi,\ D = 4$

Amplitude: $|A| = \left| -\dfrac{1}{4} \right| = \dfrac{1}{4}$

Period: $\left| \dfrac{2\pi}{C} \right| = \left| \dfrac{2\pi}{\pi} \right| = 2$

Phase shift: $\dfrac{D}{C} = \dfrac{4}{\pi}$

$y = -\dfrac{1}{4}\cos(\pi x - 4)$

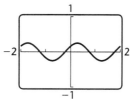

29. $y = -\cos 2x$

Shrink the graph of $y = \cos x$ horizontally by a factor of 2 and reflect it across the x-axis. Graph (b) is the correct choice.

31. $y = 2\cos\left(x + \dfrac{\pi}{2} \right) = 2\cos\left[x - \left(-\dfrac{\pi}{2} \right) \right]$

Stretch the graph of $y = \cos x$ vertically by a factor of 2 and translate it left $\pi/2$ units. Graph (h) is the correct choice.

33. $y = \sin(x - \pi) - 2$

Translate the graph of $y = \sin x$ right π units and down 2 units. Graph (a) is the correct choice.

35. $y = \dfrac{1}{3}\sin 3x$

Shrink the graph of $y = \sin x$ horizontally and vertically, both by a factor of 3. Graph (f) is the correct choice.

37. This graph has the same shape as $y = \cos x$ but with an amplitude of $\dfrac{1}{2}$ and shifted up one unit. The equation is $y = \dfrac{1}{2}\cos x + 1$.

39. This graph has the same shape as $y = \cos x$ but with a phase shift of $-\dfrac{\pi}{2}$ and also a shift of 2 units down. The equation is $y = \cos\left(x + \dfrac{\pi}{2} \right) - 2$.

41. $y = 2\cos x + \cos 2x$

Graph $y = 2\cos x$ and $y = \cos 2x$ on the same set of axes. Then graphically add some ordinates to obtain points on the graph of $y = 2\cos x + \cos 2x$.

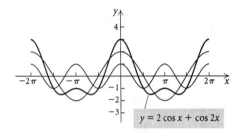

$$y = 2\cos x + \cos 2x$$

43. $y = \sin x + \cos 2x$

Graph $y = \sin x$ and $y = \cos 2x$ on the same set of axes. Then graphically add some ordinates to obtain points on the graph of $y = \sin x + \cos 2x$.

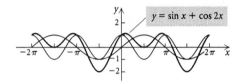

$$y = \sin x + \cos 2x$$

45. $y = \sin x - \cos x$

Graph $y = \sin x$ and $y = -\cos x$ on the same set of axes. Then graphically add some ordinates to obtain points on the graph of $y = \sin x - \cos x$.

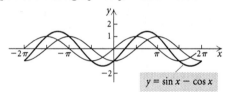

$$y = \sin x - \cos x$$

47. $y = 3\cos x + \sin 2x$

Graph $y = 3\cos x$ and $y = \sin 2x$ on the same set of axes. Then graphically add some ordinates to obtain points on the graph of $y = 3\cos x + \sin 2x$.

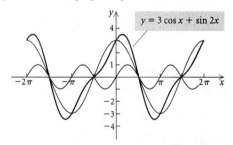

$$y = 3\cos x + \sin 2x$$

49. See the answer section in the text.

51. See the answer section in the text.

53. See the answer section in the text.

55. See the answer section in the text.

57. Discussion and Writing

59. $f(x) = 12 - x$

Solve: $0 = 12 - x$

$x = 12$

We could also graph $y = 12 - x$ on a grapher and use the Zero feature.

The zero of the function is 12.

61. The first coordinate of the x-intercept is the zero of the function. Thus from Exercise 59 we know that the x-intercept is $(12, 0)$.

63. $y = -\tan x$

Reflect the graph of $y = \tan x$ across the x-axis.

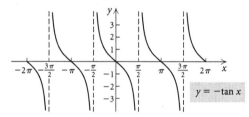

$$y = -\tan x$$

65. $y = -2 + \cot x$

Translate the graph of $y = \cot x$ down 2 units.

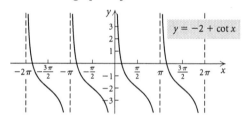

$$y = -2 + \cot x$$

67. $y = 2\tan \dfrac{1}{2}x$

Stretch the graph of $y = \tan x$ horizontally and vertically, both by a factor of 2.

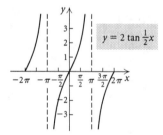

$$y = 2\tan \tfrac{1}{2}x$$

69. $y = 2\sec(x - \pi)$

Stretch the graph of $y = \sec x$ vertically by a factor of 2 and translate it to the right π units.

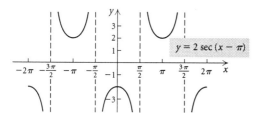

71. $y = 2\csc\left(\frac{1}{2}x - \frac{3\pi}{4}\right) = 2\csc\left[\frac{1}{2}\left(x - \frac{3\pi}{2}\right)\right]$

Stretch the graph of $y = \csc x$ horizontally and vertically, both by a factor of 2. Then translate it to the right $3\pi/2$ units.

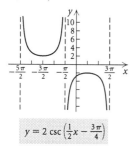

73. Graph $y = \dfrac{\sin x}{x}$ and use the Zero feature to find the zeros in the interval $[-12, 12]$. They are approximately -9.42, -6.28, -3.14, 3.14, 6.28, and 9.42.

75. Graph $y = x^3 \sin x$ and use the Zero feature to find the zeros in the interval $[-12, 12]$. They are approximately -3.14, 0, and 3.14.

77. a) $y = 101.6 + 3\sin\left(\frac{\pi}{8}x\right)$

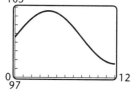

b) The maximum value occurs on day 4 when the sine function takes its maximum value, 1. It is $101.6° + 3° \cdot 1$, or $104.6°$.

The minimum value occurs on day 12 when the sine function takes its minimum value, -1. It is $101.6° + 3°(-1)$, or $98.6°$.

79. a) $y = 3000\left[\cos\frac{\pi}{45}(x - 10)\right]$

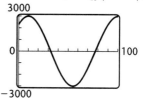

b) Amplitude: $|A| = |3000| = 3000$

Period: $\left|\dfrac{2\pi}{C}\right| = \left|\dfrac{2\pi}{\frac{\pi}{45}}\right| = 90$

Phase shift: $\dfrac{D}{C} = 10$

81. a) $y = 6e^{-0.8x}\cos(6\pi x) + 4$

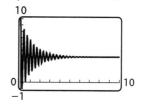

b) As t increases, $e^{-0.8t}$ decreases, and hence $6e^{-0.8t}\cos(6\pi t)$ decreases in the long run and approaches 0. Then the spring would be 4 inches from the ceiling when it stops bobbing.

Chapter 6

Trigonometric Identities, Inverse Functions, and Equations

Exercise Set 6.1

1. $(\sin x - \cos x)(\sin x + \cos x)$

$= \sin^2 x - \cos^2 x$

3. $\cos y \sin y (\sec y + \csc y)$

$= \cos y \sin y \left(\dfrac{1}{\cos y} + \dfrac{1}{\sin y} \right)$

$= \sin y + \cos y$

5. $(\sin \phi - \cos \phi)^2$

$= \sin^2 \phi - 2 \sin \phi \cos \phi + \cos^2 \phi$

$= 1 - 2 \sin \phi \cos \phi \qquad (\sin^2 \phi + \cos^2 \phi = 1)$

7. $(\sin x + \csc x)(\sin^2 x + \csc^2 x - 1)$

$= \sin^3 x + \sin x \csc^2 x - \sin x + \csc x \sin^2 x +$

$\quad \csc^3 x - \csc x$

$= \sin^3 x + \sin x \cdot \dfrac{1}{\sin^2 x} - \sin x +$

$\quad \dfrac{1}{\sin x} \cdot \sin^2 x + \csc^3 x - \dfrac{1}{\sin x}$

$= \sin^3 x + \dfrac{1}{\sin x} - \sin x + \sin x + \csc^3 x - \dfrac{1}{\sin x}$

$= \sin^3 x + \csc^3 x$

9. $\sin x \cos x + \cos^2 x$

$= \cos x (\sin x + \cos x)$

11. $\sin^4 x - \cos^4 x$

$= (\sin^2 x + \cos^2 x)(\sin^2 x - \cos^2 x)$

$= \sin^2 x - \cos^2 x$

$= (\sin x + \cos x)(\sin x - \cos x)$

13. $2 \cos^2 x + \cos x - 3$

$= (2 \cos x + 3)(\cos x - 1)$

15. $\sin^3 x + 27$

$= (\sin x)^3 + 3^3$

$= (\sin x + 3)(\sin^2 x - 3 \sin x + 9)$

17. $\dfrac{\sin^2 x \cos x}{\cos^2 x \sin x}$

$= \dfrac{\sin x}{\cos x} \cdot \dfrac{\sin x \cos x}{\sin x \cos x}$

$= \dfrac{\sin x}{\cos x}$

$= \tan x$

19. $\dfrac{\sin^2 x + 2 \sin x + 1}{\sin x + 1}$

$= \dfrac{(\sin x + 1)^2}{\sin x + 1}$

$= \sin x + 1$

21. $\dfrac{4 \tan t \sec t + 2 \sec t}{6 \tan t \sec t + 2 \sec t}$

$= \dfrac{2 \sec t (2 \tan t + 1)}{2 \sec t (3 \tan t + 1)}$

$= \dfrac{2 \tan t + 1}{3 \tan t + 1}$

23. $\dfrac{\sin^4 x - \cos^4 x}{\sin^2 x - \cos^2 x}$

$= \dfrac{(\sin^2 x + \cos^2 x)(\sin^2 x - \cos^2 x)}{\sin^2 x - \cos^2 x}$

$= 1 \qquad (\sin^2 x + \cos^2 x = 1)$

25. $\dfrac{5 \cos \phi}{\sin^2 \phi} \cdot \dfrac{\sin^2 \phi - \sin \phi \cos \phi}{\sin^2 \phi - \cos^2 \phi}$

$= \dfrac{5 \cos \phi \sin \phi (\sin \phi - \cos \phi)}{\sin^2 \phi (\sin \phi + \cos \phi)(\sin \phi - \cos \phi)}$

$= \dfrac{5 \cos \phi}{\sin \phi (\sin \phi + \cos \phi)}$

$= \dfrac{5 \cot \phi}{\sin \phi + \cos \phi}$

27. $\dfrac{1}{\sin^2 s - \cos^2 s} - \dfrac{2}{\cos s - \sin s}$

$= \dfrac{1}{\sin^2 s - \cos^2 s} - \dfrac{2}{-(-\cos s + \sin s)}$

$= \dfrac{1}{\sin^2 s - \cos^2 s} + \dfrac{2}{\sin s - \cos x}$

$= \dfrac{1}{\sin^2 s - \cos^2 s} + \dfrac{2}{\sin s - \cos s} \cdot \dfrac{\sin s + \cos s}{\sin s + \cos s}$

$= \dfrac{1 + 2 \sin s + 2 \cos s}{\sin^2 s - \cos^2 s}$

29. $\dfrac{\sin^2\theta - 9}{2\cos\theta + 1} \cdot \dfrac{10\cos\theta + 5}{3\sin\theta + 9}$

$= \dfrac{(\sin\theta + 3)(\sin\theta - 3)}{2\cos\theta + 1} \cdot \dfrac{5(2\cos\theta + 1)}{3(\sin\theta + 3)}$

$= \dfrac{5(\sin\theta - 3)}{3}$

31. $\sqrt{\sin^2 x \cos x} \cdot \sqrt{\cos x}$

$= \sqrt{\sin^2 x \cos^2 x}$

$= \sin x \cos x$

33. $\sqrt{\cos\alpha\sin^2\alpha} - \sqrt{\cos^3\alpha}$

$= \sin\alpha\sqrt{\cos\alpha} - \cos\alpha\sqrt{\cos\alpha}$

$= \sqrt{\cos\alpha}(\sin\alpha - \cos\alpha)$

35. $(1 - \sqrt{\sin y})(\sqrt{\sin y} + 1)$

$= (1 - \sqrt{\sin y})(1 + \sqrt{\sin y})$

$= 1 - \sin y \qquad [(\sqrt{\sin y})^2 = \sin y]$

37. $\sqrt{\dfrac{\sin x}{\cos x}}$

$= \sqrt{\dfrac{\sin x}{\cos x} \cdot \dfrac{\cos x}{\cos x}}$

$= \sqrt{\dfrac{\sin x \cos x}{\cos^2 x}}$

$= \dfrac{\sqrt{\sin x \cos x}}{\cos x}$

(Note that $\sqrt{\dfrac{\sin x}{\cos x}}$ could also be expressed as $\sqrt{\tan x}$.)

39. $\sqrt{\dfrac{\cos^2 y}{2\sin^2 y}} = \sqrt{\dfrac{\cot^2 y}{2} \cdot \dfrac{2}{2}}$

$= \dfrac{\sqrt{2}\cot y}{2}$

41. $\sqrt{\dfrac{\cos x}{\sin x}} = \sqrt{\dfrac{\cos x}{\sin x} \cdot \dfrac{\cos x}{\cos x}}$

$= \sqrt{\dfrac{\cos^2 x}{\sin x \cos x}}$

$= \dfrac{\cos x}{\sqrt{\sin x \cos x}}$

43. $\sqrt{\dfrac{1 + \sin y}{1 - \sin y}} = \sqrt{\dfrac{1 + \sin y}{1 - \sin y} \cdot \dfrac{1 + \sin y}{1 + \sin y}}$

$= \sqrt{\dfrac{(1 + \sin y)^2}{1 - \sin^2 y}}$

$= \dfrac{1 + \sin y}{\sqrt{\cos^2 y}}$

$= \dfrac{1 + \sin y}{\cos y}$

45. $\sqrt{a^2 - x^2} = \sqrt{a^2 - (a\sin\theta)^2}$ Substituting

$= \sqrt{a^2 - a^2\sin^2\theta}$

$= \sqrt{a^2(1 - \sin^2\theta)}$

$= \sqrt{a^2\cos^2\theta}$

$= a\cos\theta \qquad \left(a > 0 \text{ and } \cos\theta > 0 \right.$

$\left. \text{for } 0 < \theta < \dfrac{\pi}{2}\right)$

Then $\cos\theta = \dfrac{\sqrt{a^2 - x^2}}{a}$.

Also $x = a\sin\theta$, so $\sin\theta = \dfrac{x}{a}$. Then

$\tan\theta = \dfrac{\sin\theta}{\cos\theta} = \dfrac{\dfrac{x}{a}}{\dfrac{\sqrt{a^2 - x^2}}{a}} = \dfrac{x}{a} \cdot \dfrac{a}{\sqrt{a^2 - x^2}}$

$= \dfrac{x}{\sqrt{a^2 - x^2}}$.

47. $x = 3\sec\theta$

$x = \dfrac{3}{\cos\theta} \qquad \left(\sec\theta = \dfrac{1}{\cos\theta}\right)$

$\cos\theta = \dfrac{3}{x}$

$\sqrt{x^2 - 9} = \sqrt{(3\sec\theta)^2 - 9}$ Substituting

$= \sqrt{9\sec^2\theta - 9}$

$= \sqrt{9(\sec^2\theta - 1)}$

$= \sqrt{9\tan^2\theta}$

$= 3\tan\theta \qquad \left(\tan\theta > 0 \text{ for } 0 < \theta < \dfrac{\pi}{2}\right)$

Then $\tan\theta = \dfrac{\sqrt{x^2 - 9}}{3}$

$\dfrac{\sin\theta}{\cos\theta} = \dfrac{\sqrt{x^2 - 9}}{3}$

$\dfrac{\sin\theta}{\dfrac{3}{x}} = \dfrac{\sqrt{x^2 - 9}}{3} \qquad \left(\cos\theta = \dfrac{3}{x}\right)$

$\sin\theta = \dfrac{3}{x} \cdot \dfrac{\sqrt{x^2 - 9}}{3}$

$\sin\theta = \dfrac{\sqrt{x^2 - 9}}{x}$.

49. $\dfrac{x^2}{\sqrt{1 - x^2}} = \dfrac{\sin^2\theta}{\sqrt{1 - \sin^2\theta}}$ Substituting

$= \dfrac{\sin^2\theta}{\sqrt{\cos^2\theta}}$

$= \dfrac{\sin^2\theta}{\cos\theta} = \sin\theta \cdot \dfrac{\sin\theta}{\cos\theta}$

$= \sin\theta\tan\theta$

51. $\dfrac{\pi}{12} = \dfrac{3\pi}{12} - \dfrac{2\pi}{12} = \dfrac{\pi}{4} - \dfrac{\pi}{6}$

$\sin\dfrac{\pi}{12} = \sin\left(\dfrac{\pi}{4} - \dfrac{\pi}{6}\right)$

$\qquad = \sin\dfrac{\pi}{4}\cos\dfrac{\pi}{6} - \cos\dfrac{\pi}{4}\sin\dfrac{\pi}{6}$

$\qquad = \dfrac{\sqrt{2}}{2}\cdot\dfrac{\sqrt{3}}{2} - \dfrac{\sqrt{2}}{2}\cdot\dfrac{1}{2}$

$\qquad = \dfrac{\sqrt{6} - \sqrt{2}}{4}$

53. $\tan 105° = \tan(45° + 60°)$

$\qquad = \dfrac{\tan 45° + \tan 60°}{1 - \tan 45° \tan 60°}$

$\qquad = \dfrac{1 + \sqrt{3}}{1 - 1\cdot\sqrt{3}}$

$\qquad = \dfrac{1 + \sqrt{3}}{1 - \sqrt{3}}$

(This is equivalent to $\dfrac{\sqrt{6} + \sqrt{2}}{\sqrt{2} - \sqrt{6}}$ and $-2 - \sqrt{3}$.)

55. $\cos 15° = \cos(45° - 30°)$

$\qquad = \cos 45° \cos 30° + \sin 45° \sin 30°$

$\qquad = \dfrac{\sqrt{2}}{2}\cdot\dfrac{\sqrt{3}}{2} + \dfrac{\sqrt{2}}{2}\cdot\dfrac{1}{2}$

$\qquad = \dfrac{\sqrt{6} + \sqrt{2}}{4}$

57. $\quad \sin 37° \cos 22° + \cos 37° \sin 22°$

$= \sin(37° + 22°)$

$= \sin 59° \approx 0.8572$

59. $\dfrac{\tan 20° + \tan 32°}{1 - \tan 20° \tan 32°} = \tan(20° + 32°)$

$\qquad\qquad\qquad\qquad = \tan 52° \approx 1.2799$

Use the figures and function values below for Exercises 61 and 63.

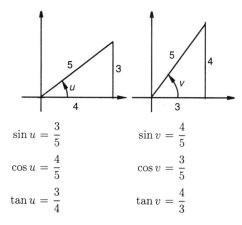

$\sin u = \dfrac{3}{5}$ $\qquad\qquad$ $\sin v = \dfrac{4}{5}$

$\cos u = \dfrac{4}{5}$ $\qquad\qquad$ $\cos v = \dfrac{3}{5}$

$\tan u = \dfrac{3}{4}$ $\qquad\qquad$ $\tan v = \dfrac{4}{3}$

61. See the figures above.

$\cos(u + v) = \cos u \cos v - \sin u \sin v$

$\qquad = \dfrac{4}{5}\cdot\dfrac{3}{5} - \dfrac{3}{5}\cdot\dfrac{4}{5}$

$\qquad = \dfrac{12}{25} - \dfrac{12}{25} = 0$

63. See the figures before Exercise 61.

$\sin(u - v) = \sin u \cos v - \cos u \sin v$

$\qquad = \dfrac{3}{5}\cdot\dfrac{3}{5} - \dfrac{4}{5}\cdot\dfrac{4}{5}$

$\qquad = -\dfrac{7}{25}$

Use the figures and function values below for Exercise 65.

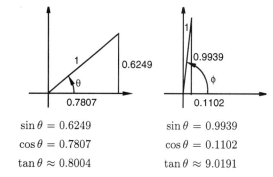

$\sin\theta = 0.6249$ $\qquad\qquad$ $\sin\theta = 0.9939$

$\cos\theta = 0.7807$ $\qquad\qquad$ $\cos\theta = 0.1102$

$\tan\theta \approx 0.8004$ $\qquad\qquad$ $\tan\theta \approx 9.0191$

65. See the figures before Exercise 64.

$\tan(\theta + \phi) = \dfrac{\tan\theta + \tan\phi}{1 - \tan\theta \tan\phi}$

$\qquad = \dfrac{0.8004 + 9.0191}{1 - 0.8004(9.0191)}$

$\qquad \approx -1.5790$

(Answer may vary slightly due to rounding differences.)

67. $\quad \sin(\alpha + \beta) + \sin(\alpha - \beta)$

$= (\sin\alpha\cos\beta + \cos\alpha\sin\beta) +$

$\qquad\qquad (\sin\alpha\cos\beta - \cos\alpha\sin\beta)$

$= 2\sin\alpha\cos\beta$

69. $\quad \cos(u + v)\cos v + \sin(u + v)\sin v$

$= (\cos u \cos v - \sin u \sin v)\cos v +$

$\qquad\qquad (\sin u \cos v + \cos u \sin v)\sin v$

$= \cos u \cos^2 v - \sin u \sin v \cos v +$

$\qquad\qquad \sin u \cos v \sin v + \cos u \sin^2 v$

$= \cos u(\cos^2 v + sin^2 v)$

$= \cos u$

71. Discussion and Writing

73. $2x - 3 = 2\left(x - \dfrac{3}{2}\right)$

$2x - 3 = 2x - 3$

$-3 = -3$ Subtracting $2x$ on both sides

The equation is true for all real numbers, so the solution set is the set of all real numbers.

75. $59°$ and $31°$ are complementary angles.   Then $\cos 59° = \sin 31° = 0.5150$, so

$$\sec 59° = \frac{1}{\cos 59°} = \frac{1}{0.5150} \approx 1.9417.$$

77. Solve each equation for y to determine the slope of each line.

$l_1: \; 2x = 3 - 2y$ 　　　　 $l_2: \; x + y = 5$

$\qquad y = -x + \dfrac{3}{2}$ 　　　　　 $y = -x + 5$

Thus $m_1 = -1$ and 　　 Thus $m_2 = -1$ and

the y-intercept is $\dfrac{3}{2}$. 　 the y-intercept is 5.

The lines are parallel, so they do not form an angle. When the formula is used, the result is $0°$.

79. Find the slope of each line.

$l_1: y = 3$ 　　　 $(y = 0x + 3)$

Thus $m = 0$.

$l_2: \; x + y = 5$

$\qquad y = -x + 5$

Thus $m = -1$.

Let ϕ be the smallest angle from l_1 to l_2.

$\tan \phi = \dfrac{-1 - 0}{1 + (-1)(0)}$ 　　 $\left(\tan \phi = \dfrac{m_2 - m_1}{1 + m_2 m_1}\right)$

$\qquad = \dfrac{-1}{1} = -1$

Since $\tan \phi$ is negative, we know that ϕ is obtuse. Thus $\phi = \dfrac{3\pi}{4}$, or $135°$.

81.

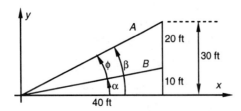

Position a set of coordinate axes as shown above. Let l_1 be the line containing the points $(0, 0)$ and $(40, 10)$. Find its slope.

$$m_1 = \tan \alpha = \frac{10}{40} = \frac{1}{4}$$

Let l_2 be the line containing the points $(0, 0)$ and $(40, 30)$. Find its slope.

$$m_2 = \tan \beta = \frac{30}{40} = \frac{3}{4}$$

Now ϕ is the smallest positive angle from l_1 to l_2.

$$\tan \phi = \frac{\dfrac{3}{4} - \dfrac{1}{4}}{1 + \dfrac{3}{4} \cdot \dfrac{1}{4}} = \frac{\dfrac{1}{2}}{\dfrac{19}{16}} = \frac{8}{19} \approx 0.4211$$

$$\phi \approx 22.83°$$

83. See the answer section in the text.

85. See the answer section in the text.

87. See the answer section in the text.

89. See the answer section in the text.

91. 　　　　　　　$\tan \phi = \dfrac{m_2 - m_1}{1 + m_2 m_1}$

$$\tan 30° = \frac{\dfrac{2}{3} - m_1}{1 + \dfrac{2}{3} m_1}$$

$$\frac{\sqrt{3}}{3} = \frac{\dfrac{2}{3} - m_1}{1 + \dfrac{2}{3} m_1}$$

$$\frac{\sqrt{3}}{3} + \frac{2\sqrt{3}}{9} m_1 = \frac{2}{3} - m_1$$

$$\frac{2\sqrt{3}}{9} m_1 + m_1 = \frac{2}{3} - \frac{\sqrt{3}}{3}$$

$$m_1 \left(\frac{2\sqrt{3}}{9} + 1\right) = \frac{2 - \sqrt{3}}{3}$$

$$m_1 \left(\frac{2\sqrt{3} + 9}{9}\right) = \frac{2 - \sqrt{3}}{3}$$

$$m_1 = \frac{2 - \sqrt{3}}{3} \cdot \frac{9}{2\sqrt{3} + 9}$$

$$m_1 = \frac{3(2 - \sqrt{3})}{2\sqrt{3} + 9} = \frac{6 - 3\sqrt{3}}{2\sqrt{3} + 9}$$

$$m_1 \approx 0.0645$$

93. Find the slope of l_1.

$m_1 = \dfrac{-2 - 7}{-3 - (-3)} = \dfrac{-9}{0}$, which is undefined so l_1 is vertical.

Find the slope of l_2.

$$m_2 = \frac{6 - (-4)}{2 - 0} = \frac{10}{2} = 5$$

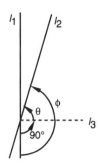

From the drawing we see that $\phi = 90° + \theta$, where θ is the smallest positive angle from the horizontal line l_3 to l_2. We find θ, recalling that the slope of a horizontal line is 0, so $m_3 = 0$.

$$\tan\theta = \frac{m_2 - m_3}{1 + m_2 m_3}$$

$$\tan\theta = \frac{5 - 0}{1 + 5 \cdot 0} = 5$$

$$\theta \approx 78.7°$$

Then $\phi \approx 90° + 78.7° \approx 168.7°$.

95.
$$\begin{aligned}
\cos 2\theta &= \cos(\theta + \theta) \\
&= \cos\theta\cos\theta - \sin\theta\sin\theta \\
&= \cos^2 - \sin^2\theta, \\
&\text{or } 1 - 2\sin^2\theta, \qquad (\cos^2\theta = 1 - \sin^2\theta) \\
&\text{or } 2\cos^2\theta - 1 \qquad (\sin^2\theta = 1 - \cos^2\theta)
\end{aligned}$$

97. See the answer section in the text.

99. See the answer section in the text.

Exercise Set 6.2

1. $\sin(3\pi/10) \approx 0.8090$, $\cos(3\pi/10) \approx 0.5878$

a) $\tan\dfrac{3\pi}{10} = \dfrac{\sin(3\pi/10)}{\cos(3\pi/10)} \approx \dfrac{0.8090}{0.5878} \approx 1.3763$,

$\csc\dfrac{3\pi}{10} = \dfrac{1}{\sin(3\pi/10)} \approx \dfrac{1}{0.8090} \approx 1.2361$,

$\sec\dfrac{3\pi}{10} = \dfrac{1}{\cos(3\pi/10)} \approx \dfrac{1}{0.5878} \approx 1.7013$,

$\cot\dfrac{3\pi}{10} = \dfrac{1}{\tan(3\pi/10)} \approx \dfrac{1}{1.3763} \approx 0.7266$

b) $\dfrac{\pi}{2} - \dfrac{3\pi}{10} = \dfrac{2\pi}{10} = \dfrac{\pi}{5}$, so $\dfrac{3\pi}{10}$ and $\dfrac{\pi}{5}$ are complements.

$\sin\dfrac{\pi}{5} = \cos\dfrac{3\pi}{10} \approx 0.5878$,

$\cos\dfrac{\pi}{5} = \sin\dfrac{3\pi}{10} \approx 0.8090$,

$\tan\dfrac{\pi}{5} = \cot\dfrac{3\pi}{10} \approx 0.7266$,

$\csc\dfrac{\pi}{5} = \sec\dfrac{3\pi}{10} \approx 1.7013$,

$\sec\dfrac{\pi}{5} = \csc\dfrac{3\pi}{10} \approx 1.2361$,

$\cot\dfrac{\pi}{5} = \tan\dfrac{3\pi}{10} \approx 1.3763$

3. We sketch a second quadrant triangle.

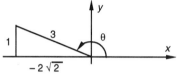

a) $\cos\theta = \dfrac{-2\sqrt{2}}{3} = -\dfrac{2\sqrt{2}}{3}$,

$\tan\theta = \dfrac{1}{-2\sqrt{2}} = -\dfrac{\sqrt{2}}{4}$,

$\csc\theta = \dfrac{3}{1} = 3$,

$\sec\theta = \dfrac{3}{-2\sqrt{2}} = -\dfrac{3\sqrt{2}}{4}$,

$\cot\theta = \dfrac{-2\sqrt{2}}{1} = -2\sqrt{2}$

b) Since θ and $\dfrac{\pi}{2} - \theta$ are complements, we have

$\sin\left(\dfrac{\pi}{2} - \theta\right) = \cos\theta = -\dfrac{2\sqrt{2}}{3}$,

$\cos\left(\dfrac{\pi}{2} - \theta\right) = \sin\theta = \dfrac{1}{3}$,

$\tan\left(\dfrac{\pi}{2} - \theta\right) = \cot\theta = -2\sqrt{2}$,

$\csc\left(\dfrac{\pi}{2} - \theta\right) = \sec\theta = -\dfrac{3\sqrt{2}}{4}$,

$\sec\left(\dfrac{\pi}{2} - \theta\right) = \csc\theta = 3$,

$\cot\left(\dfrac{\pi}{2} - \theta\right) = \tan\theta = -\dfrac{\sqrt{2}}{4}$

c) $\sin\left(\theta - \dfrac{\pi}{2}\right) = \sin\left[-\left(\dfrac{\pi}{2} - \theta\right)\right] =$

$-\sin\left(\dfrac{\pi}{2} - \theta\right) = -\left(-\dfrac{2\sqrt{2}}{3}\right) = \dfrac{2\sqrt{2}}{3},$

$\cos\left(\theta - \dfrac{\pi}{2}\right) = \cos\left[-\left(\dfrac{\pi}{2} - \theta\right)\right] = \cos\left(\dfrac{\pi}{2} - \theta\right)$

$= \dfrac{1}{3},$

$\tan\left(\theta - \dfrac{\pi}{2}\right) = \tan\left[-\left(\dfrac{\pi}{2} - \theta\right)\right] =$

$-\tan\left(\dfrac{\pi}{2} - \theta\right) = -(-2\sqrt{2}) = 2\sqrt{2},$

$\csc\left(\theta - \dfrac{\pi}{2}\right) = \dfrac{1}{\sin\left(\theta - \dfrac{\pi}{2}\right)} = \dfrac{1}{\dfrac{2\sqrt{2}}{3}} = \dfrac{3}{2\sqrt{2}} =$

$\dfrac{3\sqrt{2}}{4},$

$\sec\left(\theta - \dfrac{\pi}{2}\right) = \dfrac{1}{\cos\left(\theta - \dfrac{\pi}{2}\right)} = \dfrac{1}{\dfrac{1}{3}} = 3,$

$\cot\left(\theta - \dfrac{\pi}{2}\right) = \dfrac{1}{\tan\left(\theta - \dfrac{\pi}{2}\right)} = \dfrac{1}{2\sqrt{2}} = \dfrac{\sqrt{2}}{4}$

5. $\sec\left(x + \dfrac{\pi}{2}\right) = \dfrac{1}{\cos\left(x + \dfrac{\pi}{2}\right)} = \dfrac{1}{-\sin x} = -\csc x$

7. $\tan\left(x - \dfrac{\pi}{2}\right) = \tan\left[-\left(\dfrac{\pi}{2} - x\right)\right] =$

$\dfrac{\sin\left[-\left(\dfrac{\pi}{2} - x\right)\right]}{\cos\left[-\left(\dfrac{\pi}{2} - x\right)\right]} = \dfrac{-\sin\left(\dfrac{\pi}{2} - x\right)}{\cos\left(\dfrac{\pi}{2} - x\right)} =$

$-\tan\left(\dfrac{\pi}{2} - x\right) = -\cot x$

9. Make a drawing.

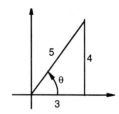

From this drawing we find that $\cos\theta = \dfrac{3}{5}$ and $\tan\theta = \dfrac{4}{3}.$

$\sin 2\theta = 2\sin\theta\cos\theta = 2 \cdot \dfrac{4}{5} \cdot \dfrac{3}{5} = \dfrac{24}{25}$

$\cos 2\theta = \cos^2\theta - \sin^2\theta = \left(\dfrac{3}{5}\right)^2 - \left(\dfrac{4}{5}\right)^2$

$= \dfrac{9}{25} - \dfrac{16}{25} = -\dfrac{7}{25}$

$\tan 2\theta = \dfrac{2\tan\theta}{1 - \tan^2\theta} = \dfrac{2 \cdot \dfrac{4}{3}}{1 - \left(\dfrac{4}{3}\right)^2} = \dfrac{\dfrac{8}{3}}{-\dfrac{7}{9}} = -\dfrac{24}{7}$

(We could have found $\tan 2\theta$ by dividing:
$\tan 2\theta = \dfrac{\sin 2\theta}{\cos 2\theta}$.)

Since $\sin 2\theta$ is positive and $\cos 2\theta$ is negative, 2θ is in quadrant II.

11. Make a drawing.

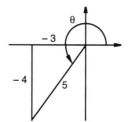

From this drawing we find that $\sin\theta = -\dfrac{4}{5}$ and $\tan\theta = \dfrac{4}{3}.$

$\sin 2\theta = 2\sin\theta\cos\theta = 2\left(-\dfrac{4}{5}\right)\left(-\dfrac{3}{5}\right) = \dfrac{24}{25}$

$\cos 2\theta = \cos^2\theta - \sin^2\theta = \left(-\dfrac{3}{5}\right)^2 - \left(-\dfrac{4}{5}\right)^2 =$

$-\dfrac{7}{25}$

$\tan 2\theta = \dfrac{\sin 2\theta}{\cos 2\theta} = -\dfrac{24}{7}$

(We could have found $\tan 2\theta$ using a double-angle identity.)

Since $\sin 2\theta$ is positive and $\cos 2\theta$ is negative, 2θ is in quadrant II.

13. Make a drawing.

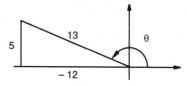

From this drawing we find that $\sin\theta = \dfrac{5}{13}$ and $\cos\theta = -\dfrac{12}{13}.$

$$\sin 2\theta = 2\sin\theta\cos\theta = 2\cdot\frac{5}{13}\left(-\frac{12}{13}\right) = -\frac{120}{169}$$

$$\cos 2\theta = \cos^2\theta - \sin^2\theta = \left(-\frac{12}{13}\right)^2 - \left(\frac{5}{13}\right)^2 = \frac{119}{169}$$

$$\tan 2\theta = \frac{\sin 2\theta}{\cos 2\theta} = \frac{-\dfrac{120}{169}}{\dfrac{119}{169}} = -\frac{120}{119}$$

(We could have found $\tan 2\theta$ using a double-angle identity.)

Since $\sin 2\theta$ is negative and $\cos 2\theta$ is positive, 2θ is in quadrant IV.

15.
$$\cos 4x$$
$$= \cos[2(2x)]$$
$$= 1 - 2\sin^2 2x$$
$$= 1 - 2(2\sin x\cos x)^2$$
$$= 1 - 8\sin^2 x\cos^2 x$$

or
$$\cos 4x$$
$$= \cos[2(2x)]$$
$$= \cos^2 2x - \sin^2 2x$$
$$= (\cos^2 x - \sin^2 x)^2 - (2\sin x\cos x)^2$$
$$= \cos^4 x - 6\sin^2 x\cos^2 x + \sin^4 x$$

or
$$\cos 4x$$
$$= \cos[2(2x)]$$
$$= 2\cos^2 2x - 1$$
$$= 2(2\cos^2 x - 1)^2 - 1$$
$$= 8\cos^4 x - 8\cos^2 x + 1$$

17. $\cos 15° = \cos\dfrac{30°}{2} = \sqrt{\dfrac{1 + \cos 30°}{2}} = \sqrt{\dfrac{1 + \sqrt{3}/2}{2}} =$

$$\sqrt{\frac{2 + \sqrt{3}}{4}} = \frac{\sqrt{2 + \sqrt{3}}}{2}$$

(We choose the positive square root since $15°$ is in quadrant I where the cosine function is positive.)

19. $\sin 112.5° = \sin\dfrac{225°}{2} = \sqrt{\dfrac{1 - \cos 225°}{2}} =$

$$\sqrt{\frac{1 - (-\sqrt{2}/2)}{2}} = \sqrt{\frac{2 + \sqrt{2}}{4}} = \frac{\sqrt{2 + \sqrt{2}}}{2}$$

(We choose the positive square root since $112.5°$ is in quadrant II where the sine function is positive.)

21. $\tan 75° = \tan\dfrac{150°}{2} = \dfrac{\sin 150°}{1 + \cos 150°} =$

$$\frac{\dfrac{1}{2}}{1 + (-\sqrt{3}/2)} = \frac{1}{2 - \sqrt{3}}, \text{ or } 2 + \sqrt{3}$$

23. First find $\cos\theta$.
$$\sin^2\theta + \cos^2\theta = 1$$
$$(0.3416)^2 + \cos^2\theta = 1$$
$$\cos^2\theta = 1 - (0.3416)^2$$
$$\cos\theta \approx 0.9398 \quad (\theta \text{ is in quadrant I.})$$
$$\sin 2\theta = 2\sin\theta\cos\theta$$
$$\approx 2(0.3416)(0.9398)$$
$$\approx 0.6421$$

25. $\sin\dfrac{\theta}{2} = \sqrt{\dfrac{1 - \cos\theta}{2}} \qquad \left(\dfrac{\theta}{2} \text{ is in quadrant I.}\right)$

$$\approx \sqrt{\frac{1 - 0.9398}{2}} \qquad \begin{array}{l}\text{We found }\cos\theta\text{ in}\\ \text{Exercise 23.}\end{array}$$
$$\approx 0.1735$$

27. Using a grapher we see that the graphs of $y_1 = \dfrac{\cos 2x}{\cos x - \sin x}$ and $y_2 = \cos x + \sin x$ are the same on $[-2\pi, 2\pi]$. The correct choice is (d). See the answer section in the text for the algebraic proof.

29. Using a grapher we see that the graphs of $y_1 = \dfrac{\sin 2x}{2\cos x}$ and $y_2 = \sin x$ are the same on $[-2\pi, 2\pi]$. The correct choice is (d). See the answer section in the text for the algebraic proof.

31. $2\cos^2\dfrac{x}{2} - 1 = \cos\left(2\cdot\dfrac{x}{2}\right) = \cos x$

33.
$$(\sin x - \cos x)^2 + (\sin 2x)$$
$$= (\sin^2 x - 2\sin x\cos x + \cos^2 x) +$$
$$\quad (2\sin x\cos x)$$
$$= \sin^2 x + \cos^2 x$$
$$= 1$$

35. $\dfrac{2 - \sec^2 x}{\sec^2 x} = \dfrac{2}{\sec^2 x} - 1 = 2\cos^2 x - 1 = \cos 2x$

37.
$$(-4\cos x \sin x + 2\cos 2x)^2 +$$
$$(2\cos 2x + 4\sin x \cos x)^2$$
$$= (16\cos^2 x \sin^2 x - 16\sin x \cos x \cos 2x +$$
$$4\cos^2 2x) + (4\cos^2 2x + 16\sin x \cos x \cos 2x +$$
$$16\sin^2 x \cos^2 x)$$
$$= 8\cos^2 2x + 32\cos^2 x \sin^2 x$$
$$= 8(\cos^2 x - \sin^2 x)^2 + 32\cos^2 x \sin^2 x$$
$$= 8(\cos^4 x - 2\cos^2 x \sin^2 x + \sin^4 x) +$$
$$32\cos^2 x \sin^2 x$$
$$= 8\cos^4 x - 16\cos^2 x \sin^2 x + 8\sin^4 x +$$
$$32\cos^2 x \sin^2 x$$
$$= 8\cos^4 x + 16\cos^2 x \sin^2 x + 8\sin^4 x$$
$$= 8(\cos^4 x + 2\cos^2 x \sin^2 x + \sin^4 x)$$
$$= 8(\cos^2 x + \sin^2 x)^2$$
$$= 8$$

39. Discussion and Writing

41. $1 - \cos^2 x = \sin^2 x \qquad (\sin^2 x + \cos^2 x = 1)$

43. $\sin^2 x - 1 = -\cos^2 x \qquad (\sin^2 x + \cos^2 x = 1)$

45. Observe that $141° = 51° + 90°$.

Find $\sin 51°$:
$$\sin^2 51° + \cos^2 51° = 1$$
$$\sin^2 51° + (0.6293)^2 = 1$$
$$\sin^2 51° = 1 - (0.6293)^2$$
$$\sin 51° \approx 0.7772 \qquad (51° \text{ is}$$
$$\text{in quadrant I.)}$$

$\sin 141° = \cos 51° \approx 0.6293$

$\cos 141° = -\sin 51° \approx -0.7772$

$\tan 141° = \dfrac{\sin 141°}{\cos 141°} \approx \dfrac{0.6293}{-0.7772} \approx -0.8097$

$\csc 141° = \dfrac{1}{\sin 141°} \approx \dfrac{1}{0.6293} \approx 1.5891$

$\sec 141° = \dfrac{1}{\cos 141°} \approx \dfrac{1}{-0.7772} \approx -1.2867$

$\cot 141° = \dfrac{1}{\tan 141°} \approx \dfrac{1}{-0.8097} \approx -1.2350$

47.
$$\cos(\pi - x) + \cot x \sin\left(x - \frac{\pi}{2}\right)$$
$$= \cos\pi \cos x + \sin\pi \sin x +$$
$$\cot x(-\cos x) \qquad \left[\sin\left(x - \frac{\pi}{2}\right) = -\cos x\right]$$
$$= -\cos x + 0 - \cot x \cos x$$
$$= -\cos x(1 + \cot x)$$

49.
$$\frac{\cos^2 y \sin\left(y + \dfrac{\pi}{2}\right)}{\sin^2 y \sin\left(\dfrac{\pi}{2} - y\right)} = \frac{\cos^2 y \cos y}{\sin^2 y \cos y}$$
$$= \frac{\cos^2 y}{\sin^2 y}$$
$$= \cot^2 y$$

51. Since $\pi < \theta \leq \dfrac{3\pi}{2}$, $\tan\theta$ is positive and $\sin\theta$ and $\cos\theta$ are negative.
$$\tan\frac{\theta}{2} = -\sqrt{\frac{1 - \cos\theta}{1 + \cos\theta}}$$
$$-\frac{5}{3} = -\sqrt{\frac{1 - \cos\theta}{1 + \cos\theta}}$$
$$\frac{25}{9} = \frac{1 - \cos\theta}{1 + \cos\theta}$$
$$25 + 25\cos\theta = 9 - 9\cos\theta$$
$$34\cos\theta = -16$$
$$\cos\theta = -\frac{16}{34} = -\frac{8}{17}$$
$$\sin\theta = -\sqrt{1 - \cos^2\theta} = -\sqrt{1 - \left(-\frac{8}{17}\right)^2} = -\frac{15}{17}$$
$$\tan\theta = \frac{\sin\theta}{\cos\theta} = \frac{-\dfrac{15}{17}}{-\dfrac{8}{17}} = \frac{15}{8}$$

53. a) Substitute $42°$ for ϕ.
$$g = 9.78049[1 + 0.005288\sin^2 42° -$$
$$0.000006\sin^2(2\cdot 42°)]$$
$$\approx 9.80359 \text{ m/sec}^2$$

b) Substitute $40°$ for ϕ.
$$g = 9.78049[1 + 0.005288\sin^2 40° -$$
$$0.000006\sin^2(2\cdot 40°)]$$
$$\approx 9.80180 \text{ m/sec}^2$$

c) $g = 9.78049[1 + 0.005288\sin^2\phi -$
$$0.000006(2\sin\phi\cos\phi)^2]$$
$$g = 9.78049(1 + 0.005288\sin^2\phi -$$
$$0.000024\sin^2\phi\cos^2\phi)$$
$$g = 9.78049[1 + 0.005288\sin^2\phi -$$
$$0.000024\sin^2\phi(1 - \sin^2\phi)]$$
$$g = 9.78049(1 + 0.005264\sin^2\phi +$$
$$0.000024\sin^4\phi)$$

d) g is the greatest at $90°$ N and $90°$ S (that is, at the poles).

g is least at $0°$(that is, at the equator).

Exercise Set 6.3

Note: Answers for the odd-numbered exercises 1-29 are in the answer section in the text.

31. Expression B completes the identity $\dfrac{\cos x + \cot x}{1 + \csc x} = \cos x$. The proof is in the answer section in the text.

33. Expression A completes the identity $\sin x \cos x + 1 = \dfrac{\sin^3 x - \cos^3 x}{\sin x - \cos x}$. The proof is in the answer section in the text.

35. Expression C completes the identity $\dfrac{1}{\cot x \sin^2 x} = \tan x + \cot x$. The proof is in the answer section in the text.

37. Discussion and Writing

39. $f(x) = 3x - 2$

    a) Find some ordered pairs.

       When $x = 0$, $f(0) = 3 \cdot 0 - 2 = 2$.

       When $x = 2$, $f(2) = 3 \cdot 2 - 2 = 4$.

       Plot these points and draw the graph.

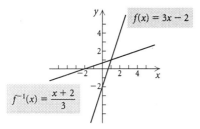

    b) Since there is no horizontal line that intersects the graph more than once, the function is one-to-one.

    c) Replace $f(x)$ with y: $\ y = 3x - 2$

       Interchange x and y: $\ x = 3y - 2$

       Solve for y: $\ y = \dfrac{x + 2}{3}$

       Replace y with $f^{-1}(x)$: $\ f^{-1}(x) = \dfrac{x + 2}{3}$

    d) Find some ordered pairs or reflect the graph of $f(x)$ across the line $y = x$. The graph is shown in part (a) above.

41. $f(x) = x^2 - 4$, $x \geq 0$

    a) Find some ordered pairs.

       When $x = 0$, $f(0) = 0^2 - 4 = -4$.

       When $x = 1$, $f(1) = 1^2 - 4 = -3$.

       When $x = 2$, $f(2) = 2^2 - 4 = 0$.

       When $x = 3$, $f(3) = 3^2 - 4 = 5$.

       Plot these points and draw the graph.

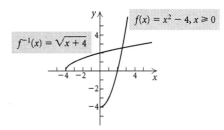

    b) Since there is no horizontal line that intersects the graph more than once, the function is one-to-one.

    c) Replace $f(x)$ with y: $\ y = x^2 - 4$

       Interchange x and y: $\ x = y^2 - 4$

       Solve for y: $\ y = \sqrt{x + 4}$ (We choose the positive square root since $y \geq 0$.)

       Replace y with $f^{-1}(x)$: $\ f^{-1}(x) = \sqrt{x + 4}$

    d) Find some ordered pairs or reflect the graph of $f(x)$ across the line $y = x$. The graph is shown in part (a) above.

43. See the answer section in the text.

45. See the answer section in the text.

47. See the answer section in the text.

Exercise Set 6.4

1. The only number in the restricted range $[-\pi/2, \pi/2]$ with a sine of $-\sqrt{3}/2$ is $-\pi/3$. Thus, $\sin^{-1}(-\sqrt{3}/2) = -\pi/3$, or $-60°$.

3. The only number in the restricted range $(-\pi/2, \pi/2)$ with a tangent of 1 is $\pi/4$. Thus, $\arctan 1 = \pi/4$, or $45°$.

5. The only number in the restricted range $[0, \pi]$ with a cosine of $\sqrt{2}/2$ is $\pi/4$. Thus, $\arccos(\sqrt{2}/2) = \pi/4$, or $45°$.

7. The only number in the restricted range $(-\pi/2, \pi/2)$ with a tangent of 0 is 0. Thus, $\tan^{-1} 0 = 0$, or $0°$.

9. The only number in the restricted range $[0, \pi]$ with a cosine of $\sqrt{3}/2$ is $\pi/6$. Thus, $\cos^{-1}(\sqrt{3}/2) = \pi/6$, or $30°$.

11. The only number in the restricted range $[-\pi/2, 0) \cup (0, \pi/2]$ with a cosecant of 2 is $\pi/6$. Thus, $\csc^{-1} 2 = \pi/6$, or $30°$.

13. The only number in the restricted range $[-\pi/2, 0) \cup (0, \pi/2]$ with a cotangent of $-\sqrt{3}$ is $-\pi/6$. Thus, $\operatorname{arccot}(-\sqrt{3}) = -\pi/6$, or $-30°$.

15. The only number in the restricted range $[-\pi/2, \pi/2]$ with a sine of $-\dfrac{1}{2}$ is $-\dfrac{\pi}{6}$. Thus, $\arcsin\left(-\dfrac{1}{2}\right) = -\dfrac{\pi}{6}$, or $-30°$.

17. The only number in the restricted range $[0, \pi]$ with a cosine of 0 is $\pi/2$. Thus, $\cos^{-1} 0 = \pi/2$, or $90°$.

19. The only number in the restricted range $[0, \pi/2) \cup (\pi/2, \pi]$ with a secant of 2 is $\pi/3$. Thus, $\operatorname{arcsec} 2 = \pi/3$, or $60°$.

21. $\arctan 0.3673 \approx 0.352$, or $20.2°$

23. $\sin^{-1} 0.9613 \approx 1.292$, or $74.0°$

25. $\cos^{-1}(-0.9810) \approx 2.946$, or $168.8°$

27. $\csc^{-1}(-6.2774) = \sin^{-1}\left(\dfrac{1}{-6.2774}\right) \approx$
-0.160, or $-9.2°$

29. $\tan^{-1} 1.091 \approx 0.829$, or $47.5°$

31. $\arcsin(-0.8192) \approx -0.960$, or $-55.0°$

33. $\sin^{-1}$: $[-1, 1]$; $\cos^{-1}$: $[-1, 1]$; $\tan^{-1}$: $(-\infty, \infty)$

35.

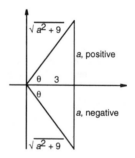

$\sin\theta = \dfrac{2000}{d}, -\dfrac{\pi}{2} \le \theta \le \dfrac{\pi}{2}$, so $\theta = \sin^{-1}\left(\dfrac{2000}{d}\right)$.

37. Since 0.3 is in the interval $[-1, 1]$, $\sin(\arcsin 0.3) = 0.3$.

39. $\cos^{-1}\left[\cos\left(-\dfrac{\pi}{4}\right)\right] = \cos^{-1}\left(\dfrac{\sqrt{2}}{2}\right) = \dfrac{\pi}{4}$

41. $\sin^{-1}\left(\sin\dfrac{\pi}{5}\right) = \dfrac{\pi}{5}$ because $\dfrac{\pi}{5}$ is in the range of the arcsine function.

43. $\tan^{-1}\left(\tan\dfrac{2\pi}{3}\right) = \tan^{-1}(-\sqrt{3}) = -\dfrac{\pi}{3}$

45. $\sin\left(\arctan\dfrac{\sqrt{3}}{3}\right) = \sin\dfrac{\pi}{6} = \dfrac{1}{2}$

47. $\tan\left(\cos^{-1}\dfrac{\sqrt{2}}{2}\right) = \tan\dfrac{\pi}{4} = 1$

49. $\arcsin\left(\cos\dfrac{\pi}{6}\right) = \arcsin\dfrac{\sqrt{3}}{2} = \dfrac{\pi}{3}$

51. Find $\tan(\arcsin 0.1)$

We wish to find the tangent of an angle whose sine is 0.1, or $\dfrac{1}{10}$.

The length of the other leg is $3\sqrt{11}$.

Thus, $\tan(\arcsin 0.1) = \dfrac{1}{3\sqrt{11}}$, or $\dfrac{\sqrt{11}}{33}$.

53. Find $\sin\left(\arctan\dfrac{a}{3}\right)$.

We draw right triangles whose legs have lengths $|a|$ and 3 so that $\tan\theta = \dfrac{a}{3}$.

$\sin\left(\arctan\dfrac{a}{3}\right) = \dfrac{a}{\sqrt{a^2+9}}$

55. Find $\cot\left(\sin^{-1}\dfrac{p}{q}\right)$.

We draw right triangles with one length of length $|p|$ and hypotenuse q so that $\sin\theta = \dfrac{p}{q}$.

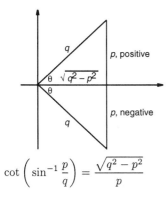

$$\cot\left(\sin^{-1}\frac{p}{q}\right) = \frac{\sqrt{q^2 - p^2}}{p}$$

57. Find $\tan\left(\arcsin\dfrac{p}{\sqrt{p^2 + 9}}\right)$.

We draw the right triangle with one leg of length $|p|$ and hypotenuse $\sqrt{p^2 + 9}$.

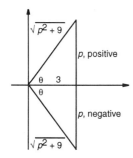

$$\tan\left(\arcsin\frac{p}{\sqrt{p^2 + 9}}\right) = \frac{p}{3}$$

59. $\cos\left(\dfrac{1}{2}\arcsin\dfrac{\sqrt{3}}{2}\right) = \cos\left(\dfrac{1}{2}\cdot 60°\right) = \dfrac{\sqrt{3}}{2}$

We could also have used a half-angle identity:

$$\cos\left(\frac{1}{2}\arcsin\frac{\sqrt{3}}{2}\right) = \sqrt{\frac{1 + \cos[\arcsin(\sqrt{3}/2)]}{2}}$$

$$= \sqrt{\frac{1 + \cos\dfrac{\pi}{3}}{2}}$$

$$= \sqrt{\frac{1 + \dfrac{1}{2}}{2}}$$

$$= \frac{\sqrt{3}}{2}$$

61. Evaluate $\cos\left(\sin^{-1}\dfrac{\sqrt{2}}{2} + \cos^{-1}\dfrac{3}{5}\right)$.

This is the cosine of a sum so we use the identity $\cos(u + v) = \cos u \cos v - \sin u \sin v$.

$$\cos\left(\sin^{-1}\frac{\sqrt{2}}{2} + \cos^{-1}\frac{3}{5}\right)$$

$$= \cos\left(\sin^{-1}\frac{\sqrt{2}}{2}\right)\cos\left(\cos^{-1}\frac{3}{5}\right) -$$

$$\sin\left(\sin^{-1}\frac{\sqrt{2}}{2}\right)\sin\left(\cos^{-1}\frac{3}{5}\right)$$

$$= \left(\cos\frac{\pi}{4}\right)\left(\frac{3}{5}\right) - \frac{\sqrt{2}}{2}\cdot\sin\left(\cos^{-1}\frac{3}{5}\right)$$

$$= \frac{\sqrt{2}}{2}\cdot\frac{3}{5} - \frac{\sqrt{2}}{2}\cdot\sin\left(\cos^{-1}\frac{3}{5}\right)$$

We draw a triangle in order to find $\sin\left(\cos^{-1}\dfrac{3}{5}\right)$.

Our expression simplifies to

$$\frac{\sqrt{2}}{2}\cdot\frac{3}{5} - \frac{\sqrt{2}}{2}\cdot\frac{4}{5} = \frac{3\sqrt{2} - 4\sqrt{2}}{10} = -\frac{\sqrt{2}}{10}.$$

63. Evaluate $\sin(\sin^{-1} x + \cos^{-1} y)$.

We will use the identity $\sin(u + v) = \sin u \cos v + \cos u \sin v$. We draw a triangle with an angle whose sine is x and another with an angle whose cosine is y.

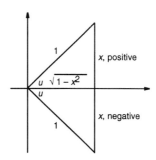

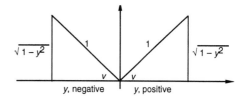

$$\sin(\sin^{-1} x + \cos^{-1} y)$$
$$= \sin(\sin^{-1}x)\cos(\cos^{-1}y) + \cos(\sin^{-1}x)\sin(\cos^{-1}y)$$
$$= x \cdot y + \frac{\sqrt{1-x^2}}{1} \cdot \frac{\sqrt{1-y^2}}{1}$$
$$= xy + \sqrt{(1-x^2)(1-y^2)}$$

65. Evaluate $\sin(\sin^{-1} 0.6032 + \cos^{-1} 0.4621)$. We will use the identity $\sin(u + v) = \sin u \cos v + \cos u \sin v$. We draw a triangle with an angle whose sine is 0.6032 and another with an angle whose cosine is 0.4621.

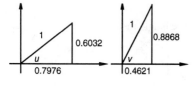

$$\sin(\sin^{-1} 0.6032 + \cos^{-1}(0.4621)$$
$$= \sin(\sin^{-1} 0.6032)\cos(\cos^{-1} 0.4621) +$$
$$\qquad \cos(\sin^{-1} 0.6032)\sin(\cos^{-1} 0.4621)$$
$$= 0.6032(0.4621) + 0.7976(0.8868)$$
$$\approx 0.9861$$

67. Discussion and Writing

69. Discussion and Writing

71.
$$2x^2 = 5x$$
$$2x^2 - 5x = 0$$
$$x(2x - 5) = 0$$
$$x = 0 \quad or \quad 2x - 5 = 0$$
$$x = 0 \quad or \qquad x = \frac{5}{2}$$

The solutions are 0 and $\frac{5}{2}$.

73. $x^4 + 5x^2 - 36 = 0$

Let $u = x^2$ and substitute.
$$u^2 + 5u - 36 = 0$$
$$(u + 9)(u - 4) = 0$$
$$u + 9 = 0 \quad or \quad u - 4 = 0$$
$$u = -9 \quad or \qquad u = 4$$
$$x^2 = -9 \quad or \qquad x^2 = 4$$
$$x = \pm 3i \quad or \qquad x = \pm 2$$

The solutions are $\pm 3i$ and ± 2.

75. $\sqrt{x - 2} = 5$
$$x - 2 = 25 \quad \text{Squaring both sides}$$
$$x = 27$$

This answer checks. The solution is 27.

77. See the answer section in the text.

79. See the answer section in the text.

81. See the answer section in the text.

83.

Let $\theta = \alpha - \beta$

$$\tan \alpha = \frac{h + y}{x}, \qquad \alpha = \arctan \frac{h + y}{x}$$
$$\tan \beta = \frac{y}{x}, \qquad \beta = \arctan \frac{y}{x}$$

Thus, $\theta = \arctan \dfrac{h + y}{x} - \arctan \dfrac{y}{x}$.

When $x = 20$ ft, $y = 7$ ft, and $h = 25$ ft we have
$$\theta = \tan^{-1} \frac{25 + 7}{20} - \tan^{-1} \frac{7}{20}$$
$$\approx 57.99° - 19.29°$$
$$= 38.7°.$$

Exercise Set 6.5

1. $\cos x = \dfrac{\sqrt{3}}{2}$

Since $\cos x$ is positive the solutions are in quadrants I and IV. They are $\dfrac{\pi}{6} + 2k\pi$ or $\dfrac{11\pi}{6} + 2k\pi$, where k is any integer. The solutions can also be expressed as $30° + k \cdot 360°$ or $330° + k \cdot 360°$, where k is any integer.

3. $\tan x = -\sqrt{3}$

Since $\tan x$ is negative the solutions are in quadrants II and IV. They are $\dfrac{2\pi}{3} + 2k\pi$ or $\dfrac{5\pi}{3} + 2k\pi$. This can be condensed as $\dfrac{2\pi}{3} + k\pi$, where k is any integer. The solutions can also be expressed as $120° + k \cdot 180°$, where k is any integer.

5. $2\cos x - 1 = -1.2814$
$$2\cos x = -0.2814$$
$$\cos x = -0.1407$$

Using a grapher we find that the reference angle, $\arccos(-0.1407)$ is $x \approx 98.09°$. Since $\cos x$ is negative, the solutions are in quadrants II and III. Thus, one solution is $98.09°$. The reference angle for $98.09°$ is $180° - 98.09°$, or $81.91°$, so the other solution in $[0°, 360°)$ is $180° + 81.91°$, or $261.91°$.

7. $2\sin x + \sqrt{3} = 0$

$\qquad 2\sin x = -\sqrt{3}$

$\qquad \sin x = -\dfrac{\sqrt{3}}{2}$

The solutions in $[0, 2\pi)$ are $\dfrac{4\pi}{3}$ and $\dfrac{5\pi}{3}$.

9. $2\cos^2 x = 1$

$\qquad \cos^2 x = \dfrac{1}{2}$

$\qquad \cos x = \pm\dfrac{1}{\sqrt{2}}, \text{ or } \pm\dfrac{\sqrt{2}}{2}$

The solutions in $[0, 2\pi)$ are $\dfrac{\pi}{4}, \dfrac{3\pi}{4}, \dfrac{5\pi}{4}$, and $\dfrac{7\pi}{4}$.

11. $\qquad 2\sin^2 x + \sin x = 1$

$\qquad 2\sin^2 x + \sin x - 1 = 0$

$\qquad (2\sin x - 1)(\sin x + 1) = 0$

$2\sin x - 1 = 0 \quad or \quad \sin x + 1 = 0$

$\qquad 2\sin x = 1 \quad or \qquad \sin x = -1$

$\qquad \sin x = \dfrac{1}{2} \quad or \qquad \sin x = -1$

The solutions in $[0, 2\pi)$ are $\dfrac{\pi}{6}, \dfrac{5\pi}{6}$, and $\dfrac{3\pi}{2}$.

13. $2\cos^2 x - \sqrt{3}\cos x = 0$

$\qquad \cos x(2\cos x - \sqrt{3}) = 0$

$\qquad \cos x = 0 \quad or \quad 2\cos - \sqrt{3} = 0$

$\qquad \cos x = 0 \quad or \qquad \cos x = \dfrac{\sqrt{3}}{2}$

The solutions in $[0, 2\pi)$ are $\dfrac{\pi}{2}, \dfrac{3\pi}{2}, \dfrac{\pi}{6}$, and $\dfrac{11\pi}{6}$.

15. $\qquad 6\cos^2 \phi + 5\cos \phi + 1 = 0$

$\qquad (3\cos \phi + 1)(2\cos \phi + 1) = 0$

$3\cos \phi + 1 = 0 \quad or \quad 2\cos \phi + 1 = 0$

$\qquad \cos \phi = -\dfrac{1}{3} \quad or \qquad \cos \phi = -\dfrac{1}{2}$

Using $\cos \phi = -\dfrac{1}{3}$, we find that $\phi = \arccos\left(-\dfrac{1}{3}\right) \approx$ $109.47°$, so one solution in $[0, 360°)$ is $109.47°$. The reference angle for this angle is $180° - 109.47°$, or $70.53°$. Thus, another solution is $180° + 70.53°$, or $250.53°$.

Using $\cos \phi = -\dfrac{1}{2}$, we find that the other solutions in $[0, 360°)$ are $120°$ and $240°$.

17. $\qquad \sin 2x \cos x - \sin x = 0$

$\qquad (2\sin x \cos x)\cos - \sin x = 0$

$\qquad 2\sin x \cos^2 x - \sin x = 0$

$\qquad 2\sin x(2\cos^2 x - 1) = 0$

$\sin x = 0 \quad or \quad 2\cos^2 x - 1 = 0$

$\sin x = 0 \quad or \qquad \cos^2 x = \dfrac{1}{2}$

$\sin x = 0 \quad or \qquad \cos x = \pm\dfrac{1}{\sqrt{2}}, \text{ or } \pm\dfrac{\sqrt{2}}{2}$

The solutions in $[0, 2\pi)$ are $0, \pi, \dfrac{\pi}{4}, \dfrac{3\pi}{4}, \dfrac{5\pi}{4}$, and $\dfrac{7\pi}{4}$.

19. $\cos^2 x + 6\cos x + 4 = 0$

$\cos x = \dfrac{-6 \pm \sqrt{36 - 16}}{2} = \dfrac{-6 \pm \sqrt{20}}{2}$

$\cos x \approx -0.7639 \quad or \quad \cos x \approx -5.2361$

Since cosine values are never less than -1, $\cos x \approx -5.2361$ has no solution. Using $\cos x = -0.7639$, we find that $x = \arccos(-0.7639) \approx 139.81°$. Thus, one solution in $[0, 360°)$ is $139.81°$. The reference angle for this angle is $180° - 139.81° = 40.19°$. Then the other solution in $[0, 360°)$ is $180° + 40.19° = 220.19°$.

21. $7 = \cot^2 x + 4\cot x$

$\quad 0 = \cot^2 x + 4\cot x - 7$

$\quad \cot x = \dfrac{-4 \pm \sqrt{16 + 28}}{2}$ $\qquad$ Using the quadratic formula

$\quad \cot x = \dfrac{-4 \pm \sqrt{44}}{2}$

$\quad \cot x \approx 1.3166 \quad or \quad \cot x \approx -5.3166$

Using $\cot x \approx 1.3166$, we find that $x = \text{arccot } 1.3166 \approx 37.22°$. Thus, two solutions in $[0, 360°)$ are $37.22°$ and $180° + 37.22°$, or $217.22°$.

Using $\cot x \approx -5.3166$ we find that $x = \text{arccot } (-5.3166) \approx -10.65°$. Then the other solutions in $[0, 360°)$ are $180° - 10.65°$, or $169.35°$, and $360° - 10.65°$, or $349.35°$.

23. $\qquad \cos 2x - \sin x = 1$

$\quad 1 - 2\sin^2 x - \sin x = 1$

$\qquad\qquad 0 = 2\sin^2 x + \sin x$

$\qquad\qquad 0 = \sin x(2\sin x + 1)$

$\sin x = 0 \quad or \quad 2\sin x + 1 = 0$

$\sin x = 0 \quad or \qquad \sin x = -\dfrac{1}{2}$

$x = 0, \pi \quad or \qquad x = \dfrac{7\pi}{6}, \dfrac{11\pi}{6}$

All values check. The solutions in $[0, 2\pi)$ are $0, \pi, \dfrac{7\pi}{6}$, and $\dfrac{11\pi}{6}$.

25. $\qquad \sin 4x - 2\sin 2x = 0$

$\qquad \sin[2(2x)] - 2\sin 2x = 0$

$\qquad 2\sin 2x \cos 2x - 2\sin 2x = 0$

$\qquad 2\sin 2x(\cos 2x - 1) = 0$

$$\sin 2x = 0 \quad \text{ or } \cos 2x - 1 = 0$$
$$\sin 2x = 0 \quad \text{ or } \quad \cos 2x = 1$$
$$2x = 0, \pi, 2\pi, 3\pi \quad \text{ or } \qquad 2x = 0, 2\pi$$
$$x = 0, \frac{\pi}{2}, \pi, \frac{3\pi}{2} \quad \text{ or } \qquad x = 0, \pi$$

All values check. The solutions in $[0, 2\pi)$ are 0, $\frac{\pi}{2}$, π, and $\frac{3\pi}{2}$.

27.
$$\sin 2x \cos x + \sin x = 0$$
$$(2 \sin x \cos x) \cos x + \sin x = 0$$
$$2 \sin x \cos^2 x + \sin x = 0$$
$$\sin x (2 \cos^2 x + 1) = 0$$
$$\sin x = 0 \quad \text{ or } \quad 2 \cos^2 x + 1 = 0$$
$$\sin x = 0 \quad \text{ or } \qquad \cos^2 x = -\frac{1}{2}$$
$$x = 0, \pi \quad \text{ or } \qquad \text{No solution}$$

Both values check. The solutions are 0 and π.

29.
$$2 \sec x \tan x + 2 \sec x + \tan x + 1 = 0$$
$$2 \sec x (\tan x + 1) + (\tan x + 1) = 0$$
$$(2 \sec x + 1)(\tan x + 1) = 0$$
$$2 \sec x + 1 = 0 \quad \text{ or } \quad \tan x + 1 = 0$$
$$\sec x = -\frac{1}{2} \quad \text{ or } \qquad \tan x = -1$$

No solution $\qquad\qquad x = \dfrac{3\pi}{4}, \dfrac{7\pi}{4}$

Both values check. The solutions in $[0, 2\pi)$ are $\frac{3\pi}{4}$ and $\frac{7\pi}{4}$.

31.
$$\sin 2x + \sin x + 2 \cos x + 1 = 0$$
$$2 \sin x \cos x + \sin x + 2 \cos x + 1 = 0$$
$$\sin x (2 \cos x + 1) + 2 \cos x + 1 = 0$$
$$(\sin x + 1)(2 \cos x + 1) = 0$$
$$\sin x + 1 = 0 \quad \text{ or } \quad 2 \cos x + 1 = 0$$
$$\sin x = -1 \quad \text{ or } \qquad \cos x = -\frac{1}{2}$$
$$x = \frac{3\pi}{2} \quad \text{ or } \qquad x = \frac{2\pi}{3}, \frac{4\pi}{3}$$

All values check. The solutions in $[0, 2\pi)$ are $\frac{2\pi}{3}$, $\frac{4\pi}{3}$, and $\frac{3\pi}{2}$.

33.
$$\sec^2 x - 2 \tan^2 x = 0$$
$$1 + \tan^2 x - 2 \tan^2 x = 0$$
$$1 - \tan^2 x = 0$$
$$\tan^2 x = 1$$
$$\tan x = \pm 1$$
$$x = \frac{\pi}{4}, \frac{3\pi}{4}, \frac{5\pi}{4}, \frac{7\pi}{4}$$

All values check. The solutions in $[0, 2\pi)$ are $\frac{\pi}{4}$, $\frac{3\pi}{4}$, $\frac{5\pi}{4}$, $\frac{7\pi}{4}$.

35.
$$2 \cos x + 2 \sin x = \sqrt{6}$$
$$\cos x + \sin x = \frac{\sqrt{6}}{2}$$
$$\cos^2 x + 2 \sin x \cos x + \sin^2 x = \frac{6}{4} \qquad \text{Squaring}$$
$$\qquad\qquad\qquad\qquad\qquad\qquad \text{both sides}$$
$$\sin 2x + 1 = \frac{3}{2}$$
$$\sin 2x = \frac{1}{2}$$
$$2x = \frac{\pi}{6}, \frac{5\pi}{6}, \frac{13\pi}{6}, \frac{17\pi}{6}$$
$$x = \frac{\pi}{12}, \frac{5\pi}{12}, \frac{13\pi}{12}, \frac{17\pi}{12}$$

The values $\frac{13\pi}{12}$ and $\frac{17\pi}{12}$ do not check, but the other values do. The solutions in $[0, 2\pi)$ are $\frac{\pi}{12}$ and $\frac{5\pi}{12}$.

37.
$$\sec^2 x + 2 \tan x = 6$$
$$1 + \tan^2 x + 2 \tan x = 6$$
$$\tan^2 x + 2 \tan x - 5 = 0$$
$$\tan x = \frac{-2 \pm \sqrt{4 + 20}}{2} = \frac{-2 \pm \sqrt{24}}{2}$$
$$\tan x \approx 1.4495 \quad \text{ or } \quad \tan x \approx -3.4495$$

Using $\tan x \approx 1.4495$, we find that $x = \arctan 1.4495 \approx 0.967$. Then two possible solutions in $[0, 2\pi)$ are 0.967 and $\pi + 0.967$, or 4.109.

Using $\tan x \approx -3.4495$, we find that $x = \arctan(-3.4495) \approx -1.289$. Thus, the other two possible solutions in $[0, 2\pi)$ are $\pi - 1.289$, or 1.853, and $2\pi - 1.289$, or 4.994.

All values check. The solutions in $[0, 2\pi)$ are 0.967, 1.853, 4.109, and 4.994. (Answers may vary slightly due to rounding differences.)

39.
$$\cos(\pi - x) + \sin\left(x - \frac{\pi}{2}\right) = 1$$
$$(-\cos x) + (-\cos x) = 1$$
$$-2\cos x = 1$$
$$\cos x = -\frac{1}{2}$$
$$x = \frac{2\pi}{3}, \frac{4\pi}{3}$$

Both values check. The solutions in $[0, 2\pi)$ are $\frac{2\pi}{3}$ and $\frac{4\pi}{3}$.

41. Find the points of intersection of $y_1 = x\sin x$ and $y_2 = 1$ or find the zeros of $y_1 = x\sin x - 1$. The solutions in $[0, 2\pi)$ are 1.114 and 2.773.

43. Find the point of intersection of $y_1 = 2\cos^2 x$ and $y_2 = x + 1$ or find the zero of $y_1 = 2\cos^2 x - x - 1$. The only solution in $[0, 2\pi)$ is 0.515.

45. Find the points of intersection of $y_1 = \cos x - 2$ and $y_2 = x^2 - 3x$ or find the zeros of $y_1 = \cos x - 2 - x^2 + 3x$. The solutions in $[0, 2\pi)$ are 0.422 and 1.756.

47. Discussion and Writing

49. $B = 90° - 55° = 35°$
$$\tan 55° = \frac{201}{b}$$
$$b = \frac{201}{\tan 55°}$$
$$b \approx 140.7$$

$$\sin 55° = \frac{201}{c}$$
$$c = \frac{201}{\sin 55°}$$
$$c \approx 245.4$$

51.
$$\frac{x}{27} = \frac{4}{3}$$
$$x = 36 \quad \text{Multiplying by 27}$$

53. $|\sin x| = \frac{\sqrt{3}}{2}$
$$\sin x = \frac{\sqrt{3}}{2} \quad or \quad \sin x = -\frac{\sqrt{3}}{2}$$
$$x = \frac{\pi}{3}, \frac{2\pi}{3} \quad or \quad x = \frac{4\pi}{3}, \frac{5\pi}{3}$$

All values check. The solutions in $[0, 2\pi)$ are $\frac{\pi}{3}$, $\frac{2\pi}{3}$, $\frac{4\pi}{3}$, and $\frac{5\pi}{3}$.

55.
$$\sqrt{\tan x} = \sqrt[4]{3}$$
$$(\sqrt{\tan x})^4 = (\sqrt[4]{3})^4$$
$$\tan^2 x = 3$$
$$\tan x = \pm\sqrt{3}$$
$$x = \frac{\pi}{3}, \frac{2\pi}{3}, \frac{4\pi}{3}, \frac{5\pi}{3}$$

Only $\frac{\pi}{3}$ and $\frac{4\pi}{3}$ check. They are the solutions in $[0, 2\pi)$.

57.
$$\ln(\cos x) = 0$$
$$\cos x = 1$$
$$x = 0$$

This value checks.

59.
$$\sin(\ln x) = -1$$
$$\ln x = \frac{3\pi}{2} + 2k\pi, \, k \text{ an integer}$$
$$x = e^{3\pi/2 + 2k\pi}, \, k \text{ an integer}$$

x is in the interval $[0, 2\pi)$ when $k \leq -1$. Thus, the possible solutions are $e^{3\pi/2 + 2k\pi}$, k an integer, $k \leq -1$. These values check and are the solutions.

61. a) Using the sine regression feature on a grapher, we get $y = 7\sin(-2.6180x + 0.5236) + 7$

b) In December, $x = 12$; when $x = 12$, $y \approx 10.5$. Thus, total sales in december are about \$10,500.

In July, $x = 7$; when $x = 7$, $y \approx 13.062$. Thus, total sales in July are about \$13,062.

63.
$$T(t) = 101.6° + 3° \sin\left(\frac{\pi}{8}t\right), 0 \leq t \leq 12$$
$$103° = 101.6° + 3° \sin\left(\frac{\pi}{8}t\right) \qquad \text{Substituting}$$
$$1.4° = 3° \sin\left(\frac{\pi}{8}t\right)$$
$$0.4667 \approx \sin\left(\frac{\pi}{8}t\right)$$
$$\frac{\pi}{8}t \approx 0.4855, 2.6561 \qquad \left(0 \leq t \leq 12, \text{ so}\right.$$
$$\left. 0 \leq \frac{\pi}{8}t \leq \frac{3\pi}{2}\right)$$
$$t \approx 1.24, 6.76$$

Both values check.

The patient's temperature was 103° at $t \approx 1.24$ days and $t \approx 6.76$ days.

65. $N(\phi) = 6066 - 31\cos 2\phi$

We consider ϕ in the interval $[0°, 90°]$ since we want latitude north.

$6040 = 6066 - 31 \cos 2\phi$ Substituting

$-26 = -31 \cos 2\phi$

$0.8387 \approx \cos 2\phi$ $(0° \le 2\phi \le 180°)$

$2\phi \approx 33.0°$

$\phi \approx 16.5°$

The value checks.

At about 16.5° N the length of a British nautical mile is found to be 6040 ft.

67. Sketch a triangle having an angle θ whose cosine is $\dfrac{3}{5}$. (See the triangle in Exercise 61, Exercise Set 5.4)

Then $\sin \theta = \dfrac{4}{5}$. Thus, $\arccos \dfrac{3}{5} = \arcsin \dfrac{4}{5}$.

$\arccos x = \arccos \dfrac{3}{5} - \arcsin \dfrac{4}{5}$

$\arccos x = 0$

$x = 1$

69. $\sin x = 5 \cos x$

$\dfrac{\sin x}{\cos x} = 5$

$\tan x = 5$

$x \approx 1.3734, 4.5150$

Then $\sin x \cos x = \sin(1.3734) \cos(1.3734) \approx 0.1923$.

(The result is the same if $x \approx 4.5150$ is used.)

Chapter 7

Triangles, Vectors, and Applications

Exercise Set 7.1

1.

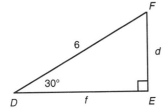

To solve this triangle find F, d, and f.

$F = 90° - 30° = 60°$

$\dfrac{d}{6} = \sin 30°$

$d = 6 \sin 30°$

$d = 3$

$\dfrac{f}{6} = \cos 30°$

$f = 6 \cos 30°$

$f = 3\sqrt{3} \approx 5.2$

3.

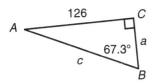

To solve this triangle, find A, a, and c.

$A = 90° - 67.3° = 22.7°$

$\cot 67.3° = \dfrac{a}{126}$

$a = 126 \cot 67.3°$

$a \approx 52.7$

$\csc 67.3° = \dfrac{c}{126}$

$c = 126 \csc 67.3°$

$c \approx 136.6$

5.

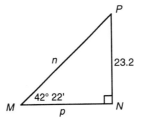

To solve this triangle, find P, p, and n.

$P = 90° - 42° 22' = 47°38'$

$\dfrac{n}{23.2} = \csc 42° 22'$

$n = 23.2 \csc 42° 22'$

$n \approx 34.4$

$\dfrac{p}{23.2} = \cot 42° 22'$

$p = 23.2 \cot 42° 22'$

$p \approx 25.4$

7.

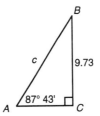

To solve this triangle, find B, b, and c.

$B = 90° - 87° 43' = 2°17'$

$\cot 87° 43' = \dfrac{b}{9.73}$

$b = 9.73 \cot 87° 43'$

$b \approx 0.39$

$\csc 87° 43' = \dfrac{c}{9.73}$

$c = 9.73 \csc 87° 43'$

$c \approx 9.74$

9.

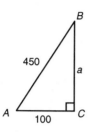

To solve this triangle, find A, B, and a.

$$\cos A = \frac{100}{450}$$
$$A \approx 77.2°$$

$$B = 90° - A \approx 90° - 77.2° \approx 12.8°$$
$$\sin A = \frac{a}{450}$$
$$\sin 77.2° \approx \frac{a}{450}$$
$$a \approx 450 \sin 77.2° \approx 439$$

11.

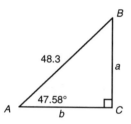

To solve this triangle, find B, a, and b.

$$B = 90° - 47.58° = 42.42°$$
$$\frac{a}{48.3} = \sin 47.58°$$
$$a = 48.3 \sin 47.58°$$
$$a \approx 35.7$$

$$\frac{b}{48.3} = \cos 47.58°$$
$$b = 48.3 \cos 47.58°$$
$$b \approx 32.6$$

13.

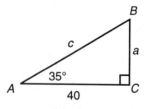

To solve this triangle find B, a, and c.

$$B = 90° - A = 90° - 35° = 55°$$

$$\tan A = \frac{a}{40}$$
$$\tan 35° = \frac{a}{40}$$
$$a = 40 \tan 35°$$
$$a \approx 28.0$$

$$\sec A = \frac{c}{40}$$
$$\sec 35° = \frac{c}{40}$$
$$c = 40 \sec 35°$$
$$c \approx 48.8$$

15.

To solve this triangle find a, A, and B.

$$\cos A = \frac{1.86}{4.02}$$
$$A \approx 62.4°$$
$$B \approx 90° - 62.4° \approx 27.6°$$
$$\sin A = \frac{a}{4.02}$$
$$\sin 62.4° \approx \frac{a}{4.02}$$
$$a \approx 4.02 \sin 62.4°$$
$$a \approx 3.56$$

17. First we find the distance from point B to the raft. We know the length of the side opposite the 50° angle and want to find the length of the hypotenuse, c. We use the sine ratio.

$$\sin 50° = \frac{40 \text{ ft}}{c}$$

$$c = \frac{40 \text{ ft}}{\sin 50°} \approx 52.2 \text{ ft}$$

Allowing 5 ft of rope at each end, Madison needs about 52.2 ft + 2 · 5 ft, or about 62.2 ft of rope.

19. We know the length of the side adjacent to the 23° angle and want to find the length of the opposite side. We use the tangent ratio.

$$\tan 23° = \frac{d}{6 \text{ ft}}$$
$$d = 6 \text{ ft} \cdot \tan 23° \approx 2.5 \text{ ft}$$

The front legs should be about 2.5 ft from the wall.

21.

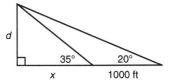

We first find the distance x, using the tangent function with 35°and 20°.

$$\tan 35° = \frac{d}{x} \qquad \text{and} \quad \tan 20° = \frac{d}{x + 800}$$
$$d = x \tan 35° \text{ and} \qquad d = (x+800)\tan 20°$$

Substitute $x \tan 35°$ for d in the second equation and solve for x.

$$x \tan 35° = (x + 800)\tan 20°$$
$$x \tan 35° = x \tan 20° + 800 \tan 20°$$
$$x(\tan 35° - \tan 20°) = 800 \tan 20°$$
$$x = \frac{800 \tan 20°}{\tan 35° - \tan 20°}$$
$$x \approx 866$$

Then we find d using the tangent function.

$$\tan 35° \approx \frac{d}{866}$$
$$d \approx 866 \tan 35° \approx 606$$

The height of the sand dune is about 606 ft.

23.

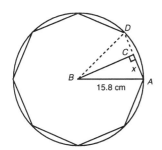

The measure of $\angle ABD$ is $\frac{1}{8} \cdot 360°$, or 45°. Then the measure of $\angle ABC$ is $\frac{1}{2} \cdot 45°$, or 22.5°. We know the length of the hypotenuse and want to find the length of the side opposite the 22.5°angle. We use the sine ratio.

$$\sin 22.5° = \frac{x}{15.8}$$
$$x = 15.8 \text{ cm} \cdot \sin 22.5° \approx 6.046 \text{ cm}$$

Then the length of each side of the octagon is about 2(6.046 cm), or 12.092 cm, and the perimeter of the octagon is about 8(12.092 cm), or about 96.7 cm.

25.

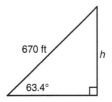

We know the length of the hypotenuse and want to find the length of the side opposite the 63.4°angle. We use the sine ratio.

$$\sin 63.4° = \frac{h}{670 \text{ ft}}$$
$$h \approx 670 \text{ ft} \cdot \sin 63.4° \approx 599 \text{ ft}$$

The kite was about 599 ft high.

27.

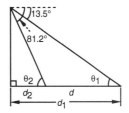

The distance to be found is d, which is $d_1 - d_2$.

From geometry we know that $\theta_1 = 13.5°$and $\theta_2 = 81.2°$.

$$\frac{d_1}{2} = \cot \theta_1 \qquad \text{and} \quad \frac{d_2}{2} = \cot \theta_2$$
$$d_1 = 2 \cot 13.5° \text{ and} \quad d_2 = 2 \cot 81.2°$$
$$d = d_1 - d_2 = 2 \cot 13.5° - 2 \cot 81.2° \approx 8$$

The towns are about 8 km apart.

29.

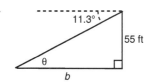

From geometry we know that $\theta = 11.3°$. We know the length of the side opposite θ and want to find the length of the side adjacent to θ. We use the tangent ratio.

$$\tan 11.3° = \frac{55 \text{ ft}}{b}$$
$$b = \frac{55 \text{ ft}}{\tan 11.3°} \approx 275 \text{ ft}$$

The boat is about 275 ft from the foot of the lighthouse.

31.

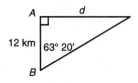

We know the length of the side adjacent to the 63° 20′ angle and want to find the length of the opposite side. We use the tangent ratio.

$$\tan 63° \, 20' = \frac{d}{12 \text{ km}}$$
$$d = 12 \text{ km} \cdot \tan 63° \, 20' \approx 24 \text{ km}$$

The lobster boat is about 24 km from the lighthouse.

33. Discussion and Writing

35. $d = \sqrt{(x_1 - x_2)^2 + (y_1 - y_2)^2}$
$d = \sqrt{[8 - (-6)]^2 + [-2 - (-4)]^2}$
$\quad = \sqrt{14^2 + 2^2}$
$\quad = \sqrt{200} = 10\sqrt{2}$
$\quad \approx 14.142$

37. $\log 0.001 = -3$ is equivalent to $\log_{10} 0.001 = -3$. Remember that the base remains the same, and the logarithm is the exponent. We have $10^{-3} = 0.001$.

39.

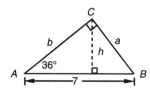

$$\cos 36° = \frac{b}{7}$$
$$b = 7 \cos 36° \approx 5.66$$
$$\sin 36° = \frac{h}{b}$$
$$\sin 36° \approx \frac{h}{5.66}$$
$$h \approx 5.66 \sin 36° \approx 3.3$$

41. $\tan \theta = \dfrac{8}{1.5}$
$\qquad \theta \approx 79.38°$

The rafters should be cut so that $\theta = 79.38°$.

43.

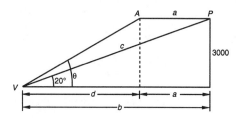

H is the perceived location of the plane;

A is the actual location of the plane when heard.

Plane's speed, 200 mph ≈ 293 ft/sec.

$$\csc 20° = \frac{c}{3000} \qquad\qquad \cot 20° = \frac{b}{3000}$$
$$c = 3000 \csc 20° \qquad\qquad b = 3000 \cot 20°$$
$$c \approx 8771 \qquad\qquad\qquad b \approx 8242$$

The time it takes the plane to fly from P to A is approximately the same time it takes the sound to travel from P to V. We use the formula distance = rate × time to find this time.

$$d = rt$$
$$8771 = 1100t$$
$$8 \approx t$$

Then the distance from P to A is given by $a = 293 \cdot 8 = 2344$ ft. Then $d = b - a = 8242 - 2344 = 5898$ ft.

$$\tan \theta = \frac{3000}{d} = \frac{3000}{5898}$$
$$\theta \approx 27°$$

Exercise Set 7.2

1.

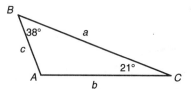

To solve this triangle find A, a, and c.

$A = 180° - (38° + 21°) = 121°$

Use the law of sines to find a and c.

Find a:

$$\frac{a}{\sin A} = \frac{b}{\sin B}$$
$$\frac{a}{\sin 121°} = \frac{24}{\sin 38°}$$
$$a = \frac{24 \sin 121°}{\sin 38°} \approx 33$$

Find c:

$$\frac{c}{\sin C} = \frac{b}{\sin B}$$

$$\frac{c}{\sin 21°} = \frac{24}{\sin 38°}$$

$$c = \frac{24 \sin 21°}{\sin 38°} \approx 14$$

3.

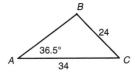

To solve this triangle find A, a, and c.

Find B:

$$\frac{b}{\sin B} = \frac{a}{\sin A}$$

$$\frac{34}{\sin B} = \frac{24}{\sin 36.5°}$$

$$\sin B = \frac{34 \sin 36.5°}{24} \approx 0.8427$$

There are two angles less than $180°$ having a sine of 0.8427. They are $57.4°$ and $122.6°$. This gives us two possible solutions.

Solution I

If $B \approx 57.4°$, then

$C \approx 180° - (36.5° + 57.4°) \approx 86.1°$.

Find c:

$$\frac{c}{\sin C} = \frac{a}{\sin A}$$

$$\frac{c}{\sin 86.1°} = \frac{24}{\sin 36.5°}$$

$$c = \frac{24 \sin 86.1°}{\sin 36.5°} \approx 40$$

Solution II

If $B \approx 122.6°$, then

$C \approx 180° - (36.5° + 122.6°) \approx 20.9°$.

Find c:

$$\frac{c}{\sin C} = \frac{a}{\sin A}$$

$$\frac{c}{\sin 20.9°} = \frac{24}{\sin 36.5°}$$

$$c = \frac{24 \sin 20.9°}{\sin 36.5°} \approx 14$$

5. Find B:

$$\frac{b}{\sin B} = \frac{c}{\sin C}$$

$$\frac{24.2}{\sin B} = \frac{30.3}{\sin 61°10'}$$

$$\sin B = \frac{24.2 \sin 61°10'}{30.3} \approx 0.6996$$

Then $B \approx 44°24'$ or $B \approx 135°36'$. An angle of $135°36'$ cannot be an angle of this triangle because it already has an angle of $61°10'$ and the two would total more than $180°$. Thus $B \approx 44°24'$.

Find A:

$A \approx 180° - (61°10' + 44°24') \approx 74°26'$

Find a:

$$\frac{a}{\sin A} = \frac{c}{\sin C}$$

$$\frac{a}{\sin 74°26'} = \frac{30.3}{\sin 61°10'}$$

$$a = \frac{30.3 \sin 74°26'}{\sin 61°10'} \approx 33.3$$

7. Find A:

$A = 180° - (37.48° + 32.16°) = 110.36°$

Find a:

$$\frac{a}{\sin A} = \frac{c}{\sin C}$$

$$\frac{a}{\sin 110.36°} = \frac{3}{\sin 32.16°}$$

$$a = \frac{3 \sin 110.36°}{\sin 32.16°} \approx 5 \text{ mi}$$

Find b:

$$\frac{b}{\sin B} = \frac{c}{\sin C}$$

$$\frac{b}{\sin 37.48°} = \frac{3}{\sin 32.16°}$$

$$b = \frac{3 \sin 37.48°}{\sin 32.16°} \approx 3 \text{ mi}$$

9. Find B:

$$\frac{b}{\sin B} = \frac{c}{\sin C}$$

$$\frac{56.78}{\sin B} = \frac{56.78}{\sin 83.78°}$$

$$\sin B = \frac{56.78 \sin 83.78°}{56.78} \approx 0.9941$$

Then $B \approx 83.78°$ or $B \approx 96.22°$. An angle of $96.22°$ cannot be an angle of this triangle because it already has an angle of $83.78°$ and the two would total $180°$. Thus, $B \approx 83.78°$.

Find A:

$A \approx 180° - (83.78° + 83.78°) \approx 12.44°$

Find a:

$$\frac{a}{\sin A} = \frac{c}{\sin C}$$

$$\frac{a}{\sin 12.44°} = \frac{56.78}{\sin 83.78°}$$

$$a = \frac{56.78 \sin 12.44°}{\sin 83.78} \approx 12.30 \text{ yd}$$

11. Find B:

$$\frac{b}{\sin B} = \frac{a}{\sin A}$$

$$\frac{10.07}{\sin B} = \frac{20.01}{\sin 30.3°}$$

$$\sin B = \frac{10.07 \sin 30.3°}{20.01} \approx 0.2539$$

Then $B \approx 14.7°$ or $B \approx 165.3°$. An angle of $165.3°$ cannot be an angle of this triangle because it already has an angle of $30.3°$ and the two would total $180°$. Thus, $B \approx 14.7°$.

Find C:

$$C \approx 180° - (30.3° + 14.7°) \approx 135.0°$$

Find c:

$$\frac{c}{\sin C} = \frac{a}{\sin A}$$

$$\frac{c}{\sin 135.0°} = \frac{20.01}{\sin 30.3°}$$

$$c = \frac{20.01 \sin 135.0°}{\sin 30.3°} \approx 28.04 \text{ cm}$$

13. Find B:

$$\frac{b}{\sin B} = \frac{a}{\sin A}$$

$$\frac{18.4}{\sin B} = \frac{15.6}{\sin 89°}$$

$$\sin B = \frac{18.4 \sin 89°}{15.6} \approx 1.1793$$

Since there is no angle having a sine greater than 1, there is no solution.

15. Find B:

$$B = 180° - (32.76° + 21.97°) = 125.27°$$

Find b:

$$\frac{b}{\sin B} = \frac{a}{\sin A}$$

$$\frac{b}{\sin 125.27°} = \frac{200}{\sin 32.76°}$$

$$b = \frac{200 \sin 125.27°}{\sin 32.76°} \approx 302 \text{ m}$$

Find c:

$$\frac{c}{\sin C} = \frac{a}{\sin A}$$

$$\frac{c}{\sin 21.97°} = \frac{200}{\sin 32.76°}$$

$$c = \frac{200 \sin 21.97°}{\sin 32.76°} \approx 138 \text{ m}$$

17. $K = \dfrac{1}{2} ac \sin B$

$K = \dfrac{1}{2}(7.2)(3.4) \sin 42°$ Substituting

$K \approx 8.2 \text{ ft}^2$

19. $K = \dfrac{1}{2} ab \sin C$

$K = \dfrac{1}{2} \cdot 4 \cdot 6 \cdot \sin 82°54'$ Substituting

$K \approx 12 \text{ yd}^2$

21. $K = \dfrac{1}{2} ac \sin B$

$K = \dfrac{1}{2}(46.12)(36.74) \sin 135.2°$

$K \approx 596.98 \text{ ft}^2$

23. $K = \dfrac{1}{2} bc \sin A$

$K = \dfrac{1}{2} \cdot 42 \cdot 53 \sin 135°$

$K \approx 787 \text{ ft}^2$

25.

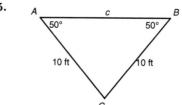

To solve this triangle find A, a, and c.

First we find C:

$$C = 180° - (50° + 50°) = 80°$$

Now we use the law of sines to find c:

$$\frac{c}{\sin C} = \frac{a}{\sin A}$$

$$\frac{c}{\sin 80°} = \frac{10}{\sin 50°}$$

$$c = \frac{10 \sin 80°}{\sin 50°} \approx 12.86$$

The speakers should be placed about 12.86 ft apart, or about 12 ft 10 in. apart.

27.

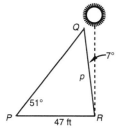

To solve this triangle find A, a, and c.

Find R: $R = 90° − 7° = 83°$

Find Q: $Q = 180° − (51° + 83°) = 46°$

Find p: (p is the length of the pole.)

$$\frac{p}{\sin P} = \frac{Q}{\sin Q}$$

$$\frac{p}{\sin 51°} = \frac{47}{\sin 46°}$$

$$p = \frac{47 \sin 51°}{\sin 46°} \approx 51$$

The pole is about 51 ft long.

29.

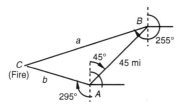

To solve this triangle find A, a, and c.

$A = 45° + (360° − 295°) = 110°$

$B = 255° − 180° − 45° = 30°$

$C = 180° − (110° + 30°) = 40°$

The distance from Tower A to the fire is b:

$$\frac{b}{\sin B} = \frac{c}{\sin C}$$

$$\frac{b}{\sin 30°} = \frac{45}{\sin 40°}$$

$$b = \frac{45 \sin 30°}{\sin 40°} \approx 35 \text{ mi}$$

The distance from Tower B to the fire is a:

$$\frac{a}{\sin A} = \frac{c}{\sin C}$$

$$\frac{a}{\sin 110°} = \frac{45}{\sin 40°}$$

$$a = \frac{45 \sin 110°}{\sin 40°} \approx 66 \text{ mi}$$

31.

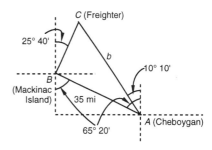

To solve this triangle find A, a, and c.

Find A: $A = 65°20′ − 10°10′ = 55°10′$

Find B: $B = 180° − 25°40′ − 65°20′ = 89°$

Find C: $C = 180° − (55°10′ + 89°) = 35°50′$

Find b:

$$\frac{b}{\sin B} = \frac{c}{\sin C}$$

$$\frac{b}{\sin 89°} = \frac{35}{\sin 35°50′}$$

$$b = \frac{35 \sin 89°}{\sin 35°50′} \approx 60$$

The freighter is about 60 mi from Cheboygan.

33. Discussion and Writing

35. $\cos A = 0.2213$

Set the grapher in Radian mode and press
$\boxed{\text{2nd}}$ $\boxed{\text{COS}}$ .2213 $\boxed{\text{ENTER}}$. The grapher returns
1.347649005, so $A \approx 1.348$ radians. Set the grapher
in Degree mode and repeat the keystrokes above.
The grapher returns 77.21460028, so $A \approx 77.2°$.

37. With the grapher set in Degree mode, enter
$18°14′20″$ as $18′14′20′$. We find that $18°14′20″ \approx 18.24°$.

39. See the answer section in the text.

41. See the answer section in the text.

Exercise Set 7.3

1.

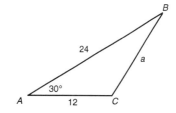

To solve this triangle find A, a, and c.

To solve this triangle find a, B, and C.

From the law of cosines,

$$a^2 = b^2 + c^2 - 2bc \cos A$$
$$a^2 = 12^2 + 24^2 - 2 \cdot 12 \cdot 24 \cos 30°$$
$$a^2 \approx 221$$
$$a \approx 15$$

Next we use the law of cosines again to find a second angle.

$$b^2 = a^2 + c^2 - 2ac \cos B$$
$$12^2 = 15^2 + 24^2 - 2(15)(24) \cos B$$
$$144 = 225 + 576 - 720 \cos B$$
$$-657 = -720 \cos B$$
$$\cos B \approx 0.9125$$
$$B \approx 24°$$

Now we find the third angle.

$$C \approx 180° - (30° + 24°) \approx 126°$$

3.

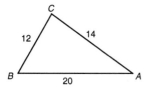

To solve this triangle find A, B, and C.

Find A:

$$a^2 = b^2 + c^2 - 2bc \cos A$$
$$12^2 = 14^2 + 20^2 - 2 \cdot 14 \cdot 20 \cos A$$
$$144 = 196 + 400 - 560 \cos A$$
$$-452 = -560 \cos A$$
$$\cos A \approx 0.8071$$

Thus $A \approx 36.18°$.

Find B:

$$b^2 = a^2 + c^2 - 2ac \cos B$$
$$14^2 = 12^2 + 20^2 - 2 \cdot 12 \cdot 20 \cos B$$
$$196 = 144 + 400 - 480 \cos B$$
$$-348 = -480 \cos B$$
$$\cos B \approx 0.7250$$

Thus $B \approx 43.53°$.

Then $C \approx 180° - (36.18° + 43.53°) \approx 100.29°$.

5. To solve this triangle find b, A, and C.

Find b:

$$b^2 = a^2 + c^2 - 2ac \cos B$$
$$b^2 = 78^2 + 16^2 - 2 \cdot 78 \cdot 16 \cos 72°40'$$
$$b^2 \approx 5596$$
$$b \approx 75 \text{ m}$$

Find A:

$$a^2 = b^2 + c^2 - 2bc \cos A$$
$$78^2 = 75^2 + 16^2 - 2 \cdot 75 \cdot 16 \cos A$$
$$6084 = 5625 + 256 - 2400 \cos A$$
$$203 = -2400 \cos A$$
$$\cos A \approx -0.0846$$
$$a \approx 94°51'$$

Then $C = 180° - (94°51' + 72°40') = 12°29'$

7. Find A:

$$a^2 = b^2 + c^2 - 2bc \cos A$$
$$16^2 = 20^2 + 32^2 - 2 \cdot 20 \cdot 32 \cos A$$
$$256 = 400 + 1024 - 1280 \cos A$$
$$-1168 = -1280 \cos A$$
$$\cos A \approx 0.9125$$
$$A \approx 24.15°.$$

Find B:

$$b^2 = a^2 + c^2 - 2ac \cos B$$
$$20^2 = 16^2 + 32^2 - 2 \cdot 16 \cdot 32 \cos B$$
$$400 = 256 + 1024 - 1024 \cos B$$
$$-880 = -1024 \cos B$$
$$\cos B \approx 0.8594$$
$$B \approx 30.75°.$$

Then $C \approx 180° - (24.15° + 30.75°) \approx 125.10°.$

9. Find A:

$$a^2 = b^2 + c^2 - 2bc \cos A$$
$$2^2 = 3^2 + 8^2 - 2 \cdot 3 \cdot 8 \cos A$$
$$4 = 9 + 64 - 48 \cos A$$
$$-69 = -48 \cos A$$
$$\cos A \approx 1.4375$$

Since there is no angle whose cosine is greater than 1, there is no solution.

11. Find A:

$$a^2 = b^2 + c^2 - 2bc \cos A$$
$$(26.12)^2 = (21.34)^2 + (19.25)^2 -$$
$$2(21.34)(19.25) \cos A$$
$$682.2544 = 455.3956 + 370.5625 - 821.59 \cos A$$
$$-143.7037 = -821.59 \cos A$$
$$\cos A \approx 0.1749$$
$$A \approx 79.93°$$

Find B:
$$b^2 = a^2 + c^2 - 2ac\cos B$$
$$(21.34)^2 = (26.12)^2 + (19.25)^2 -$$
$$2(26.12)(19.25)\cos B$$
$$455.3956 = 682.2544 + 370.5625 - 1005.62\cos B$$
$$-597.4213 = -1005.62\cos B$$
$$\cos B \approx 0.5941$$
$$B \approx 53.55°$$
$$C \approx 180° - (79.93° + 53.55°) \approx 46.52°$$

13. Find c:
$$c^2 = a^2 + b^2 - 2ab\cos C$$
$$c^2 = (60.12)^2 + (40.23)^2 -$$
$$2(60.12)(40.23)\cos 48.7°$$
$$c^2 \approx 2040$$
$$c \approx 45.17\text{ mi}$$

Find A:
$$a^2 = b^2 + c^2 - 2bc\cos A$$
$$(60.12)^2 = (40.23)^2 + (45.17)^2 -$$
$$2(40.23)(45.17)\cos A$$
$$3614.4144 = 1618.4529 + 2040.3289 -$$
$$3634.3782\cos A$$
$$-44.3674 = -3634.3782\cos A$$
$$\cos A \approx 0.0122$$
$$A \approx 89.3°$$
$$B \approx 180° - (89.3° + 48.7°) \approx 42.0°$$

15. Find a:
$$a^2 = b^2 + c^2 - 2bc\cos A$$
$$a^2 = (10.2)^2 + (17.3)^2 - 2(10.2)(17.3)\cos 53.456°$$
$$a^2 \approx 193.19$$
$$a \approx 13.9\text{ in.}$$

Find B:
$$b^2 = a^2 + c^2 - 2ac\cos B$$
$$(10.2)^2 = (13.9)^2 + (17.3)^2 - 2(13.9)(17.3)\cos B$$
$$104.04 = 193.21 + 299.29 - 480.94\cos B$$
$$-388.46 = -480.94\cos B$$
$$\cos B \approx 0.8077$$
$$B \approx 36.127°$$

Find C:
$$C \approx 180° - (53.456° + 36.127°) \approx 90.417°$$

17. We are given two sides and the angle opposite one of them. The law of sines applies.
$$C = 180° - (70° + 12°) = 98°$$

$$\frac{a}{\sin A} = \frac{b}{\sin B}$$
$$\frac{a}{\sin 70°} = \frac{21.4}{\sin 12°}$$
$$a = \frac{21.4\sin 70°}{\sin 12°} \approx 96.7$$
$$\frac{c}{\sin C} = \frac{b}{\sin B}$$
$$\frac{c}{\sin 98°} = \frac{21.4}{\sin 12°}$$
$$c = \frac{21.4\sin 98°}{\sin 12°} \approx 101.9$$

19. We are given all three sides of the triangle. The law of cosines applies.
$$a^2 = b^2 + c^2 - 2bc\cos A$$
$$(3.3)^2 = (2.7)^2 + (2.8)^2 - 2(2.7)(2.8)\cos A$$
$$10.89 = 7.29 + 7.84 - 15.12\cos A$$
$$-4.24 = -15.12\cos A$$
$$\cos A \approx 0.2804$$
$$A \approx 73.71°$$
$$b^2 = a^2 + c^2 - 2ac\cos B$$
$$(2.7)^2 = (3.3)^2 + (2.8)^2 - 2(3.3)(2.8)\cos B$$
$$7.29 = 10.89 + 7.84 - 18.48\cos B$$
$$-11.44 = -18.48\cos B$$
$$\cos B \approx 0.6190$$
$$B \approx 51.75°$$
$$C \approx 180° - (73.71° + 51.75°) \approx 54.54°$$

21. We are given the three angles of a triangle. Neither law applies. This triangle cannot be solved using the given information.

23. We are given all three sides of the triangle. The law of cosines applies.
$$a^2 = b^2 + c^2 - 2bc\cos A$$
$$(3.6)^2 = (6.2)^2 + (4.1)^2 - 2(6.2)(4.1)\cos A$$
$$12.96 = 38.44 + 16.81 - 50.84\cos A$$
$$-42.29 = -50.84\cos A$$
$$\cos A \approx 0.8318$$
$$A \approx 33.71°$$
$$b^2 = a^2 + c^2 - 2ac\cos B$$
$$(6.2)^2 = (3.6)^2 + (4.1)^2 - 2(3.6)(4.1)\cos B$$
$$38.44 = 12.96 + 16.81 - 29.52\cos B$$
$$8.67 = -29.52\cos B$$
$$\cos B \approx -0.2937$$
$$B \approx 107.08°$$
$$C \approx 180° - (33.71° + 107.08°) \approx 39.21°$$

25.

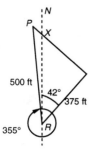

Angle $XRP = 360° - 355° = 5°$, so $R = 42° + 5° = 47°$. Use the law of cosines to find x, the distance from P to G.

$$x^2 = 500^2 + 375^2 - 2 \cdot 500 \cdot 375 \cos 47°$$

$$x^2 \approx 134,876$$

$$x \approx 367$$

The distance between the poachers and the game is about 367 ft.

27. Use the law of cosine to find the distance d from A to B.

$$d^2 = (0.5)^2 + (1.3)^2 - 2(0.5)(1.3) \cos 110°$$

$$d^2 \approx 2.38$$

$$d \approx 1.5$$

He skated about 1.5 mi.

29. 25 knots $\times$ 2 hr = 50 nautical mi

20 knots $\times$ 2 hr = 40 nautical mi

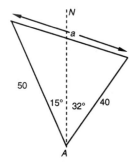

Find A: $A = 15° + 32° = 47°$

Find a:

$$a^2 = b^2 + c^2 - 2bc \cos A$$

$$a^2 = 50^2 + 40^2 - 2 \cdot 50 \cdot 40 \cos 47°$$

$$a^2 \approx 1372$$

$$a \approx 37$$

The ships are about 37 nautical mi apart.

31. 150 km/h $\times$ 3 hr = 450 km

200 km/h $\times$ 3 hr = 600 km

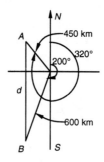

The angle opposite d is $320° - 200°$, or $120°$. Use the law of cosines to find d:

$$d^2 = 450^2 + 600^2 - 2 \cdot 450 \cdot 600 \cos 120°$$

$$d \approx 912$$

The planes are about 912 km apart.

33.

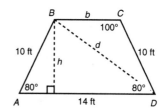

Let b represent the other base, d represent the diagonal, and h the height.

a) Find d:

Using the law of cosines,

$$d^2 = 10^2 + 14^2 - 2 \cdot 10 \cdot 14 \cos 80°$$

$$d^2 \approx 247$$

$$d \approx 16$$

The length of the diagonal is about 16 ft.

b) Find h:

$$\frac{h}{10} = \sin 80°$$

$$h = 10 \sin 80°$$

$$h \approx 9.85$$

Find $\angle CBD$:

Using the law of sines,

$$\frac{10}{\sin \angle CBD} = \frac{16}{\sin 100°}$$

$$\sin \angle CBD = \frac{10 \sin 100°}{16} \approx 0.6155$$

Thus $\angle CBD \approx 38°$.

Then $\angle CDB \approx 180° - (100° + 38°) \approx 42°$.

Find b:

Using the law of sines,

$$\frac{b}{\sin 42°} = \frac{16}{\sin 100°}$$

$$b = \frac{16 \sin 42°}{\sin 100°} \approx 10.87$$

$$\text{Area} = \frac{1}{2}h(b_1 + b_2)$$

$$\text{Area} \approx \frac{1}{2}(9.85)(10.87 + 14)$$

$$\approx 122$$

The area is about 122 ft^2. (Answers may vary due to rounding differences.)

35. Place the figure on a coordinate system as shown.

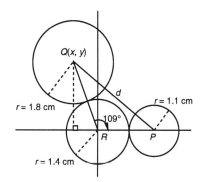

Let (x, y) be the coordinates of point Q. The coordinates of point P are $(1.4 + 1.1, 0)$, or $(2.5, 0)$. The length of QR is $1.8 + 1.4$, or 3.2.

We use the law of cosines to find d.

$$d^2 = (2.5)^2 + (3.2)^2 - 2(2.5)(3.2)\cos 109°$$

$$d^2 \approx 21.70$$

$$d \approx 4.7$$

The length PQ is about 4.7 cm.

37. Discussion and Writing

39. The distance of -5 from 0 is 5, so $|-5| = 5$.

41. $\sin 45° = \dfrac{1}{\sqrt{2}}$, or $\dfrac{\sqrt{2}}{2}$

43. $\cos\left(-\dfrac{2\pi}{3}\right) = -\cos\dfrac{\pi}{3} = -\dfrac{1}{2}$

45.

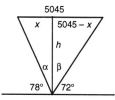

$\alpha = 90° - 78° = 12°$

$\beta = 90 - 72° = 18°$

$$\tan 12° = \frac{x}{h}, \text{ so } x = h\tan 12°$$

$$\tan 18° = \frac{5045 - x}{h}, \text{ or}$$

$$\tan 18° = \frac{5045 - h\tan 12°}{h}$$

Substituting for x

$$h\tan 18° = 5045 - h\tan 12°$$

$$h\tan 18° + h\tan 12° = 5045$$

$$h(\tan 18° + \tan 12°) = 5045$$

$$h = \frac{5045}{\tan 18° + \tan 12°}$$

$$h \approx 9386$$

The canyon is about 9386 ft deep.

47.

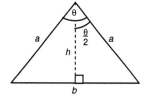

Let b represent the base, h the height, and θ the included angle.

Find h:

$$\frac{h}{a} = \cos\frac{\theta}{2}$$

$$h = a\cos\frac{\theta}{2}$$

Find b:

$$b^2 = a^2 + a^2 - 2 \cdot a \cdot a\cos\theta$$

$$b^2 = 2a^2 - 2a^2\cos\theta$$

$$b^2 = 2a^2(1 - \cos\theta)$$

$$b = \sqrt{2}a\sqrt{1 - \cos\theta}$$

Find A (area):

$$A = \frac{1}{2}bh$$

$$A = \frac{1}{2}\left(\sqrt{2}a\sqrt{1 - \cos\theta}\right)\left(a\cos\frac{\theta}{2}\right)$$

$$A = \left(a\sqrt{\frac{1 - \cos\theta}{2}}\right)\left(a\sqrt{\frac{1 + \cos\theta}{2}}\right)$$

$$A = \frac{1}{2}a^2\sqrt{1 - \cos^2\theta} = \frac{1}{2}a^2\sqrt{\sin^2\theta}$$

$$A = \frac{1}{2}a^2\sin\theta$$

The maximum area occurs when θ is 90°, because the sine function takes its maximum value, 1, at $\theta = 90°$.

Exercise Set 7.4

1.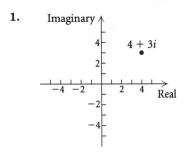

$$|4 + 3i| = \sqrt{4^2 + 3^2} = \sqrt{16 + 9} = \sqrt{25} = 5$$

3.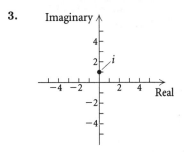

$$|i| = |0 + 1 \cdot i| = \sqrt{1^2} = 1$$

5.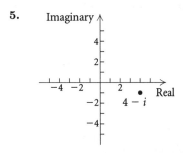

$$|4 - i| = \sqrt{4^2 + (-1)^2} = \sqrt{16 + 1} = \sqrt{17}$$

7.

$$|3| = |3 + 0i| = \sqrt{3^2} = 3$$

9. From the graph we see that standard notation for the number is $3 - 3i$.

Find trigonometric notation:

$$r = \sqrt{3^2 + (-3)^2} = \sqrt{18} = 3\sqrt{2}$$

$$\sin \theta = \frac{-3}{3\sqrt{2}} = -\frac{1}{\sqrt{2}}, \text{ or } -\frac{\sqrt{2}}{2}$$

$$\cos \theta = \frac{3}{3\sqrt{2}} = \frac{1}{\sqrt{2}}, \text{ or } \frac{\sqrt{2}}{2}$$

Thus, $\theta = \frac{7\pi}{4}$, or 315°, and we have

$$3 - 3i = 3\sqrt{2}\left(\cos \frac{7\pi}{4} + i \sin \frac{7\pi}{4} \right), \text{ or}$$

$$3 - 3i = 3\sqrt{2}(\cos 315° + i \sin 315°).$$

11. From the graph we see that standard notation for the number is $0 + 4i$, or $4i$.

Find trigonometric notation:

$$r = \sqrt{4^2} = 4$$

$\sin \theta = \frac{4}{4} = 1$, $\cos \theta = \frac{0}{4} = 0$. Thus, $\theta = \frac{\pi}{2}$, or 90°, and we have

$$4i = 4\left(\cos \frac{\pi}{2} + i \sin \frac{\pi}{2} \right), \text{ or}$$

$$4i = 4(\cos 90° + i \sin 90°).$$

13. Find trigonometric notation for $1 - i$.

$$r = \sqrt{1^2 + (-1)^2} = \sqrt{2}$$

$$\sin \theta = \frac{-1}{\sqrt{2}}, \text{ or } -\frac{\sqrt{2}}{2}, \cos \theta = \frac{1}{\sqrt{2}}, \text{ or } \frac{\sqrt{2}}{2}$$

Thus, $\theta = \frac{7\pi}{4}$, or 315°, and we have

$$1 - i = \sqrt{2}\left(\cos \frac{7\pi}{4} + i \sin \frac{7\pi}{4} \right), \text{ or}$$

$$1 - i = \sqrt{2}(\cos 315° + i \sin 315°).$$

15. Find trigonometric notation for $-3i$.

$$r = \sqrt{(-3)^2} = 3$$

$\sin \theta = \frac{-3}{3} = -1$, $\cos \theta = \frac{0}{3} = 0$

Thus, $\theta = \frac{3\pi}{2}$, or 270°, and we have

$$-3i = 3\left(\cos \frac{3\pi}{2} + i \sin \frac{3\pi}{2} \right), \text{ or}$$

$$-3i = 3(\cos 270° + i \sin 270°).$$

17. Find trigonometric notation for $\sqrt{3} + i$.

$$r\sqrt{(\sqrt{3})^2 + 1^2} = \sqrt{4} = 2$$

$\sin \theta = \frac{1}{2}$, $\cos \theta = \frac{\sqrt{3}}{2}$. Thus, $\theta = \frac{\pi}{6}$, or 30°, and we have

$$\sqrt{3} + i = 2\left(\cos \frac{\pi}{6} + i \sin \frac{\pi}{6} \right), \text{ or}$$

$$\sqrt{3} + i = 2(\cos 30° + i \sin 30°).$$

19. Find trigonometric notation for $\dfrac{2}{5}$.

$$r = \sqrt{\left(\dfrac{2}{5}\right)^2} = \dfrac{2}{5}$$

$\sin\theta = \dfrac{0}{\frac{2}{5}} = 0$, $\cos\theta = \dfrac{\frac{2}{5}}{\frac{2}{5}} = 1$. Thus, $\theta = 0$, or $0°$,

and we have

$r = \dfrac{2}{5}(\cos 0 + i\sin 0)$, or

$r = \dfrac{2}{5}(\cos 0° + i\sin 0°)$.

21. $3(\cos 30° + i\sin 30°) = 3\cos 30° + (3\sin 30°)i$

$a = 3\cos 30° = 3\cdot\dfrac{\sqrt{3}}{2} = \dfrac{3\sqrt{3}}{2}$

$b = 3\sin 30° = 3\cdot\dfrac{1}{2} = \dfrac{3}{2}$

Thus $3(\cos 30° + i\sin 30°) = \dfrac{3\sqrt{3}}{2} + \dfrac{3}{2}i$.

23. $10(\cos 270° + i\sin 270°) = 10\cos 270° + (10\sin 270°)i$

$a = 10\cos 270° = 10\cdot 0 = 0$

$b = 10\sin 270° = 10(-1) = -10$

Thus $10(\cos 270° + i\sin 270°) = 0 + (-10)i = -10i$.

25. $\sqrt{8}\left(\cos\dfrac{\pi}{4} + i\sin\dfrac{\pi}{4}\right) = \sqrt{8}\cos\dfrac{\pi}{4} + \left(\sqrt{8}\cos\dfrac{\pi}{4}\right)i$

$a = \sqrt{8}\cos\dfrac{\pi}{4} = \sqrt{8}\cdot\dfrac{\sqrt{2}}{2} = 2$

$b = \sqrt{8}\sin\dfrac{\pi}{4} = \sqrt{8}\cdot\dfrac{\sqrt{2}}{2} = 2$

Thus $\sqrt{8}\left(\cos\dfrac{\pi}{4} + i\sin\dfrac{\pi}{4}\right) = 2 + 2i$.

27. $2\left(\cos\dfrac{\pi}{2} + i\sin\dfrac{\pi}{2}\right) = 2\cos\dfrac{\pi}{2} + \left(2\sin\dfrac{\pi}{2}\right)i$

$a = 2\cos\dfrac{\pi}{2} = 2\cdot 0 = 0$

$b = 2\sin\dfrac{\pi}{2} = 2\cdot 1 = 2$

Thus $2\left(\cos\dfrac{\pi}{2} + i\sin\dfrac{\pi}{2}\right) = 0 + 2i = 2i$.

29. $\dfrac{12(\cos 48° + i\sin 48°)}{3(\cos 6° + i\sin 6°)}$

$= \dfrac{12}{3}[\cos(48° - 6°) + i\sin(48° - 6°)]$

$= 4(\cos 42° + i\sin 42°)$

31. $2.5(\cos 35° + i\sin 35°)\cdot 4.5(\cos 21° + i\sin 21°)$

$= 2.5(4.5)[\cos(35° + 21°) + i\sin(35° + 21°)]$

$= 11.25(\cos 56° + i\sin 56°)$

33. $(1 - i)(2 + 2i)$

Find trigonometric notation for $1 - i$.

$r = \sqrt{1^2 + (-1)^2} = \sqrt{2}$

$\sin\theta = \dfrac{-1}{\sqrt{2}} = -\dfrac{\sqrt{2}}{2}$, $\cos\theta = \dfrac{1}{\sqrt{2}} = \dfrac{\sqrt{2}}{2}$

Thus $\theta = \dfrac{7\pi}{4}$, or $315°$, and $1 - i = \sqrt{2}(\cos 315° + i\sin 315°)$.

Find trigonometric notation for $2 + 2i$.

$r = \sqrt{2^2 + 2^2} = \sqrt{8} = 2\sqrt{2}$

$\sin\theta = \dfrac{2}{2\sqrt{2}} = \dfrac{\sqrt{2}}{2}$, $\cos\theta = \dfrac{2}{2\sqrt{2}} = \dfrac{\sqrt{2}}{2}$

Thus $\theta = \dfrac{\pi}{4}$, or $45°$, and $2 + 2i = 2\sqrt{2}(\cos 45° + i\sin 45°)$.

$(1 - i)(2 + 2i)$

$= \sqrt{2}(\cos 315° + i\sin 315°)\cdot 2\sqrt{2}(\cos 45° + \sin 45°)$

$= \sqrt{2}\cdot 2\sqrt{2}[\cos(315° + 45°) + i\sin(315° + 45°)]$

$= 4(\cos 360° + i\sin 360°)$

$= 4(\cos 0° + i\sin 0°)$, or 4

35. $\dfrac{1 - i}{1 + i}$

Find trigonometric notation for $1 - i$.

$r = \sqrt{1^2 + (-1)^2} = \sqrt{2}$

$\sin\theta = \dfrac{-1}{\sqrt{2}} = -\dfrac{\sqrt{2}}{2}$, $\cos\theta = \dfrac{1}{\sqrt{2}} = \dfrac{\sqrt{2}}{2}$

Thus $\theta = \dfrac{7\pi}{4}$, or $315°$, and $1 - i = \sqrt{2}(\cos 315° + i\sin 315°)$

From Example 3(a) we know that $1 + i = \sqrt{2}(\cos 45° + i\sin 45°)$.

$\dfrac{1 - i}{1 + i}$

$= \dfrac{\sqrt{2}(\cos 315° + i\sin 315°)}{\sqrt{2}(\cos 45° + i\sin 45°)}$

$= \dfrac{\sqrt{2}}{\sqrt{2}}[\cos(315° - 45°) + i\sin(315° - 45°)]$

$= 1(\cos 270° + i\sin 270°)$

$= 1[0 + i(-1)]$

$= -i$

37. $(3\sqrt{3} - 3i)(2i)$

Find trigonometric notation for $3\sqrt{3} - 3i$.

$r = \sqrt{(3\sqrt{3})^2 + (-3)^2} = \sqrt{36} = 6$

$\sin\theta = \dfrac{-3}{6} = -\dfrac{1}{2}, \cos\theta = \dfrac{3\sqrt{3}}{6} = \dfrac{\sqrt{3}}{2}$

Thus $\theta = 330°$, and $3\sqrt{3} - 3i = 6(\cos 330° + i\sin 330°)$.

Find trigonometric notation for $2i$.

$r = \sqrt{2^2} = 2$

$\sin\theta = \dfrac{2}{2} = 1, \cos\theta = \dfrac{0}{2} = 0$

Thus $\theta = 90°$, and $2i = 2(\cos 90° + i\sin 90°)$.

$(3\sqrt{3} - 3i)(2i)$

$= 6(\cos 330° + i\sin 330°) \cdot 2(\cos 90° + i\sin 90°)$

$= 6 \cdot 2[\cos(330° + 90°) + i\sin(330° + 90°)]$

$= 12(\cos 420° + i\sin 420°)$

$= 12(\cos 60° + i\sin 60°)$

$= 12\left(\dfrac{1}{2} + i \cdot \dfrac{\sqrt{3}}{2}\right)$

$= 6 + 6\sqrt{3}i$

39. $\dfrac{2\sqrt{3} - 2i}{1 + \sqrt{3}i}$

Find trigonometric notation for $2\sqrt{3} - 2i$.

$r = \sqrt{(2\sqrt{3})^2 + (-2)^2} = \sqrt{16} = 4$

$\sin\theta = \dfrac{-2}{4} = -\dfrac{1}{2}, \cos\theta = \dfrac{2\sqrt{3}}{4} = \dfrac{\sqrt{3}}{2}$

Thus $\theta = 330°$, and $2\sqrt{3} - 2i = 4(\cos 330° + i\sin 330°)$

Find trigonometric notation for $1 + \sqrt{3}i$.

$r = \sqrt{1^2 + (\sqrt{3})^2} = \sqrt{4} = 2$

$\sin\theta = \dfrac{\sqrt{3}}{2}, \cos\theta = \dfrac{1}{2}$

Thus $\theta = 60°$, and $1 + \sqrt{3}i = 2(\cos 60° + i\sin 60°)$

$\dfrac{2\sqrt{3} - 2i}{1 + \sqrt{3}i}$

$= \dfrac{4(\cos 330° + i\sin 330°)}{2(\cos 60° + i\sin 60°)}$

$= \dfrac{4}{2}[\cos(330° - 60°) + i\sin(330° - 60°)]$

$= 2(\cos 270° + i\sin 270°)$

$= 2[0 + i \cdot (-1)]$

$= -2i$

41. $\left[2\left(\cos\dfrac{\pi}{3} + i\sin\dfrac{\pi}{3}\right)\right]^3$

$= 2^3\left[\cos\left(3 \cdot \dfrac{\pi}{3}\right) + i\sin\left(3 \cdot \dfrac{\pi}{3}\right)\right]$

$= 8(\cos\pi + i\sin\pi)$

43. From Exercise 35 we know that $1 + i = \sqrt{2}(\cos 45° + i\sin 45°)$.

$(1 + i)^6 = [\sqrt{2}(\cos 45° + i\sin 45°)]^6$

$\qquad = (\sqrt{2})^6[\cos(6 \cdot 45°) + i\sin(6 \cdot 45°)]$

$\qquad = 8(\cos 270° + i\sin 270°)$, or

$\qquad\qquad 8\left(\cos\dfrac{3\pi}{2} + i\sin\dfrac{3\pi}{2}\right)$

45. $[3(\cos 20° + i\sin 20°)]^3$

$= 27(\cos 60° + i\sin 60°)$

$= 27\left(\dfrac{1}{2} + i \cdot \dfrac{\sqrt{3}}{2}\right)$

$= \dfrac{27}{2} + \dfrac{27\sqrt{3}}{2}i$

47. From Exercise 13 we know that $1 - i = \sqrt{2}(\cos 315° + i\sin 315°)$.

$(1 - i)^5 = [\sqrt{2}(\cos 315° + i\sin 315°)]^5$

$\qquad = (\sqrt{2})^5(\cos 1575° + i\sin 1575°)$

$\qquad = 2^{5/2}(\cos 135° + i\sin 135°)$

$\qquad = 4\sqrt{2}\left(-\dfrac{\sqrt{2}}{2} + i \cdot \dfrac{\sqrt{2}}{2}\right)$

$\qquad = -4 + 4i$

49. Find trigonometric notation for $\dfrac{1}{\sqrt{2}} - \dfrac{1}{\sqrt{2}}i$:

$r = \sqrt{\left(\dfrac{1}{\sqrt{2}}\right)^2 + \left(-\dfrac{1}{\sqrt{2}}\right)^2} = \sqrt{1} = 1$

$\sin\theta = \dfrac{-\dfrac{1}{\sqrt{2}}}{1} = -\dfrac{1}{\sqrt{2}}, \text{ or } -\dfrac{\sqrt{2}}{2}$

$\cos\theta = \dfrac{\dfrac{1}{\sqrt{2}}}{1} = \dfrac{1}{\sqrt{2}}, \text{ or } \dfrac{\sqrt{2}}{2}$

Thus $\theta = 315°$, so $\dfrac{1}{\sqrt{2}} - \dfrac{1}{\sqrt{2}}i = 1(\cos 315° + i\sin 315°)$.

$\left(\dfrac{1}{\sqrt{2}} - \dfrac{1}{\sqrt{2}}i\right)^{12} = [1(\cos 315° + i\sin 315°)]^{12}$

$\qquad\qquad = 1(\cos 3780° + i\sin 3780°)$

$\qquad\qquad = 1(\cos 180° + i\sin 180°)$

$\qquad\qquad = 1(-1 + i \cdot 0)$

$\qquad\qquad = -1$

51. $-i = 1(\cos 270° + i \sin 270°)$

$(-i)^{1/2}$

$= [1(\cos 270° + i \sin 270)]^{1/2}$

$= 1^{1/2}\left[\cos\left(\dfrac{270°}{2} + k \cdot \dfrac{360°}{2}\right) +\right.$

$\left. i \sin\left(\dfrac{270°}{2} + k \cdot \dfrac{360°}{2}\right)\right], k = 0, 1$

$= 1[\cos(135° + k \cdot 180°) + i \sin(135° + k \cdot 180°)],$

$k = 0, 1$

The roots are

$1(\cos 135° + i \sin 135°)$, for $k = 0$

and

$1(\cos 315° + i \sin 315°)$, for $k = 1$,

or $-\dfrac{\sqrt{2}}{2} + \dfrac{\sqrt{2}}{2}i$ and $\dfrac{\sqrt{2}}{2} - \dfrac{\sqrt{2}}{2}i$.

53. $2\sqrt{2} - 2\sqrt{2}i = 4(\cos 315° + i \sin 315°)$

$(2\sqrt{2} - 2\sqrt{2}i)^{1/2}$

$= [4(\cos 315° + i \sin 315°)]^{1/2}$

$= 4^{1/2}\left[\cos\left(\dfrac{315°}{2} + k \cdot \dfrac{360°}{2}\right) +\right.$

$\left. i \sin\left(\dfrac{315°}{2} + k \cdot \dfrac{360°}{2}\right)\right], k = 0, 1$

The roots are

$2(\cos 157.5° + i \sin 157.5°)$, for $k = 0$

and

$2(\cos 337.5° + i \sin 337.5°)$, for $k = 1$.

55. $i = 1(\cos 90° + i \sin 90°)$

$i^{1/3}$

$= [1(\cos 90° + i \sin 90°)]^{1/3}$

$= 1\left[\cos\left(\dfrac{90°}{3} + k \cdot \dfrac{360°}{3}\right) + i \sin\left(\dfrac{90°}{3} + k \cdot \dfrac{360°}{3}\right)\right],$

$k = 0, 1, 2$

The roots are $1(\cos 30° + i \sin 30°)$, $1(\cos 150° + i \sin 150°)$, and $1(\cos 270° + i \sin 270°)$, or $\dfrac{\sqrt{3}}{2} + \dfrac{1}{2}i$, $-\dfrac{\sqrt{3}}{2} + \dfrac{1}{2}i$, and $-i$.

57. $(2\sqrt{3} - 2i)^{1/3}$

$= 4(\cos 330° + i \sin 330°)^{1/3}$

$= 4^{1/3}\left[\cos\left(\dfrac{330°}{3} + k \cdot \dfrac{360°}{3}\right) +\right.$

$\left. i \sin\left(\dfrac{330°}{3} + k \cdot \dfrac{360°}{3}\right)\right], k = 0, 1, 2$

The roots are $\sqrt[3]{4}(\cos 110° + i \sin 110°)$,

$\sqrt[3]{4}(\cos 230° + i \sin 230°)$, and

$\sqrt[3]{4}(\cos 350° + i \sin 350°)$.

59. $16^{1/4}$

$= [16(\cos 0° + i \sin 0°)]^{1/4}$

$= 2\left[\cos\left(\dfrac{0°}{4} + k \cdot \dfrac{360°}{4}\right) + i \sin\left(0° + k \cdot \dfrac{360°}{4}\right)\right],$

$k = 0, 1, 2, 3$

The roots are $2(\cos 0° + i \sin 0°)$, $2(\cos 90° + i \sin 90°)$, $2(\cos 180° + i \sin 180°)$, and $2(\cos 270° + i \sin 270°)$, or 2, $2i$, -2, and $-2i$.

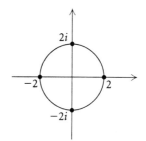

61. $(-1)^{1/5}$

$= [1(\cos 180° + i \sin 180°)]^{1/5}$

$= 1\left[\cos\left(\dfrac{180°}{5} + k \cdot \dfrac{360°}{5}\right) + i \sin\left(\dfrac{180°}{5} + k \cdot \dfrac{360°}{5}\right)\right],$

$k = 0, 1, 2, 3, 4$

The roots are $\cos 36° + i \sin 36°$; $\cos 108° + i \sin 108°$; $\cos 180° + i \sin 180°$, or -1; $\cos 252° + i \sin 252°$; and $\cos 324° + i \sin 324°$.

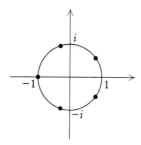

63. $8^{1/10}$

$= [8(\cos 0° + i \sin 0°)]^{1/10}$

$= \left[8^{1/10}\left(\cos\dfrac{0°}{10} + k \cdot \dfrac{360°}{10}\right) + i \sin\left(\dfrac{0°}{10} + k \cdot \dfrac{360°}{10}\right)\right],$

$k = 0, 1, 2, 3, 4, 5, 6, 7, 8, 9$

The roots are $\sqrt[10]{8}(\cos 0° + i \sin 0°)$, or $\sqrt[10]{8}$; $\sqrt[10]{8}(\cos 36° + i \sin 36°)$; $\sqrt[10]{8}(\cos 72° + i \sin 72°)$, $\sqrt[10]{8}(\cos 108° + i \sin 108°)$; $\sqrt[10]{8}(\cos 144° + i \sin 144°)$;

$\sqrt[10]{8}(\cos 180° + i \sin 180°)$, or $- \sqrt[10]{8}$;

$\sqrt[10]{8}(\cos 216° + i \sin 216°)$; $\sqrt[10]{8}(\cos 252° + i \sin 252°)$;

$\sqrt[10]{8}(\cos 288° + i \sin 288°)$; and

$\sqrt[10]{8}(\cos 324° + i \sin 324°)$.

65. $(-1)^{1/6}$

$= [1(\cos 180° + i \sin 180°)]^{1/6}$

$= 1\left[\cos\left(\dfrac{180°}{6} + k \cdot \dfrac{360°}{6}\right) + \right.$

$\left. i \sin\left(\dfrac{180°}{6} + k \cdot \dfrac{360°}{6}\right)\right], k = 0, 1, 2, 3, 4, 5$

The roots are $\cos 30° + i \sin 30°$,

$\cos 90° + i \sin 90°$, $\cos 150° + i \sin 150°$,

$\cos 210° + i \sin 210°$, $\cos 270° + i \sin 270°$,

and $\cos 330° + i \sin 330°$, or $\dfrac{\sqrt{3}}{2} + \dfrac{1}{2}i$,

$i, -\dfrac{\sqrt{3}}{2} + \dfrac{1}{2}i, -\dfrac{\sqrt{3}}{2} - \dfrac{1}{2}i, -i$, and $\dfrac{\sqrt{3}}{2} - \dfrac{1}{2}i$.

67. $x^3 = 1$

The solutions of this equation are the cube roots of 1. These were found in Example 11 in the text. They are $1, -\dfrac{1}{2} + \dfrac{\sqrt{3}}{2}i$, and $-\dfrac{1}{2} - \dfrac{\sqrt{3}}{2}i$.

69. $x^4 + i = 0$

$x^4 = -i$

Find the fourth roots of $-i$.

$(-i)^{1/4}$

$= [1(\cos 270° + i \sin 270°)]^{1/4}$

$= 1\left[\cos\left(\dfrac{270°}{4} + k \cdot \dfrac{360°}{4}\right) + \right.$

$\left. i \sin\left(\dfrac{270°}{4} + k \cdot \dfrac{360°}{4}\right)\right], k = 0, 1, 2, 3$

The solutions are $\cos 67.5° + i \sin 67.5°$,

$\cos 157.5° + i \sin 157.5°$, $\cos 247.5° + i \sin 247.5°$, and

$\cos 337.5° + i \sin 337.5°$.

71. $x^6 + 64 = 0$

$x^6 = -64$

Find the sixth roots of -64.

$(-64)^{1/6}$

$= [64(\cos 180° + i \sin 180°)]^{1/6}$

$= 2\left[\cos\left(\dfrac{180°}{6} + k \cdot \dfrac{360°}{6}\right) + \right.$

$\left. i \sin\left(\dfrac{180°}{6} + k \cdot \dfrac{360°}{6}\right)\right], k = 0, 1, 2, 3, 4, 5$

The solutions are $2(\cos 30° + i \sin 30°)$, $2(\cos 90° + i \sin 90°)$, $2(\cos 150° + i \sin 150°)$, $2(\cos 210° + i \sin 210°)$, $2(\cos 270° + i \sin 270°)$, and $2(\cos 330° + i \sin 330°)$, or $\sqrt{3}+i, 2i, -\sqrt{3}+i, -\sqrt{3}-i, -2i$, and $\sqrt{3} - i$.

73. Discussion and Writing

75. Use the Pythagorean theorem.

$r^2 = 3^2 + (-6)^2$

$r^2 = 9 + 36 = 45$

$r = \sqrt{45} = 3\sqrt{5}$

77. $x^2 + (1 - i)x + i = 0$

Use the quadratic formula.

$a = 1, b = 1 - i, c = i$

$x = \dfrac{-(1-i) \pm \sqrt{(1-i)^2 - 4 \cdot 1 \cdot i}}{2 \cdot 1}$

$= \dfrac{-1 + i \pm \sqrt{1 - 2i + i^2 - 4i}}{2}$

$= \dfrac{-1 + i \pm \sqrt{-6i}}{2}$

Now we find the square roots of $-6i$.

$(-6i)^{1/2}$

$= [6(\cos 270° + i \sin 270°)]^{1/2}$

$= 6\left[\cos\left(\dfrac{270°}{2} + k \cdot \dfrac{360°}{2}\right) + i \sin\left(\dfrac{270°}{2} + k \cdot \dfrac{360°}{5}\right)\right]$,

$k = 0, 1$

The roots are $\sqrt{6}(\cos 135° + i \sin 135°)$ and $\sqrt{6}(\cos 315° + i \sin 315°)$, or $-\sqrt{3} + \sqrt{3}i$ and $\sqrt{3} - \sqrt{3}i$.

Then the solutions of the original equation are

$x = \dfrac{-1 + i - \sqrt{3} + \sqrt{3}i}{2} = \dfrac{-1 - \sqrt{3}}{2} + \dfrac{1 + \sqrt{3}}{2}i$, or

$-\dfrac{1 + \sqrt{3}}{2} + \dfrac{1 + \sqrt{3}}{2}i$

and

$x = \dfrac{-1 + i + \sqrt{3} - \sqrt{3}i}{2} = \dfrac{-1 + \sqrt{3}}{2} + \dfrac{1 - \sqrt{3}}{2}i$, or

$-\dfrac{1 - \sqrt{3}}{2} + \dfrac{1 - \sqrt{3}}{2}i$.

79. $(\cos \theta + i \sin \theta)^{-1}$

$= \dfrac{1}{\cos \theta + i \sin \theta} \cdot \dfrac{\cos \theta - i \sin \theta}{\cos \theta - i \sin \theta}$

$= \dfrac{\cos \theta - i \sin \theta}{\cos^2 \theta + \sin^2 \theta}$

$= \cos \theta - i \sin \theta$

81. See the answer section in the text.

83. See the answer section in the text.

85. See the answer section in the text.

87.
$$z + \bar{z} = 3$$
$$(a + bi) + (a - bi) = 3$$
$$2a = 3$$
$$a = \frac{3}{2}$$

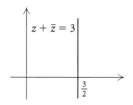

Exercise Set 7.5

1.

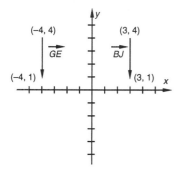

First we find the length of each vector using the distance formula.
$$|\overrightarrow{GE}| = \sqrt{[-4 - (-4)]^2 + (4 - 1)^2} = \sqrt{9} = 3$$
$$|\overrightarrow{BJ}| = \sqrt{(3 - 3)^2 + (4 - 1)^2} = \sqrt{9} = 3$$

Thus, $|\overrightarrow{GE}| = |\overrightarrow{BJ}|$.

Both vectors point down. To verify that they have the same direction we calculate the slopes of the lines that they lie on.

Slope of $\overrightarrow{GE} = \dfrac{4 - 1}{-4 - (-4)} = \dfrac{3}{0}$ (undefined)

Slope of $\overrightarrow{BJ} = \dfrac{4 - 1}{3 - 3} = \dfrac{3}{0}$ (undefined)

Both vectors have undefined slope, so they lie on vertical lines and hence have the same direction.

Since $\overrightarrow{GE}$ and $\overrightarrow{BJ}$ have the same magnitude and same direction, they are equivalent.

3.

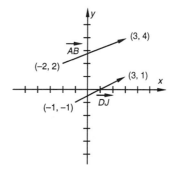

First we find the length of each vector using the distance formula.
$$|\overrightarrow{DJ}| = \sqrt{[3 - (-1)]^2 + [1 - (-1)]^2} = \sqrt{16 + 4} = \sqrt{20}$$
$$|\overrightarrow{AB}| = \sqrt{[3 - (-2)]^2 + (4 - 2)^2} = \sqrt{25 + 4} = \sqrt{29}$$

Since $|\overrightarrow{DJ}| \neq |\overrightarrow{AB}|$, the vectors are not equivalent.

5.

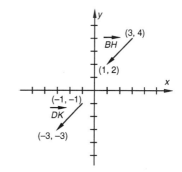

First we find the length of each vector using the distance formula.
$$|\overrightarrow{DK}| = \sqrt{[-3 - (-1)]^2 + [-3 - (-1)]^2} = \sqrt{4 + 4} = \sqrt{8}$$
$$|\overrightarrow{BH}| = \sqrt{(3 - 1)^2 + (4 - 2)^2} = \sqrt{4 + 4} = \sqrt{8}$$

Thus, $|\overrightarrow{DK}| = |\overrightarrow{BH}|$.

Both vectors point down and to the left. We calculate the slopes of the lines that they lie on.

Slope of $\overrightarrow{DK} = \dfrac{-3 - (-1)}{-3 - (-1)} = \dfrac{-2}{-2} = 1$

Slope of $\overrightarrow{BH} = \dfrac{4 - 2}{3 - 1} = \dfrac{2}{2} = 1$

The slopes are the same.

Since $\overrightarrow{DK}$ and $\overrightarrow{BH}$ have the same magnitude and same direction, they are equivalent.

7.

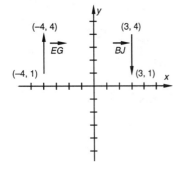

The vectors clearly have different directions, so they are not equivalent.

9.

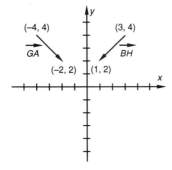

The vectors clearly have different directions, so they are not equivalent.

11.

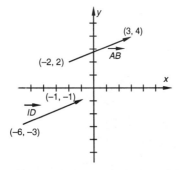

First we find the length of each vector using the distance formula.

$$|\overrightarrow{AB}| = \sqrt{[3-(-2)]^2 + (4-2)^2} = \sqrt{25+4} = \sqrt{29}$$

$$|\overrightarrow{ID}| = \sqrt{[-1-(-6)]^2 + [-1-(-3)]^2} = \sqrt{25+4} = \sqrt{29}$$

Thus, $|\overrightarrow{AB}| = |\overrightarrow{ID}|$.

Both vectors point up and to the right. We calculate the slopes of the lines that they lie on.

Slope of $\overrightarrow{AB} = \dfrac{2-4}{-2-3} = \dfrac{-2}{-5} = \dfrac{2}{5}$

Slope of $\overrightarrow{ID} = \dfrac{-3-(-1)}{-6-(-1)} = \dfrac{-2}{-5} = \dfrac{2}{5}$

The slopes are the same.

Since $\overrightarrow{AB}$ and $\overrightarrow{ID}$ have the same magnitude and same direction, they are equivalent.

13.

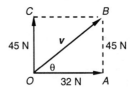

Use the Pythagorean theorem to find the magnitude of the resultant **v**.

$$|\mathbf{v}|^2 = 32^2 + 45^2$$

$$|\mathbf{v}| = \sqrt{32^2 + 45^2} \approx 55 \text{ N}$$

To find θ use the fact that triangle OAB is a right triangle.

$$\tan\theta = \frac{45}{32}$$

$$\theta \approx 55°$$

15.

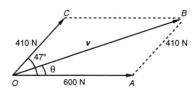

$\angle A = 180° - 47° = 133°$

Use the law of cosines to find the magnitude of the resultant **v**.

$$|\mathbf{v}|^2 = 600^2 + 410^2 - 2 \cdot 600 \cdot 410 \cos 133°$$

$$|\mathbf{v}| \approx \sqrt{863,643} \approx 929 \text{ N}$$

Use the law of sines to find θ.

$$\frac{410}{\sin\theta} = \frac{929}{\sin 133°}$$

$$\sin\theta = \frac{410 \sin 133°}{929} \approx 0.3228$$

$$\theta \approx 19°$$

17.

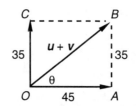

We use the Pythagorean theorem to find the magnitude of **u** + **v**.

$$|\mathbf{u} + \mathbf{v}|^2 = 35^2 + 45^2$$

$$|\mathbf{u} + \mathbf{v}| = \sqrt{35^2 + 45^2}$$

$$|\mathbf{u} + \mathbf{v}| \approx 57.0$$

To find the direction of $\mathbf{u}+\mathbf{v}$ we note that since OAB is a right triangle

$$\tan \theta = \frac{35}{45}$$

$$\theta \approx 38°.$$

19.

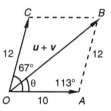

$$A = 180° - 67° = 113°$$

Use the law of cosines to find $|\mathbf{u} + \mathbf{v}|$.

$$|\mathbf{u} + \mathbf{v}|^2 = 10^2 + 12^2 - 2 \cdot 10 \cdot 12 \cos 113°$$

$$|\mathbf{u} + \mathbf{v}| \approx \sqrt{337.78} \approx 18.4$$

Use the law of sines to find θ.

$$\frac{12}{\sin \theta} = \frac{18.4}{\sin 113°}$$

$$\sin \theta = \frac{12 \sin 113°}{18.4} \approx 0.6003$$

$$\theta \approx 37°$$

21.

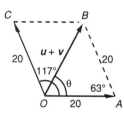

$$A = 180° - 117° = 63°$$

Use the law of cosines to find $|\mathbf{u} + \mathbf{v}|$.

$$|\mathbf{u} + \mathbf{v}|^2 = 20^2 + 20^2 - 2 \cdot 20 \cdot 20 \cos 63°$$

$$|\mathbf{u} + \mathbf{v}| \approx \sqrt{436.81} \approx 20.9$$

Triangle OAB is isosceles so θ and angle OBA have the same measure.

Thus, $\theta = \dfrac{1}{2}(180° - 63°) = \dfrac{1}{2}(117°) = 58.5° \approx 59°.$

(If we use the law of sines and $|\mathbf{u} + \mathbf{v}| \approx 20.9$ to find θ, we get $\theta \approx 58°$.)

23.

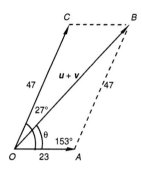

$$A = 180° - 27° = 153°$$

Use the law of cosines to find $|\mathbf{u} + \mathbf{v}|$.

$$|\mathbf{u} + \mathbf{v}|^2 = 23^2 + 47^2 - 2 \cdot 23 \cdot 47 \cos 153°$$

$$|\mathbf{u} + \mathbf{v}| \approx \sqrt{4664.36} \approx 68.3$$

Use the law of sines to find θ.

$$\frac{47}{\sin \theta} = \frac{68.3}{\sin 153°}$$

$$\sin \theta = \frac{47 \sin 153°}{68.3} \approx 0.3124$$

$$\theta \approx 18°$$

25.

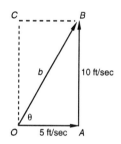

Use the Pythagorean theorem to find the speed of the balloon, $|\mathbf{b}|$.

$$|\mathbf{b}|^2 = 5^2 + 10^2$$

$$|\mathbf{b}| = \sqrt{5^2 + 10^2} \approx 11 \text{ ft/sec}$$

Use the fact that triangle OAB is a right triangle to find θ.

$$\tan \theta = \frac{10}{5} = 2$$

$$\theta \approx 63°$$

27.

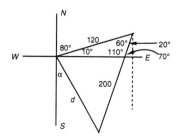

Note:

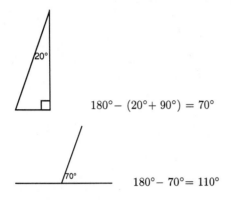

$180° - (20° + 90°) = 70°$

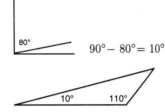

$180° - 70° = 110°$

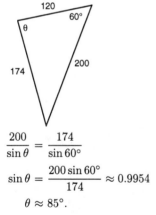

$90° - 80° = 10°$

$180° - (10° + 110°) = 60°$

Use the law of cosines to find $|\mathbf{d}|$.

$|\mathbf{d}|^2 = 120^2 + 200^2 - 2 \cdot 120 \cdot 200 \cos 60°$

$|\mathbf{d}| = \sqrt{30,400} \approx 174$

Use the law of sines to find θ in this triangle.

$\dfrac{200}{\sin \theta} = \dfrac{174}{\sin 60°}$

$\sin \theta = \dfrac{200 \sin 60°}{174} \approx 0.9954$

$\theta \approx 85°.$

Now we can find α.

$\alpha = 180° - (80° + 85°) = 15°.$

The ship is 174 nautical miles from the starting point in the direction S 15° E. (Answers may vary due to rounding differences.)

29.

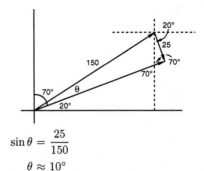

$\sin \theta = \dfrac{25}{150}$

$\theta \approx 10°$

Then the airplane's actual heading will be about $90° - (10° + 20°)$, or 60°.

31.

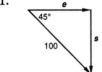

$\dfrac{|\mathbf{e}|}{100} = \cos 45°$

$|\mathbf{e}| = 100 \cos 45° \approx 70.7$

$\dfrac{|\mathbf{s}|}{100} = \sin 45°$

$|\mathbf{s}| = 100 \sin 45° \approx 70.7$

The easterly and southerly components are both 70.7.

33.

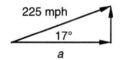

$|\mathbf{a}| = 225 \cos 17° \approx 215.17$

$|\mathbf{b}| = 225 \sin 17° \approx 65.78$

The horizontal component is about 215.17 mph forward, and the vertical component is about 65.78 mph up.

35.

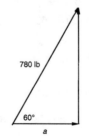

$|\mathbf{a}| = 780 \cos 60° = 390$

$|\mathbf{b}| = 780 \sin 60° \approx 675.5$

The horizontal component is about 390 lb forward, and the vertical component is about 675.5 lb up.

37.

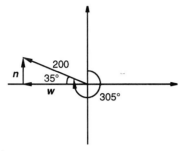

$$|\mathbf{w}| = 200\cos 35° \approx 164$$
$$|\mathbf{n}| = 200\sin 35° \approx 115$$

The westerly component is about 164 km/h, and the northerly component is about 115 km/h.

39.

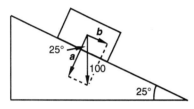

$$|\mathbf{a}| = 100\cos 25° \approx 90.6$$
$$|\mathbf{b}| = 100\sin 25° \approx 42.3$$

The magnitude of the component perpendicular to the incline is about 90.6 lb; the magnitude of the component parallel to the incline is about 42.3 lb.

41.

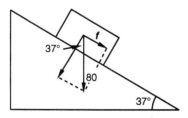

$$|\mathbf{f}| = 80\sin 37° \approx 48.1 \text{ lb}$$

43. Discussion and Writing

45. The point determined by $\dfrac{2\pi}{3}$ is a reflection across the y-axis of the point determined by $\dfrac{\pi}{3}$. The coordinates of the point determined by $\dfrac{\pi}{3}$ are $\left(\dfrac{1}{2}, \dfrac{\sqrt{3}}{2}\right)$, so the coordinates of the point determined by $\dfrac{2\pi}{3}$ are $\left(-\dfrac{1}{2}, \dfrac{\sqrt{3}}{2}\right)$. Thus,

$$\sin\frac{2\pi}{3} = y = \frac{\sqrt{3}}{2}.$$

47. The coordinates of the point determined by $\dfrac{\pi}{4}$ are $\left(\dfrac{\sqrt{2}}{2}, \dfrac{\sqrt{2}}{2}\right)$. Thus,

$$\cos\frac{\pi}{4} = x = \frac{\sqrt{2}}{2}.$$

49.

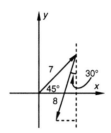

a)

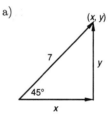

$$x = 7\cos 45° \approx 4.950$$
$$y = 7\sin 45° \approx 4.950$$

The cliff is located at $(4.950, 4.950)$.

b)

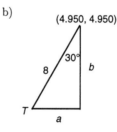

$$a = 8\sin 30° = 4$$
$$b = 8\cos 30° \approx 6.928$$

Then the coordinates of T are $(4.950 - 4, 4.950 - 6.928)$, or $(0.950, -1.978)$.

Exercise Set 7.6

1. $\overrightarrow{MN} = \langle -3 - 6, -2 - (-7)\rangle = \langle -9, 5\rangle$
$$|\overrightarrow{MN}| = \sqrt{(-9)^2 + 5^2} = \sqrt{81 + 25} = \sqrt{106}$$

3. $\overrightarrow{FE} = \langle 8 - 11, 4 - (-2)\rangle = \langle -3, 6\rangle$
$$|\overrightarrow{FE}| = \sqrt{(-3)^2 + 6^2} = \sqrt{9 + 36} = \sqrt{45}, \text{ or } 3\sqrt{5}$$

5. $\overrightarrow{KL} = \langle 8 - 4, -3 - (-3)\rangle = \langle 4, 0\rangle$
$$|\overrightarrow{KL}| = \sqrt{4^2 + 0^2} = \sqrt{16} = 4$$

7. $|\mathbf{u}| = \sqrt{(-1)^2 + 6^2} = \sqrt{1 + 36} = \sqrt{37}$

9. $\mathbf{u} + \mathbf{w} = \langle 5 + (-1), -2 + (-3) \rangle = \langle 4, -5 \rangle$

11. $|3\mathbf{w} - \mathbf{v}| = |3\langle -1, -3 \rangle - \langle -4, 7 \rangle|$
$= |\langle -3, -9 \rangle - \langle -4, 7 \rangle|$
$= |\langle -3, -(-4), -9 - 7 \rangle|$
$= |\langle 1, -16 \rangle|$
$= \sqrt{1^2 + (-16)^2}$
$= \sqrt{257}$

13. $\mathbf{v} - \mathbf{u} = \langle -4, 7 \rangle - \langle 5, -2 \rangle = \langle -4 - 5, 7 - (-2) \rangle = \langle -9, 9 \rangle$

15. $5\mathbf{u} - 4\mathbf{v} = 5\langle 5, -2 \rangle - 4\langle -4, 7 \rangle$
$= \langle 25, -10 \rangle - \langle -16, 28 \rangle$
$= \langle 25 - (-16), -10 - 28 \rangle$
$= \langle 41, -38 \rangle$

17. $|3\mathbf{u}| - |\mathbf{v}| = |3\langle 5, -2 \rangle| - |\langle -4, 7 \rangle|$
$= |\langle 15, -6 \rangle| - |\langle -4, 7 \rangle|$
$= \sqrt{15^2 + (-6)^2} - \sqrt{(-4)^2 + 7^2}$
$= \sqrt{261} - \sqrt{65}$

19. $\mathbf{v} + \mathbf{u} + 2\mathbf{w} = \langle -4, 7 \rangle + \langle 5, -2 \rangle + 2\langle -1, -3 \rangle$
$= \langle -4, 7 \rangle + \langle 5, -2 \rangle + \langle -2, -6 \rangle$
$= \langle -4 + 5 + (-2), 7 + (-2) + (-6) \rangle$
$= \langle -1, -1 \rangle$

21. $2\mathbf{v} + \mathbf{O} = 2\mathbf{v} = 2\langle -4, 7 \rangle = \langle -8, 14 \rangle$

23. $\mathbf{u} \cdot \mathbf{w} = 5(-1) + (-2)(-3) = -5 + 6 = 1$

25. $\mathbf{u} \cdot \mathbf{v} = 5(-4) + (-2)(7) = -20 - 14 = -34$

27. See the answer section in the text.

29. See the answer section in the text.

31. (a) $\mathbf{w} = \mathbf{u} + \mathbf{v}$
(b) $\mathbf{v} = \mathbf{w} - \mathbf{u}$

33. $|\langle -5, 12 \rangle| = \sqrt{(-5)^2 + 12^2} = \sqrt{169} = 13$
$\frac{1}{13}\langle -5, 12 \rangle = \left\langle -\frac{5}{13}, \frac{12}{13} \right\rangle$

35. $|\langle 1, -10 \rangle| = \sqrt{1^2 + (-10)^2} = \sqrt{101}$
$\frac{1}{\sqrt{101}}\langle 1, -10 \rangle = \left\langle \frac{1}{\sqrt{101}}, -\frac{10}{\sqrt{101}} \right\rangle$

37. $|\langle -2, -8 \rangle| = \sqrt{(-2)^2 + (-8)^2} = \sqrt{68} = 2\sqrt{17}$
$\frac{1}{2\sqrt{17}}\langle -2, -8 \rangle = \left\langle -\frac{2}{2\sqrt{17}}, -\frac{8}{2\sqrt{17}} \right\rangle = \left\langle -\frac{1}{\sqrt{17}}, -\frac{4}{\sqrt{17}} \right\rangle$

39. $\langle -4, 6 \rangle = -4\mathbf{i} + 6\mathbf{j}$

41. $\langle 2, 5 \rangle = 2\mathbf{i} + 5\mathbf{j}$

43. Horizontal component: $-3 - 4 = -7$
Vertical component: $3 - (-2) = 5$
We write the vector as $-7\mathbf{i} + 5\mathbf{j}$.

45. (a) $4\mathbf{u} - 5\mathbf{w} = 4(2\mathbf{i} + \mathbf{j}) - 5(\mathbf{i} - 5\mathbf{j})$
$= 8\mathbf{i} + 4\mathbf{j} - 5\mathbf{i} + 25\mathbf{j}$
$= 3\mathbf{i} + 29\mathbf{j}$
(b) $3\mathbf{i} + 29\mathbf{j} = \langle 3, 29 \rangle$

47. (a) $\mathbf{u} - (\mathbf{v} + \mathbf{w}) = 2\mathbf{i} + \mathbf{j} - (-3\mathbf{i} - 10\mathbf{j} + \mathbf{i} - 5\mathbf{j})$
$= 2\mathbf{i} + \mathbf{j} - (-2\mathbf{i} - 15\mathbf{j})$
$= 2\mathbf{i} + \mathbf{j} + 2\mathbf{i} + 15\mathbf{j}$
$= 4\mathbf{i} + 16\mathbf{j}$
(b) $4\mathbf{i} + 16\mathbf{j} = \langle 4, 16 \rangle$

49.
$\mathbf{u} = \left(\cos\frac{\pi}{2}\right)\mathbf{i} + \left(\sin\frac{\pi}{2}\right)\mathbf{j} = 0\mathbf{i} + 1\mathbf{j} = \mathbf{j}$, or $\langle 0, 1 \rangle$

51.
$\mathbf{u} = \left(\cos\frac{4\pi}{3}\right)\mathbf{i} + \left(\sin\frac{4\pi}{3}\right)\mathbf{j} = -\frac{1}{2}\mathbf{i} - \frac{\sqrt{3}}{2}\mathbf{j}$, or $\left\langle -\frac{1}{2}, -\frac{\sqrt{3}}{2} \right\rangle$

53. $\tan\theta = \frac{-5}{-2} = \frac{5}{2}$
$\theta = \tan^{-1}\frac{5}{2}$
The vector is in the third quadrant, so θ is a third-quadrant angle. The reference angle is
$\tan^{-1}\frac{5}{2} \approx 68°$
and $\theta \approx 180° + 68°$, or $248°$.

55.
$$\tan\theta = \frac{2}{1} = 2$$
$$\theta = \tan^{-1} 2$$

The vector is in the first quadrant, so θ is a first-quadrant angle. Then
$$\theta = \tan^{-1} 2 \approx 63°.$$

57.
$$\tan\theta = \frac{6}{5}$$
$$\theta = \tan^{-1} \frac{6}{5}$$

The vector is in the first quadrant, so θ is a first-quadrant angle. Then
$$\theta = \tan^{-1} \frac{6}{5} \approx 50°.$$

59.
$$|\mathbf{u}| = \sqrt{(3\cos 45°)^2 + (3\sin 45°)^2}$$
$$= \sqrt{9\cos^2 45° + 9\sin^2 45°}$$
$$= \sqrt{9(\cos^2 45° + \sin^2 45°)}$$
$$= \sqrt{9 \cdot 1} = \sqrt{9} = 3$$

The vector is given in terms of the direction angle, 45°.

61.
$$|\mathbf{v}| = \sqrt{\left(-\frac{1}{2}\right)^2 + \left(\frac{\sqrt{3}}{2}\right)^2} = \sqrt{\frac{1}{4} + \frac{3}{4}} = \sqrt{1} = 1$$

$$\tan\theta = \frac{\frac{\sqrt{3}}{2}}{-\frac{1}{2}} = \frac{\sqrt{3}}{2}\left(-\frac{2}{1}\right) = -\sqrt{3}$$

$$\theta = \tan^{-1}(-\sqrt{3})$$

The vector is in the second quadrant, so θ is a second-quadrant angle. The reference angle is
$$\tan^{-1}\sqrt{3} = 60°$$
and $\theta = 180° - 60°$, or $120°$.

63. $\mathbf{u} = \langle 2, -5 \rangle$, $\mathbf{v} = \langle 1, 4 \rangle$
$$\mathbf{u} \cdot \mathbf{v} = 2 \cdot 1 + (-5)(4) = -18$$
$$|\mathbf{u}| = \sqrt{2^2 + (-5)^2} = \sqrt{29}$$
$$|\mathbf{v}| = \sqrt{1^2 + 4^2} = \sqrt{17}$$

$$\cos\alpha = \frac{\mathbf{u} \cdot \mathbf{v}}{|\mathbf{u}|\,|\mathbf{v}|} = \frac{-18}{\sqrt{29}\sqrt{17}}$$
$$\alpha = \cos^{-1}\frac{-18}{\sqrt{29}\sqrt{17}}$$
$$\alpha \approx 144.2°$$

65. $\mathbf{w} = \langle 3, 5 \rangle$, $\mathbf{r} = \langle 5, 5 \rangle$
$$\mathbf{w} \cdot \mathbf{r} = 3 \cdot 5 + 5 \cdot 5 = 40$$
$$|\mathbf{w}| = \sqrt{3^2 + 5^2} = \sqrt{34}$$

$$|\mathbf{r}| = \sqrt{5^2 + 5^2} = \sqrt{50}$$

$$\cos\alpha = \frac{\mathbf{w} \cdot \mathbf{r}}{|\mathbf{w}|\,|\mathbf{r}|} = \frac{40}{\sqrt{34}\sqrt{50}}$$
$$\alpha = \cos^{-1}\frac{40}{\sqrt{34}\sqrt{50}}$$
$$\alpha \approx 14.0°$$

67. $\mathbf{a} = \mathbf{i} + \mathbf{j}$, $\mathbf{t} = 2\mathbf{i} - 3\mathbf{j}$
$$\mathbf{a} \cdot \mathbf{b} = 1 \cdot 2 + 1(-3) = -1$$
$$|\mathbf{a}| = \sqrt{1^2 + 1^2} = \sqrt{2}$$
$$|\mathbf{b}| = \sqrt{2^2 + (-3)^2} = \sqrt{13}$$

$$\cos\alpha = \frac{\mathbf{a} \cdot \mathbf{b}}{|\mathbf{a}|\,|\mathbf{b}|} = \frac{-1}{\sqrt{2}\sqrt{13}}$$
$$\alpha = \cos^{-1}\frac{-1}{\sqrt{2}\sqrt{13}}$$
$$\alpha \approx 101.3°$$

69. For $\theta = \frac{\pi}{6}$, $\mathbf{u} = \left(\cos\frac{\pi}{6}\right)\mathbf{i} + \left(\sin\frac{\pi}{6}\right)\mathbf{j} = \frac{\sqrt{3}}{2}\mathbf{i} + \frac{1}{2}\mathbf{j}$.

For $\theta = \frac{3\pi}{4}$,
$$\mathbf{u} = \left(\cos\frac{3\pi}{4}\right)\mathbf{i} + \left(\sin\frac{3\pi}{4}\right)\mathbf{j} = -\frac{\sqrt{2}}{2}\mathbf{i} + \frac{\sqrt{2}}{2}\mathbf{j}.$$

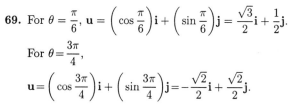

71. $\mathbf{u} = (\cos\theta)\mathbf{i} + (\sin\theta)\mathbf{j}$ where $\theta = \frac{\pi}{2} + \frac{3\pi}{4}$, or $\frac{5\pi}{4}$.
Then $\mathbf{u} = \left(\cos\frac{5\pi}{4}\right)\mathbf{i} + \left(\sin\frac{5\pi}{4}\right)\mathbf{j} = -\frac{\sqrt{2}}{2}\mathbf{i} - \frac{\sqrt{2}}{2}\mathbf{j}.$

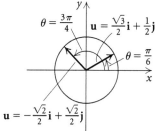

73. Find the magnitude of $-\mathbf{i} + 3\mathbf{j}$.
$$\sqrt{(-1)^2 + 3^2} = \sqrt{10}$$

Then the desired unit vector is
$$\frac{-\mathbf{i} + 3\mathbf{j}}{\sqrt{10}} = -\frac{1}{\sqrt{10}}\mathbf{i} + \frac{3}{\sqrt{10}}\mathbf{j}, \text{ or } -\frac{\sqrt{10}}{10}\mathbf{i} + \frac{3\sqrt{10}}{10}\mathbf{j}.$$

75. Find the magnitude of $2\mathbf{i} - 3\mathbf{j}$.

$$\sqrt{2^2 + (-3)^2} = \sqrt{13}$$

Then the desired vector is $\sqrt{13}\left(\dfrac{2\mathbf{i} - 3\mathbf{j}}{\sqrt{13}}\right) =$

$\sqrt{13}\left(\dfrac{2}{\sqrt{13}}\mathbf{i} - \dfrac{3}{\sqrt{13}}\mathbf{j}\right)$, or $\sqrt{13}\left(\dfrac{2\sqrt{13}}{13}\mathbf{i} - \dfrac{3\sqrt{13}}{13}\mathbf{j}\right)$.

77. See the answer section in the text.

79. Refer to the drawing in this manual accompanying the solution for Exercise 27, Exercise Set 7.5. The vector representing the first part of the ship's trip can be given by

$$120(\cos 10°)\mathbf{i} + 120(\sin 10°)\mathbf{j}.$$

The vector representing the second part of the trip can be given by

$$200(\cos 250°)\mathbf{i} + 200(\sin 250°)\mathbf{j}.$$

Then the resultant is

$$120(\cos 10°)\mathbf{i} + 120(\sin 10°)\mathbf{j} + 200(\cos 250°)\mathbf{i} +$$
$$200(\sin 250°)\mathbf{j}$$

$$= [120(\cos 10°) + 200(\cos 250°)]\mathbf{i} +$$
$$[120(\sin 10°) + 200(\sin 250°)]\mathbf{j}$$

$$\approx 49.77\mathbf{i} - 167.10\mathbf{j}$$

Then the distance from the starting point is

$$\sqrt{(49.77)^2 + (-167.10)^2} \approx 174 \text{ nautical miles.}$$

Now we find the direction angle of the resultant.

$$\tan \theta = \frac{-167.10}{49.77}$$

$$\theta = \tan^{-1}\frac{-167.10}{49.77}$$

θ is a fourth-quadrant angle. The reference angle is

$$\tan^{-1}\frac{167.10}{49.77} \approx 73°$$

and $\theta \approx 360° - 73°$, or $287°$.

Thus, the ship's bearing is about S17°E. (This answer differs from the answer found in Section 7.5 due to rounding differences.)

81. Refer to the drawing in this manual accompanying the solution for Exercise 29, Exercise Set 7.5. We can use the Pythagorean theorem to find the magnitude of the vector representing the desired velocity of the airplane:

$$\sqrt{150^2 - 25^2} \approx 148$$

Then the vector representing the desired velocity can be given by

$$148(\cos 20°)\mathbf{i} + 148(\sin 20°)\mathbf{j}.$$

The vector representing the wind can be given by

$$25(\cos 290°)\mathbf{i} + 25(\sin 290°)\mathbf{j}.$$

The vector representing the desired velocity is the resultant of the vectors representing the airplane's actual velocity and the wind, so we subtract to find the vector representing the airplane's actual velocity.

$$[148(\cos 20°)\mathbf{i} + 148(\sin 20°)\mathbf{j}] -$$
$$[25(\cos 290°)\mathbf{i} + 25(\sin 290°)\mathbf{j}]$$
$$= [148(\cos 20°) - 25(\cos 290°)]\mathbf{i} +$$
$$[148(\sin 20°) - 25(\sin 290°)]\mathbf{j}$$
$$\approx 130.52\mathbf{i} + 74.11\mathbf{j}$$

Now we find the direction angle of this vector. Note that α is a first-quadrant angle.

$$\tan \alpha = \frac{74.11}{130.52}$$

$$\alpha = \tan^{-1}\frac{74.11}{130.52} \approx 30°$$

Thus, the airplane's actual bearing is $60°$.

83. We draw a force diagram with the initial point of each vector at the origin.

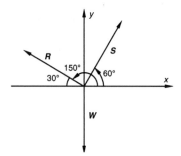

We express each vector in terms of its magnitude and direction angle.

$$\mathbf{R} = |\mathbf{R}|[(\cos 150°)\mathbf{i} + (\sin 150°)\mathbf{j}]$$
$$\mathbf{S} = |\mathbf{S}|[(\cos 60°)\mathbf{i} + (\sin 60°)\mathbf{j}]$$
$$\mathbf{W} = 1000(\cos 270°)\mathbf{i} + 1000(\sin 270°)\mathbf{j} = -1000\mathbf{j}$$

Substituting for $\mathbf{R}$, $\mathbf{S}$, and $\mathbf{W}$ in $\mathbf{R} + \mathbf{S} + \mathbf{W} = \mathbf{O}$, we have

$|\mathbf{R}|[(\cos 150°)\mathbf{i} + (\sin 150°)\mathbf{j}] +$
$|\mathbf{S}|[(\cos 60°)\mathbf{i} + (\sin 60°)\mathbf{j}] - 1000\mathbf{j} = 0\mathbf{i} + 0\mathbf{j}.$

This gives us a system of equations.

$$|\mathbf{R}|(\cos 150°) + |\mathbf{S}|(\cos 60°) = 0,$$

$$|\mathbf{R}|(\sin 150°) + |\mathbf{S}|(\sin 60°) = 1000$$

Solving this system, we get

$$|\mathbf{R}| = 500, \quad |\mathbf{S}| \approx 866.$$

The tension in the cable on the left is 500 lb and in the cable on the right is about 866 lb.

85. We draw a force diagram with the initial point of each vector at the origin.

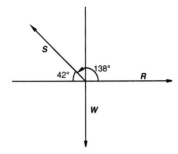

We express each vector in terms of its magnitude and direction angle.

$$\mathbf{R} = |\mathbf{R}|[(\cos 0°)\mathbf{i} + (\sin 0°)\mathbf{j}] = |\mathbf{R}|\mathbf{i}$$

$$\mathbf{S} = |\mathbf{S}|[(\cos 138°)\mathbf{i} + (\sin 138°)\mathbf{j}]$$

$$\mathbf{W} = 150(\cos 270°)\mathbf{i} + 150(\sin 270°)\mathbf{j} = -150\mathbf{j}$$

Substituting for $\mathbf{R}$, $\mathbf{S}$, and $\mathbf{W}$ in $\mathbf{R} + \mathbf{S} + \mathbf{W} = \mathbf{O}$, we have

$$|\mathbf{R}|\mathbf{i} + |\mathbf{S}|[(\cos 138°)\mathbf{i} + (\sin 138°)\mathbf{j}] - 150\mathbf{j} = 0\mathbf{i} + 0\mathbf{j}.$$

This gives us a system of equations.

$$|\mathbf{R}| + |\mathbf{S}|(\cos 138°) = 0,$$

$$|\mathbf{S}|(\sin 138°) = 150$$

Solving this system, we get

$$|\mathbf{R}| \approx 167, \ |\mathbf{S}| \approx 224.$$

The tension in the cable is about 224 lb, and the compression in the boom is about 167 lb. (Answers may vary slightly due to rounding differences.)

87. $\mathbf{u} + \mathbf{v} = \langle u_1, u_2 \rangle + \langle v_1, v_2 \rangle$

$\quad\quad = \langle u_1 + v_1, u_2 + v_2 \rangle$

$\quad\quad = \langle v_1 + u_1, v_2 + u_2 \rangle$

$\quad\quad = \langle v_1, v_2 \rangle + \langle u_1, u_2 \rangle$

$\quad\quad = \mathbf{v} + \mathbf{u}$

89. Discussion and Writing

91. $-\dfrac{1}{5}x - y = 15$

$\quad\quad -y = \dfrac{1}{5}x + 15$

$\quad\quad y = -\dfrac{1}{5}x - 15 \quad\quad$ Multiplying by -1

With the equation written in the form $y = mx + b$ we see that the slope is $-\dfrac{1}{5}$ and the y-intercept is $(0, -15)$.

93. $x^3 - 4x^2 = 0$

$\quad x^2(x - 4) = 0$

$\quad x^2 = 0 \ \ or \ \ x - 4 = 0$

$\quad x = 0 \ \ or \quad\quad x = 4$

The zeros are 0 and 4.

We could also graph $y = x^3 - 4x^2$ and use the Zero feature on a grapher to find the zeros.

95. (a) Assume neither $|\mathbf{u}|$ nor $|\mathbf{v}|$ is zero.

If $\mathbf{u} \cdot \mathbf{v} = |\mathbf{u}||\mathbf{v}|\cos\theta = 0$, then $\cos\theta = 0$, or $\theta = 90°$ and the vectors are perpendicular.

(b) Answers may vary.

Let $\mathbf{u} = \mathbf{i}$ and $\mathbf{v} = \mathbf{j}$. Then $\mathbf{u} \cdot \mathbf{v} = 1 \cdot 0 + 0 \cdot 1 = 0$.

97. Find the magnitude of $\mathbf{u} = \langle 3, -4 \rangle$:

$$|\mathbf{u}| = \sqrt{3^2 + (-4)^2} = \sqrt{25} = 5$$

The unit vector in the same direction as $\mathbf{u}$ is

$$\frac{3\mathbf{i} - 4\mathbf{j}}{5} = \frac{3}{5}\mathbf{i} - \frac{4}{5}\mathbf{j}.$$

The unit vector in the opposite direction to $\mathbf{u}$ is

$$-\left(\frac{3\mathbf{i} - 4\mathbf{j}}{5}\right), \text{ or } -\frac{3}{5}\mathbf{i} + \frac{4}{5}\mathbf{j}.$$

These are the only unit vectors parallel to $\langle 3, -4 \rangle$.

99. B has coordinates (x, y) such that

$$x - 2 = 3, \text{ or } x = 5$$

and

$$y - 9 = -1, \text{ or } y = 8.$$

Thus point B is $(5,8)$.

Chapter 8

Systems of Equations and Matrices

Exercise Set 8.1

1. Graph (c) is the graph of this system.

3. Graph (f) is the graph of this system.

5. Graph (b) is the graph of this system.

7. Graph $y_1 = 2 - x$ and $y_2 = -3x$ in the same window and find the point of intersection.

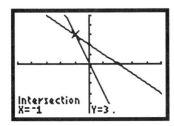

The solution is $(-1, 3)$.

The solution is $(3, -2)$.

9. Graph $y_1 = -\frac{1}{2}x + \frac{3}{2}$ and $y_2 = -\frac{1}{4}x + \frac{13}{8}$ in the same window and find the point of intersection.

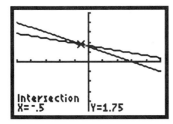

The solution is $(-0.5, 1.75)$.

11. Graph $y_1 = 2x - 1$ and $y_2 = 2x + 1$ in the same window and find the point of intersection.

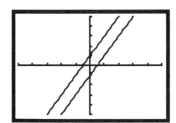

The lines do not intersect, so there is no solution.

13.
$$x + y = 9, \quad (1)$$
$$2x - 3y = -2 \quad (2)$$

Solve equation (1) for either x or y. We choose to solve for y.

$$y = 9 - x$$

Then substitute $9 - x$ for y in equation (2) and solve the resulting equation.

$$2x - 3(9 - x) = -2$$
$$2x - 27 + 3x = -2$$
$$5x - 27 = -2$$
$$5x = 25$$
$$x = 5$$

Now substitute 5 for x in either equation (1) or (2) and solve for y.

$$5 + y = 9 \quad \text{Using equation (1)}$$
$$y = 4$$

The solution is $(5, 4)$. The grapher confirms that this is correct.

15.
$$x - 2y = 7, \quad (1)$$
$$x = y + 4 \quad (2)$$

Use equation (2) and substitute $y + 4$ for x in equation (1). Then solve for y.

$$y + 4 - 2y = 7$$
$$-y + 4 = 7$$
$$-y = 3$$
$$y = -3$$

Substitute -3 for y in equation (2) to find x.

$$x = -3 + 4 = 1$$

The solution is $(1, -3)$. The grapher confirms that this is correct.

17.
$$y = 2x - 6, \quad (1)$$
$$5x - 3y = 16 \quad (2)$$

Use equation (1) and substitute $2x - 6$ for y in equation (2). Then solve for x.

$$5x - 3(2x - 6) = 16$$
$$5x - 6x + 18 = 16$$
$$-x + 18 = 16$$
$$-x = -2$$
$$x = 2$$

Substitute 2 for x in equation (1) to find y.

$$y = 2 \cdot 2 - 6 = 4 - 6 = -2$$

The solution is $(2, -2)$. The grapher confirms that this is correct.

19. $x - 5y = 4$,   (1)

$y = 7 - 2x$ (2)

Use equation (2) and substitute $7 - 2x$ for y in equation (1). Then solve for x.

$$x - 5(7 - 2x) = 4$$
$$x - 35 + 10x = 4$$
$$11x - 35 = 4$$
$$11x = 39$$
$$x = \frac{39}{11}$$

Substitute $\frac{39}{11}$ for x in equation (2) to find y.

$$y = 7 - 2 \cdot \frac{39}{11} = 7 - \frac{78}{11} = -\frac{1}{11}$$

The solution is $\left(\frac{39}{11}, -\frac{1}{11}\right)$. The grapher confirms that this is correct.

21. $2x - 3y = 5$,   (1)

$5x + 4y = 1$ (2)

Solve one equation for either x or y. We choose to solve equation (1) for x.

$$2x - 3y = 5$$
$$2x = 3y + 5$$
$$x = \frac{3}{2}y + \frac{5}{2}$$

Substitute $\frac{3}{2}y + \frac{5}{2}$ for x in equation (2) and solve for y.

$$5\left(\frac{3}{2}y + \frac{5}{2}\right) + 4y = 1$$
$$\frac{15}{2}y + \frac{25}{2} + 4y = 1$$
$$\frac{23}{2}y + \frac{25}{2} = 1$$
$$\frac{23}{2}y = -\frac{23}{2}$$
$$y = -1$$

Substitute -1 for y in either equation (1) or (2) and solve for x.

$$2x - 3(-1) = 5 \quad \text{Using equation (1)}$$
$$2x + 3 = 5$$
$$2x = 2$$
$$x = 1$$

The solution is $(1, -1)$. The grapher confirms that this is correct.

23. $x + 2y = 7$,   (1)

$x - 2y = -5$ (2)

We add the equations to eliminate y.

$$\begin{array}{r} x + 2y = 7 \\ \underline{x - 2y = -5} \\ 2x \phantom{{}+2y} = 2 \quad \text{Adding} \\ x = 1 \end{array}$$

Back-substitute in either equation and solve for y.

$$1 + 2y = 7 \quad \text{Using equation (1)}$$
$$2y = 6$$
$$y = 3$$

The solution is $(1, 3)$. The grapher confirms that this is correct.

Since the system of equations has exactly one solution it is consistent and independent.

25. $x - 3y = 2$,   (1)

$6x + 5y = -34$ (2)

Multiply equation (1) by -6 and add it to equation (2) to eliminate x.

$$\begin{array}{r} -6x + 18y = -12 \\ \underline{6x + 5y = -34} \\ 23y = -46 \\ y = -2 \end{array}$$

Back-substitute to find x.

$$x - 3(-2) = 2 \quad \text{Using equation (1)}$$
$$x + 6 = 2$$
$$x = -4$$

The solution is $(-4, -2)$. The grapher confirms that this is correct.

Since the system of equations has exactly one solution it is consistent and independent.

27. $0.3x - 0.2y = -0.9$,

$0.2x - 0.3y = -0.6$

First, multiply each equation by 10 to clear the decimals.

$$3x - 2y = -9 \quad (1)$$
$$2x - 3y = -6 \quad (2)$$

Now multiply equation (1) by 3 and equation (2) by -2 and add to eliminate y.

$$\begin{array}{r} 9x - 6y = -27 \\ \underline{-4x + 6y = 12} \\ 5x \phantom{{}+6y} = -15 \\ x = -3 \end{array}$$

Back-substitute to find y.

$$3(-3) - 2y = -9 \quad \text{Using equation (1)}$$
$$-9 - 2y = -9$$
$$-2y = 0$$
$$y = 0$$

The solution is $(-3, 0)$. The grapher confirms that this is correct. Since the system of equations has exactly one solution it is consistent and independent.

29. $3x - 12y = 6,$ (1)

   $2x - 8y = 4$ (2)

Multiply equation (1) by 2 and equation (2) by -3 and add.

$$6x - 24y = 12$$
$$\underline{-6x + 24y = -12}$$
$$0 = 0$$

The equation $0 = 0$ is true for all values of x and y. Thus, the system of equations has infinitely many solutions. Solving either equation for y, we can write $y = \frac{1}{4}x - \frac{1}{2}$ so the solutions are ordered pairs of the form $\left(x, \frac{1}{4}x - \frac{1}{2}\right)$. Equivalently, if we solve either equation for x we get $x = 4y + 2$ so the solutions can also be expressed as $(4y + 2, y)$. The grapher confirms that the graphs of the equations coincide.

Since there are infinitely many solutions, the system of equations is consistent and dependent.

31. $\frac{1}{5}x + \frac{1}{2}y = 6,$ (1)

   $\frac{3}{5}x - \frac{1}{2}y = 2$ (2)

We could multiply both equations by 10 to clear fractions, but since the y-coefficients differ only by sign we will just add to eliminate y.

$$\frac{1}{5}x + \frac{1}{2}y = 6$$
$$\underline{\frac{3}{5}x - \frac{1}{2}y = 2}$$
$$\frac{4}{5}x \qquad = 8$$
$$x = 10$$

Back-substitute to find y.

$$\frac{1}{5} \cdot 10 + \frac{1}{2}y = 6 \quad \text{Using equation (1)}$$
$$2 + \frac{1}{2}y = 6$$
$$\frac{1}{2}y = 4$$
$$y = 8$$

The solution is $(10, 8)$. The grapher confirms that this is correct.

Since the system of equations has exactly one solution it is consistent and independent.

33. $2x = 5 - 3y,$ (1)

   $4x = 11 - 7y$ (2)

We rewrite the equations.

   $2x + 3y = 5,$ (1a)

   $4x + 7y = 11$ (2a)

Multiply equation (2a) by -2 and add to eliminate x.

$$-4x - 6y = -10$$
$$\underline{4x + 7y = 11}$$
$$y = 1$$

Back-substitute to find x.

$$2x = 5 - 3 \cdot 1 \quad \text{Using equation (1)}$$
$$2x = 2$$
$$x = 1$$

The solution is $(1, 1)$. The grapher confirms that this is correct.

Since the system of equations has exactly one solution it is consistent and independent.

35. *Familiarize*. Let $x =$ the number of video rentals and $y =$ the number of boxes of popcorn given away. Then the total cost of the rentals is $1 \cdot x$, or x, and the total cost of the popcorn is $2y$.

***Translate*.** The number of new members is the same as the number of incentives so we have one equation:

$$x + y = 48.$$

The total cost of the incentive was \$86. This gives us another equation:

$$x + 2y = 86.$$

***Carry out*.** We solve the system

   $x + y = 48,$ (1)

   $x + 2y = 86.$ (2)

Multiply equation (1) by -1 and add.

$$-x - y = -48$$
$$\underline{x + 2y = 86}$$
$$y = 38$$

Back-substitute to find x.

$$x + 38 = 48 \quad \text{Using equation (1)}$$
$$x = 10$$

***Check*.** When 10 rentals and 38 boxes of popcorn are given away, a total of 48 new members have signed up. Ten rentals cost the store \$10 and 38 boxes of popcorn cost $2 \cdot 38$, or \$76, so the total cost of the incentives was $\$10 + \$76$, or \$86. The solution checks.

State. Ten rentals and 38 boxes of popcorn were given away.

37. Familiarize. Let $x =$ the cost of an adult's admission and $y =$ the cost of a child's admission. Then the total cost for 4 adults was $4x$ and for 16 children was $16y$.

Translate. Each adult's admission costs twice as much as each child's admission, so we have one equation:

$$x = 2y.$$

The total admission cost was \$72, so we have:

$$4x + 16y = 72$$

Solve. We solve the system

$$x = 2y, \quad (1)$$
$$4x + 16y = 72. \quad (2)$$

Substitute $2y$ for x in equation (2) and solve for y.

$$4 \cdot 2y + 16y = 72$$
$$24y = 72$$
$$y = 3$$

Back-substitute in equation (1) to find x.

$$x = 2 \cdot 3 = 6$$

Check. The cost of an adult's admission, \$6, is twice the cost of a child's admission, \$3. The total admission cost was $4 \cdot \$6 + 16 \cdot \$3 = \$24 + \$48 = \$72$. The solution checks.

State. Each adult's admission is \$6 and each child's admission is \$3.

39. Familiarize and Translate. We use the system of equations given in the problem.

$$y = 70 + 2x \quad (1)$$
$$y = 175 - 5x, \quad (2)$$

Carry out. Substitute $175 - 5x$ for y in equation (1) and solve for x.

$$175 - 5x = 70 + 2x$$
$$105 = 7x \quad \text{Adding } 5x \text{ and subtracting } 70$$
$$15 = x$$

Back-substitute in either equation to find y. We choose equation (1).

$$y = 70 + 2 \cdot 15 = 70 + 30 = 100$$

Check. Substituting 15 for x and 100 for y in both of the original equations yields true equations, so the solution checks. We could also check graphically by finding the point of intersection of equations (1) and (2).

State. The equilibrium point is $(15, \$100)$.

41. Familiarize and Translate. We find the value of x for which $C = R$, where

$$C = 14x + 350,$$
$$R = 16.5x.$$

Carry out. When $C = R$ we have:

$$14x + 350 = 16.5x$$
$$350 = 2.5x$$
$$140 = x$$

Check. When $x = 140$, $C = 14 \cdot 140 + 350$, or 2310 and $R = 16.5(140)$, or 2310. Since $C = R$, the solution checks.

State. 140 units must be produced and sold in order to break even.

43. Familiarize and Translate. We find the value of x for which $C = R$, where

$$C = 15x + 12,000,$$
$$R = 18x - 6000.$$

Carry out. When $C = R$ we have:

$$15x + 12,000 = 18x - 6000$$
$$18,000 = 3x \qquad \text{Subtracting } 15x \text{ and}$$
$$\qquad\qquad\qquad\qquad \text{adding } 6000$$
$$6000 = x$$

Check. When $x = 6000$, $C = 15 \cdot 6000 + 12,000$, or 102,000 and $R = 18 \cdot 6000 - 6000$, or 102,000. Since $C = R$, the solution checks.

State. 6000 units must be produced and sold in order to break even.

45. Familiarize. Let $x =$ the number of servings of spaghetti and meatballs required and $y =$ the number of servings of iceberg lettuce required. Then x servings of spaghetti contain $260x$ Cal and $32x$ g of carbohydrates; y servings of lettuce contain $5y$ Cal and $1 \cdot y$ or y, g of carbohydrates.

Translate. One equation comes from the fact that 400 Cal are desired:

$$260x + 5y = 400.$$

A second equation comes from the fact that 50g of carbohydrates are required:

$$32x + y = 50.$$

Solve. We solve the system

$$260x + 5y = 400, \quad (1)$$
$$32x + \ y = 50. \quad (2)$$

Multiply equation (2) by -5 and add.

$$260x + 5y = 400$$
$$\underline{-160x - 5y = -250}$$
$$100x \qquad = 150$$
$$x = 1.5$$

Back-substitute to find y.

$$32(1.5) + y = 50 \quad \text{Using equation (2)}$$
$$48 + y = 50$$
$$y = 2$$

Check. 1.5 servings of spaghetti contain $260(1.5)$, or 390 Cal and $32(1.5)$, or 48 g of carbohydrates; 2 servings of lettuce contain $5 \cdot 2$, or 10 Cal and $1 \cdot 2$, or 2 g of carbohydrates. Together they contain $390 + 10$, or 400 Cal and $48 + 2$, or 50 g of carbohydrates. The solution checks.

State. 1.5 servings of spaghetti and meatballs and 2 servings of iceberg lettuce are required.

47. Familiarize. It helps to make a drawing. Then organize the information in a table. Let $x =$ the speed of the boat and $y =$ the speed of the stream. The speed upstream is $x - y$. The speed downstream is $x + y$.

46 km	2 hr	$(x + y)$ km/h

Downstream

51 km	3 hr	$(x - y)$ km/h

Upstream

	Distance	Speed	Time
Downstream	46	$x + y$	2
Upstream	51	$x - y$	3

Translate. Using $d = rt$ in each row of the table, we get a system of equations.

$$46 = (x + y)2 \qquad x + y = 23, \quad (1)$$
$$\text{or}$$
$$51 = (x - y)3 \qquad x - y = 17 \quad (2)$$

Carry out. We begin by adding equations (1) and (2).

$$x + y = 23$$
$$\underline{x - y = 17}$$
$$2x \quad = 40$$
$$x = 20$$

Back-substitute to find y.

$$20 + y = 23 \quad \text{Using equation (1)}$$
$$y = 23$$

Check. The speed downstream is $20 + 3$, or 23 km/h. The distance traveled downstream in 2 hr is $23 \cdot 2$, or 46 km. The speed upstream is $20 - 3$, or 17 km/h.

The distance traveled upstream in 3 hr is $17 \cdot 3$, or 51 km. The solution checks.

State. The speed of the boat is 20 km/h. The speed of the stream is 3 km/h.

49. Familiarize. Let $x =$ the amount invested at 7% and $y =$ the amount invested at 9%. Then the interest from the investments is 7%x and 9%y, or $0.07x$ and $0.09y$.

Translate.

The total investment is $15,000.

$$x + y = 15,000$$

The total interest is $1230.

$$0.07x + 0.09y = 1230$$

We have a system of equations:

$$x + \quad y = 15,000,$$
$$0.07x + 0.09y = 1230$$

Multiplying the second equation by 100 to clear the decimals, we have:

$$x + \quad y = 15,000, \quad (1)$$
$$7x + 9y = 123,000. \quad (2)$$

Carry out. We begin by multiplying equation (1) by -7 and adding.

$$-7x - 7y = -105,000$$
$$\underline{7x + 9y = 123,000}$$
$$2y = 18,000$$
$$y = 9000$$

Back-substitute to find x.

$$x + 9000 = 15,000 \quad \text{Using equation (1)}$$
$$x = 6000$$

Check. The total investment is $6000 + $9000, or $15,000. The total interest is $0.07(\$6000) + 0.09(\$9000)$, or $420 + $810, or $1230. The solution checks.

State. $6000 was invested at 7% and $9000 was invested at 9%.

51. Familiarize. Let $x =$ the number of pounds of French roast coffee used and $y =$ the number of pounds of Kenyan coffee. We organize the information in a table.

	French roast	Kenyan	Mixture
Amount	x	y	10 lb
Price per pound	$9.00	$7.50	$8.40
Total cost	$9x$	$7.50y$	$8.40(10), or $84

Translate. The first and third rows of the table give us a system of equations.

$$x + y = 10,$$
$$9x + 7.5y = 84$$

Multiply the second equation by 10 to clear the decimals.

$$x + y = 10, \quad (1)$$
$$90x + 75y = 840 \quad (2)$$

Carry out. Begin by multiplying equation (1) by -75 and adding.

$$-75x - 75y = -750$$
$$\underline{90x + 75y = 840}$$
$$15x \qquad = 90$$
$$x = 6$$

Back-substitute to find y.

$$6 + y = 10 \quad \text{Using equation (1)}$$
$$y = 4$$

Check. The total amount of coffee in the mixture is $6 + 4$, or 10 lb. The total value of the mixture is $6(\$9) + 4(\$7.50)$, or $\$54 + \$30$, or $\$84$. The solution checks.

State. 6 lb of French roast coffee and 4 lb of Kenyan coffee should be used.

53. *Familiarize*. Let $x =$ the monthly sales, $C =$ the earnings with the straight-commission plan, and $S =$ the earnings with the salary-plus-commission plan. Then 8% of sales is represented by 8%x, or $0.08x$, and 1% of sales is represented by 1%x, or $0.01x$.

Translate.

Straight commission pays 8% of sales.

$$C = 0.08x$$

Salary plus commission pays $1500 plus 1% of sales.

$$S = 1500 + 0.01x$$

Carry out. We find the value of x for which $C = S$. We have:

$$0.08x = 1500 + 0.01x$$
$$0.07x = 1500$$
$$x \approx 21,428.57$$

Check. When $x \approx 21,428.57$, $C \approx 0.08(21,428.57) \approx 1714.29$ and $S \approx 1500 + 0.01(21,428.57) \approx 1714.29$. Since $C = S$, the solution checks.

State. The two plans pay the same amount for monthly sales of about $21,428.57.

55. a) $t(x) = 1.69541779x + 44.15363881$,
$r(x) = -0.4528301887x + 68.83018868$

b) Graph $y_1 = t(x)$ and $y_2 = r(x)$ and find the first coordinate of the point of intersection of the graphs. It is approximately 11.5, so the number of take-out meals purchased was equal to the number of restaurant meals purchased about 11.5 years after 1984.

57. Discussion and Writing

59. $2x - 5 = 0$
$$2x = 5$$
$$x = \frac{5}{2}$$

We could also graph $y = 2x - 5$ and use the Zero feature.

The solution is $\frac{5}{2}$, or 2.5.

61. $x^2 - 4x + 3 = 0$
$$(x - 1)(x - 3) = 0$$
$$x - 1 = 0 \quad or \quad x - 3 = 0$$
$$x = 1 \quad or \qquad x = 3$$

We could also graph $y = x^2 - 4x + 3$ and use the Zero feature twice.

The solutions are 1 and 3.

63. *Familiarize*. Let $x =$ the time spent jogging and $y =$ the time spent walking. Then Nancy jogs $8x$ km and walks $4y$ km. We organize the information in a table.

Translate.

The total time is 1 hr.

$$x + y = 1$$

The total distance is 6 km.

$$8x + 4y = 6$$

Carry out. Solve the system

$$x + y = 1, \quad (1)$$
$$8x + 4y = 6. \quad (2)$$

Multiply equation (1) by -4 and add.

$$-4x - 4y = -4$$
$$\underline{8x + 4y = 6}$$
$$4x \qquad = 2$$
$$x = \frac{1}{2}$$

This is the time we need to find the distance spent jogging, so we could stop here. However, we will not be able to check the solution unless we find y also so we continue. We back-substitute.

$$\frac{1}{2} + y = 1 \quad \text{Using equation (1)}$$
$$y = \frac{1}{2}$$

Then the distance jogged is $8 \cdot \dfrac{1}{2}$, or 4 km.

Check. The total time is $\dfrac{1}{2}$ hr $+\dfrac{1}{2}$ hr, or 1 hr. The total distance is 4 km $+2$ km, or 6 km. The solution checks.

State. Nancy jogged 4 km on each trip.

65. Familiarize and Translate. We let x and y represent the speeds of the trains. Organize the information in a table. Using $d = rt$, we let $3x$, $2y$, $1.5x$, and $3y$ represent the distances the trains travel.

First situation:

3 hours	x km/h	y km/h	2 hours
Union			Central

├──────── 216 km ────────┤

Second situation:

1.5 hours	x km/h	y km/h	3 hours
Union			Central

├──────── 216 km ────────┤

	Distance traveled in first situation	Distance traveled in second situation
Train$_1$ (from Union to Central)	$3x$	$1.5x$
Train$_2$ (from Central to Union)	$2y$	$3y$
Total	216	216

The total distance in each situation is 216 km. Thus, we have a system of equations.

$$3x + 2y = 216, \quad (1)$$
$$1.5x + 3y = 216 \quad (2)$$

Carry out. Multiply equation (2) by -2 and add.

$$\begin{aligned} 3x + 2y &= 216 \\ \underline{-3x - 6y} &= \underline{-432} \\ -4y &= -216 \\ y &= 54 \end{aligned}$$

Back-substitute to find x.

$$3x + 2 \cdot 54 = 216 \quad \text{Using equation (1)}$$
$$3x + 108 = 216$$
$$3x = 108$$
$$x = 36$$

Check. If $x = 36$ and $y = 54$, the total distance the trains travel in the first situation is $3 \cdot 36 + 2 \cdot 54$, or 216 km. The total distance they travel in the second

situation is $1.5 \cdot 36 + 3 \cdot 54$, or 216 km. The solution checks.

State. The speed of the first train is 36 km/h. The speed of the second train is 54 km/h.

67. Substitute the given solutions in the equation $Ax + By = 1$ to get a system of equations.

$$3A - B = 1, \quad (1)$$
$$-4A - 2B = 1 \quad (2)$$

Multiply equation (1) by -2 and add.

$$\begin{aligned} -6A + 2B &= -2 \\ \underline{-4A - 2B} &= \underline{1} \\ -10A &= -1 \\ A &= \frac{1}{10} \end{aligned}$$

Back-substitute to find B.

$$3\left(\frac{1}{10}\right) - B = 1 \qquad \text{Using equation (1)}$$
$$\frac{3}{10} - B = 1$$
$$-B = \frac{7}{10}$$
$$B = -\frac{7}{10}$$

We have $A = \dfrac{1}{10}$ and $B = -\dfrac{7}{10}$.

69. Familiarize. Let $x =$ the number of city miles driven and $y =$ the number of highway miles driven. The $x/19$ and $y/28$ represent the number of gallons of gasoline used driving x mi and y mi, respectively.

Translate.

The car was driven a total of 405 mi.

$$x + y = 405$$

A total of 23 gal of gasoline was used.

$$\frac{x}{19} + \frac{y}{28} = 18$$

Carry out. We solve the system

$$x + y = 405,$$
$$\frac{x}{19} + \frac{y}{28} = 18.$$

Multiply the second equation by 532 to clear fractions.

$$x + y = 405, \quad (1)$$
$$28x + 19y = 9576 \quad (2)$$

Now multiply equation (1) by -19 and add.

$$\begin{aligned} -19x - 19y &= -7695 \\ \underline{28x + 19y} &= \underline{9576} \\ 9x &= 1881 \\ x &= 209 \end{aligned}$$

Back-substitute to find y.

$$209 + y = 405 \quad \text{Using equation (1)}$$
$$y = 196$$

Check. A total of $209 + 196$, or 405 mi was driven. City driving used $209/19$, or 11 gal and highway driving used $196/28$, or 7 gal. Thus, a total of $11 + 7$, or 18 gal of gasoline was used. The solution checks.

State. 209 mi were driven in the city and 196 mi were driven on the highway.

Exercise Set 8.2

1.
$$\begin{aligned} x + y + z &= 2, &\quad (1)\\ 6x - 4y + 5z &= 31, &\quad (2)\\ 5x + 2y + 2z &= 13 &\quad (3) \end{aligned}$$

Multiply equation (1) by -6 and add it to equation (2). We also multiply equation (1) by -5 and add it to equation (3).
$$\begin{aligned} x + y + z &= 2 &\quad (1)\\ -10y - z &= 19 &\quad (4)\\ -3y - 3z &= 3 &\quad (5) \end{aligned}$$

Multiply the last equation by 10 to make the y-coefficient a multiple of the y-coefficient in equation (4).
$$\begin{aligned} x + y + z &= 2 &\quad (1)\\ -10y - z &= 19 &\quad (4)\\ -30y - 30z &= 30 &\quad (6) \end{aligned}$$

Multiply equation (4) by -3 and add it to equation (6).
$$\begin{aligned} x + y + z &= 2 &\quad (1)\\ -10y - z &= 19 &\quad (4)\\ -27z &= -27 &\quad (7) \end{aligned}$$

Solve equation (7) for z.
$$-27z = -27$$
$$z = 1$$

Back-substitute 1 for z in equation (4) and solve for y.
$$-10y - 1 = 19$$
$$-10y = 20$$
$$y = -2$$

Back-substitute 1 for z for -2 and y in equation (1) and solve for x.
$$x + (-2) + 1 = 2$$
$$x - 1 = 2$$
$$x = 3$$

The solution is $(3, -2, 1)$.

3.
$$\begin{aligned} x - y + 2z &= -3 &\quad (1)\\ x + 2y + 3z &= 4 &\quad (2)\\ 2x + y + z &= -3 &\quad (3) \end{aligned}$$

Multiply equation (1) by -1 and add it to equation (2). We also multiply equation (1) by -2 and add it to equation (3).
$$\begin{aligned} x - y + 2z &= -3 &\quad (1)\\ 3y + z &= 7 &\quad (4)\\ 3y - 3z &= 3 &\quad (5) \end{aligned}$$

Multiply equation (4) by -1 and add it to equation (5).
$$\begin{aligned} x - y + 2z &= -3 &\quad (1)\\ 3y + z &= 7 &\quad (4)\\ -4z &= -4 &\quad (6) \end{aligned}$$

Solve equation (6) for z.
$$-4z = -4$$
$$z = 1$$

Back-substitute 1 for z in equation (4) and solve for y.
$$3y + 1 = 7$$
$$3y = 6$$
$$y = 2$$

Back-substitute 1 for z and 2 for y in equation (1) and solve for x.
$$x - 2 + 2 \cdot 1 = -3$$
$$x = -3$$

The solution is $(-3, 2, 1)$.

5.
$$\begin{aligned} x + 2y - z &= 5, &\quad (1)\\ 2x - 4y + z &= 0, &\quad (2)\\ 3x + 2y + 2z &= 3 &\quad (3) \end{aligned}$$

Multiply equation (1) by -2 and add it to equation (2). Also, multiply equation (1) by -3 and add it to equation (3).
$$\begin{aligned} x + 2y - z &= 5, &\quad (1)\\ -8y + 3z &= -10, &\quad (4)\\ -4y + 5z &= -12 &\quad (5) \end{aligned}$$

Multiply equation (5) by 2 to make the y-coefficient a multiple of the y-coefficient of equation (4).
$$\begin{aligned} x + 2y - z &= 5, &\quad (1)\\ -8y + 3z &= -10, &\quad (4)\\ -8y + 10z &= -24 &\quad (6) \end{aligned}$$

Multiply equation (4) by -1 and add it to equation (6).

$$x + 2y - z = 5, \qquad (1)$$
$$-8y + 3z = -10, \qquad (4)$$
$$7z = -14 \qquad (7)$$

Solve equation (7) for z.

$$7z = -14$$
$$z = -2$$

Back-substitute -2 for z in equation (4) and solve for y.

$$-8y + 3(-2) = -10$$
$$-8y - 6 = -10$$
$$-8y = -4$$
$$y = \frac{1}{2}$$

Back-substitute $\frac{1}{2}$ for y and -2 for z in equation (1) and solve for x.

$$x + 2 \cdot \frac{1}{2} - (-2) = 5$$
$$x + 1 + 2 = 5$$
$$x = 2$$

The solution is $\left(2, \frac{1}{2}, -2 \right)$.

7.
$$x + 2y - z = -8, \quad (1)$$
$$2x - y + z = 4, \quad (2)$$
$$8x + y + z = 2 \quad (3)$$

Multiply equation (1) by -2 and add it to equation (2). Also, multiply equation (1) by -8 and add it to equation (3).

$$x + 2y - z = -8, \quad (1)$$
$$-5y + 3z = 20, \quad (4)$$
$$-15y + 9z = 66 \quad (5)$$

Multiply equation (4) by -3 and add it to equation (5).

$$x + 2y - z = -8, \quad (1)$$
$$-5y + 3z = 20, \quad (4)$$
$$0 = 6 \quad (6)$$

Equation (6) is false, so the system of equations has no solution.

9.
$$2x + y - 3z = 1, \quad (1)$$
$$x - 4y + z = 6, \quad (2)$$
$$4x - 7y - z = 13 \quad (3)$$

Interchange equations (1) and (2).

$$x - 4y + z = 6, \quad (2)$$
$$2x + y - 3z = 1, \quad (1)$$
$$4x - 7y - z = 13 \quad (3)$$

Multiply equation (2) by -2 and add it to equation (1). Also, multiply equation (2) by -4 and add it to equation (3).

$$x - 4y + z = 6, \quad (2)$$
$$9y - 5z = -11, \quad (4)$$
$$9y - 5z = -11 \quad (5)$$

Multiply equation (4) by -1 and add it to equation (5).

$$x - 4y + z = 6, \quad (1)$$
$$9y - 5z = -11, \quad (4)$$
$$0 = 0 \quad (6)$$

The equation $0 = 0$ tells us that equation (3) of the original system is dependent on the first two equations. The system of equations has infinitely many solutions and is equivalent to

$$2x + y - 3z = 1, \quad (1)$$
$$x - 4y + z = 6. \quad (2)$$

To find an expression for the solutions, we first solve equation (4) for either y or z. We choose to solve for z.

$$9y - 5z = -11$$
$$-5z = -9y - 11$$
$$z = \frac{9y + 11}{5}$$

Back-substitute in equation (2) to find an expression for x in terms of y.

$$x - 4y + \frac{9y + 11}{5} = 6$$
$$x - 4y + \frac{9}{5}y + \frac{11}{5} = 6$$
$$x = \frac{11}{5}y + \frac{19}{5} = \frac{11y + 19}{5}$$

The solutions are given by $\left(\dfrac{11y + 19}{5}, y, \dfrac{9y + 11}{5} \right)$, where y is any real number.

11.
$$4a + 9b = 8, \quad (1)$$
$$8a + 6c = -1, \quad (2)$$
$$6b + 6c = -1 \quad (3)$$

Multiply equation (1) by -2 and add it to equation (2).

$$4a + 9b = 8, \quad (1)$$
$$-18b + 6c = -17, \quad (4)$$
$$6b + 6c = -1 \quad (3)$$

Multiply equation (3) by 3 to make the b-coefficient a multiple of the b-coefficient in equation (4).

$$4a + 9b = 8, \quad (1)$$
$$-18b + 6c = -17, \quad (4)$$
$$18b + 18c = -3 \quad (5)$$

Add equation (4) to equation (5).

$$4a + 9b \qquad = 8, \qquad (1)$$
$$-18b + 6c = -17, \quad (4)$$
$$24c = -20 \quad (6)$$

Solve equation (6) for c.

$$24c = -20$$
$$c = -\frac{20}{24} = -\frac{5}{6}$$

Back-substitute $-\dfrac{5}{6}$ for c in equation (4) and solve for b.

$$-18b + 6c = -17$$
$$-18b + 6\left(-\frac{5}{6}\right) = -17$$
$$-18b - 5 = -17$$
$$-18b = -12$$
$$b = \frac{12}{18} = \frac{2}{3}$$

Back-substitute $\dfrac{2}{3}$ for b in equation (1) and solve for a.

$$4a + 9b = 8$$
$$4a + 9 \cdot \frac{2}{3} = 8$$
$$4a + 6 = 8$$
$$4a = 2$$
$$a = \frac{1}{2}$$

The solution is $\left(\dfrac{1}{2}, \dfrac{2}{3}, -\dfrac{5}{6}\right)$.

13.
$$w + x + y + z = 2 \qquad (1)$$
$$w + 2x + 2y + 4z = 1 \qquad (2)$$
$$-w + x - y - z = -6 \qquad (3)$$
$$-w + 3x + y - z = -2 \qquad (4)$$

Multiply equation (1) by -1 and add to equation (2). Add equation (1) to equation (3) and to equation (4).

$$w + x + y + z = 2 \qquad (1)$$
$$x + y + 3z = -1 \qquad (5)$$
$$2x \qquad = -4 \qquad (6)$$
$$4x + 2y \qquad = 0 \qquad (7)$$

Solve equation (6) for x.

$$2x = -4$$
$$x = -2$$

Back-substitute -2 for x in equation (7) and solve for y.

$$4(-2) + 2y = 0$$
$$-8 + 2y = 0$$
$$2y = 8$$
$$y = 4$$

Back-substitute -2 for x and 4 for y in equation (5) and solve for z.

$$-2 + 4 + 3z = -1$$
$$3z = -3$$
$$z = -1$$

Back-substitute -2 for x, 4 for y, and -1 for z in equation (1) and solve for w.

$$w - 2 + 4 - 1 = 2$$
$$w = 1$$

The solution is $(1, -2, 4, -1)$.

15. *Familiarize*. Let $x =$ the number of orders under 10 lb, $y =$ the number of orders from 10 lb up to 15 lb, and $z =$ the number of orders of 15 lb or more. Then the total shipping charges for each category of order are \$3x, \$5y, and \$7.50z.

***Translate*.**

The total number of orders was 150.

$$x + y + z = 150$$

Total shipping charges were \$680.

$$3x + 5y + 7.5z = 680$$

The number of orders under 10 lb was three times the number of orders weighing 15 lb or more.

$$x = 3z$$

We have a system of equations

$$x + y + z = 150, \qquad x + y + z = 150,$$
$$3x + 5y + 7.5z = 680, \quad \text{or} \quad 30x + 50y + 75z = 6800,$$
$$x = 3z \qquad x - 3z = 0$$

***Carry out*.** Solving the system of equations, we get $(60, 70, 20)$.

***Check*.** The total number of orders is $60 + 70 + 20$, or 150. The total shipping charges are $\$3 \cdot 60 + \$5 \cdot 70 + \$7.50(20)$, or \$680. The number of orders under 10 lb, 60, is three times the number of orders weighing over 15 lb, 20. The solution checks.

***State*.** There were 60 packages under 10 lb, 70 packages from 10 lb up to 15 lb, and 20 packages weighing 15 lb or more.

17. *Familiarize*. Let x, y, and z represent the number of servings of ground beef, baked potato, and strawberries required, respectively. One serving of ground beef contains $245x$ Cal, $0x$ or 0 g of carbohydrates, and $9x$ mg of calcium. One baked potato contains

145y Cal, 34y g of carbohydrates, and 8y mg of calcium. One serving of strawberries contains 45z Cal, 10z g of carbohydrates, and 21z mg of calcium.

Translate.

The total number of calories is 485.

$$245x + 145y + 45z = 485$$

A total of 41.5 g of carbohydrates is required.

$$34y + 10z = 41.5$$

A total of 35 mg of calcium is required.

$$9x + 8y + 21z = 35$$

We have a system of equations.

$$245x + 145y + 45z = 485,$$
$$34y + 10z = 41.5,$$
$$9x + 8y + 21z = 35$$

Carry out. Solving the system of equations, we get $(1.25, 1, 0.75)$.

Check. 1.25 servings of ground beef contains 306.25 Cal, no carbohydrates, and 11.25 mg of calcium; 1 baked potato contains 145 Cal, 34 g of carbohydrates, and 8 mg of calcium; 0.75 servings of strawberries contains 33.75 Cal, 7.5 g of carbohydrates, and 15.75 mg of calcium. Thus, there are a total of $306.25 + 145 + 33.75$, or 485 Cal, $34 + 7.5$, or 41.5 g of carbohydrates, and $11.25 + 8 + 15.75$, or 35 mg of calcium. The solution checks.

State. 1.25 servings of ground beef, 1 baked potato, and 0.75 serving of strawberries are required.

19. *Familiarize*. Let x, y, and z represent the amounts invested at 4%, 6%, and 7%, respectively. Then the annual interest from the investments is $4\%x$, $6\%y$, and $7\%z$, or $0.04x$, $0.06y$, and $0.07z$.

Translate.

A total of $5000 was invested.

$$x + y + z = 5000$$

The total interest is $302.

$$0.04x + 0.06y + 0.07z = 302$$

The amount invested at 7% is $1500 more than the amount invested at 4%.

$$z = x + 1500$$

We have a system of equations.

$$x + y + z = 5000,$$
$$0.04x + 0.06y + 0.07z = 302,$$
$$z = x + 1500$$
or
$$x + y + z = 5000,$$
$$4x + 6y + 7z = 30,200,$$
$$-x + z = 1500$$

Carry out. Solving the system of equations, we get $(1300, 900, 2800)$.

Check. The total investment was $1300 + $900 + $2800, or $5000. The total interest was $0.04(\$1300) + 0.06(\$900) + 0.07(\$2800) = \$52 + \$54 + \$196$, or $302. The amount invested at 7%, $2800, is $1500 more than the amount invested at 4%, $1300. The solution checks.

State. $1300 was invested at 4%, $900 at 6%, and $2800 at 7%.

21. *Familiarize*. Let x, y, and z represent the prices of orange juice, a raisin bagel, and a cup of coffee, respectively. The new price for orange juice is $x + 50\%x$, or $x + 0.5x$, or $1.5x$; the new price of a bagel is $y + 20\%y$, or $y + 0.2y$, or $1.2y$.

Translate.

Orange juice, a raisin bagel, and a cup of coffee cost $3.

$$x + y + x = 3$$

After the price increase, orange juice, a raisin bagel, and a cup of coffee will cost $3.75.

$$1.5x + 1.2y + z = 3.75$$

After the price increases, orange juice will cost twice as much as coffee.

$$1.5x = 2z$$

We have a system of equations.

$$x + y + z = 3, \qquad x + y + z = 3,$$
$$1.5x + 1.2y + z = 3.75, \text{ or } 150x + 120y + 100z = 375,$$
$$1.5x = 2z \qquad 15x - 20z = 0$$

Carry out. Solving the system of equations, we get $(1, 1.25, 0.75)$.

Check. If orange juice costs $1, a bagel costs $1.25, and a cup of coffee costs $0.75, then together they cost $1 + $1.25 + $0.75, or $3. After the price increases orange juice will cost 1.5($1), or $1.50 and a bagel will cost 1.2($1.25) or $1.50. Then orange juice, a bagel, and coffee will cost $1.50 + $1.50 + $0.75, or $3.75. After the price increase the price of orange juice, $1.50, will be twice the price of coffee, $0.75. The solution checks.

State. Before the increase orange juice costs $1, a raisin bagel cost $1.25, and a cup of coffee cost $0.75.

23. *Familiarize*. Let x, y, and z represent the volume of passenger traffic on domestic airways, by bus, and by railroads, respectively, in billions of passenger-miles.

Translate.

The total volume of passenger traffic was 405 billion passenger-miles.

$$x + y + z = 405$$

The volume of bus traffic was 10 billion passenger-miles more than the volume of railroad traffic.

$$y = z + 10$$

The volume of railroad traffic was 329 billion passenger-miles less than the volume of traffic on domestic airways.

$$z = x - 329$$

We have a system of equations.

$$
\begin{aligned}
x + y + z &= 405, & x + y + z &= 405, \\
y &= z + 10, & \text{or} \qquad y - z &= 10, \\
z &= x - 329 & -x \quad + z &= -329
\end{aligned}
$$

Carry out. Solving the system of equations, we get $(351, 32, 22)$.

Check. The total volume is $351 + 32 + 22$, or 405 billion passenger-miles. The volume of bus traffic, 32 billion passenger-miles, is 10 billion passenger-miles more than the volume of railroad traffic, 22 billion passenger-miles. The volume of railroad traffic, 22 billion passenger-miles, is 329 billion passenger-miles less than the volume of traffic on domestic airways, 351 billion passenger-miles. The solution checks.

State. The volume of traffic on domestic airways was 351 billion passenger-miles, by bus was 32 billion passenger-miles, and by railroad was 22 billion passenger-miles.

25. Familiarize. Let x, y, and z represent the number of par-3, par-4, and par-5 holes, respectively. A golfer who shoots par on every hole has a score of $3x$ from the par-3 holes, $4y$ from the par-4 holes, and $5z$ from the par-5 holes.

Translate.

The total number of holes is 18.

$$x + y + z = 18$$

A golfer who shoots par on every hole has a score of 72.

$$3x + 4y + 5z = 72$$

The sum of the number of par-3 holes and the number of par-5 holes is 8.

$$x + z = 8$$

We have a system of equations.

$$
\begin{aligned}
x + y + z &= 18, \\
3x + 4y + 5z &= 72, \\
x \quad + z &= 8
\end{aligned}
$$

Carry out. We solve the system. The solution is $(4, 10, 4)$.

Check. The total number of holes is $4 + 10 + 4$, or 18. A golfer who shoots par on every hole has a score

of $3 \cdot 4 + 4 \cdot 10 + 5 \cdot 4$, or 72. The sum of the number of par-3 holes and the number of par-5 holes is $4 + 4$, or 8. The solution checks.

State. There are 4 par-3 holes, 10 par-4 holes, and 4 par-5 holes.

27. a) Substitute the data points $(0, 211)$, $(10, 237)$, and $(16, 203)$ in the function $f(x) = ax^2 + bx + c$.

$$211 = a \cdot 0^2 + b \cdot 0 + c$$
$$237 = a \cdot 10^2 + b \cdot 10 + c$$
$$203 = a \cdot 16^2 + b \cdot 16 + c$$

We have a system of equations.

$$
\begin{aligned}
c &= 211, \\
100a + 10b + c &= 237, \\
256a + 16b + c &= 203
\end{aligned}
$$

Solving the system of equations, we get
$$\left(-\frac{31}{60}, \frac{233}{30}, 211 \right), \text{ so}$$
$$f(x) = -\frac{31}{60}x^2 + \frac{233}{30}x + 211.$$

b) In 2005, $x = 2005 - 1980$, or 25.
$$f(25) = -\frac{31}{60}(25)^2 + \frac{233}{30}(25) + 211 \approx 82$$

There will be about 82 thousand marriages in California in 2005.

29. a) Substitute the data points $(0, 594)$, $(5, 567)$, and $(11, 576)$ in the function $f(x) = ax^2 + bx + c$.

$$594 = a \cdot 0^2 + b \cdot 0 + c$$
$$567 = a \cdot 5^2 + b \cdot 5 + c$$
$$576 = a \cdot 11^2 + b \cdot 11 + c$$

We have a system of equations.

$$
\begin{aligned}
c &= 594, \\
25a + 5b + c &= 567, \\
121a + 11b + c &= 576
\end{aligned}
$$

Solving the system of equations, we get
$$\left(\frac{69}{110}, -\frac{939}{110}, 594 \right), \text{ so}$$
$$f(x) = \frac{69}{110}x^2 - \frac{939}{110}x + 594.$$

b) In 2005, $x = 2005 - 1985$, or 20.
$$f(20) = \frac{69}{110}(20)^2 - \frac{939}{110}(20) + 594 \approx 674.$$

The per capita milk consumption in 2005 will be about 674 lb.

31. Discussion and Writing

33. $(3 - 4i) - (-2 - i) = 3 - 4i + 2 + i =$
$$(3 + 2) + (-4 + 1)i = 5 - 3i$$

35. $(1 - 2i)(6 + 2i) = 6 + 2i - 12i - 4i^2 =$
$6 + 2i - 12i + 4 = 10 - 10i$

37. $\dfrac{2}{x} - \dfrac{1}{y} - \dfrac{3}{z} = -1,$

$\dfrac{2}{x} - \dfrac{1}{y} + \dfrac{1}{z} = -9,$

$\dfrac{1}{x} + \dfrac{2}{y} - \dfrac{4}{z} = 17$

First substitute u for $\dfrac{1}{x}$, v for $\dfrac{1}{y}$, and w for $\dfrac{1}{z}$ and solve for u, v, and w.

$2u - v - 3w = -1,$

$2u - v + w = -9,$

$u + 2v - 4w = 17$

Solving this system we get $(-1, 5, -2)$.

If $u = -1$, and $u = \dfrac{1}{x}$, then $-1 = \dfrac{1}{x}$, or $x = -1$.

If $v = 5$ and $v = \dfrac{1}{y}$, then $5 = \dfrac{1}{y}$, or $y = \dfrac{1}{5}$.

If $w = -2$ and $w = \dfrac{1}{z}$, then $-2 = \dfrac{1}{z}$, or $z = -\dfrac{1}{2}$.

The solution of the original system is $\left(-1, \dfrac{1}{5}, -\dfrac{1}{2} \right)$.

39. Label the angle measures at the tips of the stars a, b, c, d, and e. Also label the angles of the pentagon p, q, r, s, and t.

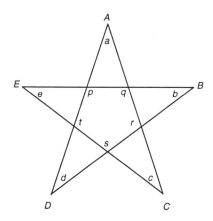

Using the geometric fact that the sum of the angle measures of a triangle is $180°$, we get 5 equations.

$p + b + d = 180$

$q + c + e = 180$

$r + a + d = 180$

$s + b + e = 180$

$t + a + c = 180$

Adding these equations, we get

$(p + q + r + s + t) + 2a + 2b + 2c + 2d + 2e = 5(180).$

The sum of the angle measures of any convex polygon with n sides is given by the formula $S = (n - 2)180$. Thus $p + q + r + s + t = (5 - 2)180$, or 540. We substitute and solve for $a + b + c + d + e$.

$540 + 2(a + b + c + d + e) = 900$

$2(a + b + c + d + e) = 360$

$a + b + c + d + e = 180$

The sum of the angle measures at the tips of the star is $180°$.

41. Substituting, we get

$A + \dfrac{3}{4}B + 3C = 12,$

$\dfrac{4}{3}A + B + 2C = 12,$

$2A + B + C = 12,$ or

$4A + 3B + 12C = 48,$

$4A + 3B + 6C = 36,$ Clearing fractions

$2A + B + C = 12.$

Solving the system of equations, we get $(3, 4, 2)$. The equation is $3x + 4y + 2z = 12$.

43. Substituting, we get

$59 = a(-2)^3 + b(-2)^2 + c(-2) + d,$

$13 = a(-1)^3 + b(-1)^2 + c(-1) + d,$

$-1 = a \cdot 1^3 + b \cdot 1^2 + c \cdot 1 + d,$

$-17 = a \cdot 2^3 + b \cdot 2^2 + c \cdot 2 + d,$ or

$-8a + 4b - 2c + d = 59,$

$-a + b - c + d = 13,$

$a + b + c + d = -1,$

$8a + 4b + 2c + d = -17.$

Solving the system of equations, we get $(-4, 5, -3, 1)$, so $y = -4x^3 + 5x^2 - 3x + 1$.

45. *Familiarize and Translate*. Let a, s, and c represent the number of adults, students, and children in attendance, respectively.

The total attendance was 100.

$a + s + c = 100$

The total amount of money taken in was $100.

(Express 50 cents as $\dfrac{1}{2}$ dollar.)

$10a + 3s + \dfrac{1}{2}c = 100$

The resulting system is

$$a + s + c = 100,$$
$$10a + 3s + \frac{1}{2}c = 100.$$

Carry out. Multiply the first equation by -3 and add it to the second equation to obtain $7a - \frac{5}{2}c = -200$ or $a = \frac{5}{14}(c - 80)$ where $(c-80)$ is a positive multiple of 14 (because a must be a positive integer). That is $(c - 80) = k \cdot 14$ or $c = 80 + k \cdot 14$, where k is a positive integer. If $k > 1$, then $c > 100$. This is impossible since the total attendance is 100. Thus $k = 1$, so $c = 80 + 1 \cdot 14 = 94$. Then $a = \frac{5}{14}(94 - 80) = \frac{5}{14} \cdot 14 = 5$, and $5 + s + 94 = 10$, or $s = 1$.

Check. The total attendance is $5 + 1 + 94$, or 100.

The total amount of money taken in was $\$10 \cdot 5 + \$3 \cdot 1 + \$\frac{1}{2} \cdot 94 = \$100$. The result checks.

State. There were 5 adults, 1 student, and 94 children in attendance.

Exercise Set 8.3

1. The matrix has 3 rows and 2 columns, so its order is 3×2.

3. The matrix has 1 row and 4 columns, so its order is 1×4.

5. The matrix has 3 rows and 3 columns, so its order is 3×3.

7. We omit the variables and replace the equals signs with a vertical line.

$$\left[\begin{array}{rr|r} 2 & -1 & 7 \\ 1 & 4 & -5 \end{array} \right]$$

9. We omit the variables, writing zeros for the missing terms, and replace the equals signs with a vertical line.

$$\left[\begin{array}{rrr|r} 1 & -2 & 3 & 12 \\ 2 & 0 & -4 & 8 \\ 0 & 3 & 1 & 7 \end{array} \right]$$

11. Insert variables and replace the vertical line with equals signs.
$$3x - 5y = 1,$$
$$x + 4y = -2$$

13. Insert variables and replace the vertical line with equals signs.
$$2x + y - 4z = 12,$$
$$3x \quad\quad + 5z = -1,$$
$$x - y + z = 2$$

15. $4x + 2y = 11,$
$$3x - y = 2$$

Write the augmented matrix. We will use Gaussian elimination.

$$\left[\begin{array}{rr|r} 4 & 2 & 11 \\ 3 & -1 & 2 \end{array} \right]$$

Multiply row 2 by 4 to make the first number in row 2 a multiple of 4.

$$\left[\begin{array}{rr|r} 4 & 2 & 11 \\ 12 & -4 & 8 \end{array} \right]$$

Multiply row 1 by -3 and add it to row 2.

$$\left[\begin{array}{rr|r} 4 & 2 & 11 \\ 0 & -10 & -25 \end{array} \right]$$

Multiply row 1 by $\frac{1}{4}$ and row 2 by $-\frac{1}{10}$.

$$\left[\begin{array}{rr|r} 1 & \frac{1}{2} & \frac{11}{4} \\ 0 & 1 & \frac{5}{2} \end{array} \right]$$

Write the system of equations that corresponds to the last matrix.
$$x + \frac{1}{2}y = \frac{11}{4}, \quad (1)$$
$$y = \frac{5}{2} \quad (2)$$

Back-substitute in equation (1) and solve for x.
$$x + \frac{1}{2} \cdot \frac{5}{2} = \frac{11}{4}$$
$$x + \frac{5}{4} = \frac{11}{4}$$
$$x = \frac{6}{4} = \frac{3}{2}$$

The solution is $\left(\frac{3}{2}, \frac{5}{2} \right)$.

17. $5x - 2y = -3,$
$$2x + 5y = -24$$

Write the augmented matrix. We will use Gaussian elimination.

$$\left[\begin{array}{rr|r} 5 & -2 & -3 \\ 2 & 5 & -24 \end{array}\right]$$

Multiply row 2 by 5 to make the first number in row 2 a multiple of 5.

$$\left[\begin{array}{rr|r} 5 & -2 & -3 \\ 10 & 25 & -120 \end{array}\right]$$

Multiply row 1 by -2 and add it to row 2.

$$\left[\begin{array}{rr|r} 5 & -2 & -3 \\ 0 & 29 & -114 \end{array}\right]$$

Multiply row 1 by $\dfrac{1}{5}$ and row 2 by $\dfrac{1}{29}$.

$$\left[\begin{array}{rr|r} 1 & -\dfrac{2}{5} & -\dfrac{3}{5} \\ 0 & 1 & -\dfrac{114}{29} \end{array}\right]$$

Write the system of equations that corresponds to the last matrix.

$$x - \frac{2}{5}y = -\frac{3}{5}, \quad (1)$$
$$y = -\frac{114}{29} \quad (2)$$

Back-substitute in equation (1) and solve for x.

$$x - \frac{2}{5}\left(-\frac{114}{29}\right) = -\frac{3}{5}$$
$$x + \frac{228}{145} = -\frac{3}{5}$$
$$x = -\frac{315}{145} = -\frac{63}{29}$$

The solution is $\left(-\dfrac{63}{29}, -\dfrac{114}{29}\right)$.

19. $3x + 4y = 7,$
$-5x + 2y = 10$

Write the augmented matrix. We will use Gaussian elimination.

$$\left[\begin{array}{rr|r} 3 & 4 & 7 \\ -5 & 2 & 10 \end{array}\right]$$

Multiply row 2 by 3 to make the first number in row 2 a multiple of 3.

$$\left[\begin{array}{rr|r} 3 & 4 & 7 \\ -15 & 6 & 30 \end{array}\right]$$

Multiply row 1 by 5 and add it to row 2.

$$\left[\begin{array}{rr|r} 3 & 4 & 7 \\ 0 & 26 & 65 \end{array}\right]$$

Multiply row 1 by $\dfrac{1}{3}$ and row 2 by $\dfrac{1}{26}$.

$$\left[\begin{array}{rr|r} 1 & \dfrac{4}{3} & \dfrac{7}{3} \\ 0 & 1 & \dfrac{5}{2} \end{array}\right]$$

Write the system of equations that corresponds to the last matrix.

$$x + \frac{4}{3}y = \frac{7}{3}, \quad (1)$$
$$y = \frac{5}{2} \quad (2)$$

Back-substitute in equation (1) and solve for x.

$$x + \frac{4}{3} \cdot \frac{5}{2} = \frac{7}{3}$$
$$x + \frac{10}{3} = \frac{7}{3}$$
$$x = -\frac{3}{3} = -1$$

The solution is $\left(-1, \dfrac{5}{2}\right)$.

21. $3x + 2y = 6,$
$2x - 3y = -9$

Write the augmented matrix. We will use Gauss-Jordan elimination.

$$\left[\begin{array}{rr|r} 3 & 2 & 6 \\ 2 & -3 & -9 \end{array}\right]$$

Multiply row 2 by 3 to make the first number in row 2 a multiple of 3.

$$\left[\begin{array}{rr|r} 3 & 2 & 6 \\ 6 & -9 & -27 \end{array}\right]$$

Multiply row 1 by -2 and add it to row 2.

$$\left[\begin{array}{rr|r} 3 & 2 & 6 \\ 0 & -13 & -39 \end{array}\right]$$

Multiply row 2 by $-\dfrac{1}{13}$.

$$\begin{bmatrix} 3 & 2 & | & 6 \\ 0 & 1 & | & 3 \end{bmatrix}$$

Multiply row 2 by -2 and add it to row 1.

$$\begin{bmatrix} 3 & 0 & | & 0 \\ 0 & 1 & | & 3 \end{bmatrix}$$

Multiply row 1 by $\frac{1}{3}$.

$$\begin{bmatrix} 1 & 0 & | & 0 \\ 0 & 1 & | & 3 \end{bmatrix}$$

We have $x = 0$, $y = 3$. The solution is $(0, 3)$.

23. $x - 3y = 8,$
       $2x - 6y = 3$

Write the augmented matrix.

$$\begin{bmatrix} 1 & -3 & | & 8 \\ 2 & -6 & | & 3 \end{bmatrix}$$

Multiply row 1 by -2 and add it to row 2.

$$\begin{bmatrix} 1 & -3 & | & 8 \\ 0 & 0 & | & -13 \end{bmatrix}$$

The last row corresponds to the false equation $0 = -13$, so there is no solution.

25. $-2x + 6y = 4,$
       $3x - 9y = -6$

Write the augmented matrix.

$$\begin{bmatrix} -2 & 6 & | & 4 \\ 3 & -9 & | & -6 \end{bmatrix}$$

Multiply row 1 by $-\frac{1}{2}$.

$$\begin{bmatrix} 1 & -3 & | & -2 \\ 3 & -9 & | & -6 \end{bmatrix}$$

Multiply row 1 by -3 and add it to row 2.

$$\begin{bmatrix} 1 & -3 & | & -2 \\ 0 & 0 & | & 0 \end{bmatrix}$$

The last row corresponds to the equation $0 = 0$ which is true for all values of x and y. Thus, the system of

equations is dependent and is equivalent to the first equation $-2x + 6y = 4$, or $x - 3y = -2$. Solving for x, we get $x = 3y - 2$. Then the solutions are of the form $(3y - 2, y)$, where y is any real number.

27. $x + 2y - 3z = 9,$
       $2x - y + 2z = -8,$
       $3x - y - 4z = 3$

Write the augmented matrix. We will use Gauss-Jordan elimination.

$$\begin{bmatrix} 1 & 2 & -3 & | & 9 \\ 2 & -1 & 2 & | & -8 \\ 3 & -1 & -4 & | & 3 \end{bmatrix}$$

Multiply row 1 by -2 and add it to row 2. Also, multiply row 1 by -3 and add it to row 3.

$$\begin{bmatrix} 1 & 2 & -3 & | & 9 \\ 0 & -5 & 8 & | & -26 \\ 0 & -7 & 5 & | & -24 \end{bmatrix}$$

Multiply row 2 by $-\frac{1}{5}$ to get a 1 in the second row, second column.

$$\begin{bmatrix} 1 & 2 & -3 & | & 9 \\ 0 & 1 & -\frac{8}{5} & | & \frac{26}{5} \\ 0 & -7 & 5 & | & -24 \end{bmatrix}$$

Multiply row 2 by -2 and add it to row 1. Also, multiply row 2 by 7 and add it to row 3.

$$\begin{bmatrix} 1 & 0 & \frac{1}{5} & | & -\frac{7}{5} \\ 0 & 1 & -\frac{8}{5} & | & \frac{26}{5} \\ 0 & 0 & -\frac{31}{5} & | & \frac{62}{5} \end{bmatrix}$$

Multiply row 3 by $-\frac{5}{31}$ to get a 1 in the third row, third column.

$$\begin{bmatrix} 1 & 0 & \frac{1}{5} & | & -\frac{7}{5} \\ 0 & 1 & -\frac{8}{5} & | & \frac{26}{5} \\ 0 & 0 & 1 & | & -2 \end{bmatrix}$$

Multiply row 3 by $-\frac{1}{5}$ and add it to row 1. Also, multiply row 3 by $\frac{8}{5}$ and add it to row 2.

$$\begin{bmatrix} 1 & 0 & 0 & | & -1 \\ 0 & 1 & 0 & | & 2 \\ 0 & 0 & 1 & | & -2 \end{bmatrix}$$

We have $x = -1$, $y = 2$, $z = -2$. The solution is $(-1, 2, -2)$.

29. $4x - y - 3z = 1,$
$8x + y - z = 5,$
$2x + y + 2z = 5$

Write the augmented matrix. We will use Gauss-Jordan elimination.

$$\begin{bmatrix} 4 & -1 & -3 & | & 1 \\ 8 & 1 & -1 & | & 5 \\ 2 & 1 & 2 & | & 5 \end{bmatrix}$$

First interchange rows 1 and 3 so that each number below the first number in the first row is a multiple of that number.

$$\begin{bmatrix} 2 & 1 & 2 & | & 5 \\ 8 & 1 & -1 & | & 5 \\ 4 & -1 & -3 & | & 1 \end{bmatrix}$$

Multiply row 1 by -4 and add it to row 2. Also, multiply row 1 by -2 and add it to row 3.

$$\begin{bmatrix} 2 & 1 & 2 & | & 5 \\ 0 & -3 & -9 & | & -15 \\ 0 & -3 & -7 & | & -9 \end{bmatrix}$$

Multiply row 2 by -1 and add it to row 3.

$$\begin{bmatrix} 2 & 1 & 2 & | & 5 \\ 0 & -3 & -9 & | & -15 \\ 0 & 0 & 2 & | & 6 \end{bmatrix}$$

Multiply row 2 by $-\dfrac{1}{3}$ to get a 1 in the second row, second column.

$$\begin{bmatrix} 2 & 1 & 2 & | & 5 \\ 0 & 1 & 3 & | & 5 \\ 0 & 0 & 2 & | & 6 \end{bmatrix}$$

Multiply row 2 by -1 and add it to row 1.

$$\begin{bmatrix} 2 & 0 & -1 & | & 0 \\ 0 & 1 & 3 & | & 5 \\ 0 & 0 & 2 & | & 6 \end{bmatrix}$$

Multiply row 3 by $\dfrac{1}{2}$ to get a 1 in the third row, third column.

$$\begin{bmatrix} 2 & 0 & -1 & | & 0 \\ 0 & 1 & 3 & | & 5 \\ 0 & 0 & 1 & | & 3 \end{bmatrix}$$

Add row 3 to row 1. Also multiply row 3 by -3 and add it to row 2.

$$\begin{bmatrix} 2 & 0 & 0 & | & 3 \\ 0 & 1 & 0 & | & -4 \\ 0 & 0 & 1 & | & 3 \end{bmatrix}$$

Finally, multiply row 1 by $\dfrac{1}{2}$.

$$\begin{bmatrix} 1 & 0 & 0 & | & \dfrac{3}{2} \\ 0 & 1 & 0 & | & -4 \\ 0 & 0 & 1 & | & 3 \end{bmatrix}$$

We have $x = \dfrac{3}{2}$, $y = -4$, $z = 3$. The solution is $\left(\dfrac{3}{2}, -4, 3\right)$.

31. $x - 2y + 3z = 4,$
$3x + y - z = 0,$
$2x + 3y - 5z = 1$

Write the augmented matrix. We will use Gaussian elimination.

$$\begin{bmatrix} 1 & -2 & 3 & | & -4 \\ 3 & 1 & -1 & | & 0 \\ 2 & 3 & -5 & | & 1 \end{bmatrix}$$

Multiply row 1 by -3 and add it to row 2. Also, multiply row 1 by -2 and add it to row 3.

$$\begin{bmatrix} 1 & -2 & 3 & | & -4 \\ 0 & 7 & -10 & | & 12 \\ 0 & 7 & -11 & | & 9 \end{bmatrix}$$

Multiply row 2 by -1 and add it to row 3.

$$\begin{bmatrix} 1 & -2 & 3 & | & -4 \\ 0 & 7 & -10 & | & 12 \\ 0 & 0 & -1 & | & -3 \end{bmatrix}$$

Multiply row 2 by $\dfrac{1}{7}$ and multiply row 3 by -1.

$$\begin{bmatrix} 1 & -2 & 3 & | & -4 \\ 0 & 1 & -\dfrac{10}{7} & | & \dfrac{12}{7} \\ 0 & 0 & 1 & | & 3 \end{bmatrix}$$

Now write the system of equations that corresponds to the last matrix.

$x - 2y + 3z = -4,$ (1)

$y - \dfrac{10}{7}z = \dfrac{12}{7},$ (2)

$z = 3$ (3)

Back-substitute 3 for z in equation (2) and solve for y.

$$y - \frac{10}{7} \cdot 3 = \frac{12}{7}$$

$$y - \frac{30}{7} = \frac{12}{7}$$

$$y = \frac{42}{7} = 6$$

Back-substitute 6 for y and 3 for z in equation (1) and solve for x.

$$x - 2 \cdot 6 + 3 \cdot 3 = -4$$

$$x - 3 = -4$$

$$x = -1$$

The solution is $(-1, 6, 3)$.

33. $2x - 4y - 3z = 3,$
$\quad x + 3y + \ z = -1,$
$\quad 5x + \ y - 2z = 2$

Write the augmented matrix.

$$\begin{bmatrix} 2 & -4 & -3 & 3 \\ 1 & 3 & 1 & -1 \\ 5 & 1 & -2 & 2 \end{bmatrix}$$

Interchange the first two rows to get a 1 in the first row, first column.

$$\begin{bmatrix} 1 & 3 & 1 & -1 \\ 2 & -4 & -3 & 3 \\ 5 & 1 & -2 & 2 \end{bmatrix}$$

Multiply row 1 by -2 and add it to row 2. Also, multiply row 1 by -5 and add it to row 3.

$$\begin{bmatrix} 1 & 3 & 1 & -1 \\ 0 & -10 & -5 & 5 \\ 0 & -14 & -7 & 7 \end{bmatrix}$$

Multiply row 2 by $-\frac{1}{10}$ to get a 1 in the second row, second column.

$$\begin{bmatrix} 1 & 3 & 1 & -1 \\ 0 & 1 & \frac{1}{2} & -\frac{1}{2} \\ 0 & -14 & -7 & 7 \end{bmatrix}$$

Multiply row 2 by 14 and add it to row 3.

$$\begin{bmatrix} 1 & 3 & 1 & -1 \\ 0 & 1 & \frac{1}{2} & -\frac{1}{2} \\ 0 & 0 & 0 & 0 \end{bmatrix}$$

The last row corresponds to the equation $0 = 0$. This indicates that the system of equations is dependent. It is equivalent to

$$x + 3y + \ z = -1,$$

$$y + \frac{1}{2}z = -\frac{1}{2}$$

We solve the second equation for y.

$$y = -\frac{1}{2}z - \frac{1}{2}$$

Substitute for y in the first equation and solve for x.

$$x + 3\left(-\frac{1}{2}z - \frac{1}{2}\right) + z = -1$$

$$x - \frac{3}{2}z - \frac{3}{2} + z = -1$$

$$x = \frac{1}{2}z + \frac{1}{2}$$

The solution is $\left(\frac{1}{2}z + \frac{1}{2}, -\frac{1}{2}z - \frac{1}{2}, z\right)$, where z is any real number.

35. $p + \ q + \ r = 1,$
$\quad p + 2q + 3r = 4,$
$\quad 4p + 5q + 6r = 7$

Write the augmented matrix.

$$\begin{bmatrix} 1 & 1 & 1 & 1 \\ 1 & 2 & 3 & 4 \\ 4 & 5 & 6 & 7 \end{bmatrix}$$

Multiply row 1 by -1 and add it to row 2. Also, multiply row 1 by -4 and add it to row 3.

$$\begin{bmatrix} 1 & 1 & 1 & 1 \\ 0 & 1 & 2 & 3 \\ 0 & 1 & 2 & 3 \end{bmatrix}$$

Multiply row 2 by -1 and add it to row 3.

$$\begin{bmatrix} 1 & 1 & 1 & 1 \\ 0 & 1 & 2 & 3 \\ 0 & 0 & 0 & 0 \end{bmatrix}$$

The last row corresponds to the equation $0 = 0$. This indicates that the system of equations is dependent. It is equivalent to

$$p + q + \ r = 1,$$

$$q + 2r = 3.$$

We solve the second equation for q.

$$q = -2r + 3$$

Substitute for y in the first equation and solve for p.

$$p - 2r + 3 + r = 1$$
$$p - r + 3 = 1$$
$$p = r - 2$$

The solution is $(r-2, -2r+3, r)$, where r is any real number.

37. $a + b - c = 7,$
$a - b + c = 5,$
$3a + b - c = -1$

Write the augmented matrix.

$$\begin{bmatrix} 1 & 1 & -1 & 7 \\ 1 & -1 & 1 & 5 \\ 3 & 1 & -1 & -1 \end{bmatrix}$$

Multiply row 1 by -1 and add it to row 2. Also, multiply row 1 by -3 and add it to row 3.

$$\begin{bmatrix} 1 & 1 & -1 & 7 \\ 0 & -2 & 2 & -2 \\ 0 & -2 & 2 & -22 \end{bmatrix}$$

Multiply row 2 by -1 and add it to row 3.

$$\begin{bmatrix} 1 & 1 & -1 & 7 \\ 0 & -2 & 2 & -2 \\ 0 & 0 & 0 & -20 \end{bmatrix}$$

The last row corresponds to the false equation $0 = -20$. Thus, the system of equations has no solution.

39. $-2w + 2x + 2y - 2z = -10,$
$w + x + y + z = -5,$
$3w + x - y + 4z = -2,$
$w + 3x - 2y + 2z = -6$

Write the augmented matrix. We will use Gaussian elimination.

$$\begin{bmatrix} -2 & 2 & 2 & -2 & -10 \\ 1 & 1 & 1 & 1 & -5 \\ 3 & 1 & -1 & 4 & -2 \\ 1 & 3 & -2 & 2 & -6 \end{bmatrix}$$

Interchange rows 1 and 2.

$$\begin{bmatrix} 1 & 1 & 1 & 1 & -5 \\ -2 & 2 & 2 & -2 & -10 \\ 3 & 1 & -1 & 4 & -2 \\ 1 & 3 & -2 & 2 & -6 \end{bmatrix}$$

Multiply row 1 by 2 and add it to row 2. Multiply row 1 by -3 and add it to row 3. Multiply row 1 by -1 and add it to row 4.

$$\begin{bmatrix} 1 & 1 & 1 & 1 & -5 \\ 0 & 4 & 4 & 0 & -20 \\ 0 & -2 & -4 & 1 & 13 \\ 0 & 2 & -3 & 1 & -1 \end{bmatrix}$$

Interchange rows 2 and 3.

$$\begin{bmatrix} 1 & 1 & 1 & 1 & -5 \\ 0 & -2 & -4 & 1 & 13 \\ 0 & 4 & 4 & 0 & -20 \\ 0 & 2 & -3 & 1 & -1 \end{bmatrix}$$

Multiply row 2 by 2 and add it to row 3. Add row 2 to row 4.

$$\begin{bmatrix} 1 & 1 & 1 & 1 & -5 \\ 0 & -2 & -4 & 1 & 13 \\ 0 & 0 & -4 & 2 & 6 \\ 0 & 0 & -7 & 2 & 12 \end{bmatrix}$$

Multiply row 4 by 4.

$$\begin{bmatrix} 1 & 1 & 1 & 1 & -5 \\ 0 & -2 & -4 & 1 & 13 \\ 0 & 0 & -4 & 2 & 6 \\ 0 & 0 & -28 & 8 & 48 \end{bmatrix}$$

Multiply row 3 by -7 and add it to row 4.

$$\begin{bmatrix} 1 & 1 & 1 & 1 & -5 \\ 0 & -2 & -4 & 1 & 13 \\ 0 & 0 & -4 & 2 & 6 \\ 0 & 0 & 0 & -6 & 6 \end{bmatrix}$$

Multiply row 2 by $-\dfrac{1}{2}$, row 3 by $-\dfrac{1}{4}$, and row 6 by $-\dfrac{1}{6}$.

$$\begin{bmatrix} 1 & 1 & 1 & 1 & -5 \\ 0 & 1 & 2 & -\dfrac{1}{2} & -\dfrac{13}{2} \\ 0 & 0 & 1 & -\dfrac{1}{2} & -\dfrac{3}{2} \\ 0 & 0 & 0 & 1 & -1 \end{bmatrix}$$

Write the system of equations that corresponds to the last matrix.

$$w + x + y + z = -5, \quad (1)$$
$$x + 2y - \frac{1}{2}z = -\frac{13}{2}, \quad (2)$$
$$y - \frac{1}{2}z = -\frac{3}{2}, \quad (3)$$
$$z = -1 \quad (4)$$

Back-substitute in equation (3) and solve for y.

$$y - \frac{1}{2}(-1) = -\frac{3}{2}$$

$$y + \frac{1}{2} = -\frac{3}{2}$$

$$y = -2$$

Back-substitute in equation (2) and solve for x.

$$x + 2(-2) - \frac{1}{2}(-1) = -\frac{13}{2}$$

$$x - 4 + \frac{1}{2} = -\frac{13}{2}$$

$$x = -3$$

Back-substitute in equation (1) and solve for w.

$$w - 3 - 2 - 1 = -5$$

$$w = 1$$

The solution is $(1, -3, -2, -1)$.

41. *Familiarize.* Let x = the number of hours the Houlihans were out before 11 P.M. and y = the number of hours after 11 P.M. Then they pay the babysitter $\$5x$ before 11 P.M. and $\$7.50y$ after 11 P.M.

Translate.

The Houlihans were out for a total of 5 hr.

$$x + y = 5$$

They paid the sitter a total of $30.

$$5x + 7.5y = 30$$

Carry out. Use Gaussian elimination or Gauss-Jordan elimination to solve the system of equations.

$$x + \quad y = 5,$$
$$5x + 7.5y = 30.$$

The solution is $(3, 2)$. The coordinate $y = 2$ indicates that the Houlihans were out 2 hr after 11 P.M., so they came home at 1 A.M.

Check. The total time is $3 + 2$, or 5 hr. The total pay is $\$5 \cdot 3 + \$7.50(2)$, or $\$15 + \$15$, or $\$30$. The solution checks.

State. The Houlihans came home at 1 A.M.

43. *Familiarize*. Let x, y, and z represent the amounts borrowed at 8%, 10%, and 12%, respectively. Then the annual interest is $8\%x$, $10\%y$, and $12\%z$, or $0.08x$, $0.1y$, and $0.12z$.

Translate.

The total amount borrowed was $30,000.

$$x + y + z = 30,000$$

The total annual interest was $3040.

$$0.08x + 0.1y + 0.12z = 3040$$

The total amount borrowed at 8% and 10% was twice the amount borrowed at 12%.

$$x + y = 2z$$

We have a system of equations.

$$x + y + z = 30,000,$$
$$0.08x + 0.1y + 0.12z = 3040,$$
$$x + y = 2z, \text{ or}$$

$$x + \quad y + \quad z = 30,000,$$
$$0.08x + 0.1y + 0.12z = 3040,$$
$$x + \quad y - \quad 2z = 0$$

Carry out. Using Gaussian elimination or Gauss-Jordan elimination, we find that the solution is $(8000, 12,000, 10,000)$.

Check. The total amount borrowed was $\$8000 + \$12,000 + \$10,000$, or $\$30,000$. The total annual interest was $0.08(\$8000) + 0.1(\$12,000) + 0.12(\$10,000)$, or $\$640 + \$1200 + \$1200$, or $\$3040$. The total amount borrowed at 8% and 10%, $\$8000 + \$12,000$ or $\$20,000$, was twice the amount borrowed at 12%, $\$10,000$. The solution checks.

State. The amounts borrowed at 8%, 10%, and 12% were $8000, $12,000 and $10,000, respectively.

45. Discussion and Writing

47.
$$2x^2 + x = 7$$
$$2x^2 + x - 7 = 0$$
$$a = 2, b = 1, c = -7$$
$$x = \frac{-b \pm \sqrt{b^2 - 4ac}}{2a}$$
$$= \frac{-1 \pm \sqrt{1^2 - 4 \cdot 2 \cdot (-7)}}{2 \cdot 2} = \frac{-1 \pm \sqrt{1 + 56}}{4}$$
$$= \frac{-1 \pm \sqrt{57}}{4}$$

The solutions are $\dfrac{-1 + \sqrt{57}}{4}$ and $\dfrac{-1 - \sqrt{57}}{4}$, or $\dfrac{-1 \pm \sqrt{57}}{4}$.

We could find approximate solutions by graphing $y_1 = 2x^2 + x$ and $y_2 = 7$ and using the Intersect feature twice to find the first coordinates of the points of intersection of the graphs.

49.
$$\sqrt{2x+1} - 1 = \sqrt{2x-4}$$
$$(\sqrt{2x+1} - 1)^2 = (\sqrt{2x-4})^2 \quad \text{Squaring both sides}$$
$$2x + 1 - 2\sqrt{2x+1} + 1 = 2x - 4$$
$$2x + 2 - 2\sqrt{2x+1} = 2x - 4$$
$$2 - 2\sqrt{2x+1} = -4 \quad \text{Subtracting } 2x$$
$$-2\sqrt{2x+1} = -6 \quad \text{Subtracting } 2$$
$$\sqrt{2x+1} = 3 \quad \text{Dividing by } -2$$
$$(\sqrt{2x+1})^2 = 3^2 \quad \text{Squaring both sides}$$
$$2x + 1 = 9$$
$$2x = 8$$
$$x = 4$$

The number 4 checks. It is the solution.

51. Substitute to find three equations.
$$12 = a(-3)^2 + b(-3) + c$$
$$-7 = a(-1)^2 + b(-1) + c$$
$$-2 = a \cdot 1^2 + b \cdot 1 + c$$

We have a system of equations.
$$9a - 3b + c = 12,$$
$$a - b + c = -7,$$
$$a + b + c = -2$$

Write the augmented matrix. We will use Gaussian elimination.
$$\begin{bmatrix} 9 & -3 & 1 & | & 12 \\ 1 & -1 & 1 & | & -7 \\ 1 & 1 & 1 & | & -2 \end{bmatrix}$$

Interchange the first two rows.
$$\begin{bmatrix} 1 & -1 & 1 & | & -7 \\ 9 & -3 & 1 & | & 12 \\ 1 & 1 & 1 & | & -2 \end{bmatrix}$$

Multiply row 1 by -9 and add it to row 2. Also, multiply row 1 by -1 and add it to row 3.
$$\begin{bmatrix} 1 & -1 & 1 & | & -7 \\ 0 & 6 & -8 & | & 75 \\ 0 & 2 & 0 & | & 5 \end{bmatrix}$$

Interchange row 2 and row 3.
$$\begin{bmatrix} 1 & -1 & 1 & | & -7 \\ 0 & 2 & 0 & | & 5 \\ 0 & 6 & -8 & | & 75 \end{bmatrix}$$

Multiply row 2 by -3 and add it to row 3.

$$\begin{bmatrix} 1 & -1 & 1 & | & -7 \\ 0 & 2 & 0 & | & 5 \\ 0 & 0 & -8 & | & 60 \end{bmatrix}$$

Multiply row 2 by $\frac{1}{2}$ and row 3 by $-\frac{1}{8}$.
$$\begin{bmatrix} 1 & -1 & 1 & | & -7 \\ 0 & 1 & 0 & | & \frac{5}{2} \\ 0 & 0 & 1 & | & -\frac{15}{2} \end{bmatrix}$$

Write the system of equations that corresponds to the last matrix.
$$x - y + z = -7,$$
$$y = \frac{5}{2},$$
$$z = -\frac{15}{2}$$

Back-substitute $\frac{5}{2}$ for y and $-\frac{15}{2}$ for z in the first equation and solve for x.
$$x - \frac{5}{2} - \frac{15}{2} = -7$$
$$x - 10 = -7$$
$$x = 3$$

The solution is $\left(3, \frac{5}{2}, -\frac{15}{2}\right)$, so the equation is
$$y = 3x^2 + \frac{5}{2}x - \frac{15}{2}.$$

53.
$$\begin{bmatrix} 1 & 5 \\ 3 & 2 \end{bmatrix}$$

Multiply row 1 by -3 and add it to row 2.
$$\begin{bmatrix} 1 & 5 \\ 0 & -13 \end{bmatrix}$$

Multiply row 2 by $-\frac{1}{13}$.
$$\begin{bmatrix} 1 & 5 \\ 0 & 1 \end{bmatrix} \quad \text{Row-echelon form}$$

Multiply row 2 by -5 and add it to row 1.
$$\begin{bmatrix} 1 & 0 \\ 0 & 1 \end{bmatrix} \quad \text{Reduced row-echelon form}$$

55. $y = x + z,$

     $3y + 5z = 4,$

     $x + 4 = y + 3z,$ or

     $x - y + z = 0,$

         $3y + 5z = 4,$

     $x - y - 3z = -4$

Write the augmented matrix. We will use Gauss-Jordan elimination.

$$\left[\begin{array}{rrr|r} 1 & -1 & 1 & 0 \\ 0 & 3 & 5 & 4 \\ 1 & -1 & -3 & -4 \end{array}\right]$$

Multiply row 1 by -1 and add it to row 3.

$$\left[\begin{array}{rrr|r} 1 & -1 & 1 & 0 \\ 0 & 3 & 5 & 4 \\ 0 & 0 & -4 & -4 \end{array}\right]$$

Multiply row 3 by $-\dfrac{1}{4}$.

$$\left[\begin{array}{rrr|r} 1 & -1 & 1 & 0 \\ 0 & 3 & 5 & 4 \\ 0 & 0 & 1 & 1 \end{array}\right]$$

Multiply row 3 by -1 and add it to row 1. Also, multiply row 3 by -5 and add it to row 2.

$$\left[\begin{array}{rrr|r} 1 & -1 & 0 & -1 \\ 0 & 3 & 0 & -1 \\ 0 & 0 & 1 & 1 \end{array}\right]$$

Multiply row 2 by $\dfrac{1}{3}$.

$$\left[\begin{array}{rrr|r} 1 & -1 & 0 & -1 \\ 0 & 1 & 0 & -\dfrac{1}{3} \\ 0 & 0 & 1 & 1 \end{array}\right]$$

Add row 2 to row 1.

$$\left[\begin{array}{rrr|r} 1 & 0 & 0 & -\dfrac{4}{3} \\ 0 & 1 & 0 & -\dfrac{1}{3} \\ 0 & 0 & 1 & 1 \end{array}\right]$$

Read the solution from the last matrix. It is

$$\left(-\frac{4}{3}, -\frac{1}{3}, 1\right).$$

57. $x - 4y + 2z = 7,$

     $3x + y + 3z = -5$

Write the augmented matrix.

$$\left[\begin{array}{rrr|r} 1 & -4 & 2 & 7 \\ 3 & 1 & 3 & -5 \end{array}\right]$$

Multiply row 1 by -3 and add it to row 2.

$$\left[\begin{array}{rrr|r} 1 & -4 & 2 & 7 \\ 0 & 13 & -3 & -26 \end{array}\right]$$

Multiply row 2 by $\dfrac{1}{13}$.

$$\left[\begin{array}{rrr|r} 1 & -4 & 2 & 7 \\ 0 & 1 & -\dfrac{3}{13} & -2 \end{array}\right]$$

Write the system of equations that corresponds to the last matrix.

$$x - 4y + 2z = 7,$$
$$y - \frac{3}{13}z = -2$$

Solve the second equation for y.

$$y = \frac{3}{13}z - 2$$

Substitute in the first equation and solve for x.

$$x - 4\left(\frac{3}{13}z - 2\right) + 2z = 7$$
$$x - \frac{12}{13}z + 8 + 2z = 7$$
$$x = -\frac{14}{13}z - 1$$

The solution is $\left(-\dfrac{14}{13}z - 1, \dfrac{3}{13}z - 2, z\right)$, where z is any real number.

59. $4x + 5y = 3,$

     $-2x + y = 9,$

     $3x - 2y = -15$

Write the augmented matrix.

$$\left[\begin{array}{rr|r} 4 & 5 & 3 \\ -2 & 1 & 9 \\ 3 & -2 & -15 \end{array}\right]$$

Multiply row 2 by 2 and row 3 by 4.

$$\left[\begin{array}{rr|r} 4 & 5 & 3 \\ -4 & 2 & 18 \\ 12 & -8 & -60 \end{array}\right]$$

Add row 1 to row 2. Also, multiply row 1 by -3 and add it to row 3.

$$\left[\begin{array}{rr|r} 4 & 5 & 3 \\ 0 & 7 & 21 \\ 0 & -23 & -69 \end{array}\right]$$

Multiply row 2 by $\frac{1}{7}$ and row 3 by $-\frac{1}{23}$.

$$\left[\begin{array}{rr|r} 4 & 5 & 3 \\ 0 & 1 & 3 \\ 0 & 1 & 3 \end{array}\right]$$

Multiply row 2 by -1 and add it to row 3.

$$\left[\begin{array}{rr|r} 4 & 5 & 3 \\ 0 & 1 & 3 \\ 0 & 0 & 0 \end{array}\right]$$

The last row corresponds to the equation $0 = 0$. Thus we have a dependent system that is equivalent to

$$4x + 5y = 3, \quad (1)$$
$$y = 3. \quad (2)$$

Back-substitute in equation (1) to find x.

$$4x + 5 \cdot 3 = 3$$
$$4x + 15 = 3$$
$$4x = -12$$
$$x = -3$$

The solution is $(-3, 3)$.

Exercise Set 8.4

1. $\begin{bmatrix} 5 & x \end{bmatrix} = \begin{bmatrix} y & -3 \end{bmatrix}$

Corresponding entries of the two matrices must be equal. Thus we have $5 = y$ and $x = -3$.

3. $\begin{bmatrix} 3 & 2x \\ y & -8 \end{bmatrix} = \begin{bmatrix} 3 & -2 \\ 1 & -8 \end{bmatrix}$

Corresponding entries of the two matrices must be equal. Thus, we have:

$$2x = -2 \quad \text{and} \quad y = 1$$
$$x = -1 \quad \text{and} \quad y = 1$$

5. $\mathbf{A} + \mathbf{B} = \begin{bmatrix} 1 & 2 \\ 4 & 3 \end{bmatrix} + \begin{bmatrix} -3 & 5 \\ 2 & -1 \end{bmatrix}$

$$= \begin{bmatrix} 1 + (-3) & 2 + 5 \\ 4 + 2 & 3 + (-1) \end{bmatrix}$$

$$= \begin{bmatrix} -2 & 7 \\ 6 & 2 \end{bmatrix}$$

7. $\mathbf{E} + \mathbf{O} = \begin{bmatrix} 1 & 3 \\ 2 & 6 \end{bmatrix} + \begin{bmatrix} 0 & 0 \\ 0 & 0 \end{bmatrix}$

$$= \begin{bmatrix} 1 + 0 & 3 + 0 \\ 2 + 0 & 6 + 0 \end{bmatrix}$$

$$= \begin{bmatrix} 1 & 3 \\ 2 & 6 \end{bmatrix}$$

9. $3\mathbf{F} = 3 \begin{bmatrix} 3 & 3 \\ -1 & -1 \end{bmatrix}$

$$= \begin{bmatrix} 3 \cdot 3 & 3 \cdot 3 \\ 3 \cdot (-1) & 3 \cdot (-1) \end{bmatrix}$$

$$= \begin{bmatrix} 9 & 9 \\ -3 & -3 \end{bmatrix}$$

11. $3\mathbf{F} = 3 \begin{bmatrix} 3 & 3 \\ -1 & -1 \end{bmatrix} = \begin{bmatrix} 9 & 9 \\ -3 & -3 \end{bmatrix}$,

$$2\mathbf{A} = 2 \begin{bmatrix} 1 & 2 \\ 4 & 3 \end{bmatrix} = \begin{bmatrix} 2 & 4 \\ 8 & 6 \end{bmatrix}$$

$$3\mathbf{F} + 2\mathbf{A} = \begin{bmatrix} 9 & 9 \\ -3 & -3 \end{bmatrix} + \begin{bmatrix} 2 & 4 \\ 8 & 6 \end{bmatrix}$$

$$= \begin{bmatrix} 9 + 2 & 9 + 4 \\ -3 + 8 & -3 + 6 \end{bmatrix}$$

$$= \begin{bmatrix} 11 & 13 \\ 5 & 3 \end{bmatrix}$$

13. $\mathbf{B} - \mathbf{A} = \begin{bmatrix} -3 & 5 \\ 2 & -1 \end{bmatrix} - \begin{bmatrix} 1 & 2 \\ 4 & 3 \end{bmatrix}$

$$= \begin{bmatrix} -3 & 5 \\ 2 & -1 \end{bmatrix} + \begin{bmatrix} -1 & -2 \\ -4 & -3 \end{bmatrix}$$

$$[\mathbf{B} - \mathbf{A} = \mathbf{B} + (-\mathbf{A})]$$

$$= \begin{bmatrix} -3 + (-1) & 5 + (-2) \\ 2 + (-4) & -1 + (-3) \end{bmatrix}$$

$$= \begin{bmatrix} -4 & 3 \\ -2 & -4 \end{bmatrix}$$

15. $\mathbf{BA} = \begin{bmatrix} -3 & 5 \\ 2 & -1 \end{bmatrix} \begin{bmatrix} 1 & 2 \\ 4 & 3 \end{bmatrix}$

$$= \begin{bmatrix} -3 \cdot 1 + 5 \cdot 4 & -3 \cdot 2 + 5 \cdot 3 \\ 2 \cdot 1 + (-1)4 & 2 \cdot 2 + (-1)3 \end{bmatrix}$$

$$= \begin{bmatrix} 17 & 9 \\ -2 & 1 \end{bmatrix}$$

17. $\mathbf{CD} = \begin{bmatrix} 1 & -1 \\ -1 & 1 \end{bmatrix} \begin{bmatrix} 1 & 1 \\ 1 & 1 \end{bmatrix}$

$$= \begin{bmatrix} 1 \cdot 1 + (-1) \cdot 1 & 1 \cdot 1 + (-1) \cdot 1 \\ -1 \cdot 1 + 1 \cdot 1 & -1 \cdot 1 + 1 \cdot 1 \end{bmatrix}$$

$$= \begin{bmatrix} 0 & 0 \\ 0 & 0 \end{bmatrix}$$

19. $\mathbf{AI} = \begin{bmatrix} 1 & 2 \\ 4 & 3 \end{bmatrix} \begin{bmatrix} 1 & 0 \\ 0 & 1 \end{bmatrix}$

$\quad\quad = \begin{bmatrix} 1\cdot 1 + 2\cdot 0 & 1\cdot 0 + 2\cdot 1 \\ 4\cdot 1 + 3\cdot 0 & 4\cdot 0 + 3\cdot 1 \end{bmatrix}$

$\quad\quad = \begin{bmatrix} 1 & 2 \\ 4 & 3 \end{bmatrix}$

21. a) $\mathbf{B} = \begin{bmatrix} 150 & 80 & 40 \end{bmatrix}$

b) $\$150 + 5\% \cdot \$150 = 1.05(\$150) = \$157.50$

$\$80 + 5\% \cdot \$80 = 1.05(\$80) = \$84$

$\$40 + 5\% \cdot \$40 = 1.05(\$40) = \$42$

We write the matrix that corresponds to these amounts.

$\mathbf{R} = \begin{bmatrix} 157.5 & 84 & 42 \end{bmatrix}$

c) $\mathbf{B} + \mathbf{R} = \begin{bmatrix} 150 & 80 & 40 \end{bmatrix} + \begin{bmatrix} 157.5 & 84 & 42 \end{bmatrix}$

$\quad\quad = \begin{bmatrix} 307.5 & 164 & 82 \end{bmatrix}$

The entries represent the total budget in each type of expenditure for June and July.

23. a) $\mathbf{C} = \begin{bmatrix} 140 & 27 & 3 & 13 & 64 \end{bmatrix}$

$\mathbf{P} = \begin{bmatrix} 180 & 4 & 11 & 24 & 662 \end{bmatrix}$

$\mathbf{B} = \begin{bmatrix} 50 & 5 & 1 & 82 & 20 \end{bmatrix}$

b) $\quad \mathbf{C} + 2\mathbf{P} + 3\mathbf{B}$

$\quad = \begin{bmatrix} 140 & 27 & 3 & 13 & 64 \end{bmatrix} +$
$\quad\quad \begin{bmatrix} 360 & 8 & 22 & 48 & 1324 \end{bmatrix} +$
$\quad\quad \begin{bmatrix} 150 & 15 & 3 & 246 & 60 \end{bmatrix}$

$\quad = \begin{bmatrix} 650 & 50 & 28 & 307 & 1448 \end{bmatrix}$

The entries represent the total nutritional value of one serving of chicken, 1 cup of potato salad, and 3 broccoli spears.

25. Use a grapher.

$\begin{bmatrix} -1 & 0 & 7 \\ 3 & -5 & 2 \end{bmatrix} \begin{bmatrix} 6 \\ -4 \\ 1 \end{bmatrix} = \begin{bmatrix} 1 \\ 40 \end{bmatrix}$

27. Use a grapher.

$\begin{bmatrix} -2 & 4 \\ 5 & 1 \\ -1 & -3 \end{bmatrix} \begin{bmatrix} 3 & -6 \\ -1 & 4 \end{bmatrix} = \begin{bmatrix} -10 & 28 \\ 14 & -26 \\ 0 & -6 \end{bmatrix}$

29. $\begin{bmatrix} 1 \\ -5 \\ 3 \end{bmatrix} \begin{bmatrix} -6 & 5 & 8 \\ 0 & 4 & -1 \end{bmatrix}$

The grapher produces an error message when this multiplication is attempted. This product is not defined because the number of columns of the first matrix, 1, is not equal to the number of rows of the second matrix, 2.

31. Use a grapher.

$\begin{bmatrix} 1 & -4 & 3 \\ 0 & 8 & 0 \\ -2 & -1 & 5 \end{bmatrix} \begin{bmatrix} 3 & 0 & 0 \\ 0 & -4 & 0 \\ 0 & 0 & 1 \end{bmatrix} =$

$\begin{bmatrix} 3 & 16 & 3 \\ 0 & -32 & 0 \\ -6 & 4 & 5 \end{bmatrix}$

33. a) $\mathbf{M} = \begin{bmatrix} 45.29 & 6.63 & 10.94 & 7.42 & 8.01 \\ 53.78 & 4.95 & 9.83 & 6.16 & 12.56 \\ 47.13 & 8.47 & 12.66 & 8.29 & 9.43 \\ 51.64 & 7.12 & 11.57 & 9.35 & 10.72 \end{bmatrix}$

b) $\mathbf{N} = \begin{bmatrix} 65 & 48 & 93 & 57 \end{bmatrix}$

c) Use a grapher.

$\quad \mathbf{NM} =$

$\begin{bmatrix} 12,851.86 & 1862.1 & 3019.81 & 2081.9 & 2611.56 \end{bmatrix}$

d) The entries of $\mathbf{NM}$ represent the total cost, in cents, of each item for the day's meals.

35. a) $\mathbf{S} = \begin{bmatrix} 8 & 15 \\ 6 & 10 \\ 4 & 3 \end{bmatrix}$

b) $\mathbf{C} = \begin{bmatrix} 3 & 1.5 & 2 \end{bmatrix}$

c) $\mathbf{CS} = \begin{bmatrix} 41 & 66 \end{bmatrix}$

d) The entries of $\mathbf{CS}$ represent the total cost, in dollars, of ingredients for each coffee shop.

37. a) $\mathbf{P} = \begin{bmatrix} 6 & 4.5 & 5.2 \end{bmatrix}$

b) $\quad \mathbf{PS} = \begin{bmatrix} 6 & 4.5 & 5.2 \end{bmatrix} \begin{bmatrix} 8 & 15 \\ 6 & 10 \\ 4 & 3 \end{bmatrix}$

$\quad\quad = \begin{bmatrix} 95.8 & 150.6 \end{bmatrix}$

The profit from Mugsey's Coffee Shop is $95.80, and the profit from The Coffee Club is $150.60.

39. $2x - 3y = 7,$

$\quad x + 5y = -6$

Write the coefficients on the left in a matrix. Then write the product of that matrix and the column matrix containing the variables, and set the result equal to the column matrix containing the constants on the right.

$\begin{bmatrix} 2 & -3 \\ 1 & 5 \end{bmatrix} \begin{bmatrix} x \\ y \end{bmatrix} = \begin{bmatrix} 7 \\ -6 \end{bmatrix}$

41. $x + y - 2z = 6,$

$\quad 3x - y + z = 7,$

$\quad 2x + 5y - 3z = 8$

Write the coefficients on the left in a matrix. Then write the product of that matrix and the column matrix containing the variables, and set the result equal to the column matrix containing the constants on the right.

$$\begin{bmatrix} 1 & 1 & -2 \\ 3 & -1 & 1 \\ 2 & 5 & -3 \end{bmatrix} \begin{bmatrix} x \\ y \\ z \end{bmatrix} = \begin{bmatrix} 6 \\ 7 \\ 8 \end{bmatrix}$$

43. $3x - 2y + 4z = 17,$

$2x + y - 5z = 13$

Write the coefficients on the left in a matrix. Then write the product of that matrix and the column matrix containing the variables, and set the result equal to the column matrix containing the constants on the right.

$$\begin{bmatrix} 3 & -2 & 4 \\ 2 & 1 & -5 \end{bmatrix} \begin{bmatrix} x \\ y \\ z \end{bmatrix} = \begin{bmatrix} 17 \\ 13 \end{bmatrix}$$

45. $-4w + x - y + 2z = 12,$

$w + 2x - y - z = 0,$

$-w + x + 4y - 3z = 1,$

$2w + 3x + 5y - 7z = 9$

Write the coefficients on the left in a matrix. Then write the product of that matrix and the column matrix containing the variables, and set the result equal to the column matrix containing the constants on the right.

$$\begin{bmatrix} -4 & 1 & -1 & 2 \\ 1 & 2 & -1 & -1 \\ -1 & 1 & 4 & -3 \\ 2 & 3 & 5 & -7 \end{bmatrix} \begin{bmatrix} w \\ x \\ y \\ z \end{bmatrix} = \begin{bmatrix} 12 \\ 0 \\ 1 \\ 9 \end{bmatrix}$$

47. Discussion and Writing

49. $f(x) = x^2 - 3x - 10$

a) $-\dfrac{b}{2a} = -\dfrac{-3}{2 \cdot 1} = \dfrac{3}{2}$

$f\left(\dfrac{3}{2}\right) = \left(\dfrac{3}{2}\right)^2 - 3\left(\dfrac{3}{2}\right) - 10 = -\dfrac{49}{4}$

The vertex is $\left(\dfrac{3}{2}, -\dfrac{49}{4}\right).$

b) The line of symmetry is $x = \dfrac{3}{2}.$

c) Since the coefficient of x^2 is positive, the function has a minimum value. It is the second coordinate of the vertex, $-\dfrac{49}{4}.$

51. $f(x) = -x^2 - 3x + 5$

a) $-\dfrac{b}{2a} = -\dfrac{-3}{2(-1)} = -\dfrac{3}{2}$

$f\left(-\dfrac{3}{2}\right) = -\left(-\dfrac{3}{2}\right)^2 - 3\left(-\dfrac{3}{2}\right) + 5 = \dfrac{29}{4}$

The vertex is $\left(-\dfrac{3}{2}, \dfrac{29}{4}\right).$

b) The line of symmetry is $x = -\dfrac{3}{2}.$

c) Since the coefficient of x^2 is negative, the function has a maximum value. It is the second coordinate of the vertex, $\dfrac{29}{4}.$

53. $\mathbf{A} = \begin{bmatrix} -1 & 0 \\ 2 & 1 \end{bmatrix},\ \mathbf{B} = \begin{bmatrix} 1 & -1 \\ 0 & 2 \end{bmatrix}$

$(\mathbf{A} + \mathbf{B})(\mathbf{A} - \mathbf{B}) = \begin{bmatrix} 0 & -1 \\ 2 & 3 \end{bmatrix} \begin{bmatrix} -2 & 1 \\ 2 & -1 \end{bmatrix}$

$= \begin{bmatrix} -2 & 1 \\ 2 & -1 \end{bmatrix}$

$\mathbf{A}^2 - \mathbf{B}^2$

$= \begin{bmatrix} -1 & 0 \\ 2 & 1 \end{bmatrix} \begin{bmatrix} -1 & 0 \\ 2 & 1 \end{bmatrix} - \begin{bmatrix} 1 & -1 \\ 0 & 2 \end{bmatrix} \begin{bmatrix} 1 & -1 \\ 0 & 2 \end{bmatrix}$

$= \begin{bmatrix} 1 & 0 \\ 0 & 1 \end{bmatrix} - \begin{bmatrix} 1 & -3 \\ 0 & 4 \end{bmatrix}$

$= \begin{bmatrix} 0 & 3 \\ 0 & -3 \end{bmatrix}$

Thus $(\mathbf{A} + \mathbf{B})(\mathbf{A} - \mathbf{B}) \neq \mathbf{A}^2 - \mathbf{B}^2.$

55. In Exercise 53 we found that $(\mathbf{A} + \mathbf{B})(\mathbf{A} - \mathbf{B}) = \begin{bmatrix} -2 & 1 \\ 2 & -1 \end{bmatrix}$

and we also found $\mathbf{A}^2$ and $\mathbf{B}^2.$

$\mathbf{BA} = \begin{bmatrix} 1 & -1 \\ 0 & 2 \end{bmatrix} \begin{bmatrix} -1 & 0 \\ 2 & 1 \end{bmatrix} = \begin{bmatrix} -3 & -1 \\ 4 & 2 \end{bmatrix}$

$\mathbf{AB} = \begin{bmatrix} -1 & 0 \\ 2 & 1 \end{bmatrix} \begin{bmatrix} 1 & -1 \\ 0 & 2 \end{bmatrix} = \begin{bmatrix} -1 & 1 \\ 2 & 0 \end{bmatrix}$

$\mathbf{A}^2 + \mathbf{BA} - \mathbf{AB} - \mathbf{B}^2$

$= \begin{bmatrix} 1 & 0 \\ 0 & 1 \end{bmatrix} + \begin{bmatrix} -3 & -1 \\ 4 & 2 \end{bmatrix} - \begin{bmatrix} -1 & 1 \\ 2 & 0 \end{bmatrix} -$

$\begin{bmatrix} 1 & -3 \\ 0 & 4 \end{bmatrix}$

$= \begin{bmatrix} -2 & 1 \\ 2 & -1 \end{bmatrix}$

Thus $(\mathbf{A} + \mathbf{B})(\mathbf{A} - \mathbf{B}) = \mathbf{A}^2 + \mathbf{BA} - \mathbf{AB} - \mathbf{B}^2.$

57. See the answer section in the text.

59. See the answer section in the text.

61. See the answer section in the text.

Exercise Set 8.5

1. $\mathbf{BA} = \begin{bmatrix} 7 & 3 \\ 2 & 1 \end{bmatrix} \begin{bmatrix} 1 & -3 \\ -2 & 7 \end{bmatrix} = \begin{bmatrix} 1 & 0 \\ 0 & 1 \end{bmatrix}$

$\mathbf{AB} = \begin{bmatrix} 1 & -3 \\ -2 & 7 \end{bmatrix} \begin{bmatrix} 7 & 3 \\ 2 & 1 \end{bmatrix} = \begin{bmatrix} 1 & 0 \\ 0 & 1 \end{bmatrix}$

Since $\mathbf{BA} = \mathbf{I} = \mathbf{AB}$, $\mathbf{B}$ is the inverse of $\mathbf{A}$.

3. $\mathbf{BA} = \begin{bmatrix} 2 & 3 & 2 \\ 3 & 3 & 4 \\ 1 & 1 & 1 \end{bmatrix} \begin{bmatrix} -1 & -1 & 6 \\ 1 & 0 & -2 \\ 1 & 0 & -3 \end{bmatrix} =$

$\begin{bmatrix} 3 & -2 & 0 \\ 4 & -3 & 0 \\ 1 & -1 & 1 \end{bmatrix}$

Since $\mathbf{BA} \neq \mathbf{I}$, $\mathbf{B}$ is not the inverse of $\mathbf{A}$.

5. $\mathbf{A} = \begin{bmatrix} 3 & 2 \\ 5 & 3 \end{bmatrix}$

Write the augmented matrix.

$\begin{bmatrix} 3 & 2 & | & 1 & 0 \\ 5 & 3 & | & 0 & 1 \end{bmatrix}$

Multiply row 2 by 3.

$\begin{bmatrix} 3 & 2 & | & 1 & 0 \\ 15 & 9 & | & 0 & 3 \end{bmatrix}$

Multiply row 1 by -5 and add it to row 2.

$\begin{bmatrix} 3 & 2 & | & 1 & 0 \\ 0 & -1 & | & -5 & 3 \end{bmatrix}$

Multiply row 2 by 2 and add it to row 1.

$\begin{bmatrix} 3 & 0 & | & -9 & 6 \\ 0 & -1 & | & -5 & 3 \end{bmatrix}$

Multiply row 1 by $\frac{1}{3}$ and row 2 by -1.

$\begin{bmatrix} 1 & 0 & | & -3 & 2 \\ 0 & 1 & | & 5 & -3 \end{bmatrix}$

Then $\mathbf{A}^{-1} = \begin{bmatrix} -3 & 2 \\ 5 & -3 \end{bmatrix}$.

7. $\mathbf{A} = \begin{bmatrix} 6 & 9 \\ 4 & 6 \end{bmatrix}$

Write the augmented matrix.

$\begin{bmatrix} 6 & 9 & | & 1 & 0 \\ 4 & 6 & | & 0 & 1 \end{bmatrix}$

Multiply row 2 by 3.

$\begin{bmatrix} 6 & 9 & | & 1 & 0 \\ 12 & 18 & | & 0 & 3 \end{bmatrix}$

Multiply row 1 by -2 and add it to row 2.

$\begin{bmatrix} 6 & 9 & | & 1 & 0 \\ 0 & 0 & | & -2 & 3 \end{bmatrix}$

We cannot obtain the identity matrix on the left since the second row contains only zeros to the left of the vertical line. Thus, $\mathbf{A}^{-1}$ does not exist.

9. $\mathbf{A} = \begin{bmatrix} 3 & 1 & 0 \\ 1 & 1 & 1 \\ 1 & -1 & 2 \end{bmatrix}$

Write the augmented matrix.

$\begin{bmatrix} 3 & 1 & 0 & | & 1 & 0 & 0 \\ 1 & 1 & 1 & | & 0 & 1 & 0 \\ 1 & -1 & 2 & | & 0 & 0 & 1 \end{bmatrix}$

Interchange the first two rows.

$\begin{bmatrix} 1 & 1 & 1 & | & 0 & 1 & 0 \\ 3 & 1 & 0 & | & 1 & 0 & 0 \\ 1 & -1 & 2 & | & 0 & 0 & 1 \end{bmatrix}$

Multiply row 1 by -3 and add it to row 2. Also, multiply row 1 by -1 and add it to row 3.

$\begin{bmatrix} 1 & 1 & 1 & | & 0 & 1 & 0 \\ 0 & -2 & -3 & | & 1 & -3 & 0 \\ 0 & -2 & 1 & | & 0 & -1 & 1 \end{bmatrix}$

Multiply row 2 by $-\frac{1}{2}$.

$\begin{bmatrix} 1 & 1 & 1 & | & 0 & 1 & 0 \\ 0 & 1 & \frac{3}{2} & | & -\frac{1}{2} & \frac{3}{2} & 0 \\ 0 & -2 & 1 & | & 0 & -1 & 1 \end{bmatrix}$

Multiply row 2 by -1 and add it to row 1. Also, multiply row 2 by 2 and add it to row 3.

$\begin{bmatrix} 1 & 0 & -\frac{1}{2} & | & \frac{1}{2} & -\frac{1}{2} & 0 \\ 0 & 1 & \frac{3}{2} & | & -\frac{1}{2} & \frac{3}{2} & 0 \\ 0 & 0 & 4 & | & -1 & 2 & 1 \end{bmatrix}$

Multiply row 3 by $\frac{1}{4}$.

$\begin{bmatrix} 1 & 0 & -\frac{1}{2} & | & \frac{1}{2} & -\frac{1}{2} & 0 \\ 0 & 1 & \frac{3}{2} & | & -\frac{1}{2} & \frac{3}{2} & 0 \\ 0 & 0 & 1 & | & -\frac{1}{4} & \frac{1}{2} & \frac{1}{4} \end{bmatrix}$

Multiply row 3 by $\frac{1}{2}$ and add it to row 1. Also, multiply row 3 by $-\frac{3}{2}$ and add it to row 2.

$\begin{bmatrix} 1 & 0 & 0 & | & \frac{3}{8} & -\frac{1}{4} & \frac{1}{8} \\ 0 & 1 & 0 & | & -\frac{1}{8} & \frac{3}{4} & -\frac{3}{8} \\ 0 & 0 & 1 & | & -\frac{1}{4} & \frac{1}{2} & \frac{1}{4} \end{bmatrix}$

Then $\mathbf{A}^{-1} = \begin{bmatrix} \frac{3}{8} & -\frac{1}{4} & \frac{1}{8} \\ -\frac{1}{8} & \frac{3}{4} & -\frac{3}{8} \\ -\frac{1}{4} & \frac{1}{2} & \frac{1}{4} \end{bmatrix}$.

11. $\mathbf{A} = \begin{bmatrix} 1 & -4 & 8 \\ 1 & -3 & 2 \\ 2 & -7 & 10 \end{bmatrix}$

Write the augmented matrix.

$\begin{bmatrix} 1 & -4 & 8 & | & 1 & 0 & 0 \\ 1 & -3 & 2 & | & 0 & 1 & 0 \\ 2 & -7 & 10 & | & 0 & 0 & 1 \end{bmatrix}$

Multiply row 1 by -1 and add it to row 2. Also, multiply row 1 by -2 and add it to row 3.

$\begin{bmatrix} 1 & -4 & 8 & | & 1 & 0 & 0 \\ 0 & 1 & -6 & | & -1 & 1 & 0 \\ 0 & 1 & -6 & | & -2 & 0 & 1 \end{bmatrix}$

Since the second and third rows are identical left of the vertical line, it will not be possible to obtain the identity matrix on the left side. Thus, $\mathbf{A}^{-1}$ does not exist.

13. Use a grapher.

$\mathbf{A}^{-1} = \begin{bmatrix} 0.4 & -0.6 \\ 0.2 & -0.8 \end{bmatrix}$

15. Use a grapher.

$\mathbf{A}^{-1} = \begin{bmatrix} -1 & -1 & -6 \\ 1 & 0 & 2 \\ 0 & 1 & 3 \end{bmatrix}$

17. Use a grapher.

$\mathbf{A}^{-1} = \begin{bmatrix} 1 & 1 & 2 \\ 1 & 1 & 1 \\ 2 & 3 & 4 \end{bmatrix}$

19. The grapher produces an error message. Thus, $\mathbf{A}^{-1}$ does not exist.

21. Use a grapher.

$\mathbf{A}^{-1} = \begin{bmatrix} 1 & -2 & 3 & 8 \\ 0 & 1 & -3 & 1 \\ 0 & 0 & 1 & -2 \\ 0 & 0 & 0 & -1 \end{bmatrix}$

23. Use a grapher.

$\mathbf{A}^{-1} = \begin{bmatrix} 0.25 & 0.25 & 1.25 & -0.25 \\ 0.5 & 1.25 & 1.75 & -1 \\ -0.25 & -0.25 & -0.75 & 0.75 \\ 0.25 & 0.5 & 0.75 & -0.5 \end{bmatrix}$

25. $4x + 3y = 2,$
$\quad x - 2y = 6$

Write an equivalent matrix equation, $\mathbf{AX} = \mathbf{B}$.

$\begin{bmatrix} 4 & 3 \\ 1 & -2 \end{bmatrix} \begin{bmatrix} x \\ y \end{bmatrix} = \begin{bmatrix} 2 \\ 6 \end{bmatrix}$

Then $\mathbf{X} = \mathbf{A}^{-1}\mathbf{B} = \begin{bmatrix} 2 \\ -2 \end{bmatrix}$.

The solution is $(2, -2)$.

27. $5x + y = 2,$
$\quad 3x - 2y = -4$

Write an equivalent matrix equation, $\mathbf{AX} = \mathbf{B}$.

$\begin{bmatrix} 5 & 1 \\ 3 & -2 \end{bmatrix} \begin{bmatrix} x \\ y \end{bmatrix} = \begin{bmatrix} 2 \\ -4 \end{bmatrix}$

Then $\mathbf{X} = \mathbf{A}^{-1}\mathbf{B} = \begin{bmatrix} 0 \\ 2 \end{bmatrix}$.

The solution is $(0, 2)$.

29. $x \quad + z = 1,$
$\quad 2x + y \quad = 3,$
$\quad x - y + z = 4$

Write an equivalent matrix equation, $\mathbf{AX} = \mathbf{B}$.

$\begin{bmatrix} 1 & 0 & 1 \\ 2 & 1 & 0 \\ 1 & -1 & 1 \end{bmatrix} \begin{bmatrix} x \\ y \\ z \end{bmatrix} = \begin{bmatrix} 1 \\ 3 \\ 4 \end{bmatrix}$

Then $\mathbf{X} = \mathbf{A}^{-1}\mathbf{B} = \begin{bmatrix} 3 \\ -3 \\ -2 \end{bmatrix}$.

The solution is $(3, -3, -2)$.

31. $2x + 3y + 4z = 2,$
$\quad x - 4y + 3z = 2,$
$\quad 5x + y + z = -4$

Write an equivalent matrix equation, $\mathbf{AX} = \mathbf{B}$.

$\begin{bmatrix} 2 & 3 & 4 \\ 1 & -4 & 3 \\ 5 & 1 & 1 \end{bmatrix} \begin{bmatrix} x \\ y \\ z \end{bmatrix} = \begin{bmatrix} 2 \\ 2 \\ -4 \end{bmatrix}$

Then $\mathbf{X} = \mathbf{A}^{-1}\mathbf{B} = \begin{bmatrix} -1 \\ 0 \\ 1 \end{bmatrix}$.

The solution is $(-1, 0, 1)$.

33. $2w - 3x + 4y - 5z = 0,$
$\quad 3w - 2x + 7y - 3z = 2,$
$\quad w + x - y + z = 1,$
$\quad -w - 3x - 6y + 4z = 6$

Write an equivalent matrix equation, $\mathbf{AX} = \mathbf{B}$.

$\begin{bmatrix} 2 & -3 & 4 & -5 \\ 3 & -2 & 7 & -3 \\ 1 & 1 & -1 & 1 \\ -1 & -3 & -6 & 4 \end{bmatrix} \begin{bmatrix} w \\ x \\ y \\ z \end{bmatrix} = \begin{bmatrix} 0 \\ 2 \\ 1 \\ 6 \end{bmatrix}$

Then $\mathbf{X} = \mathbf{A}^{-1}\mathbf{B} = \begin{bmatrix} 1 \\ -1 \\ 0 \\ 1 \end{bmatrix}$.

The solution is $(1, -1, 0, 1)$.

35. *Familiarize*. Let x = the number of hot dogs sold and y = the number of sausages.

Translate.

The total number of items sold was 145.

$$x + y = 145$$

The number of hot dogs sold is 45 more than the number of sausages.

$$x = y + 45$$

We have a system of equations:

$$x + y = 145, \qquad x + y = 145,$$
$$\text{or}$$
$$x = y + 45, \qquad x - y = 45.$$

Carry out. Write an equivalent matrix equation, $\mathbf{AX} = \mathbf{B}$.

$$\begin{bmatrix} 1 & 1 \\ 1 & -1 \end{bmatrix} \begin{bmatrix} x \\ y \end{bmatrix} = \begin{bmatrix} 145 \\ 45 \end{bmatrix}$$

Then $\mathbf{X} = \mathbf{A}^{-1}\mathbf{B} = \begin{bmatrix} 95 \\ 50 \end{bmatrix}$, so the solution is $(95, 50)$.

Check. The total number of items is $95 + 50$, or 145. The number of hot dogs, 95, is 45 more than the number of sausages. The solution checks.

State. Stefan sold 95 hot dogs and 50 Italian sausages.

37. *Familiarize*. Let x, y, and z represent the prices of one ton of topsoil, mulch, and pea gravel, respectively.

Translate.

Four tons of topsoil, 3 tons of mulch, and 6 tons of pea gravel costs $2825.

$$4x + 3y + 6z = 2825$$

Five tons of topsoil, 2 tons of mulch, and 5 tons of pea gravel costs $2663.

$$5x + 2y + 5z = 2663$$

Pea gravel costs $17 less per ton than topsoil.

$$z = x - 17$$

We have a system of equations.

$$4x + 3y + 6z = 2825,$$
$$5x + 2y + 5z = 2663,$$
$$z = x - 17, \text{ or}$$

$$4x + 3y + 6z = 2825,$$
$$5x + 2y + 5z = 2663,$$
$$x \qquad\quad - z = 17$$

Carry out. Write an equivalent matrix equation, $\mathbf{AX} = \mathbf{B}$.

$$\begin{bmatrix} 4 & 3 & 6 \\ 5 & 2 & 5 \\ 1 & 0 & -1 \end{bmatrix} \begin{bmatrix} x \\ y \\ z \end{bmatrix} = \begin{bmatrix} 2825 \\ 2663 \\ 17 \end{bmatrix}$$

Then $\mathbf{X} = \mathbf{A}^{-1}\mathbf{B} = \begin{bmatrix} 239 \\ 179 \\ 222 \end{bmatrix}$, so the solution is $(239, 179, 222)$.

Check. Four tons of topsoil, 3 tons of mulch, and 6 tons of pea gravel costs $4 \cdot \$239 + 3 \cdot \$179 + 6 \cdot \$222$, or $\$956 + \$537 + \$1332$, or $\$2825$. Five tons of topsoil, 2 tons of mulch, and 5 tons of pea gravel costs $5 \cdot \$239 + 2 \cdot \$179 + 5 \cdot \$222$, or $\$1195 + \$358 + \$1110$, or $\$2663$. The price of pea gravel, $222, is $17 less than the price of topsoil, $239. The solution checks.

State. The price of topsoil is $239 per ton, of mulch is $179 per ton, and of pea gravel is $222 per ton.

39. Discussion and Writing

41.

$$\begin{array}{r|rrrr} -2 & 1 & -6 & 4 & -8 \\ & & -2 & 16 & -40 \\ \hline & 1 & -8 & 20 & -48 \end{array}$$

$f(-2) = -48$

43. $f(x) = x^3 - 3x^2 - 6x + 8$

We use synthetic division to find one factor. We first try $x - 1$.

$$\begin{array}{r|rrrr} 1 & 1 & -3 & -6 & 8 \\ & & 1 & -2 & -8 \\ \hline & 1 & -2 & -8 & 0 \end{array}$$

Since $f(1) = 0$, $x - 1$ is a factor of $f(x)$. We have $f(x) = (x - 1)(x^2 - 2x - 8)$. Factoring the trinomial we get $f(x) = (x - 1)(x - 4)(x + 2)$.

45. $\mathbf{A} = [x]$

Write the augmented matrix.

$$\begin{bmatrix} x & | & 1 \end{bmatrix}$$

Multiply by $\dfrac{1}{x}$.

$$\begin{bmatrix} 1 & | & \dfrac{1}{x} \end{bmatrix}$$

Then $\mathbf{A}^{-1}$ exists if and only if $x \neq 0$. $\mathbf{A}^{-1} = \begin{bmatrix} \dfrac{1}{x} \end{bmatrix}$.

47. $\mathbf{A} = \begin{bmatrix} 0 & 0 & x \\ 0 & y & 0 \\ z & 0 & 0 \end{bmatrix}$

Write the augmented matrix.

$$\left[\begin{array}{ccc|ccc} 0 & 0 & x & 1 & 0 & 0 \\ 0 & y & 0 & 0 & 1 & 0 \\ z & 0 & 0 & 0 & 0 & 1 \end{array} \right]$$

Interchange row 1 and row 3.

$$\begin{bmatrix} z & 0 & 0 & 0 & 0 & 1 \\ 0 & y & 0 & 0 & 1 & 0 \\ 0 & 0 & x & 1 & 0 & 0 \end{bmatrix}$$

Multiply row 1 by $\dfrac{1}{z}$, row 2 by $\dfrac{1}{y}$, and row 3 by $\dfrac{1}{x}$.

$$\begin{bmatrix} 1 & 0 & 0 & 0 & 0 & \dfrac{1}{z} \\ 0 & 1 & 0 & 0 & \dfrac{1}{y} & 0 \\ 0 & 0 & 1 & \dfrac{1}{x} & 0 & 0 \end{bmatrix}$$

Then $\mathbf{A}^{-1}$ exists if and only if $x \neq 0$ and $y \neq 0$ and $z \neq 0$, or if and only if $xyz \neq 0$.

$$\mathbf{A}^{-1} = \begin{bmatrix} 0 & 0 & \dfrac{1}{z} \\ 0 & \dfrac{1}{y} & 0 \\ \dfrac{1}{x} & 0 & 0 \end{bmatrix}$$

Exercise Set 8.6

1. Graph (f) is the graph of $y > x$.

3. Graph (h) is the graph of $y \leq x - 3$.

5. Graph (g) is the graph of $2x + y < 4$.

7. Graph (b) is the graph of $2x - 5y > 10$.

9. Graph: $y > 2x$

   1. We first graph the related equation $y = 2x$. We draw the line dashed since the inequality symbol is $>$.

   2. To determine which half-plane to shade, test a point not on the line. We try $(1,1)$ and substitute:

   $$\begin{array}{c|c} \multicolumn{2}{c}{y > 2x} \\ \hline 1 \ ? \ 2 \cdot 1 & \\ 1 & 2 \quad \text{FALSE} \end{array}$$

   Since $1 > 2$ is false, $(1,1)$ is not a solution, nor are any points in the half-plane containing $(1,1)$. The points in the opposite half-plane are solutions, so we shade that half-plane and obtain the graph.

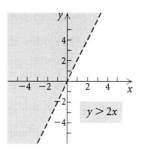

11. Graph: $y + x \geq 0$

   1. First graph the related equation $y + x = 0$. Draw the line solid since the inequality is $\geq$.

   2. Next determine which half-plane to shade by testing a point not on the line. Here we use $(2,2)$ as a check.

   $$\begin{array}{c|c} \multicolumn{2}{c}{y + x \geq 0} \\ \hline 2 + 2 \ ? \ 0 & \\ 4 & 0 \quad \text{TRUE} \end{array}$$

   Since $4 \geq 0$ is true, $(2,2)$ is a solution. Thus shade the half-plane containing $(2,2)$.

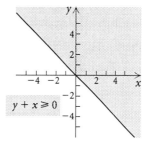

13. Graph: $y > x - 3$

   1. We first graph the related equation $y = x - 3$. Draw the line dashed since the inequality symbol is $>$.

   2. To determine which half-plane to shade, test a point not on the line. We try $(0,0)$.

   $$\begin{array}{c|c} \multicolumn{2}{c}{y > x - 3} \\ \hline 0 \ ? \ 0 - 3 & \\ 0 & -3 \quad \text{TRUE} \end{array}$$

   Since $0 > -3$ is true, $(0,0)$ is a solution. Thus we shade the half-plane containing $(0,0)$.

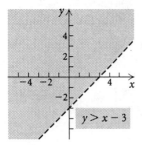

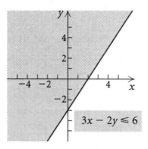

15. Graph: $x + y < 4$

    1. First graph the related equation $x + y = 4$. Draw the line dashed since the inequality is $<$.

    2. To determine which half-plane to shade, test a point not on the line. We try $(0, 0)$.

$$\frac{x + y < 4}{\begin{array}{c|c} 0 + 0 \ ? \ 4 \\ \hline 0 & 4 \quad \text{TRUE} \end{array}}$$

    Since $0 < 4$ is true, $(0, 0)$ is a solution. Thus shade the half-plane containing $(0, 0)$.

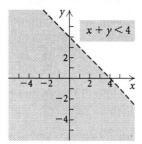

17. Graph: $3x - 2y \leq 6$

    1. First graph the related equation $3x - 2y = 6$. Draw the line solid since the inequality is $\leq$.

    2. To determine which half-plane to shade, test a point not on the line. We try $(0, 0)$.

$$\frac{3x - 2y \leq 6}{\begin{array}{c|c} 3(0) - 2(0) \ ? \ 6 \\ \hline 0 & 6 \quad \text{TRUE} \end{array}}$$

    Since $0 \leq 6$ is true, $(0, 0)$ is a solution. Thus shade the half-plane containing $(0, 0)$.

19. Graph: $3y + 2x \geq 6$

    1. First graph the related equation $3y + 2x = 6$. Draw the line solid since the inequality is $\geq$.

    2. To determine which half-plane to shade, test a point not on the line. We try $(0, 0)$.

$$\frac{3y + 2x \geq 6}{\begin{array}{c|c} 3 \cdot 0 + 2 \cdot 0 \ ? \ 6 \\ \hline 0 & 6 \quad \text{FALSE} \end{array}}$$

    Since $0 \geq 6$ is false, $(0, 0)$ is not a solution. We shade the half-plane which does not contain $(0, 0)$.

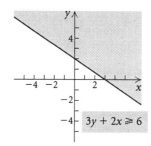

21. Graph: $3x - 2 \leq 5x + y$

$$-2 \leq 2x + y \quad \text{Adding } -3x$$

    1. First graph the related equation $2x + y = -2$. Draw the line solid since the inequality is $\leq$.

    2. To determine which half-plane to shade, test a point not on the line. We try $(0, 0)$.

$$\frac{2x + y \geq -2}{\begin{array}{c|c} 2(0) + 0 \ ? \ -2 \\ \hline 0 & -2 \quad \text{TRUE} \end{array}}$$

    Since $0 \geq -2$ is true, $(0, 0)$ is a solution. Thus shade the half-plane containing the origin.

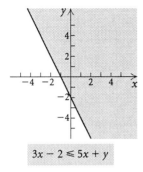

$$3x - 2 \leqslant 5x + y$$

23. Graph: $x < -4$

1. We first graph the related equation $x = -4$. Draw the line dashed since the inequality is $<$.

2. To determine which half-plane to shade, test a point not on the line. We try $(0, 0)$.

$$\frac{x < -4}{0 \ ? \ -4 \ \text{FALSE}}$$

Since $0 < -4$ is false, $(0, 0)$ is not a solution. Thus, we shade the half-plane which does not contain the origin.

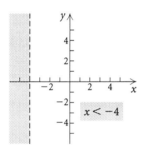

$$x < -4$$

25. Graph: $y > -3$

1. First we graph the related equation $y = -3$. Draw the line dashed since the inequality is $>$.

2. To determine which half-plane to shade we test a point not on the line. We try $(0, 0)$.

$$\frac{y > -3}{0 \ ? \ -3 \ \text{TRUE}}$$

Since $0 > -3$ is true, $(0, 0)$ is a solution. We shade the half-plane containing $(0, 0)$.

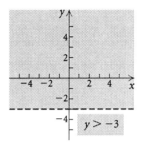

$$y > -3$$

27. Graph: $-4 < y < -1$

This is a conjunction of two inequalities

$$4 < y \quad and \quad y < -1.$$

We can graph $-4 < y$ and $y < -1$ separately and then graph the intersection, or region in both solution sets.

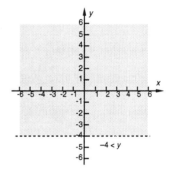

$$-4 < y$$

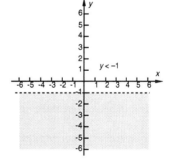

$$y < -1$$

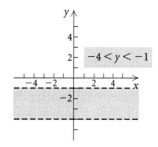

$$-4 < y < -1$$

29. Graph: $y \geq |x|$

  1. Graph the related equation $y = |x|$. Draw the line solid since the inequality symbol is $\geq$.

  2. To determine the region to shade, observe that the solution set consists of all ordered pairs (x, y) where the second coordinate is greater than or equal to the absolute value of the first coordinate. We see that the solutions are the points on or above the graph of $y = |x|$.

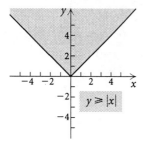

31. Graph (f) is the correct graph.

33. Graph (a) is the correct graph.

35. Graph (b) is the correct graph.

37. Graph: $y \leq x$,

   $y \geq 3 - x$

We graph the related equations $y = x$ and $y = 3 - x$ using solid lines. The arrows at the ends of the lines indicate the half-plane containing the solution set for each inequality. Shade the region common to both solution sets.

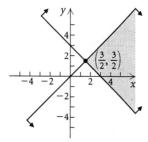

We find the vertex $\left(\dfrac{3}{2}, \dfrac{3}{2} \right)$ by solving the system

   $y = x$,
   $y = 3 - x$.

39. Graph: $y \geq x$,

   $y \leq x - 4$

We graph the related equations $y = x$ and $y = x - 4$ using solid lines. The arrows at the ends of the lines indicate the half-plane containing the solution set for

each inequality. There is not a region common to both solution sets, so there are no vertices.

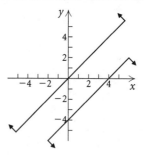

41. Graph: $y \geq -3$,

   $x \geq 1$

We graph the related equations $y = -3$ and $x = 1$ using solid lines. The arrows at the ends of the lines indicate the half-plane containing the solution set for each inequality. Shade the region common to both solution sets.

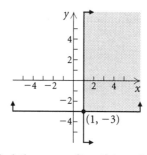

We find the vertex $(1, -3)$ by solving the system

   $y = -3$,
   $x = 1$.

43. Graph: $x \leq 3$,

   $y \geq 2 - 3x$

We graph the related equations $x = 3$ and $y = 2 - 3x$ using solid lines. The arrows at the ends of the lines indicate the half-plane containing the solution set for each inequality. Shade the region common to both solution sets.

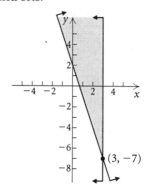

We find the vertex $(3, -7)$ by solving the system

$$x = 3,$$
$$y = 2 - 3x.$$

45. Graph: $x + y \leq 1,$
$$x - y \leq 2$$

We graph the related equations $x + y = 1$ and $x - y = 2$ using solid lines. The arrows at the ends of the lines indicate the half-plane containing the solution set for each inequality. Shade the region common to both solution sets.

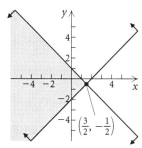

We find the vertex $\left(\dfrac{3}{2}, -\dfrac{1}{2}\right)$ by solving the system

$$x + y = 1,$$
$$x - y = 2.$$

47. Graph: $2y - x \leq 2,$
$$y + 3x \geq -1$$

We graph the related equations $2y - x = 2$ and $y + 3x = -1$ using solid lines. The arrows at the ends of the lines indicate the half-plane containing the solution set for each inequality. Shade the region common to both solution sets.

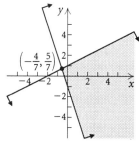

We find the vertex $\left(-\dfrac{4}{7}, \dfrac{5}{7}\right)$ by solving the system

$$2y - x = 2,$$
$$y + 3x = -1.$$

49. Graph: $x - y \leq 2,$
$$x + 2y \geq 8,$$
$$y \leq 4$$

We graph the related equations $x - y = 2$, $x + 2y = 8$, and $y = 4$ using solid lines. The arrows at the ends of the lines indicate the half-plane containing the solution set for each inequality. Shade the region common to all three solution sets.

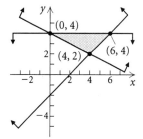

We find the vertex $(0, 4)$ by solving the system

$$x + 2y = 8,$$
$$y = 4.$$

We find the vertex $(6, 4)$ by solving the system

$$x - y = 2,$$
$$y = 4.$$

We find the vertex $(4, 2)$ by solving the system

$$x - y = 2,$$
$$x + 2y = 8.$$

51. Graph: $4x - 3y \geq -12,$
$$4x + 3y \geq -36,$$
$$y \leq 0,$$
$$x \leq 0$$

Shade the intersection of the graphs of the four inequalities.

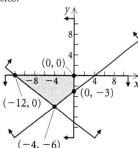

We find the vertex $(-12, 0)$ by solving the system

$$4y + 3x = -36,$$
$$y = 0.$$

We find the vertex $(0, 0)$ by solving the system

$$y = 0,$$
$$x = 0.$$

We find the vertex $(0, -3)$ by solving the system

$$4y - 3x = -12,$$
$$x = 0.$$

We find the vertex $(-4, -6)$ by solving the system

$$4y - 3x = -12,$$
$$4y + 3x = -36.$$

53. Graph: $3x + 4y \geq 12,$
$$5x + 6y \leq 30,$$
$$1 \leq x \leq 3$$

Shade the intersection of the graphs of the given inequalities.

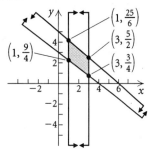

We find the vertex $\left(1, \frac{25}{6}\right)$ by solving the system

$$5x + 6y = 30,$$
$$x = 1.$$

We find the vertex $\left(3, \frac{5}{2}\right)$ by solving the system

$$5x + 6y = 30,$$
$$x = 3.$$

We find the vertex $\left(3, \frac{3}{4}\right)$ by solving the system

$$3x + 4y = 12,$$
$$x = 3.$$

We find the vertex $\left(1, \frac{9}{4}\right)$ by solving the system

$$3x + 4y = 12,$$
$$x = 1.$$

55. Find the maximum and minimum values of
$$P = 17x - 3y + 60, \text{ subject to}$$
$$6x + 8y \leq 48,$$
$$0 \leq y \leq 4,$$
$$0 \leq x \leq 3.$$

Graph the system of inequalities and determine the vertices.

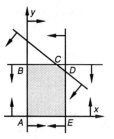

Vertex A: $(0, 0)$

Vertex B:
We solve the system $x = 0$ and $y = 4$. The coordinates of point B are $(0, 4)$.

Vertex C:
We solve the system $6x + 8y = 48$ and $y = 4$. The coordinates of point C are $\left(\frac{8}{3}, 4\right)$.

Vertex D:
We solve the system $6x + 8y = 48$ and $y = 7$. The coordinates of point D are $\left(7, \frac{3}{4}\right)$.

Vertex E:
We solve the system $x = 7$ and $y = 0$. The coordinates of point E are $(7, 0)$.

Evaluate the objective function P at each vertex.

Vertex	$P = 17x - 3y + 60$
$A(0, 0)$	$17 \cdot 0 - 3 \cdot 0 + 60 = 60$
$B(0, 4)$	$17 \cdot 0 - 3 \cdot 4 + 60 = 48$
$C\left(\frac{8}{3}, 3\right)$	$17 \cdot \frac{8}{3} - 3 \cdot 4 + 60 = 66\frac{2}{3}$
$D\left(7, \frac{3}{4}\right)$	$17 \cdot 7 - 3 \cdot \frac{3}{4} + 60 = 176\frac{3}{4}$
$E(7, 0)$	$17 \cdot 7 - 3 \cdot 0 + 60 = 179$

The maximum value of P is 179 when $x = 7$ and $y = 0$.

The minimum value of P is 48 when $x = 0$ and $y = 4$.

57. Find the maximum and minimum values of
$$F = 5x + 36y, \text{ subject to}$$
$$5x + 3y \leq 34,$$
$$3x + 5y \leq 30,$$
$$x \geq 0,$$
$$y \geq 0.$$

Graph the system of inequalities and find the vertices.

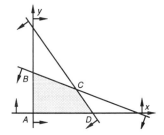

Vertex A: $(0,0)$

Vertex B:

We solve the system $3x + 5y = 30$ and $x = 0$. The coordinates of point B are $(0,6)$.

Vertex C:

We solve the system $5x + 3y = 34$ and $3x + 5y = 30$. The coordinates of point C are $(5,3)$.

Vertex D:

We solve the system $5x + 3y = 34$ and $y = 0$. The coordinates of point D are $\left(\frac{34}{5}, 0\right)$.

Evaluate the objective function F at each vertex.

Vertex	$F = 5x + 36y$
$A(0,0)$	$5 \cdot 0 + 36 \cdot 0+ = 0$
$B(0,6)$	$5 \cdot 0 + 36 \cdot 6 = 216$
$C(5,3)$	$5 \cdot 5 + 39 \cdot 3 = 133$
$D\left(\frac{34}{5}, 0\right)$	$5 \cdot \frac{34}{5} + 36 \cdot 0 = 34$

The maximum value of F is 216 when $x = 0$ and $y = 6$.

The minimum value of F is 0 when $x = 0$ and $y = 0$.

59. Let $x =$ the number of jumbo biscuits and $y =$ the number of regular biscuits to be made per day. The income I is given by

$$I = 0.10x + 0.08y$$

subject to the constrains

$$x + y \leq 200,$$
$$2x + y \leq 300,$$
$$x \geq 0,$$
$$y \geq 0.$$

We graph the system of inequalities, determine the vertices, and find the value if I at each vertex.

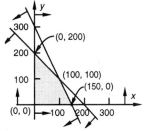

Vertex	$I = 0.10x + 0.08y$
$(0,0)$	$0.10(0) + 0.08(0) = 0$
$(0,200)$	$0.10(0) + 0.08(200) = 16$
$(100,100)$	$0.10(100) + 0.08(100) = 18$
$(150,0)$	$0.10(150) + 0.08(0) = 15$

The company will have a maximum income of $18 when 100 of each type of biscuit is made.

61. Let $x =$ the number of units of lumber and $y =$ the number of units of plywood produced per week. The profit P is given by

$$P = 20x + 30y$$

subject to the constraints

$$x + y \leq 400,$$
$$x \geq 100,$$
$$y \geq 150.$$

We graph the system of inequalities, determine the vertices and find the value of P at each vertex.

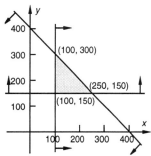

Vertex	$P = 20x + 30y$
$(100,150)$	$20 \cdot 100 + 30 \cdot 150 = 6500$
$(100,300)$	$20 \cdot 100 + 30 \cdot 300 = 11,000$
$(250,150)$	$20 \cdot 250 + 30 \cdot 150 = 9500$

The maximum profit of $11,000 is achieved by producing 100 units of lumber and 300 units of plywood.

63. Let $x =$ the number of sacks of soybean meal to be used and $y =$ the number of sacks of oats. The minimum cost is given by

$$C = 15x + 5y$$

subject to the constraints

$$50x + 15y \geq 120,$$
$$8x + 5y \geq 24,$$
$$5x + y \geq 10,$$
$$x \geq 0,$$
$$y \geq 0.$$

Graph the system of inequalities, determine the vertices, and find the value of C at each vertex.

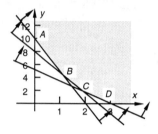

Vertex	$C = 15x + 5y$
$A(0,10)$	$15 \cdot 0 + 5 \cdot 10 = 50$
$B\left(\dfrac{6}{5},4\right)$	$15 \cdot \dfrac{6}{5} + 5 \cdot 4 = 38$
$C\left(\dfrac{24}{13},\dfrac{24}{13}\right)$	$15 \cdot \dfrac{24}{13} + 5 \cdot \dfrac{24}{13} = 36\dfrac{12}{13}$
$D(3,0)$	$15 \cdot 3 + 5 \cdot 0 = 45$

The minimum cost of $\$36\dfrac{12}{13}$ is achieved by using $\dfrac{24}{13}$, or $1\dfrac{11}{13}$ sacks of soybean meal and $\dfrac{24}{13}$, or $1\dfrac{11}{13}$ sacks of oats.

65. Let $x =$ the amount invested in corporate bonds and $y =$ the amount invested in municipal bonds. The income I is given by

$$I = 0.08x + 0.075y$$

subject to the constrains

$$x + y \leq 40,000,$$
$$6000 \leq x \leq 22,000,$$
$$y \leq 30,000.$$

We graph the system of inequalities, determine the vertices, and find the value of I at each vertex.

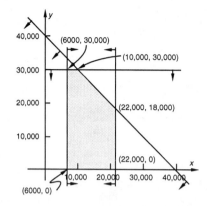

Vertex	$I = 0.08x + 0.075y$
$(6000,0)$	480
$(6000,30,000)$	2730
$(10,000,30,000)$	3050
$(22,000,18,000)$	3110
$(22,000,0)$	1760

The maximum income of $\$3110$ occurs when $\$22,000$ is invested in corporate bonds and $\$18,000$ is invested in municipal bonds.

67. Let $x =$ the number of P_1 airplanes and $y =$ the number of P_2 airplanes to be used. The operating cost C, in thousands of dollars, is given by

$$C = 12x + 10y$$

subject to the constraints

$$40x + 80y \geq 2000,$$
$$40x + 30y \geq 1500,$$
$$120x + 40y \geq 2400,$$
$$x \geq 0,$$
$$y \geq 0.$$

Graph the system of inequalities, determine the vertices, and find the value of C at each vertex.

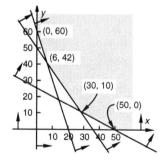

Vertex	$C = 12x + 10y$
$(0, 60)$	$12 \cdot 0 + 10 \cdot 60 = 600$
$(6, 42)$	$12 \cdot 6 + 10 \cdot 42 = 492$
$(30, 10)$	$12 \cdot 30 + 10 \cdot 10 = 460$
$(50, 0)$	$12 \cdot 50 + 10 \cdot 0 = 600$

The minimum cost of $460 thousand is achieved using 30 P_1's and 10 P_2's.

69. Let $x =$ the number of knit suits and $y =$ the number of worsted suits made. The profit is given by

$$P = 34x + 31y$$

subject to

$$2x + 4y \leq 20,$$
$$4x + 2y \leq 16,$$
$$x \geq 0,$$
$$y \geq 0.$$

Graph the system of inequalities, determine the vertices, and find the value of P at each vertex.

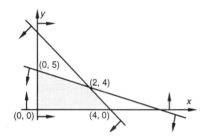

Vertex	$P = 34x + 31y$
$(0, 0)$	$34 \cdot 0 + 31 \cdot 0 = 0$
$(0, 5)$	$34 \cdot 0 + 31 \cdot 5 = 155$
$(2, 4)$	$34 \cdot 2 + 31 \cdot 4 = 192$
$(4, 0)$	$34 \cdot 4 + 31 \cdot 0 = 136$

The maximum profit per day is $192 when 2 knit suits and 4 worsted suits are made.

71. Let $x =$ the number of pounds of meat and $y =$ the number of pounds of cheese in the diet in a week. The cost is given by

$$C = 3.50x + 4.60y$$

subject to

$$2x + 3y \geq 12,$$
$$2x + y \geq 6,$$
$$x \geq 0,$$
$$y \geq 0.$$

Graph the system of inequalities, determine the vertices, and find the value of C at each vertex.

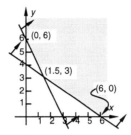

Vertex	$C = 3.50x + 4.60y$
$(0, 6)$	$3.50(0) + 4.60(6) = 27.60$
$(1.5, 3)$	$3.50(1.5) + 4.60(3) = 19.05$
$(6, 0)$	$3.50(6) + 4.60(0) = 21.00$

The minimum weekly cost of $19.05 is achieved when 1.5 lb of meat and 3 lb of cheese are used.

73. Let $x =$ the number of animal A and $y =$ the number of animal B. The total number of animals is given by

$$T = x + y$$

subject to

$$x + 0.2y \leq 600,$$
$$0.5x + y \leq 525,$$
$$x \geq 0,$$
$$y \geq 0.$$

Graph the system of inequalities, determine the vertices, and find the value of T at each vertex.

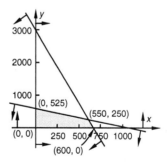

Vertex	$T = x + y$
$(0, 0)$	$0 + 0 = 0$
$(0, 525)$	$0 + 525 = 525$
$(550, 250)$	$550 + 250 = 800$
$(600, 0)$	$600 + 0 = 600$

The maximum total number of 800 is achieved when there are 550 of A and 250 of B.

75. Discussion and Writing

77. $-5 \le x + 2 < 4$

$-7 \le x < 2$ Subtracting 2

The solution set is $\{x | -7 \le x < 2\}$, or $[-7, 2)$.

79. $x^2 - 2x \le 3$ Polynomial inequality

$x^2 - 2x - 3 \le 0$

$x^2 - 2x - 3 = 0$ Related equation

$(x + 1)(x - 3) = 0$ Factoring

Using the principle of zero products or by observing the graph of $y = x^2 - 2x - 3$, we see that the solutions of the related equation are -1 and 3. These numbers divide the x-axis into the intervals $(-\infty, -1)$, $(-1, 3)$, and $(3, \infty)$. We let $f(x) = x^2 - 2x - 3$ and test a value in each interval.

$(-\infty, -1)$: $f(-2) = 5 > 0$

$(-1, 3)$: $f(0) = -3 < 0$

$(3, \infty)$: $f(4) = 5 > 0$

Function values are negative on $(-1, 3)$. This can also be determined from the graph of $y = x^2 - 2x - 3$. Since the inequality symbol is $\le$, the endpoints of the interval must be included in the solution set. It is $[-1, 3]$.

81. Graph: $y \ge x^2 - 2$,

$y \le 2 - x^2$

First graph the related equations $y = x^2 - 2$ and $y = 2 - x^2$ using solid lines. The solution set consists of the region above the graph of $y = x^2 - 2$ and below the graph of $y = 2 - x^2$.

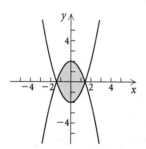

83. See the answer section in the text.

85. See the answer section in the text.

87. Let $x =$ the number of less expensive speaker assemblies and $y =$ the number of more expensive assemblies. The income is given by

$$I = 350x + 600y$$

subject to

$y \le 44$

$x + y \le 60,$

$x + 2y \le 90,$

$x \ge 0,$

$y \ge 0.$

Graph the system of inequalities, determine the vertices, and find the value of I at each vertex.

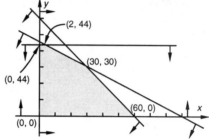

Vertex	$I = 350x + 600y$
$(0,0)$	$350 \cdot 0 + 600 \cdot 0 = 0$
$(0,44)$	$350 \cdot 0 + 600 \cdot 44 = 26,400$
$(2,44)$	$350 \cdot 2 + 600 \cdot 44 = 27,100$
$(30,30)$	$350 \cdot 30 + 600 \cdot 30 = 28,500$
$(60,0)$	$350 \cdot 60 + 600 \cdot 0 = 21,000$

The maximum income of \$28,500 is achieved when 30 less expensive and 30 more expensive assemblies are made.

Exercise Set 8.7

1. $\dfrac{x + 7}{(x - 3)(x + 2)} = \dfrac{A}{x - 3} + \dfrac{B}{x + 2}$

$\dfrac{x + 7}{(x - 3)(x + 2)} = \dfrac{A(x + 2) + B(x - 3)}{(x - 3)(x + 2)}$ Adding

Equate the numerators:

$x + 7 = A(x + 2) + B(x - 3)$

Let $x + 2 = 0$, or $x = -2$. Then we get

$-2 + 7 = 0 + B(-2 - 3)$

$5 = -5B$

$-1 = B$

Next let $x - 3 = 0$, or $x = 3$. Then we get

$3 + 7 = A(3 + 2) + 0$

$10 = 5A$

$2 = A$

The decomposition is as follows:

$$\frac{2}{x - 3} - \frac{1}{x + 2}$$

3. $\dfrac{7x-1}{6x^2-5x+1}$

$= \dfrac{7x-1}{(3x-1)(2x-1)}$ Factoring the denominator

$= \dfrac{A}{3x-1} + \dfrac{B}{2x-1}$

$= \dfrac{A(2x-1)+B(3x-1)}{(3x-1)(2x-1)}$ Adding

Equate the numerators:

$7x-1 = A(2x-1)+B(3x-1)$

Let $2x-1=0$, or $x=\dfrac{1}{2}$. Then we get

$$7\left(\frac{1}{2}\right)-1 = 0 + B\left(3\cdot\frac{1}{2}-1\right)$$

$$\frac{5}{2} = \frac{1}{2}B$$

$$5 = B$$

Next let $3x-1=0$, or $x=\dfrac{1}{3}$. We get

$$7\left(\frac{1}{3}\right)-1 = A\left(2\cdot\frac{1}{3}-1\right)$$

$$\frac{7}{3}-1 = A\left(\frac{2}{3}-1\right)$$

$$\frac{4}{3} = -\frac{1}{3}A$$

$$-4 = A$$

The decomposition is as follows:

$$-\frac{4}{3x-1} + \frac{5}{2x-1}$$

5. $\dfrac{3x^2-11x-26}{(x^2-4)(x+1)}$

$= \dfrac{3x^2-11x-26}{(x+2)(x-2)(x+1)}$ Factoring the denominator

$= \dfrac{A}{x+2} + \dfrac{B}{x-2} + \dfrac{C}{x+1}$

$= \dfrac{A(x-2)(x+1)+B(x+2)(x+1)+C(x+2)(x-2)}{(x+2)(x-2)(x+1)}$

    Adding

Equate the numerators:

$3x^2-11x-26 = A(x-2)(x+1)+$
$\qquad\qquad B(x+2)(x+1)+C(x+2)(x-2)$

Let $x+2=0$ or $x=-2$. Then we get

$3(-2)^2-11(-2)-26 = A(-2-2)(-2+1)+0+0$

$12+22-26 = A(-4)(-1)$

$8 = 4A$

$2 = A$

Next let $x-2=0$, or $x=2$. Then, we get

$3\cdot 2^2-11\cdot 2-26 = 0 + B(2+2)(2+1)+0$

$12-22-26 = B\cdot 4\cdot 3$

$-36 = 12B$

$-3 = B$

Finally let $x+1=0$, or $x=-1$. We get

$3(-1)^2-11(-1)-26 = 0+0+C(-1+2)(-1-2)$

$3+11-26 = C(1)(-3)$

$-12 = -3C$

$4 = C$

The decomposition is as follows:

$$\frac{2}{x+2} - \frac{3}{x-2} + \frac{4}{x+1}$$

7. $\dfrac{9}{(x+2)^2(x-1)}$

$= \dfrac{A}{x+2} + \dfrac{B}{(x+2)^2} + \dfrac{C}{x-1}$

$= \dfrac{A(x+2)(x-1)+B(x-1)+C(x+2)^2}{(x+2)^2(x-1)}$

    Adding

Equate the numerators:

$9 = A(x+2)(x-1)+B(x-1)+C(x+2)^2$ (1)

Let $x-1=0$, or $x=1$. Then, we get

$9 = 0+0+C(1+2)^2$

$9 = 9C$

$1 = C$

Next let $x+2=0$, or $x=-2$. Then, we get

$9 = 0+B(-2-1)+0$

$9 = -3B$

$-3 = B$

To find A we first simplify equation (1).

$9 = A(x^2+x-2)+B(x-1)+C(x^2+4x+4)$

$\quad = Ax^2+Ax-2A+Bx-B+Cx^2+4Cx+4C$

$\quad = (A+C)x^2+(A+B+4C)x+(-2A-B+4C)$

Then we equate the coefficients of x^2.

$0 = A+C$

$0 = A+1$ Substituting 1 for C

$-1 = A$

The decomposition is as follows:

$$-\frac{1}{x+2} - \frac{3}{(x+2)^2} + \frac{1}{x-1}$$

9.
$$\frac{2x^2 + 3x + 1}{(x^2 - 1)(2x - 1)}$$

$$= \frac{2x^2 + 3x + 1}{(x + 1)(x - 1)(2x - 1)} \quad \text{Factoring the denominator}$$

$$= \frac{A}{x + 1} + \frac{B}{x - 1} + \frac{C}{2x - 1}$$

$$= \frac{A(x-1)(2x-1) + B(x+1)(2x-1) + C(x+1)(x-1)}{(x + 1)(x - 1)(2x - 1)}$$
$$\text{Adding}$$

Equate the numerators:
$$2x^2 + 3x + 1 = A(x - 1)(2x - 1) +$$
$$B(x + 1)(2x - 1) + C(x + 1)(x - 1)$$

Let $x + 1 = 0$, or $x = -1$. Then, we get
$$2(-1)^2 + 3(-1) + 1 = A(-1 - 1)[2(-1) - 1] + 0 + 0$$
$$2 - 3 + 1 = A(-2)(-3)$$
$$0 = 6A$$
$$0 = A$$

Next let $x - 1 = 0$, or $x = 1$. Then, we get
$$2 \cdot 1^2 + 3 \cdot 1 + 1 = 0 + B(1 + 1)(2 \cdot 1 - 1) + 0$$
$$2 + 3 + 1 = B \cdot 2 \cdot 1$$
$$6 = 2B$$
$$3 = B$$

Finally we let $2x - 1 = 0$, or $x = \frac{1}{2}$. We get
$$2\left(\frac{1}{2}\right)^2 + 3\left(\frac{1}{2}\right) + 1 = 0 + 0 + C\left(\frac{1}{2} + 1\right)\left(\frac{1}{2} - 1\right)$$
$$\frac{1}{2} + \frac{3}{2} + 1 = C \cdot \frac{3}{2} \cdot \left(-\frac{1}{2}\right)$$
$$3 = -\frac{3}{4}C$$
$$-4 = C$$

The decomposition is as follows:
$$\frac{3}{x - 1} - \frac{4}{2x - 1}$$

11.
$$\frac{x^4 - 3x^3 - 3x^2 + 10}{(x + 1)^2(x - 3)}$$

$$= \frac{x^4 - 3x^3 - 3x^2 + 10}{x^3 - x^2 - 5x - 3} \quad \text{Multiplying the denominator}$$

Since the degree of the numerator is greater than the degree of the denominator, we divide.

$$
\begin{array}{r}
x - 2 \\
x^3 - x^2 - 5x - 3 \overline{)\, x^4 - 3x^3 - 3x^2 + 0x + 10} \\
\underline{x^4 - x^3 - 5x^2 - 3x} \\
-2x^3 + 2x^2 + 3x + 10 \\
\underline{-2x^3 + 2x^2 + 10x + 6} \\
-7x + 4
\end{array}
$$

The original expression is thus equivalent to the following:
$$x - 2 + \frac{-7x + 4}{x^3 - x^2 - 5x - 3}$$

We proceed to decompose the fraction.
$$\frac{-7x + 4}{(x + 1)^2(x - 3)}$$

$$= \frac{A}{x + 1} + \frac{B}{(x + 1)^2} + \frac{C}{x - 3}$$

$$= \frac{A(x + 1)(x - 3) + B(x - 3) + C(x + 1)^2}{(x + 1)^2(x - 3)}$$
$$\text{Adding}$$

Equate the numerators:
$$-7x + 4 = A(x + 1)(x - 3) + B(x - 3) +$$
$$C(x + 1)^2 \qquad (1)$$

Let $x - 3 = 0$, or $x = 3$. Then, we get
$$-7 \cdot 3 + 4 = 0 + 0 + C(3 + 1)^2$$
$$-17 = 16C$$
$$-\frac{17}{16} = C$$

Let $x + 1 = 0$, or $x = -1$. Then, we get
$$-7(-1) + 4 = 0 + B(-1 - 3) + 0$$
$$11 = -4B$$
$$-\frac{11}{4} = B$$

To find A we first simplify equation (1).
$$-7x + 4$$
$$= A(x^2 - 2x - 3) + B(x - 3) + C(x^2 + 2x + 1)$$
$$= Ax^2 - 2Ax - 3A + Bx - 3B + Cx^2 - 2Cx + C$$
$$= (A + C)x^2 + (-2A + B - 2C)x + (-3A - 3B + C)$$

Then equate the coefficients of x^2.
$$0 = A + C$$

Substituting $-\frac{17}{16}$ for C, we get $A = \frac{17}{16}$.

The decomposition is as follows:
$$\frac{17/16}{x + 1} - \frac{11/4}{(x + 1)^2} - \frac{17/16}{x - 3}$$

The original expression is equivalent to the following:
$$x - 2 + \frac{17/16}{x + 1} - \frac{11/4}{(x + 1)^2} - \frac{17/16}{x - 3}$$

13.
$$\frac{-x^2 + 2x - 13}{(x^2 + 2)(x - 1)}$$

$$= \frac{Ax + B}{x^2 + 2} + \frac{C}{x - 1}$$

$$= \frac{(Ax + B)(x - 1) + C(x^2 + 2)}{(x^2 + 2)(x - 1)} \quad \text{Adding}$$

Equate the numerators:

$-x^2 + 2x - 13 = (Ax + B)(x - 1) + C(x^2 + 2)$ (1)

Let $x - 1 = 0$, or $x = 1$. Then we get

$$-1^2 + 2 \cdot 1 - 13 = 0 + C(1^2 + 2)$$
$$-1 + 2 - 13 = C(1 + 2)$$
$$-12 = 3C$$
$$-4 = C$$

To find A and B we first simplify equation (1).

$$-x^2 + 2x - 13$$
$$= Ax^2 - Ax + Bx - B + Cx^2 + 2C$$
$$= (A + C)x^2 + (-A + B)x + (-B + 2C)$$

Equate the coefficients of x^2:

$-1 = A + C$

Substituting -4 for C, we get $A = 3$.

Equate the constant terms:

$-13 = -B + 2C$

Substituting -4 for C, we get $B = 5$.

The decomposition is as follows:

$$\frac{3x + 5}{x^2 + 2} - \frac{4}{x - 1}$$

15. $\dfrac{6 + 26x - x^2}{(2x - 1)(x + 2)^2}$

$$= \frac{A}{2x - 1} + \frac{B}{x + 2} + \frac{C}{(x + 2)^2}$$

$$= \frac{A(x + 2)^2 + B(2x - 1)(x + 2) + C(2x - 1)}{(2x - 1)(x + 2)^2}$$

                                        Adding

Equate the numerators:

$6 + 26x - x^2 = A(x + 2)^2 + B(2x - 1)(x + 2) +$
                        $C(2x - 1)$ (1)

Let $2x - 1 = 0$, or $x = \dfrac{1}{2}$. Then, we get

$$6 + 26 \cdot \frac{1}{2} - \left(\frac{1}{2}\right)^2 = A\left(\frac{1}{2} + 2\right)^2 + 0 + 0$$
$$6 + 13 - \frac{1}{4} = A\left(\frac{5}{2}\right)^2$$
$$\frac{75}{4} = \frac{25}{4}A$$
$$3 = A$$

Let $x + 2 = 0$, or $x = -2$. We get

$$6 + 26(-2) - (-2)^2 = 0 + 0 + C[2(-2) - 1]$$
$$6 - 52 - 4 = -5C$$
$$-50 = -5C$$
$$10 = C$$

To find B we first simplify equation (1).

$6 + 26x - x^2$
$$= A(x^2 + 4x + 4) + B(2x^2 + 3x - 2) + C(2x - 1)$$
$$= Ax^2 + 4Ax + 4A + 2Bx^2 + 3Bx - 2B + 2Cx - C$$
$$= (A + 2B)x^2 + (4A + 3B + 2C)x + (4A - 2B - C)$$

Equate the coefficients of x^2:

$-1 = A + 2B$

Substituting 3 for A, we obtain $B = -2$.

The decomposition is as follows:

$$\frac{3}{2x - 1} - \frac{2}{x + 2} + \frac{10}{(x + 2)^2}$$

17. $\dfrac{6x^3 + 5x^2 + 6x - 2}{2x^2 + x - 1}$

Since the degree of the numerator is greater than the degree of the denominator, we divide.

$$
\begin{array}{r}
3x + 1 \\
2x^2 + x - 1 \overline{\smash{)}6x^3 + 5x^2 + 6x - 2} \\
\underline{6x^3 + 3x^2 - 3x} \\
2x^2 + 9x - 2 \\
\underline{2x^2 + x - 1} \\
8x - 1
\end{array}
$$

The original expression is equivalent to

$$3x + 1 + \frac{8x - 1}{2x^2 + x - 1}$$

We proceed to decompose the fraction.

$$\frac{8x - 1}{2x^2 + x - 1} = \frac{8x - 1}{(2x - 1)(x + 1)} \quad \text{Factoring the denominator}$$

$$= \frac{A}{2x - 1} + \frac{B}{x + 1}$$

$$= \frac{A(x + 1) + B(2x - 1)}{(2x - 1)(x + 1)} \quad \text{Adding}$$

Equate the numerators:

$8x - 1 = A(x + 1) + B(2x - 1)$

Let $x + 1 = 0$, or $x = -1$. Then we get

$$8(-1) - 1 = 0 + B[2(-1) - 1]$$
$$-8 - 1 = B(-2 - 1)$$
$$-9 = -3B$$
$$3 = B$$

Next let $2x - 1 = 0$, or $x = \dfrac{1}{2}$. We get

$$8\left(\frac{1}{2}\right) - 1 = A\left(\frac{1}{2} + 1\right) + 0$$
$$4 - 1 = A\left(\frac{3}{2}\right)$$
$$3 = \frac{3}{2}A$$
$$2 = A$$

The decomposition is

$$\frac{2}{2x-1} + \frac{3}{x+1}.$$

The original expression is equivalent to

$$3x + 1 + \frac{2}{2x-1} + \frac{3}{x+1}.$$

19.
$$\frac{2x^2 - 11x + 5}{(x-3)(x^2 + 2x - 5)}$$
$$= \frac{A}{x-3} + \frac{Bx + C}{x^2 + 2x - 5}$$
$$= \frac{A(x^2 + 2x - 5) + (Bx + C)(x - 3)}{(x-3)(x^2 + 2x - 5)} \quad \text{Adding}$$

Equate the numerators:

$$2x^2 - 11x + 5 = A(x^2 + 2x - 5) +$$
$$(Bx + C)(x - 3) \qquad (1)$$

Let $x - 3 = 0$, or $x = 3$. Then, we get

$$2 \cdot 3^2 - 11 \cdot 3 + 5 = A(3^2 + 2 \cdot 3 - 5) + 0$$
$$18 - 33 + 5 = A(9 + 6 - 5)$$
$$-10 = 10A$$
$$-1 = A$$

To find B and C, we first simplify equation (1).

$$2x^2 - 11x + 5 = Ax^2 + 2Ax - 5A + Bx^2 - 3Bx +$$
$$Cx - 3C$$
$$= (A + B)x^2 + (2A - 3B + C)x +$$
$$(-5A - 3C)$$

Equate the coefficients of x^2:

$2 = A + B$

Substituting -1 for A, we get $B = 3$.

Equate the constant terms:

$5 = -5A - 3C$

Substituting -1 for A, we get $C = 0$.

The decomposition is as follows:

$$-\frac{1}{x-3} + \frac{3x}{x^2 + 2x - 5}$$

21. $\dfrac{-4x^2 - 2x + 10}{(3x + 5)(x + 1)^2}$

The decomposition looks like

$$\frac{A}{3x + 5} + \frac{B}{x + 1} + \frac{C}{(x + 1)^2}.$$

Add and equate the numerators.

$$-4x^2 - 2x + 10$$
$$= A(x + 1)^2 + B(3x + 5)(x + 1) + C(3x + 5)$$
$$= A(x^2 + 2x + 1) + B(3x^2 + 8x + 5) + C(3x + 5)$$
or

$$-4x^2 - 2x + 10$$
$$= (A + 3B)x^2 + (2A + 8B + 3C)x + (A + 5B + 5C)$$

Then equate corresponding coefficients.

$$-4 = A + 3B \qquad \text{Coefficients of } x^2\text{-terms}$$
$$-2 = 2A + 8B + 3C \qquad \text{Coefficients of } x\text{-terms}$$
$$10 = A + 5B + 5C \qquad \text{Constant terms}$$

We solve this system of three equations and find $A = 5$, $B = -3$, $C = 4$.

The decomposition is

$$\frac{5}{3x + 5} - \frac{3}{x + 1} + \frac{4}{(x + 1)^2}.$$

23. $\dfrac{36x + 1}{12x^2 - 7x - 10} = \dfrac{36x + 1}{(4x - 5)(3x + 2)}$

The decomposition looks like

$$\frac{A}{4x - 5} + \frac{B}{3x + 2}.$$

Add and equate the numerators.

$$36x + 1 = A(3x + 2) + B(4x - 5)$$
$$\text{or } 36x + 1 = (3A + 4B)x + (2A - 5B)$$

Then equate corresponding coefficients.

$$36 = 3A + 4B \qquad \text{Coefficients of } x\text{-terms}$$
$$1 = 2A - 5B \qquad \text{Constant terms}$$

We solve this system of equations and find

$A = 8$ and $B = 3$.

The decomposition is

$$\frac{8}{4x - 5} + \frac{3}{3x + 2}.$$

25. $\dfrac{-4x^2 - 9x + 8}{(3x^2 + 1)(x - 2)}$

The decomposition looks like

$$\frac{Ax + B}{3x^2 + 1} + \frac{C}{x - 2}.$$

Add and equate the numerators.

$$-4x^2 - 9x + 8$$
$$= (Ax + B)(x - 2) + C(3x^2 + 1)$$
$$= Ax^2 - 2Ax + Bx - 2B + 3Cx^2 + C$$
or

$$-4x^2 - 9x + 8$$
$$= (A + 3C)x^2 + (-2A + B)x + (-2B + C)$$

Then equate corresponding coefficients.

$$-4 = A + 3C \qquad \text{Coefficients of } x^2\text{-terms}$$
$$-9 = -2A + B \qquad \text{Coefficients of } x\text{-terms}$$
$$8 = -2B + C \qquad \text{Constant terms}$$

We solve this system of equations and find

$A = 2$, $B = -5$, $C = -2$.

The decomposition is

$$\frac{2x-5}{3x^2+1} - \frac{2}{x-2}.$$

27. Discussion and Writing

29. Discussion and Writing

31. $f(x) = x^3 + x^2 - 3x - 2$

We use synthetic division to factor the polynomial. Using the possibilities found by the rational zeros theorem we find that $x+2$ is a factor:

$$\begin{array}{r|rrrr} -2 & 1 & 1 & -3 & -2 \\ & & -2 & 2 & 2 \\ \hline & 1 & -1 & -1 & 0 \end{array}$$

We have $x^3 + x^2 - 3x - 2 = (x+2)(x^2-x-1)$.

$$x^3 + x^2 - 3x - 2 = 0$$
$$(x+2)(x^2-x-1) = 0$$
$$x+2 = 0 \quad or \quad x^2 - x - 1 = 0$$

The solution of the first equation is -2. We use the quadratic formula to solve the second equation.

$$x = \frac{-b \pm \sqrt{b^2 - 4ac}}{2a}$$
$$= \frac{-(-1) \pm \sqrt{(-1)^2 - 4 \cdot 1 \cdot (-1)}}{2 \cdot 1}$$
$$= \frac{1 \pm \sqrt{5}}{2}$$

The solutions are -2, $\dfrac{1+\sqrt{5}}{2}$ and $\dfrac{1-\sqrt{5}}{2}$.

33. $f(x) = x^3 + 5x^2 + 5x - 3$

$$\begin{array}{r|rrrr} -3 & 1 & 5 & 5 & -3 \\ & & -3 & -6 & 3 \\ \hline & 1 & 2 & -1 & 0 \end{array}$$

$$x^3 + 5x^2 + 5x - 3 = 0$$
$$(x+3)(x^2 + 2x - 1) = 0$$
$$x+3 = 0 \quad or \quad x^2 + 2x - 1 = 0$$

The solution of the first equation is -3. We use the quadratic formula to solve the second equation.

$$x = \frac{-b \pm \sqrt{b^2 - 4ac}}{2a}$$
$$= \frac{-2 \pm \sqrt{2^2 - 4 \cdot 1 \cdot (-1)}}{2 \cdot 1} = \frac{-2 \pm \sqrt{8}}{2}$$
$$= \frac{-2 \pm 2\sqrt{2}}{2} = \frac{2(-1 \pm \sqrt{2})}{2}$$
$$= -1 \pm \sqrt{2}$$

The solutions are -3, $-1 + \sqrt{2}$, and $-1 - \sqrt{2}$.

35. $\dfrac{x}{x^4 - a^4}$

$$= \frac{x}{(x^2 + a^2)(x+a)(x-a)} \quad \text{Factoring the denominator}$$
$$= \frac{Ax + B}{x^2 + a^2} + \frac{C}{x+a} + \frac{D}{x-a}$$
$$= [(Ax+B)(x+a)(x-a) + C(x^2+a^2)(x-a) + D(x^2+a^2)(x+a)]/[(x^2+a^2)(x+a)(x-a)]$$

Equate the numerators:

$$x = (Ax+B)(x+a)(x-a) + C(x^2+a^2)(x-a) + D(x^2+a^2)(x+a)$$

Let $x - a = 0$, or $x = a$. Then, we get

$$a = 0 + 0 + D(a^2 + a^2)(a+a)$$
$$a = D(2a^2)(2a)$$
$$a = 4a^3 D$$
$$\frac{1}{4a^2} = D$$

Let $x + a = 0$, or $x = -a$. We get

$$-a = 0 + C[(-a)^2 + a^2](-a - a) + 0$$
$$-a = C(2a^2)(-2a)$$
$$-a = -4a^3 C$$
$$\frac{1}{4a^2} = C$$

Equate the coefficients of x^3:

$$0 = A + C + D$$

Substituting $\dfrac{1}{4a^2}$ for C and for D, we get

$$A = -\frac{1}{2a^2}.$$

Equate the constant terms:

$$0 = -Ba^2 - Ca^3 + Da^3$$

Substitute $\dfrac{1}{4a^2}$ for C and for D. Then solve for B.

$$0 = -Ba^2 - \frac{1}{4a^2} \cdot a^3 + \frac{1}{4a^2} \cdot a^3$$
$$0 = -Ba^2$$
$$0 = B$$

The decomposition is as follows:

$$-\frac{\frac{1}{2a^2}x}{x^2 + a^2} + \frac{\frac{1}{4a^2}}{x+a} + \frac{\frac{1}{4a^2}}{x-a}$$

37. $\dfrac{1 + \ln x^2}{(\ln x + 2)(\ln x - 3)^2} = \dfrac{1 + 2\ln x}{(\ln x + 2)(\ln x - 3)^2}$

Let $u = \ln x$. Then we have:

$$\frac{1+2u}{(u+2)(u-3)^2}$$

$$= \frac{A}{u+2} + \frac{B}{u-3} + \frac{C}{(u-3)^2}$$

$$= \frac{A(u3)^2 + B(u+2)(u-3) + C(u+2)}{(u+2)(u-3)^2}$$

Equate the numerators:

$$1+2u = A(u-3)^2 + B(u+2)(u-3) + C(u+2)$$

Let $u-3=0$, or $u=3$.

$$1+2\cdot 3 = 0 + 0 + C(5)$$

$$7 = 5C$$

$$\frac{7}{5} = C$$

Let $u+2=0$, or $u=-2$.

$$1+2(-2) = A(-2-3)^2 + 0 + 0$$

$$-3 = 25A$$

$$-\frac{3}{25} = A$$

To find B, we equate the coefficients of u^2:

$$0 = A + B$$

Substituting $-\dfrac{3}{25}$ for A and solving for B, we get

$B = \dfrac{3}{25}$.

The decomposition of $\dfrac{1+2u}{(u+2)(u-3)^2}$ is as follows:

$$-\frac{3}{25(u+2)} + \frac{3}{25(u-3)} + \frac{7}{5(u-3)^2}$$

Substituting $\ln x$ for u we get

$$-\frac{3}{25(\ln x + 2)} + \frac{3}{25(\ln x - 3)} + \frac{7}{5(\ln x - 3)^2}.$$

Chapter 9

Analytic Geometry Topics

Exercise Set 9.1

1. Graph (f) is the graph of $x^2 = 8y$.

3. Graph (b) is the graph of $(y-2)^2 = -3(x+4)$.

5. Graph (d) is the graph of $13x^2 - 8y - 9 = 0$.

7. $x^2 = 20y$

$\quad x^2 = 4 \cdot 5 \cdot y \qquad$ Writing $x^2 = 4py$

$\quad$ Vertex: $(0,0)$

$\quad$ Focus: $(0,5) \qquad\qquad [(0,p)]$

$\quad$ Directrix: $y = -5 \quad (y = -p)$

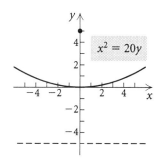

9. $y^2 = -6x$

$\quad y^2 = 4\left(-\dfrac{3}{2}\right)x \qquad$ Writing $y^2 = 4px$

$\quad$ Vertex: $(0,0)$

$\quad$ Focus: $\left(-\dfrac{3}{2},0\right) \qquad\qquad [(p,0)]$

$\quad$ Directrix: $x = -\left(-\dfrac{3}{2}\right) = \dfrac{3}{2} \quad (x = -p)$

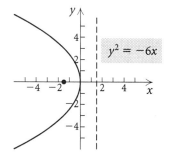

11. $x^2 - 4y = 0$

$\quad x^2 = 4y$

$\quad x^2 = 4 \cdot 1 \cdot y \qquad$ Writing $x^2 = 4py$

$\quad$ Vertex: $(0,0)$

$\quad$ Focus: $(0,1) \qquad\qquad [(0,p)]$

$\quad$ Directrix: $y = -1 \quad (y = -p)$

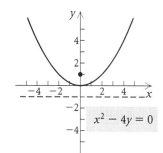

13. $x = 2y^2$

$\quad y^2 = \dfrac{1}{2}x$

$\quad y^2 = 4 \cdot \dfrac{1}{8} \cdot x \qquad$ Writing $y^2 = 4px$

$\quad$ Vertex: $(0,0)$

$\quad$ Focus: $\left(\dfrac{1}{8},0\right)$

$\quad$ Directrix: $x = -\dfrac{1}{8}$

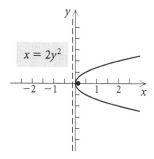

15. Since the directrix, $x = -4$, is a vertical line, the equation is of the form $(y-k)^2 = 4p(x-h)$. The focus, $(4,0)$, is on the x-axis so the line of symmetry is the x-axis and $p = 4$. The vertex, (h,k), is the point on the x-axis midway between the directrix and the focus. Thus, it is $(0,0)$. We have

$$(y - k)^2 = 4p(x - h)$$
$$(y - 0)^2 = 4 \cdot 4(x - 0) \quad \text{Substituting}$$
$$y^2 = 16x.$$

17. Since the directrix, $y = \pi$, is a horizontal line, the equation is of the form $(x - h)^2 = 4p(y - k)$. The focus, $(0, -\pi)$, is on the y-axis so the line of symmetry is the y-axis and $p = -\pi$. The vertex (h, k) is the point on the y-axis midway between the directrix and the focus. Thus, it is $(0, 0)$. We have

$$(x - h)^2 = 4p(y - k)$$
$$(x - 0)^2 = 4(-\pi)(y - 0) \quad \text{Substituting}$$
$$x^2 = -4\pi y$$

19. Since the directrix, $x = -4$, is a vertical line, the equation is of the form $(y - k)^2 = 4p(x - h)$. The focus, $(3, 2)$, is on the horizontal line $y = 2$, so the line of symmetry is $y = 2$. The vertex is the point on the line $y = 2$ that is midway between the directrix and the focus. That is, it is the midpoint of the segment from $(-4, 2)$ to $(3, 2)$: $\left(\dfrac{-4 + 3}{2}, \dfrac{2 + 2}{2} \right)$, or $\left(-\dfrac{1}{2}, 2 \right)$. Then $h = -\dfrac{1}{2}$ and the directrix is $x = h - p$, so we have

$$x = h - p$$
$$-4 = -\frac{1}{2} - p$$
$$-\frac{7}{2} = -p$$
$$\frac{7}{2} = p.$$

Now we find the equation of the parabola.
$$(y - k)^2 = 4p(x - h)$$
$$(y - 2)^2 = 4\left(\frac{7}{2}\right)\left[x - \left(-\frac{1}{2}\right)\right]$$
$$(y - 2)^2 = 14\left(x + \frac{1}{2}\right)$$

21. $\quad (x + 2)^2 = -6(y - 1)$

$[x - (-2)]^2 = 4\left(-\dfrac{3}{2}\right)(y - 1) \quad [(x-h)^2 = 4p(y-k)]$

Vertex: $(-2, 1) \qquad\qquad\qquad [(h, k)]$

Focus: $\left(-2, 1 + \left(-\dfrac{3}{2} \right) \right)$, or $\left(-2, -\dfrac{1}{2} \right)$
$$[(h, k + p)]$$

Directrix: $y = 1 - \left(-\dfrac{3}{2} \right) = \dfrac{5}{2} \quad (y = k - p)$

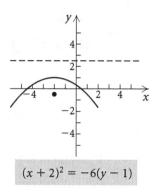

$(x + 2)^2 = -6(y - 1)$

23. $x^2 + 2x + 2y + 7 = 0$
$$x^2 + 2x = -2y - 7$$
$$(x^2 + 2x + 1) = -2y - 7 + 1 = -2y - 6$$
$$(x + 1)^2 = -2(y + 3)$$
$$[x - (-1)]^2 = 4\left(-\frac{1}{2} \right)[y - (-3)]$$
$$[(x - h)^2 = 4p(y - k)]$$

Vertex: $(-1, -3) \qquad [(h, k)]$

Focus: $\left(-1, -3 + \left(-\dfrac{1}{2} \right) \right)$, or $\left(-1, -\dfrac{7}{2} \right)$
$$[(h, k+p)]$$

Directrix: $y = -3 - \left(-\dfrac{1}{2} \right) = -\dfrac{5}{2} \quad (y = k - p)$

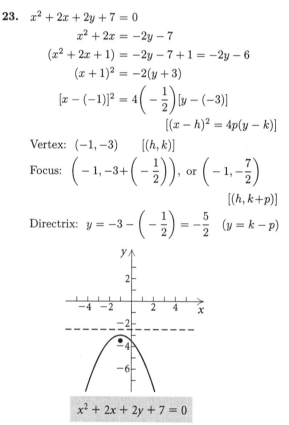

$x^2 + 2x + 2y + 7 = 0$

25. $x^2 - y - 2 = 0$
$$x^2 = y + 2$$
$$(x - 0)^2 = 4 \cdot \frac{1}{4} \cdot [y - (-2)]$$
$$[(x - h)^2 = 4p(y - k)]$$

Vertex: $(0, -2) \qquad [(h, k)]$

Focus: $\left(0, -2 + \dfrac{1}{4} \right)$, or $\left(0, -\dfrac{7}{4} \right) \; [(h, k + p)]$

Directrix: $y = -2 - \dfrac{1}{4} = -\dfrac{9}{4} \quad (y = k - p)$

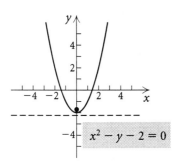

$$x^2 - y - 2 = 0$$

27.
$$y = x^2 + 4x + 3$$
$$y - 3 = x^2 + 4x$$
$$y - 3 + 4 = x^2 + 4x + 4$$
$$y + 1 = (x + 2)^2$$
$$4 \cdot \frac{1}{4} \cdot [y - (-1)] = [x - (-2)]^2$$
$$[(x - h)^2 = 4p(y - k)]$$

Vertex: $(-2, -1)$ $[(h, k)]$

Focus: $\left(-2, -1 + \frac{1}{4} \right)$, or $\left(-2, -\frac{3}{4} \right)$ $[(h, k + p)]$

Directrix: $y = -1 - \frac{1}{4} = -\frac{5}{4}$ $(y = k - p)$

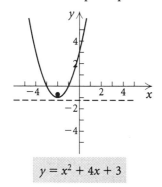

$$y = x^2 + 4x + 3$$

29.
$$y^2 - y - x + 6 = 0$$
$$y^2 - y = x - 6$$
$$y^2 - y + \frac{1}{4} = x - 6 + \frac{1}{4}$$
$$\left(y - \frac{1}{2} \right)^2 = x - \frac{23}{4}$$
$$\left(y - \frac{1}{2} \right)^2 = 4 \cdot \frac{1}{4} \left(x - \frac{23}{4} \right)$$
$$[(y - k)^2 = 4p(x - h)]$$

Vertex: $\left(\frac{23}{4}, \frac{1}{2} \right)$ $[(h, k)]$

Focus: $\left(\frac{23}{4} + \frac{1}{4}, \frac{1}{2} \right)$, or $\left(6, \frac{1}{2} \right)$ $[(h + p, k)]$

Directrix: $x = \frac{23}{4} - \frac{1}{4} = \frac{22}{4}$ or $\frac{11}{2}$ $(x = h - p)$

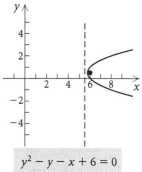

$$y^2 - y - x + 6 = 0$$

31. a) The vertex is $(0, 0)$. The focus is $(4, 0)$, so $p = 4$. The parabola has a horizontal axis of symmetry so the equation is of the form $y^2 = 4px$. We have
$$y^2 = 4px$$
$$y^2 = 4 \cdot 4 \cdot x$$
$$y^2 = 16x$$

   b) We make a drawing.

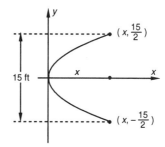

The depth of the satellite dish at the vertex is x where $\left(x, \frac{15}{2} \right)$ is a point on the parabola.
$$y^2 = 16x$$
$$\left(\frac{15}{2} \right)^2 = 16x \qquad \text{Substituting } \frac{15}{2} \text{ for } y$$
$$\frac{225}{4} = 16x$$
$$\frac{225}{64} = x, \text{ or}$$
$$3\frac{33}{64} = x$$

The depth of the satellite dish at the vertex is $3\frac{33}{64}$ ft.

33. We position a coordinate system with the origin at the vertex and the x-axis on the parabola's axis of symmetry.

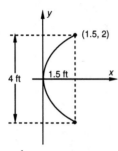

The parabola is of the form $y^2 = 4px$ and a point on the parabola is $\left(1.5, \dfrac{4}{2}\right)$, or $(1.5, 2)$.

$$y^2 = 4px$$
$$2^2 = 4 \cdot p \cdot (1.5) \quad \text{Substituting}$$
$$4 = 6p$$
$$\frac{4}{6} = p, \text{ or}$$
$$\frac{2}{3} = p$$

Since the focus is at $(p, 0)$, or $\left(\dfrac{2}{3}, 0\right)$, the focus is $\dfrac{2}{3}$ ft, or 8 in., from the vertex.

35. Discussion and Writing

37. $x^2 + 10x = x^2 + 10x + 25 \quad (\frac{1}{2} \cdot 10 = 5 \text{ and } 5^2 = 25)$
$\qquad\qquad = (x+5)^2$

39. $\quad (x-1)^2 + (y+2)^2 = 9$
$\quad (x-1)^2 + [y-(-2)]^2 = 3^2$

Center: $(1, -2)$

Radius: 3

41. A parabola with a vertical axis of symmetry has an equation of the type $(x-h)^2 = 4p(y-k)$.

Solve for p substituting $(-1, 2)$ for (h, k) and $(-3, 1)$ for (x, y).
$$[-3 - (-1)]^2 = 4p(1-2)$$
$$4 = -4p$$
$$-1 = p$$
The equation of the parabola is
$$[x - (-1)]^2 = 4(-1)(y-2), \text{ or}$$
$$(x+1)^2 = -4(y-2).$$

43. Vertex: $(0.867, 0.348)$

Focus: $(0.867, -0.191)$

Directrix: $y = 0.887$

45. Position a coordinate system as shown below with the y-axis on the parabola's axis of symmetry.

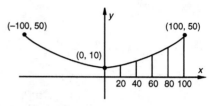

The equation of the parabola is of the form $(x-h)^2 = 4p(y-k)$. Substitute 100 for x, 50 for y, 0 for h, and 10 for k and solve for p.

$$(x-h)^2 = 4p(y-k)$$
$$(100-0)^2 = 4p(50-10)$$
$$10{,}000 = 160p$$
$$\frac{250}{4} = p$$

Then the equation is
$$x^2 = 4\left(\frac{250}{4}\right)(y-10), \text{ or}$$
$$x^2 = 250(y-10).$$

To find the lengths of the vertical cables, find y when $x = 0, 20, 40, 60, 80,$ and 100.

When $x = 0$: $0^2 = 250(y-10)$
$$0 = y - 10$$
$$10 = y$$

When $x = 20$: $20^2 = 250(y-10)$
$$400 = 250(y-10)$$
$$1.6 = y - 10$$
$$11.6 = y$$

When $x = 40$: $40^2 = 250(y-10)$
$$1600 = 250(y-10)$$
$$6.4 = y - 10$$
$$16.4 = y$$

When $x = 60$: $60^2 = 250(y-10)$
$$3600 = 250(y-10)$$
$$14.4 = y - 10$$
$$24.4 = y$$

When $x = 80$: $80^2 = 250(y-10)$
$$6400 = 250(y-10)$$
$$25.6 = y - 10$$
$$35.6 = y$$

When $x = 100$, we know from the given information that $y = 50$.

The lengths of the vertical cables are 10 ft, 11.6 ft, 16.4 ft, 24.4 ft, 35.6 ft, and 50 ft.

Exercise Set 9.2

1. Graph (b) is the graph of $x^2 + y^2 = 5$.

3. Graph (d) is the graph of $x^2 + y^2 - 6x + 2y = 6$.

5. Graph (a) is the graph of $x^2 + y^2 - 5x + 3y = 0$.

7. Complete the square twice.
$$x^2 + y^2 - 14x + 4y = 11$$
$$x^2 - 14x + y^2 + 4y = 11$$
$$x^2 - 14x + 49 + y^2 + 4y + 4 = 11 + 49 + 4$$
$$(x - 7)^2 + (y + 2)^2 = 64$$
$$(x - 7)^2 + [y - (-2)]^2 = 8^2$$

Center: $(7, -2)$

Radius: 8

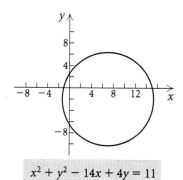

$$x^2 + y^2 - 14x + 4y = 11$$

9. Complete the square twice.
$$x^2 + y^2 + 4x - 6y - 12 = 0$$
$$x^2 + 4x + y^2 - 6y = 12$$
$$x^2 + 4x + 4 + y^2 - 6y + 9 = 12 + 4 + 9$$
$$(x + 2)^2 + (y - 3)^2 = 25$$
$$[x - (-2)]^2 + (y - 3)^2 = 5^2$$

Center: $(-2, 3)$

Radius: 5

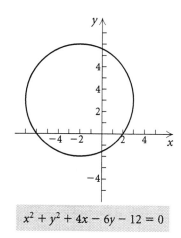

$$x^2 + y^2 + 4x - 6y - 12 = 0$$

11. Complete the square twice.
$$x^2 + y^2 + 6x - 10y = 0$$
$$x^2 + 6x + y^2 - 10y = 0$$
$$x^2 + 6x + 9 + y^2 - 10y + 25 = 0 + 9 + 25$$
$$(x + 3)^2 + (y - 5)^2 = 34$$
$$[x - (-3)]^2 + (y - 5)^2 = (\sqrt{34})^2$$

Center: $(-3, 5)$

Radius: $\sqrt{34}$

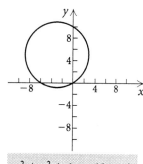

$$x^2 + y^2 + 6x - 10y = 0$$

13. Complete the square twice.
$$x^2 + y^2 - 9x = 7 - 4y$$
$$x^2 - 9x + y^2 + 4y = 7$$
$$x^2 - 9x + \frac{81}{4} + y^2 + 4y + 4 = 7 + \frac{81}{4} + 4$$
$$\left(x - \frac{9}{2}\right)^2 + (y + 2)^2 = \frac{125}{4}$$
$$\left(x - \frac{9}{2}\right)^2 + [y - (-2)]^2 = \left(\frac{5\sqrt{5}}{2}\right)^2$$

Center: $\left(\frac{9}{2}, -2\right)$

Radius: $\frac{5\sqrt{5}}{2}$

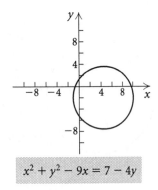

$$x^2 + y^2 - 9x = 7 - 4y$$

15. Graph (c) is the graph of $16x^2 + 4y^2 = 64$.

17. Graph (d) is the graph of $x^2 + 9y^2 - 6x + 90y = -225$.

19. $\dfrac{x^2}{4} + \dfrac{y^2}{1} = 1$

$\dfrac{x^2}{2^2} + \dfrac{y^2}{1^2} = 1$ Standard form

$a = 2,\ b = 1$

The major axis is horizontal, so the vertices are $(-2, 0)$ and $(2, 0)$. Since we know that $c^2 = a^2 - b^2$, we have $c^2 = 4 - 1 = 3$, so $c = \sqrt{3}$ and the foci are $(-\sqrt{3}, 0)$ and $(\sqrt{3}, 0)$.

To graph the ellipse, plot the vertices. Note also that since $b = 1$, the y-intercepts are $(0, -1)$ and $(0, 1)$. Plot these points as well and connect the four plotted points with a smooth curve.

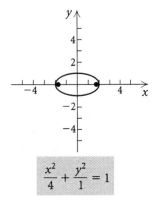

$$\dfrac{x^2}{4} + \dfrac{y^2}{1} = 1$$

21. $16x^2 + 9y^2 = 144$

$\dfrac{x^2}{9} + \dfrac{y^2}{16} = 1$ Dividing by 144

$\dfrac{x^2}{3^2} + \dfrac{y^2}{4^2} = 1$ Standard form

$a = 4,\ b = 3$

The major axis is vertical, so the vertices are $(0, -4)$ and $(0, 4)$. Since $c^2 = a^2 - b^2$, we have $c^2 = 16 - 9 = 7$, so $c = \sqrt{7}$ and the foci are $(0, -\sqrt{7})$ and $(0, \sqrt{7})$.

To graph the ellipse, plot the vertices. Note also that since $b = 3$, the x-intercepts are $(-3, 0)$ and $(3, 0)$. Plot these points as well and connect the four plotted points with a smooth curve.

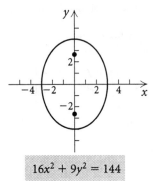

$$16x^2 + 9y^2 = 144$$

23. $2x^2 + 3y^2 = 6$

$\dfrac{x^2}{3} + \dfrac{y^2}{2} = 1$

$\dfrac{x^2}{(\sqrt{3})^2} + \dfrac{y^2}{(\sqrt{2})^2} = 1$

$a = \sqrt{3},\ b = \sqrt{2}$

The major axis is horizontal, so the vertices are $(-\sqrt{3}, 0)$ and $(\sqrt{3}, 0)$. Since $c^2 = a^2 - b^2$, we have $c^2 = 3 - 2 = 1$, so $c = 1$ and the foci are $(-1, 0)$ and $(1, 0)$.

To graph the ellipse, plot the vertices. Note also that since $b = \sqrt{2}$, the y-intercepts are $(0, -\sqrt{2})$ and $(0, \sqrt{2})$. Plot these points as well and connect the four plotted points with a smooth curve.

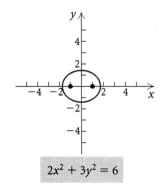

$$2x^2 + 3y^2 = 6$$

25.
$$4x^2 + 9y^2 = 1$$
$$\frac{x^2}{\frac{1}{4}} + \frac{y^2}{\frac{1}{9}} = 1$$
$$\frac{x^2}{\left(\frac{1}{2}\right)^2} + \frac{y^2}{\left(\frac{1}{3}\right)^2} = 1$$
$$a = \frac{1}{2},\, b = \frac{1}{3}$$

The major axis is horizontal, so the vertices are $\left(-\frac{1}{2}, 0\right)$ and $\left(\frac{1}{2}, 0\right)$. Since $c^2 = a^2 - b^2$, we have $c^2 = \frac{1}{4} - \frac{1}{9} = \frac{5}{36}$, so $c = \frac{\sqrt{5}}{6}$ and the foci are $\left(-\frac{\sqrt{5}}{6}, 0\right)$ and $\left(\frac{\sqrt{5}}{6}, 0\right)$.

To graph the ellipse, plot the vertices. Note also that since $b = \frac{1}{3}$, the y-intercepts are $\left(0, -\frac{1}{3}\right)$ and $\left(0, \frac{1}{3}\right)$. Plot these points as well and connect the four plotted points with a smooth curve.

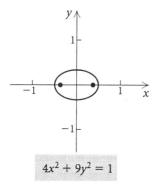

$$4x^2 + 9y^2 = 1$$

27. The vertices are on the x-axis, so the major axis is horizontal. We have $a = 7$ and $c = 3$, so we can find b^2:
$$c^2 = a^2 - b^2$$
$$3^2 = 7^2 - b^2$$
$$b^2 = 49 - 9 = 40$$

Write the equation:
$$\frac{x^2}{a^2} + \frac{y^2}{b^2} = 1$$
$$\frac{x^2}{49} + \frac{y^2}{40} = 1$$

29. The vertices, $(0, -8)$ and $(0, 8)$, are on the y-axis, so the major axis is vertical and $a = 8$. Since the vertices are equidistant from the origin, the center of the ellipse is at the origin. The length of the minor axis is 10, so $b = 10/2$, or 5.

Write the equation:
$$\frac{x^2}{b^2} + \frac{y^2}{a^2} = 1$$
$$\frac{x^2}{5^2} + \frac{y^2}{8^2} = 1$$
$$\frac{x^2}{25} + \frac{y^2}{64} = 1$$

31. The foci, $(-2, 0)$ and $(2, 0)$ are on the x-axis, so the major axis is horizontal and $c = 2$. Since the foci are equidistant from the origin, the center of the ellipse is at the origin. The length of the major axis is 6, so $a = 6/2$, or 3. Now we find b^2:
$$c^2 = a^2 - b^2$$
$$2^2 = 3^2 - b^2$$
$$4 = 9 - b^2$$
$$b^2 = 5$$

Write the equation:
$$\frac{x^2}{a^2} + \frac{y^2}{b^2} = 1$$
$$\frac{x^2}{9} + \frac{y^2}{5} = 1$$

33.
$$\frac{(x-1)^2}{9} + \frac{(y-2)^2}{4} = 1$$
$$\frac{(x-1)^2}{3^2} + \frac{(y-2)^2}{2^2} = 1 \quad \text{Standard form}$$

The center is $(1, 2)$. Note that $a = 3$ and $b = 2$. The major axis is horizontal so the vertices are 3 units left and right of the center:

$(1 - 3, 2)$ and $(1 + 3, 2)$, or $(-2, 2)$ and $(4, 2)$.

We know that $c^2 = a^2 - b^2$, so $c^2 = 9 - 4 = 5$ and $c = \sqrt{5}$. Then the foci are $\sqrt{5}$ units left and right of the center:

$(1 - \sqrt{5}, 2)$ and $(1 + \sqrt{5}, 2)$.

To graph the ellipse, plot the vertices. Since $b = 2$, two other points on the graph are 2 units below and above the center:

$(1, 2 - 2)$ and $(1, 2 + 2)$ or $(1, 0)$ and $(1, 4)$

Plot these points also and connect the four plotted points with a smooth curve.

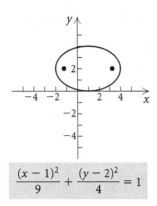

$$\frac{(x-1)^2}{9} + \frac{(y-2)^2}{4} = 1$$

35. $\dfrac{(x+3)^2}{25} + \dfrac{(y-5)^2}{36} = 1$

$\dfrac{[x-(-3)]^2}{5^2} + \dfrac{(y-5)^2}{6^2} = 1$ Standard form

The center is $(-3,5)$. Note that $a=6$ and $b=5$. The major axis is vertical so the vertices are 6 units below and above the center:

$(-3, 5-6)$ and $(-3, 5+6)$, or $(-3,-1)$ and $(-3, 11)$.

We know that $c^2 = a^2 - b^2$, so $c^2 = 36 - 25 = 11$ and $c = \sqrt{11}$. Then the foci are $\sqrt{11}$ units below and above the vertex:

$(-3, 5-\sqrt{11})$ and $(-3, 5+\sqrt{11})$.

To graph the ellipse, plot the vertices. Since $b=5$, two other points on the graph are 5 units left and right of the center:

$(-3-5, 5)$ and $(-3+5, 5)$, or $(-8, 5)$ and $(2, 5)$

Plot these points also and connect the four plotted points with a smooth curve.

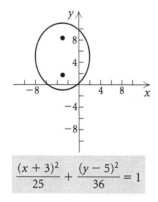

$$\frac{(x+3)^2}{25} + \frac{(y-5)^2}{36} = 1$$

37. $3(x+2)^2 + 4(y-1)^2 = 192$

$\dfrac{(x+2)^2}{64} + \dfrac{(y-1)^2}{48} = 1$ Dividing by 192

$\dfrac{[x-(-2)]^2}{8^2} + \dfrac{(y-1)^2}{(\sqrt{48})^2} = 1$ Standard form

The center is $(-2, 1)$. Note that $a = 8$ and $b = \sqrt{48}$, or $4\sqrt{3}$. The major axis is horizontal so the vertices are 8 units left and right of the center:

$(-2-8, 1)$ and $(-2+8, 1)$, or $(-10, 1)$ and $(6, 1)$.

We know that $c^2 = a^2 - b^2$, so $c^2 = 64 - 48 = 16$ and $c = 4$. Then the foci are 4 units left and right of the center:

$(-2-4, 1)$ and $(-2+4, 1)$ or $(-6, 1)$ and $(2, 1)$.

To graph the ellipse, plot the vertices. Since $b = 4\sqrt{3} \approx 6.928$, two other points on the graph are about 6.928 units below and above the center:

$(-2, 1-6.928)$ and $(-2, 1+6.928)$, or

$(-2, -5.928)$ and $(-2, 7.928)$.

Plot these points also and connect the four plotted points with a smooth curve.

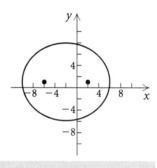

$$3(x+2)^2 + 4(y-1)^2 = 192$$

39. Begin by completing the square twice.

$4x^2 + 9y^2 - 16x + 18y - 11 = 0$

$4x^2 - 16x + 9y^2 + 18y = 11$

$4(x^2 - 4x) + 9(y^2 + 2y) = 11$

$4(x^2 - 4x + 4) + 9(y^2 + 2y + 1) = 11 + 4 \cdot 4 + 9 \cdot 1$

$4(x-2)^2 + 9(y+1)^2 = 36$

$\dfrac{(x-2)^2}{9} + \dfrac{(y+1)^2}{4} = 1$

$\dfrac{(x-2)^2}{3^2} + \dfrac{[y-(-1)]^2}{2^2} = 1$

The center is $(2, -1)$. Note that $a = 3$ and $b = 2$. The major axis is horizontal so the vertices are 3 units left and right of the center:

$(2-3, -1)$ and $(2+3, -1)$, or $(-1, -1)$ and $(5, -1)$.

We know that $c^2 = a^2 - b^2$, so $c^2 = 9 - 4 = 5$ and $c = \sqrt{5}$. Then the foci are $\sqrt{5}$ units left and right of the center:

$(2-\sqrt{5}, -1)$ and $(2+\sqrt{5}, -1)$.

To graph the ellipse, plot the vertices. Since $b = 2$, two other points on the graph are 2 units below and above the center:

$(2, -1-2)$ and $(2, -1+2)$, or $(2, -3)$ and $(2, 1)$.

Plot these points also and connect the four plotted points with a smooth curve.

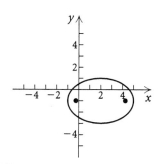

$$4x^2 + 9y^2 - 16x + 18y - 11 = 0$$

41. Begin by completing the square twice.
$$4x^2 + y^2 - 8x - 2y + 1 = 0$$
$$4x^2 - 8x + y^2 - 2y = -1$$
$$4(x^2 - 2x) + y^2 - 2y = -1$$
$$4(x^2 - 2x + 1) + y^2 - 2y + 1 = -1 + 4 \cdot 1 + 1$$
$$4(x-1)^2 + (y-1)^2 = 4$$
$$\frac{(x-1)^2}{1} + \frac{(y-1)^2}{4} = 1$$
$$\frac{(x-1)^2}{1^2} + \frac{(y-1)^2}{2^2} = 1$$

The center is $(1, 1)$. Note that $a = 2$ and $b = 1$. The major axis is vertical so the vertices are 2 units below and above the center:

$(1, 1-2)$ and $(1, 1+2)$, or $(1, -1)$ and $(1, 3)$.

We know that $c^2 = a^2 - b^2$, so $c^2 = 4 - 1 = 3$ and $c = \sqrt{3}$. Then the foci are $\sqrt{3}$ units below and above the center:

$(1, 1-\sqrt{3})$ and $(1, 1+\sqrt{3})$.

To graph the ellipse, plot the vertices. Since $b = 1$, two other points on the graph are 1 unit left and right of the center:

$(1-1, 1)$ and $(1+1, 1)$ or $(0, 1)$ and $(2, 1)$.

Plot these points also and connect the four plotted points with a smooth curve.

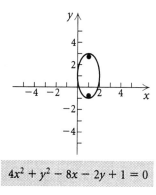

$$4x^2 + y^2 - 8x - 2y + 1 = 0$$

43. The ellipse in Example 4 is flatter than the one in Example 2, so the ellipse in Example 2 has the smaller eccentricity.

We compute the the eccentricities: In Example 2, $c = 3$ and $a = 5$, so $e = c/a = 3/5 = 0.6$. In Example 4, $c = 2\sqrt{3}$ and $a = 4$, so $e = c/a = 2\sqrt{3}/4 \approx 0.866$. These computations confirm that the ellipse in Example 2 has the smaller eccentricity.

45. Since the vertices, $(0, -4)$ and $(0, 4)$ are on the y-axis and are equidistant from the origin, we know that the major axis of the ellipse is vertical, its center is at the origin, and $a = 4$. Use the information that $e = 1/4$ to find c:
$$e = \frac{c}{a}$$
$$\frac{1}{4} = \frac{c}{4} \quad \text{Substituting}$$
$$c = 1$$
Now $c^2 = a^2 - b^2$, so we can find b^2:
$$1^2 = 4^2 - b^2$$
$$1 = 16 - b^2$$
$$b^2 = 15$$
Write the equation of the ellipse:
$$\frac{x^2}{b^2} + \frac{y^2}{a^2} = 1$$
$$\frac{x^2}{15} + \frac{y^2}{16} = 1$$

47. From the figure in the text we see that the center of the ellipse is $(0, 0)$, the major axis is horizontal, the vertices are $(-50, 0)$ and $(50, 0)$, and one y-intercept is $(0, 12)$. Then $a = 50$ and $b = 12$. The equation is
$$\frac{x^2}{a^2} + \frac{y^2}{b^2} = 1$$
$$\frac{x^2}{50^2} + \frac{y}{12^2} = 1$$
$$\frac{x^2}{2500} + \frac{y^2}{144} = 1.$$

49. Position a coordinate system as shown below where $1 \text{ unit} = 10^7 \text{ mi}$.

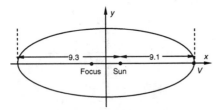

The length of the major axis is $9.3 + 9.1$, or 18.4. Then the distance from the center of the ellipse (the origin) to V is $18.4/2$, or 9.2. Since the distance from the sun to V is 9.1, the distance from the sun to the center is $9.2 - 9.1$, or 0.1. Then the distance from the sun to the other focus is twice this distance:

$$2(0.1 \times 10^7 \text{ mi}) = 0.2 \times 10^7 \text{ mi}$$
$$= 2 \times 10^6 \text{ mi}$$

51. Discussion and Writing

53. a) The graph of $f(x) = 2x - 3$ is shown below.

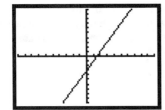

Since there is no horizontal line that crosses the graph more than once, the function is one-to-one.

b) Replace $f(x)$ with y: $y = 2x - 3$

Interchange x and y: $x = 2y - 3$

Solve for y: $x + 3 = 2y$

$$\frac{x+3}{2} = y$$

Replace y with $f^{-1}(x)$: $f^{-1}(x) = \dfrac{x+3}{2}$

55. a) The graph of $f(x) = \dfrac{5}{x-1}$ is shown below.

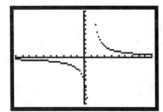

Since there is no horizontal line that crosses the graph more than once, the function is one-to-one.

b) Replace $f(x)$ with y: $y = \dfrac{5}{x-1}$

Interchange x and y: $x = \dfrac{5}{y-1}$

Solve for y: $x(y-1) = 5$

$$y - 1 = \frac{5}{x}$$
$$y = \frac{5}{x} + 1$$

Replace y with $f^{-1}(x)$: $f^{-1} = \dfrac{5}{x} + 1$, or $\dfrac{5+x}{x}$

57. The center of the ellipse is the midpoint of the segment connecting the vertices:

$$\left(\frac{3+3}{2}, \frac{-4+6}{2}\right), \text{ or } (3, 1).$$

Now a is the distance from the origin to a vertex. We use the vertex $(3, 6)$.

$$a = \sqrt{(3-3)^2 + (6-1)^2} = 5$$

Also b is one-half the length of the minor axis.

$$b = \frac{\sqrt{(5-1)^2 + (1-1)^2}}{2} = \frac{4}{2} = 2$$

The vertices lie on the vertical line $x = 3$, so the major axis is vertical. We write the equation of the ellipse.

$$\frac{(x-h)^2}{b^2} + \frac{(y-k)^2}{a^2} = 1$$

$$\frac{(x-3)^2}{4} + \frac{(y-1)^2}{25} = 1$$

59. The center is the midpoint of the segment connecting the vertices:

$$\left(\frac{-3+3}{0}, \frac{0+0}{0}\right), \text{ or } (0, 0).$$

Then $a = 3$ and since the vertices are on the x-axis, the major axis is horizontal. The equation is of the form $\dfrac{x^2}{a^2} + \dfrac{y^2}{b^2} = 1$.

Substitute 3 for a, 2 for x, and $\dfrac{22}{3}$ for y and solve for b^2.

$$\frac{4}{9} + \frac{\frac{484}{9}}{b^2} = 1$$

$$\frac{4}{9} + \frac{484}{9b^2} = 1$$

$$4b^2 + 484 = 9b^2$$

$$484 = 5b^2$$

$$\frac{484}{5} = b^2$$

Then the equation is $\dfrac{x^2}{9} + \dfrac{y^2}{484/5} = 1$.

61. Center: $(2.003, -1.005)$

Vertices: $(-1.017, -1.005)$, $(5.023, -1.005)$

63. Position a coordinate system as shown.

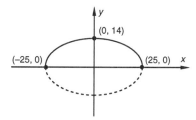

The equation of the ellipse is

$$\frac{x^2}{25^2} + \frac{y^2}{14^2} = 1$$

$$\frac{x^2}{625} + \frac{y^2}{196} = 1.$$

A point 6 ft from the riverbank corresponds to $(25 - 6, 0)$, or $(19, 0)$ or to $(-25 + 6, 0)$, or $(-19, 0)$. Substitute either 19 or -19 for x and solve for y, the clearance.

$$\frac{19^2}{625} + \frac{y^2}{196} = 1$$

$$\frac{y^2}{196} = 1 - \frac{361}{625}$$

$$y^2 = 196\left(1 - \frac{361}{625}\right)$$

$$y \approx 9.1$$

The clearance 6 ft from the riverbank is about 9.1 ft.

Exercise Set 9.3

1. Graph (b) is the graph of $\dfrac{x^2}{25} - \dfrac{y^2}{9} = 1$.

3. Graph (c) is the graph of $\dfrac{(y-1)^2}{16} - \dfrac{(x+3)^2}{1} = 1$.

5. Graph (a) is the graph of $25x^2 - 16y^2 = 400$.

7. The vertices are equidistant from the origin and are on the y-axis, so the center is at the origin and the transverse axis is vertical. Since $c^2 = a^2 + b^2$, we have $5^2 = 3^2 + b^2$ so $b^2 = 16$.

The equation is of the form $\dfrac{y^2}{a^2} - \dfrac{x^2}{b^2} = 1$, so we have $\dfrac{y^2}{9} - \dfrac{x^2}{16} = 1$.

9. The asymptotes pass through the origin, so the center is the origin. The given vertex is on the x-axis, so the transverse axis is horizontal. Since $\dfrac{b}{a}x = \dfrac{3}{2}x$

and $a = 2$, we have $b = 3$. The equation is of the form $\dfrac{x^2}{a^2} - \dfrac{y^2}{b^2} = 1$, so we have $\dfrac{x^2}{2^2} - \dfrac{y^2}{3^2} = 1$, or $\dfrac{x^2}{4} - \dfrac{y^2}{9} = 1$.

11. $\dfrac{x^2}{4} - \dfrac{y^2}{4} = 1$

$\dfrac{x^2}{2^2} - \dfrac{y^2}{2^2} = 1$ Standard form

The center is $(0, 0)$; $a = 2$ and $b = 2$. The transverse axis is horizontal so the vertices are $(-2, 0)$ and $(2, 0)$. Since $c^2 = a^2 + b^2$, we have $c^2 = 4 + 4 = 8$ and $c = \sqrt{8}$, or $2\sqrt{2}$. Then the foci are $(-2\sqrt{2}, 0)$ and $(2\sqrt{2}, 0)$.

Find the asymptotes:

$$y = \frac{b}{a}x \text{ and } y = -\frac{b}{a}x$$

$$y = \frac{2}{2}x \text{ and } y = -\frac{2}{2}x$$

$$y = x \quad \text{ and } y = -x$$

To draw the graph sketch the asymptotes, plot the vertices, and draw the branches of the hyperbola outward from the vertices toward the asymptotes.

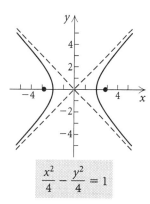

$$\boxed{\frac{x^2}{4} - \frac{y^2}{4} = 1}$$

13. $\dfrac{(x-2)^2}{9} - \dfrac{(y+5)^2}{1} = 1$

$\dfrac{(x-2)^2}{3^2} - \dfrac{[y-(-5)]^2}{1^2} = 1$ Standard form

The center is $(2, -5)$; $a = 3$ and $b = 1$. The transverse axis is horizontal, so the vertices are 3 units left and right of the center:

$(2 - 3, -5)$ and $(2 + 3, -5)$, or $(-1, -5)$ and $(5, -5)$.

Since $c^2 = a^2 + b^2$, we have $c^2 = 9 + 1 = 10$ and $c = \sqrt{10}$. Then the foci are $\sqrt{10}$ units left and right of the center:

$$(2 - \sqrt{10}, -5) \text{ and } (2 + \sqrt{10}, -5).$$

Find the asymptotes:

$$y - k = \frac{b}{a}(x-h) \quad \text{and} \quad y - k = -\frac{b}{a}(x-h)$$

$$y - (-5) = \frac{1}{3}(x-2) \quad \text{and} \quad y - (-5) = -\frac{1}{3}(x-2)$$

$$y + 5 = \frac{1}{3}(x-2) \quad \text{and} \quad y + 5 = -\frac{1}{3}(x-2), \text{ or}$$

$$y = \frac{1}{3}x - \frac{17}{3} \quad \text{and} \quad y = -\frac{1}{3}x - \frac{13}{3}$$

Sketch the asymptotes, plot the vertices, and draw the graph.

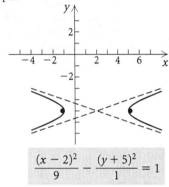

$$\boxed{\frac{(x-2)^2}{9} - \frac{(y+5)^2}{1} = 1}$$

15.
$$\frac{(y+3)^2}{4} - \frac{(x+1)^2}{16} = 1$$

$$\frac{[y-(-3)]^2}{2^2} - \frac{[x-(-1)]^2}{4^2} = 1 \quad \text{Standard form}$$

The center is $(-1,-3)$; $a = 2$ and $b = 4$. The transverse axis is vertical, so the vertices are 2 units below and above the center:

$(-1, -3-2)$ and $(1, -3+2)$, or $(-1,-5)$ and $(-1,-1)$.

Since $c^2 = a^2 + b^2$, we have $c^2 = 4 + 16 = 20$ and $c = \sqrt{20}$, or $2\sqrt{5}$. Then the foci are $2\sqrt{5}$ units below and above of the center:

$(-1, -3 - 2\sqrt{5})$ and $(-1, -3 + 2\sqrt{5})$.

Find the asymptotes:

$$y - k = \frac{a}{b}(x-h) \quad \text{and} \quad y - k = -\frac{a}{b}(x-h)$$

$$y - (-3) = \frac{2}{4}(x-(-1)) \quad \text{and} \quad y - (-3) = -\frac{2}{4}(x-(-1))$$

$$y + 3 = \frac{1}{2}(x+1) \quad \text{and} \quad y + 3 = -\frac{1}{2}(x+1), \text{ or}$$

$$y = \frac{1}{2}x - \frac{5}{2} \quad \text{and} \quad y = -\frac{1}{2}x - \frac{7}{2}$$

Sketch the asymptotes, plot the vertices, and draw the graph.

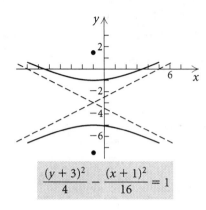

$$\boxed{\frac{(y+3)^2}{4} - \frac{(x+1)^2}{16} = 1}$$

17. $x^2 - 4y^2 = 4$

$$\frac{x^2}{4} - \frac{y^2}{1} = 1$$

$$\frac{x^2}{2^2} - \frac{y^2}{1^2} = 1 \quad \text{Standard form}$$

The center is $(0,0)$; $a = 2$ and $b = 1$. The transverse axis is horizontal, so the vertices are $(-2,0)$ and $(2,0)$. Since $c^2 = a^2 + b^2$, we have $c^2 = 4 + 1 = 5$ and $c = \sqrt{5}$. Then the foci are $(-\sqrt{5}, 0)$ and $(\sqrt{5}, 0)$.

Find the asymptotes:

$$y = \frac{b}{a}x \quad \text{and} \quad y = -\frac{b}{a}x$$

$$y = \frac{1}{2}x \quad \text{and} \quad y = -\frac{1}{2}x$$

Sketch the asymptotes, plot the vertices, and draw the graph.

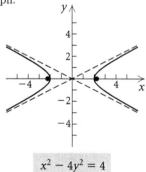

$$\boxed{x^2 - 4y^2 = 4}$$

19. $9y^2 - x^2 = 81$

$$\frac{y^2}{9} - \frac{x^2}{81} = 1$$

$$\frac{y^2}{3^2} - \frac{x^2}{9^2} = 1 \quad \text{Standard form}$$

The center is $(0,0)$; $a = 3$ and $b = 9$. The transverse axis is vertical, so the vertices are $(0,-3)$ and $(0,3)$. Since $c^2 = a^2 + b^2$, we have $c^2 = 9 + 81 = 90$ and $c = \sqrt{90}$, or $3\sqrt{10}$. Then the foci are $(0, -3\sqrt{10})$ and $(0, 3\sqrt{10})$.

Find the asymptotes:

$$y = \frac{a}{b}x \quad \text{and} \quad y = -\frac{a}{b}x$$

$$y = \frac{3}{9}x \quad \text{and} \quad y = -\frac{3}{9}x$$

$$y = \frac{1}{3}x \quad \text{and} \quad y = -\frac{1}{3}x$$

Sketch the asymptotes, plot the vertices, and draw the graph.

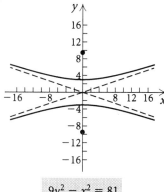

$$9y^2 - x^2 = 81$$

21.

$$x^2 - y^2 = 2$$

$$\frac{x^2}{2} - \frac{y^2}{2} = 1$$

$$\frac{x^2}{(\sqrt{2})^2} - \frac{y^2}{(\sqrt{2})^2} = 1 \quad \text{Standard form}$$

The center is $(0,0)$; $a = \sqrt{2}$ and $b = \sqrt{2}$. The transverse axis is horizontal, so the vertices are $(-\sqrt{2}, 0)$ and $(\sqrt{2}, 0)$. Since $c^2 = a^2 + b^2$, we have $c^2 = 2 + 2 = 4$ and $c = 2$. Then the foci are $(-2, 0)$ and $(2, 0)$.

Find the asymptotes:

$$y = \frac{b}{a}x \quad \text{and} \quad y = -\frac{b}{a}x$$

$$y = \frac{\sqrt{2}}{\sqrt{2}}x \quad \text{and} \quad y = -\frac{\sqrt{2}}{\sqrt{2}}x$$

$$y = x \quad \text{and} \quad y = -x$$

Sketch the asymptotes, plot the vertices, and draw the graph.

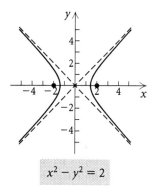

$$x^2 - y^2 = 2$$

23.

$$y^2 - x^2 = \frac{1}{4}$$

$$\frac{y^2}{1/4} - \frac{x^2}{1/4} = 1$$

$$\frac{y^2}{(1/2)^2} - \frac{x^2}{(1/2)^2} = 1 \quad \text{Standard form}$$

The center is $(0,0)$; $a = \frac{1}{2}$ and $b = \frac{1}{2}$. The transverse axis is vertical, so the vertices are $\left(0, -\frac{1}{2}\right)$ and $\left(0, \frac{1}{2}\right)$. Since $c^2 = a^2 + b^2$, we have $c^2 = \frac{1}{4} + \frac{1}{4} = \frac{1}{2}$ and $c = \sqrt{\frac{1}{2}}$, or $\frac{\sqrt{2}}{2}$. Then the foci are $\left(0, -\frac{\sqrt{2}}{2}\right)$ and $\left(0, \frac{\sqrt{2}}{2}\right)$.

Find the asymptotes:

$$y = \frac{a}{b}x \quad \text{and} \quad y = -\frac{a}{b}x$$

$$y = \frac{1/2}{1/2}x \quad \text{and} \quad y = -\frac{1/2}{1/2}x$$

$$y = x \quad \text{and} \quad y = -x$$

Sketch the asymptotes, plot the vertices, and draw the graph.

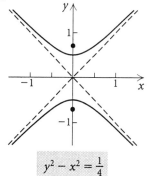

$$y^2 - x^2 = \frac{1}{4}$$

25. Begin by completing the square twice.

$$x^2 - y^2 - 2x - 4y - 4 = 0$$
$$(x^2 - 2x) - (y^2 + 4y) = 4$$
$$(x^2 - 2x + 1) - (y^2 + 4y + 4) = 4 + 1 - 1 \cdot 4$$
$$(x - 1)^2 - (y + 2)^2 = 1$$
$$\frac{(x-1)^2}{1^2} - \frac{[y-(-2)]^2}{1^2} = 1 \quad \text{Standard form}$$

The center is $(1, -2)$; $a = 1$ and $b = 1$. The transverse axis is horizontal, so the vertices are 1 unit left and right of the center:

$(1 - 1, -2)$ and $(1 + 1, -2)$ or $(0, -2)$ and $(2, -2)$

Since $c^2 = a^2 + b^2$, we have $c^2 = 1 + 1 = 2$ and $c = \sqrt{2}$. Then the foci are $\sqrt{2}$ units left and right of the center:

$$(1 - \sqrt{2}, -2) \text{ and } (1 + \sqrt{2}, -2).$$

Find the asymptotes:

$$y - k = \frac{b}{a}(x - h) \text{ and } y - k = -\frac{b}{a}(x - h)$$
$$y - (-2) = \frac{1}{1}(x - 1) \text{ and } y - (-2) = -\frac{1}{1}(x - 1)$$
$$y + 2 = x - 1 \quad \text{and} \quad y + 2 = -(x - 1), \text{ or}$$
$$y = x - 3 \quad \text{and} \quad y = -x - 1$$

Sketch the asymptotes, plot the vertices, and draw the graph.

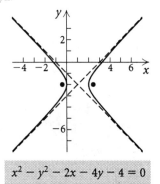

$$x^2 - y^2 - 2x - 4y - 4 = 0$$

27. Begin by completing the square twice.

$$36x^2 - y^2 - 24x + 6y - 41 = 0$$
$$(36x^2 - 24x) - (y^2 - 6y) = 41$$
$$36\left(x^2 - \frac{2}{3}x\right) - (y^2 - 6y) = 41$$
$$36\left(x^2 - \frac{2}{3}x + \frac{1}{9}\right) - (y^2 - 6y + 9) = 41 + 36 \cdot \frac{1}{9} - 1 \cdot 9$$
$$36\left(x - \frac{1}{3}\right)^2 - (y - 3)^2 = 36$$
$$\frac{\left(x - \frac{1}{3}\right)^2}{1} - \frac{(y-3)^2}{36} = 1$$
$$\frac{\left(x - \frac{1}{3}\right)^2}{1^2} - \frac{(y-3)^2}{6^2} = 1 \quad \text{Standard form}$$

The center is $\left(\frac{1}{3}, 3\right)$; $a = 1$ and $b = 6$. The transverse axis is horizontal, so the vertices are 1 unit left and right of the center:

$\left(\frac{1}{3} - 1, 3\right)$ and $\left(\frac{1}{3} + 1, 3\right)$ or $\left(-\frac{2}{3}, 3\right)$ and $\left(\frac{4}{3}, 3\right)$.

Since $c^2 = a^2 + b^2$, we have $c^2 = 1 + 36 = 37$ and $c = \sqrt{37}$. Then the foci are $\sqrt{37}$ units left and right of the center:

$$\left(\frac{1}{3} - \sqrt{37}, 3\right) \text{ and } \left(\frac{1}{3} + \sqrt{37}, 3\right).$$

Find the asymptotes:

$$y - k = \frac{b}{a}(x - h) \quad \text{and } y - k = -\frac{b}{a}(x - h)$$
$$y - 3 = \frac{6}{1}\left(x - \frac{1}{3}\right) \quad \text{and } y - 3 = -\frac{6}{1}\left(x - \frac{1}{3}\right)$$
$$y - 3 = 6\left(x - \frac{1}{3}\right) \quad \text{and } y - 3 = -6\left(x - \frac{1}{3}\right), \text{ or}$$
$$y = 6x + 1 \quad \text{and} \quad y = -6x + 5$$

Sketch the asymptotes, plot the vertices, and draw the graph.

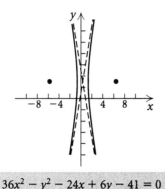

$$36x^2 - y^2 - 24x + 6y - 41 = 0$$

29. Begin by completing the square twice.
$$9y^2 - 4x^2 - 18y + 24x - 63 = 0$$
$$9(y^2 - 2y) - 4(x^2 - 6x) = 63$$
$$9(y^2 - 2y + 1) - 4(x^2 - 6x + 9) = 63 + 9 \cdot 1 - 4 \cdot 9$$
$$9(y - 1)^2 - 4(x - 3)^2 = 36$$
$$\frac{(y-1)^2}{4} - \frac{(x-3)^2}{9} = 1$$
$$\frac{(y-1)^2}{2^2} - \frac{(x-3)^2}{3^2} = 1 \quad \text{Standard} \atop \text{form}$$

The center is $(3, 1)$; $a = 2$ and $b = 3$. The transverse axis is vertical, so the vertices are 2 units below and above the center:

$(3, 1 - 2)$ and $(3, 1 + 2)$, or $(3, -1)$ and $(3, 3)$.

Since $c^2 = a^2 + b^2$, we have $c^2 = 4 + 9 = 13$ and $c = \sqrt{13}$. Then the foci are $\sqrt{13}$ units below and above the center:

$(3, 1 - \sqrt{13})$ and $(3, 1 + \sqrt{13})$.

Find the asymptotes:
$$y - k = \frac{a}{b}(x - h) \text{ and } y - k = -\frac{a}{b}(x - h)$$
$$y - 1 = \frac{2}{3}(x - 3) \text{ and } y - 1 = -\frac{2}{3}(x - 3), \text{ or}$$
$$y = \frac{2}{3}x - 1 \quad \text{and} \quad y = -\frac{2}{3}x + 3$$

Sketch the asymptotes, plot the vertices, and draw the graph.

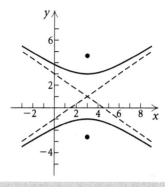

$$9y^2 - 4x^2 - 18y + 24x - 63 = 0$$

31. Begin by completing the square twice.
$$x^2 - y^2 - 2x - 4y = 4$$
$$(x^2 - 2x + 1) - (y^2 + 4y + 4) = 4 + 1 - 4$$
$$(x - 1)^2 - (y + 2)^2 = 1$$
$$\frac{(x-1)^2}{1^2} - \frac{[y-(-2)]^2}{1^2} = 1 \quad \text{Standard} \atop \text{form}$$

The center is $(1, -2)$; $a = 1$ and $b = 1$. The transverse axis is horizontal, so the vertices are 1 unit left and right of the center:

$(1 - 1, -2)$ and $(1 + 1, -2)$, or $(0, -2)$ and $(2, -2)$.

Since $c^2 = a^2 + b^2$, we have $c^2 = 1 + 1 = 2$ and $c = \sqrt{2}$. Then the foci are $\sqrt{2}$ units left and right of the center:

$(1 - \sqrt{2}, -2)$ and $(1 + \sqrt{2}, -2)$.

Find the asymptotes:
$$y - k = \frac{b}{a}(x - h) \text{ and } \quad y - k = -\frac{b}{a}(x - h)$$
$$y - (-2) = \frac{1}{1}(x - 1) \text{ and } y - (-2) = -\frac{1}{1}(x - 1)$$
$$y + 2 = x - 1 \quad \text{and} \quad y + 2 = -(x - 1), \text{ or}$$
$$y = x - 3 \quad \text{and} \quad y = -x - 1$$

Sketch the asymptotes, plot the vertices, and draw the graph.

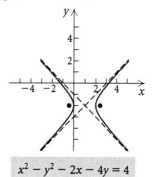

$$x^2 - y^2 - 2x - 4y = 4$$

33. Begin by completing the square twice.
$$y^2 - x^2 - 6x - 8y - 29 = 0$$
$$(y^2 - 8y + 16) - (x^2 + 6x + 9) = 29 + 16 - 9$$
$$(y - 4)^2 - (x + 3)^2 = 36$$
$$\frac{(y-4)^2}{36} - \frac{(x+3)^2}{36} = 1$$
$$\frac{(y-4)^2}{6^2} - \frac{[x-(-3)]^2}{6^2} = 1 \quad \begin{array}{l}\text{Standard} \\ \text{form}\end{array}$$

The center is $(-3, 4)$; $a = 6$ and $b = 6$. The transverse axis is vertical, so the vertices are 6 units below and above the center:

$(-3, 4-6)$ and $(-3, 4+6)$, or $(-3, -2)$ and $(-3, 10)$.

Since $c^2 = a^2 + b^2$, we have $c^2 = 36 + 36 = 72$ and $c = \sqrt{72}$, or $6\sqrt{2}$. Then the foci are $6\sqrt{2}$ units below and above the center:

$$(-3, 4 - 6\sqrt{2}) \text{ and } (-3, 4 + 6\sqrt{2}).$$

Find the asymptotes:

$$y - k = \frac{a}{b}(x - h) \qquad \text{and} \quad y - k = -\frac{a}{b}(x - h)$$

$$y - 4 = \frac{6}{6}(x - (-3)) \text{ and } y - 4 = -\frac{6}{6}(x - (-3))$$

$$y - 4 = x + 3 \qquad \text{and} \quad y - 4 = -(x + 3), \text{ or}$$

$$y = x + 7 \qquad \text{and} \qquad y = -x + 1$$

Sketch the asymptotes, plot the vertices, and draw the graph.

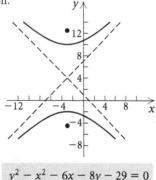

$$y^2 - x^2 - 6x - 8y - 29 = 0$$

35. The hyperbola in Example 3 is wider than the one in Example 2, so the hyperbola in Example 3 has the larger eccentricity.

Compute the eccentricities: In Example 2, $c = 5$ and $a = 4$, so $e = 5/4$, or 1.25. In Example 3, $c = \sqrt{5}$ and $a = 1$, so $e = \sqrt{5}/1 \approx 2.24$. These computations confirm that the hyperbola in Example 3 has the larger eccentricity.

37. The center is the midpoint of the segment connecting the vertices:

$$\left(\frac{3-3}{2}, \frac{7+7}{2}\right), \text{ or } (0, 7).$$

The vertices are on the horizontal line $y = 7$, so the transverse axis is horizontal. Since the vertices are 3 units left and right of the center, $a = 3$.

Find c:

$$e = \frac{c}{a} = \frac{5}{3}$$

$$\frac{c}{3} = \frac{5}{3} \quad \text{Substituting 3 for } a$$

$$c = 5$$

Now find b^2:

$$c^2 = a^2 + b^2$$
$$5^2 = 3^2 + b^2$$
$$16 = b^2$$

Write the equation:

$$\frac{(x-h)^2}{a^2} - \frac{(y-k)^2}{b^2} = 1$$

$$\frac{x^2}{9} - \frac{(y-7)^2}{16} = 1$$

39.

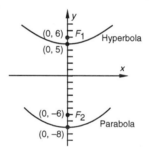

One focus is 6 units above the center of the hyperbola, so $c = 6$. One vertex is 5 units above the center, so $a = 5$. Find b^2:

$$c^2 = a^2 + b^2$$
$$6^2 = 5^2 + b^2$$
$$11 = b^2$$

Write the equation:

$$\frac{y^2}{a^2} - \frac{x^2}{b^2} = 1$$

$$\frac{y^2}{25} - \frac{x^2}{11} = 1$$

41. Discussion and Writing

43.
$$\begin{array}{ll} x + y = 5, & (1) \\ \underline{x - y = 7} & (2) \\ 2x \quad\;\; = 12 & \text{Adding} \\ \quad\; x = 6 \end{array}$$

Back-substitute in either equation (1) or (2) and solve for y. We use equation (1).

$$6 + y = 5$$
$$y = -1$$

The solution is $(6, -1)$.

45. $2x - 3y = 7,$ (1)

 $3x + 5y = 1$ (2)

Multiply equation (1) by 5 and equation (2) by 3 and add to eliminate y.

$$10x - 15y = 35$$
$$\underline{9x + 15y = 3}$$
$$19x = 38$$
$$x = 2$$

Back-substitute and solve for y.

$$3 \cdot 2 + 5y = 1 \quad \text{Using equation (2)}$$
$$5y = -5$$
$$y = -1$$

The solution is $(2, -1)$.

47. The center is the midpoint of the segment connecting $(3, -8)$ and $(3, -2)$:

$$\left(\frac{3+3}{2}, \frac{-8-2}{2}\right), \text{ or } (3, -5).$$

The vertices are on the vertical line $x = 3$ and are 3 units above and below the center so the transverse axis is vertical and $a = 3$. Use the equation of an asymptote to find b:

$$y - k = \frac{a}{b}(x - h)$$

$$y + 5 = \frac{3}{b}(x - 3)$$

$$y = \frac{3}{b}x - \frac{9}{b} - 5$$

This equation corresponds to the asymptote $y = 3x - 14$, so $\frac{3}{b} = 3$ and $b = 1$.

Write the equation of the hyperbola:

$$\frac{(y-k)^2}{a^2} - \frac{(x-h)^2}{b^2} = 1$$

$$\frac{(y+5)^2}{9} - \frac{(x-3)^2}{1} = 1$$

49. Center: $(-1.460, -0.957)$

Vertices: $(-2.360, -0.957), (-0.560, -0.957)$

Asymptotes: $y = -1.429x - 3.043, y = 1.429x + 1.129$

51. S and T are the foci of the hyperbola, so $c = 300/2 = 150$.

$200 \text{ microseconds} \cdot \dfrac{0.186 \text{ mi}}{1 \text{ microsecond}} = 37.2 \text{ mi}$, the difference of the ships' distances from the foci. That is, $2a = 37.2$, so $a = 18.6$.

Find b^2:

$$c^2 = a^2 + b^2$$
$$150^2 = 18.6^2 + b^2$$
$$22,154.04 = b^2$$

Then the equation of the hyperbola is

$$\frac{x^2}{18.6^2} - \frac{y^2}{22,154.04} = 1, \text{ or } \frac{x^2}{345.96} - \frac{y^2}{22,154.04} = 1.$$

Exercise Set 9.4

1. The correct graph is (e).

3. The correct graph is (c).

5. The correct graph is (b).

7. $x^2 + y^2 = 25,$ (1)

 $y - x = 1$ (2)

First solve equation (2) for y.

$$y = x + 1 \quad (3)$$

Then substitute $x + 1$ for y in equation (1) and solve for x.

$$x^2 + y^2 = 25$$
$$x^2 + (x + 1)^2 = 25$$
$$x^2 + x^2 + 2x + 1 = 25$$
$$2x^2 + 2x - 24 = 0$$
$$x^2 + x - 12 = 0 \quad \text{Multiplying by } \frac{1}{2}$$
$$(x + 4)(x - 3) = 0 \quad \text{Factoring}$$
$$x + 4 = 0 \quad \text{or} \quad x - 3 = 0 \quad \text{Principle of zero}$$
$$\phantom{x + 4 = 0 \quad \text{or} \quad x - 3 = 0 \quad} \text{products}$$
$$x = -4 \quad \text{or} \quad x = 3$$

Now substitute these numbers into equation (3) and solve for y.

$$y = -4 + 1 = -3$$
$$y = 3 + 1 = 4$$

The pairs $(-4, -3)$ and $(3, 4)$ check, so they are the solutions.

9. $4x^2 + 9y^2 = 36,$ (1)

 $3y + 2x = 6$ (2)

First solve equation (2) for y.

$$3y = -2x + 6$$

$$y = -\frac{2}{3}x + 2 \quad (3)$$

Then substitute $-\dfrac{2}{3}x + 2$ for y in equation (1) and solve for x.

$$4x^2 + 9y^2 = 36$$

$$4x^2 + 9\left(-\frac{2}{3}x + 2\right)^2 = 36$$

$$4x^2 + 9\left(\frac{4}{9}x^2 - \frac{8}{3}x + 4\right) = 36$$

$$4x^2 + 4x^2 - 24x + 36 = 36$$

$$8x^2 - 24x = 0$$

$$x^2 - 3x = 0$$

$$x(x - 3) = 0$$

$x = 0$ or $x = 3$

Now substitute these numbers in equation (3) and solve for y.

$$y = -\frac{2}{3} \cdot 0 + 2 = 2$$

$$y = -\frac{2}{3} \cdot 3 + 2 = 0$$

The pairs $(0, 2)$ and $(3, 0)$ check, so they are the solutions.

11. $\quad x^2 + y^2 = 25, \qquad (1)$

$\qquad y^2 = x + 5 \qquad\qquad (2)$

We substitute $x + 5$ for y^2 in equation (1) and solve for x.

$$x^2 + y^2 = 25$$

$$x^2 + (x + 5) = 25$$

$$x^2 + x - 20 = 0$$

$$(x + 5)(x - 4) = 0$$

$$x + 5 = 0 \quad \text{or} \quad x - 4 = 0$$

$$x = -5 \quad \text{or} \qquad x = 4$$

We substitute these numbers for x in either equation (1) or equation (2) and solve for y. Here we use equation (2).

$y^2 = -5 + 5 = 0$ and $y = 0$.

$y^2 = 4 + 5 = 9$ and $y = \pm 3$.

The pairs $(-5, 0)$, $(4, 3)$ and $(4, -3)$ check. They are the solutions.

13. $\quad x^2 + y^2 = 9, \qquad (1)$

$\qquad x^2 - y^2 = 9 \qquad\qquad (2)$

Here we use the elimination method.

$$
\begin{array}{rll}
x^2 + y^2 = & 9 & (1) \\
x^2 - y^2 = & 9 & (2) \\
\hline
2x^2 = & 18 & \text{Adding} \\
x^2 = & 9 & \\
x = & \pm 3 &
\end{array}
$$

If $x = 3$, $x^2 = 9$, and if $x = -3$, $x^2 = 9$, so substituting 3 or -3 in equation (1) gives us

$$x^2 + y^2 = 9$$

$$9 + y^2 = 9$$

$$y^2 = 0$$

$$y = 0.$$

The pairs $(3, 0)$ and $(-3, 0)$ check. They are the solutions.

15. $\quad y^2 - x^2 = 9 \qquad (1)$

$\qquad 2x - 3 = y \qquad\quad (2)$

Substitute $2x - 3$ for y in equation (1) and solve for x.

$$y^2 - x^2 = 9$$

$$(2x - 3)^2 - x^2 = 9$$

$$4x^2 - 12x + 9 - x^2 = 9$$

$$3x^2 - 12x = 0$$

$$x^2 - 4x = 0$$

$$x(x - 4) = 0$$

$x = 0$ or $x = 4$

Now substitute these numbers into equation (2) and solve for y.

If $x = 0$, $y = 2 \cdot 0 - 3 = -3$.

If $x = 4$, $y = 2 \cdot 4 - 3 = 5$.

The pairs $(0, -3)$ and $(4, 5)$ check. They are the solutions.

17. $\quad y^2 = x + 3, \qquad (1)$

$\qquad 2y = x + 4 \qquad\quad (2)$

First solve equation (2) for x.

$$2y - 4 = x \qquad (3)$$

Then substitute $2y - 4$ for x in equation (1) and solve for y.

$$y^2 = x + 3$$

$$y^2 = (2y - 4) + 3$$

$$y^2 = 2y - 1$$

$$y^2 - 2y + 1 = 0$$

$$(y - 1)(y - 1) = 0$$

$$y - 1 = 0 \quad \text{or} \quad y - 1 = 0$$

$$y = 1 \quad \text{or} \qquad y = 1$$

Now substitute 1 for y in equation (3) and solve for x.

$$2 \cdot 1 - 4 = x$$

$$-2 = x$$

The pair $(-2, 1)$ checks. It is the solution.

19. $\quad x^2 + y^2 = 25, \quad (1)$

$\qquad xy = 12 \qquad\qquad (2)$

First we solve equation (2) for y.

$$xy = 12$$

$$y = \frac{12}{x}$$

Then we substitute $\frac{12}{x}$ for y in equation (1) and solve for x.

$$x^2 + y^2 = 25$$

$$x^2 + \left(\frac{12}{x}\right)^2 = 25$$

$$x^2 + \frac{144}{x^2} = 25$$

$$x^4 + 144 = 25x^2 \quad \text{Multiplying by } x^2$$

$$x^4 - 25x^2 + 144 = 0$$

$$u^2 - 25u + 144 = 0 \qquad \text{Letting } u = x^2$$

$$(u - 9)(u - 16) = 0$$

$$u = 9 \quad \text{or} \quad u = 16$$

We now substitute x^2 for u and solve for x.

$$x^2 = 9 \quad \text{or} \quad x^2 = 16$$

$$x = \pm 3 \quad \text{or} \quad x = \pm 4$$

Since $y = 12/x$, if $x = 3$, $y = 4$; if $x = -3$, $y = -4$; if $x = 4$, $y = 3$; and if $x = -4$, $y = -3$. The pairs $(3, 4)$, $(-3, -4)$, $(4, 3)$, and $(-4, -3)$ check. They are the solutions.

21. $x^2 + y^2 = 4,$ (1)

     $16x^2 + 9y^2 = 144$ (2)

$$\begin{array}{ll} -9x^2 - 9y^2 = -36 & \text{Multiplying (1) by } -9 \\ \underline{16x^2 + 9y^2 = 144} & \\ 7x^2 \qquad\quad = 108 & \text{Adding} \end{array}$$

$$x^2 = \frac{108}{7}$$

$$x = \pm\sqrt{\frac{108}{7}} = \pm 6\sqrt{\frac{3}{7}}$$

$$x = \pm\frac{6\sqrt{21}}{7} \quad \begin{array}{l}\text{Rationalizing the de-}\\\text{nominator}\end{array}$$

Substituting $\frac{6\sqrt{21}}{7}$ or $-\frac{6\sqrt{21}}{7}$ for x in equation (1) gives us

$$\frac{36 \cdot 21}{49} + y^2 = 4$$

$$y^2 = 4 - \frac{108}{7}$$

$$y^2 = -\frac{80}{7}$$

$$y = \pm\sqrt{-\frac{80}{7}} = \pm 4i\sqrt{\frac{5}{7}}$$

$$y = \pm\frac{4i\sqrt{35}}{7}. \quad \begin{array}{l}\text{Rationalizing the}\\\text{denominator}\end{array}$$

The pairs $\left(\dfrac{6\sqrt{21}}{7}, \dfrac{4i\sqrt{35}}{7}\right)$,

$\left(\dfrac{6\sqrt{21}}{7}, -\dfrac{4i\sqrt{35}}{7}\right)$, $\left(-\dfrac{6\sqrt{21}}{7}, \dfrac{4i\sqrt{35}}{7}\right)$, and

$\left(-\dfrac{6\sqrt{21}}{7}, -\dfrac{4i\sqrt{35}}{7}\right)$ check. They are the solutions.

23. $x^2 + 4y^2 = 25,$ (1)

     $x + 2y = 7$ (2)

First solve equation (2) for x.

$$x = -2y + 7 \qquad (3)$$

Then substitute $-2y + 7$ for x in equation (1) and solve for y.

$$x^2 + 4y^2 = 25$$

$$(-2y + 7)^2 + 4y^2 = 25$$

$$4y^2 - 28y + 49 + 4y^2 = 25$$

$$8y^2 - 28y + 24 = 0$$

$$2y^2 - 7y + 6 = 0$$

$$(2y - 3)(y - 2) = 0$$

$$y = \frac{3}{2} \text{ or } y = 2$$

Now substitute these numbers in equation (3) and solve for x.

$$x = -2 \cdot \frac{3}{2} + 7 = 4$$

$$x = -2 \cdot 2 + 7 = 3$$

The pairs $\left(4, \dfrac{3}{2}\right)$ and $(3, 2)$ check, so they are the solutions.

25. $x^2 - xy + 3y^2 = 27,$ (1)

     $x - y = 2$ (2)

First solve equation (2) for y.

$$x - 2 = y \qquad\qquad (3)$$

Then substitute $x - 2$ for y in equation (1) and solve for x.

$$x^2 - xy + 3y^2 = 27$$

$$x^2 - x(x - 2) + 3(x - 2)^2 = 27$$

$$x^2 - x^2 + 2x + 3x^2 - 12x + 12 = 27$$

$$3x^2 - 10x - 15 = 0$$

$$x = \frac{-(-10) \pm \sqrt{(-10)^2 - 4(3)(-15)}}{2 \cdot 3}$$

$$x = \frac{10 \pm \sqrt{100 + 180}}{6} = \frac{10 \pm \sqrt{280}}{6}$$

$$x = \frac{10 \pm 2\sqrt{70}}{6} = \frac{5 \pm \sqrt{70}}{3}$$

Now substitute these numbers in equation (3) and solve for y.

$y = \dfrac{5 + \sqrt{70}}{3} - 2 = \dfrac{-1 + \sqrt{70}}{3}$

$y = \dfrac{5 - \sqrt{70}}{3} - 2 = \dfrac{-1 - \sqrt{70}}{3}$

The pairs $\left(\dfrac{5 + \sqrt{70}}{3}, \dfrac{-1 + \sqrt{70}}{3}\right)$ and

$\left(\dfrac{5 - \sqrt{70}}{3}, \dfrac{-1 - \sqrt{70}}{3}\right)$ check, so they are the solutions.

27. $\quad x^2 + y^2 = 16, \qquad x^2 + y^2 = 16, \quad (1)$
$\qquad\qquad\qquad\qquad \text{or}$
$\quad y^2 - 2x^2 = 10 \qquad -2x^2 + y^2 = 10 \quad (2)$

Here we use the elimination method.

$\begin{aligned} 2x^2 + 2y^2 &= 32 \quad \text{Multiplying (1) by 2} \\ -2x^2 + y^2 &= 10 \\ \hline 3y^2 &= 42 \quad \text{Adding} \\ y^2 &= 14 \\ y &= \pm\sqrt{14} \end{aligned}$

Substituting $\sqrt{14}$ or $-\sqrt{14}$ for y in equation (1) gives us

$x^2 + 14 = 16$

$x^2 = 2$

$x = \pm\sqrt{2}$

The pairs $(-\sqrt{2}, -\sqrt{14})$, $(-\sqrt{2}, \sqrt{14})$, $(\sqrt{2}, -\sqrt{14})$, and $(\sqrt{2}, \sqrt{14})$ check. They are the solutions.

29. $\quad x^2 + y^2 = 5, \quad (1)$
$\qquad xy = 2 \qquad\quad (2)$

First we solve equation (2) for y.

$xy = 2$

$y = \dfrac{2}{x}$

Then we substitute $\dfrac{2}{x}$ for y in equation (1) and solve for x.

$x^2 + y^2 = 5$

$x^2 + \left(\dfrac{2}{x}\right)^2 = 5$

$x^2 + \dfrac{4}{x^2} = 5$

$x^4 + 4 = 5x^2 \quad \text{Multiplying by } x^2$

$x^4 - 5x^2 + 4 = 0$

$u^2 - 5u + 4 = 0 \quad \text{Letting } u = x^2$

$(u - 4)(u - 1) = 0$

$u = 4 \quad \text{or} \quad u = 1$

We now substitute x^2 for u and solve for x.

$x^2 = 4 \quad \text{or} \quad x^2 = 1$

$x = \pm 2 \qquad\quad x = \pm 1$

Since $y = 2/x$, if $x = 2$, $y = 1$; if $x = -2$, $y = -1$; if $x = 1$, $y = 2$; and if $x = -1$, $y = -2$. The pairs $(2, 1)$, $(-2, -1)$, $(1, 2)$, and $(-1, -2)$ check. They are the solutions.

31. $\quad 3x + y = 7 \qquad (1)$
$\qquad 4x^2 + 5y = 56 \quad (2)$

First solve equation (1) for y.

$3x + y = 7$

$y = 7 - 3x \quad (3)$

Next substitute $7 - 3x$ for y in equation (2) and solve for x.

$4x^2 + 5y = 56$

$4x^2 + 5(7 - 3x) = 56$

$4x^2 + 35 - 15x = 56$

$4x^2 - 15x - 21 = 0$

Using the quadratic formula, we find that

$x = \dfrac{15 - \sqrt{561}}{8} \quad \text{or} \quad x = \dfrac{15 + \sqrt{561}}{8}.$

Now substitute these numbers into equation (3) and solve for y.

If $x = \dfrac{15 - \sqrt{561}}{8}$, $y = 7 - 3\left(\dfrac{15 - \sqrt{561}}{8}\right)$, or

$\dfrac{11 + 3\sqrt{561}}{8}.$

If $x = \dfrac{15 + \sqrt{561}}{8}$, $y = 7 - 3\left(\dfrac{15 + \sqrt{561}}{8}\right)$, or

$\dfrac{11 - 3\sqrt{561}}{8}.$

The pairs $\left(\dfrac{15 - \sqrt{561}}{8}, \dfrac{11 + 3\sqrt{561}}{8}\right)$ and

$\left(\dfrac{15 + \sqrt{561}}{8}, \dfrac{11 - 3\sqrt{561}}{8}\right)$ check and are the solutions.

33. $\quad a + b = 7, \quad (1)$
$\qquad ab = 4 \qquad\quad (2)$

First solve equation (1) for a.

$a = -b + 7 \quad (3)$

Then substitute $-b + 7$ for a in equation (2) and solve for b.

$(-b + 7)b = 4$

$-b^2 + 7b = 4$

$0 = b^2 - 7b + 4$

$b = \dfrac{-(-7) \pm \sqrt{(-7)^2 - 4 \cdot 1 \cdot 4}}{2 \cdot 1}$

$b = \dfrac{7 \pm \sqrt{33}}{2}$

Now substitute these numbers in equation (3) and solve for a.

$$a = -\left(\frac{7 + \sqrt{33}}{2}\right) + 7 = \frac{7 - \sqrt{33}}{2}$$

$$a = -\left(\frac{7 - \sqrt{33}}{2}\right) + 7 = \frac{7 + \sqrt{33}}{2}$$

The pairs $\left(\frac{7 - \sqrt{33}}{2}, \frac{7 + \sqrt{33}}{2}\right)$ and

$\left(\frac{7 + \sqrt{33}}{2}, \frac{7 - \sqrt{33}}{2}\right)$ check, so they are the solutions.

35. $x^2 + y^2 = 13$, $\qquad$ (1)

$\qquad xy = 6$ $\qquad\qquad$ (2)

First we solve Equation (2) for y.

$$xy = 6$$

$$y = \frac{6}{x}$$

Then we substitute $\frac{6}{x}$ for y in equation (1) and solve for x.

$$x^2 + y^2 = 13$$

$$x^2 + \left(\frac{6}{x}\right)^2 = 13$$

$$x^2 + \frac{36}{x^2} = 13$$

$$x^4 + 36 = 13x^2 \qquad \text{Multiplying by } x^2$$

$$x^4 - 13x^2 + 36 = 0$$

$$u^2 - 13u + 36 = 0 \qquad \text{Letting } u = x^2$$

$$(u - 9)(u - 4) = 0$$

$$u = 9 \quad \text{or} \quad u = 4$$

We now substitute x^2 for u and solve for x.

$$x^2 = 9 \quad \text{or} \quad x^2 = 4$$

$$x = \pm 3 \quad \text{or} \quad x = \pm 2$$

Since $y = 6/x$, if $x = 3$, $y = 2$; if $x = -3$, $y = -2$; if $x = 2$, $y = 3$; and if $x = -2$, $y = -3$. The pairs $(3, 2)$, $(-3, -2)$, $(2, 3)$, and $(-2, -3)$ check. They are the solutions.

37. $x^2 + y^2 + 6y + 5 = 0$ $\qquad$ (1)

$\qquad x^2 + y^2 - 2x - 8 = 0$ $\qquad$ (2)

Using the elimination method, multiply equation (2) by -1 and add the result to equation (1).

$$\begin{array}{ll} x^2 + y^2 + 6y + 5 = 0 & (1) \\ \underline{-x^2 - y^2 + 2x + 8 = 0} & (2) \\ 2x + 6y + 13 = 0 & (3) \end{array}$$

Solve equation (3) for x.

$$2x + 6y + 13 = 0$$

$$2x = -6y - 13$$

$$x = \frac{-6y - 13}{2}$$

Substitute $\frac{-6y - 13}{2}$ for x in equation (1) and solve for y.

$$x^2 + y^2 + 6y + 5 = 0$$

$$\left(\frac{-6y - 13}{2}\right)^2 + y^2 + 6y + 5 = 0$$

$$\frac{36y^2 + 156y + 169}{4} + y^2 + 6y + 5 = 0$$

$$36y^2 + 156y + 169 + 4y^2 + 24y + 20 = 0$$

$$40y^2 + 180y + 189 = 0$$

Using the quadratic formula, we find that

$y = \frac{-45 \pm 3\sqrt{15}}{20}$. Substitute $\frac{-45 \pm 3\sqrt{15}}{20}$ for y in

$x = \frac{-6y - 13}{2}$ and solve for x.

If $y = \frac{-45 + 3\sqrt{15}}{20}$, then

$$x = \frac{-6\left(\frac{-45 + 3\sqrt{15}}{20}\right) - 13}{2} = \frac{5 - 9\sqrt{15}}{20}.$$

If $y = \frac{-45 - 3\sqrt{15}}{20}$, then

$$x = \frac{-6\left(\frac{-45 - 3\sqrt{15}}{20}\right) - 13}{2} = \frac{5 + 9\sqrt{15}}{20}.$$

The pairs $\left(\frac{5 + 9\sqrt{15}}{20}, \frac{-45 - 3\sqrt{15}}{20}\right)$ and

$\left(\frac{5 - 9\sqrt{15}}{20}, \frac{-45 + 3\sqrt{15}}{20}\right)$ check and are the solutions.

39. $2a + b = 1$, $\qquad$ (1)

$\qquad b = 4 - a^2$ $\qquad$ (2)

Equation (2) is already solved for b. Substitute $4 - a^2$ for b in equation (1) and solve for a.

$$2a + 4 - a^2 = 1$$

$$0 = a^2 - 2a - 3$$

$$0 = (a - 3)(a + 1)$$

$$a = 3 \quad \text{or} \quad a = -1$$

Substitute these numbers in equation (2) and solve for b.

$$b = 4 - 3^2 = -5$$

$$b = 4 - (-1)^2 = 3$$

The pairs $(3, -5)$ and $(-1, 3)$ check. They are the solutions.

41. $a^2 + b^2 = 89,$ (1)

     $a - b = 3$ (2)

First solve equation (2) for a.

     $a = b + 3$ (3)

Then substitute $b + 3$ for a in equation (1) and solve for b.

$$(b+3)^2 + b^2 = 89$$
$$b^2 + 6b + 9 + b^2 = 89$$
$$2b^2 + 6b - 80 = 0$$
$$b^2 + 3b - 40 = 0$$
$$(b+8)(b-5) = 0$$
$$b = -8 \text{ or } b = 5$$

Substitute these numbers in equation (3) and solve for a.

$$a = -8 + 3 = -5$$
$$a = 5 + 3 = 8$$

The pairs $(-5, -8)$ and $(8, 5)$ check. They are the solutions.

43. $xy - y^2 = 2,$ (1)

    $2xy - 3y^2 = 0$ (2)

$$\begin{array}{rl} -2xy + 2y^2 = -4 & \text{Multiplying (1) by } -2 \\ \underline{2xy - 3y^2 = 0} & \\ -y^2 = -4 & \text{Adding} \\ y^2 = 4 & \\ y = \pm 2 & \end{array}$$

We substitute for y in equation (1) and solve for x.

When $y = 2$: $x \cdot 2 - 2^2 = 2$

$$2x - 4 = 2$$
$$2x = 6$$
$$x = 3$$

When $y = -2$: $x(-2) - (-2)^2 = 2$

$$-2x - 4 = 2$$
$$-2x = 6$$
$$x = -3$$

The pairs $(3, 2)$ and $(-3, -2)$ check. They are the solutions.

45. $m^2 - 3mn + n^2 + 1 = 0,$ (1)

    $3m^2 - mn + 3n^2 = 13$ (2)

   $m^2 - 3mn + n^2 = -1$ (3)   Rewriting (1)

   $3m^2 - mn + 3n^2 = 13$ (2)

$$\begin{array}{rl} -3m^2 + 9mn - 3n^2 = 3 & \text{Multiplying (3) by } -3 \\ \underline{3m^2 - mn + 3n^2 = 13} & \\ 8mn = 16 & \\ mn = 2 & \\ n = \dfrac{2}{m} & (4) \end{array}$$

Substitute $\dfrac{2}{m}$ for n in equation (1) and solve for m.

$$m^2 - 3m\left(\frac{2}{m}\right) + \left(\frac{2}{m}\right)^2 + 1 = 0$$
$$m^2 - 6 + \frac{4}{m^2} + 1 = 0$$
$$m^2 - 5 + \frac{4}{m^2} = 0$$
$$m^4 - 5m^2 + 4 = 0 \quad \begin{array}{l}\text{Multiplying} \\ \text{by } m^2\end{array}$$

Substitute u for m^2.

$$u^2 - 5u + 4 = 0$$
$$(u - 4)(u - 1) = 0$$
$$u = 4 \quad or \quad u = 1$$
$$m^2 = 4 \quad or \quad m^2 = 1$$
$$m = \pm 2 \quad or \quad m = \pm 1$$

Substitute for m in equation (4) and solve for n.

When $m = 2$, $n = \dfrac{2}{2} = 1$.

When $m = -2$, $n = \dfrac{2}{-2} = -1$.

When $m = 1$, $n = \dfrac{2}{1} = 2$.

When $m = -1$, $n = \dfrac{2}{-1} = -2$.

The pairs $(2, 1)$, $(-2, -1)$, $(1, 2)$, and $(-1, -2)$ check. They are the solutions.

47. $x^2 + y^2 = 5,$ (1)

    $x - y = 8$ (2)

First solve equation (2) for x.

$$x = y + 8 \quad (3)$$

Then substitute $y + 8$ for x in equation (1) and solve for y.

$$(y+8)^2 + y^2 = 5$$
$$y^2 + 16y + 64 + y^2 = 5$$
$$2y^2 + 16y + 59 = 0$$
$$y = \frac{-16 \pm \sqrt{(16)^2 - 4(2)(59)}}{2 \cdot 2}$$
$$y = \frac{-16 \pm \sqrt{-216}}{4}$$
$$y = \frac{-16 \pm 6i\sqrt{6}}{4}$$
$$y = -4 \pm \frac{3}{2}i\sqrt{6}$$

Now substitute these numbers in equation (3) and solve for x.

$$x = -4 + \frac{3}{2}i\sqrt{6} + 8 = 4 + \frac{3}{2}i\sqrt{6}$$

$$x = -4 - \frac{3}{2}i\sqrt{6} + 8 = 4 - \frac{3}{2}i\sqrt{6}$$

The pairs $\left(4 + \frac{3}{2}i\sqrt{6}, -4 + \frac{3}{2}i\sqrt{6}\right)$ and

$\left(4 - \frac{3}{2}i\sqrt{6}, -4 - \frac{3}{2}i\sqrt{6}\right)$ check. They are the solutions.

49. $a^2 + b^2 = 14,$ (1)

$ab = 3\sqrt{5}$ (2)

Solve equation (2) for b.

$$b = \frac{3\sqrt{5}}{a}$$

Substitute $\frac{3\sqrt{5}}{a}$ for b in equation (1) and solve for a.

$$a^2 + \left(\frac{3\sqrt{5}}{a}\right)^2 = 14$$

$$a^2 + \frac{45}{a^2} = 14$$

$$a^4 + 45 = 14a^2$$

$$a^4 - 14a^2 + 45 = 0$$

$$u^2 - 14u + 45 = 0 \qquad \text{Letting } u = a^2$$

$$(u - 9)(u - 5) = 0$$

$$u = 9 \quad or \quad u = 5$$

$$a^2 = 9 \quad or \quad a^2 = 5$$

$$a = \pm 3 \quad or \quad a = \pm\sqrt{5}$$

Since $b = 3\sqrt{5}/a$, if $a = 3$, $b = \sqrt{5}$; if $a = -3$, $b = -\sqrt{5}$; if $a = \sqrt{5}$, $b = 3$; and if $a = -\sqrt{5}$, $b = -3$. The pairs $(3, \sqrt{5})$, $(-3, -\sqrt{5})$, $(\sqrt{5}, 3)$, $(-\sqrt{5}, -3)$ check. They are the solutions.

51. $x^2 + y^2 = 25,$ (1)

$9x^2 + 4y^2 = 36$ (2)

$$-4x^2 - 4y^2 = -100 \qquad \text{Multiplying (1) by } -4$$

$$\frac{9x^2 + 4y^2 = 36}{5x^2 \qquad = -64}$$

$$x^2 = -\frac{64}{5}$$

$$x = \pm\sqrt{\frac{-64}{5}} = \pm\frac{8i}{\sqrt{5}}$$

$$x = \pm\frac{8i\sqrt{5}}{5} \qquad \text{Rationalizing the denominator}$$

Substituting $\frac{8i\sqrt{5}}{5}$ or $-\frac{8i\sqrt{5}}{5}$ for x in equation (1) and solving for y gives us

$$-\frac{64}{5} + y^2 = 25$$

$$y^2 = \frac{189}{5}$$

$$y = \pm\sqrt{\frac{189}{5}} = \pm 3\sqrt{\frac{21}{5}}$$

$$y = \pm\frac{3\sqrt{105}}{5}. \qquad \text{Rationalizing the denominator}$$

The pairs $\left(\frac{8i\sqrt{5}}{5}, \frac{3\sqrt{105}}{5}\right)$, $\left(-\frac{8i\sqrt{5}}{5}, \frac{3\sqrt{105}}{5}\right)$, $\left(\frac{8i\sqrt{5}}{5}, -\frac{3\sqrt{105}}{5}\right)$, and $\left(-\frac{8i\sqrt{5}}{5}, -\frac{3\sqrt{105}}{5}\right)$ check.

They are the solutions.

53. $5y^2 - x^2 = 1,$ (1)

$xy = 2$ (2)

Solve equation (2) for x.

$$x = \frac{2}{y}$$

Substitute $\frac{2}{y}$ for x in equation (1) and solve for y.

$$5y^2 - \left(\frac{2}{y}\right)^2 = 1$$

$$5y^2 - \frac{4}{y^2} = 1$$

$$5y^4 - 4 = y^2$$

$$5y^4 - y^2 - 4 = 0$$

$$5u^2 - u - 4 = 0 \qquad \text{Letting } u = y^2$$

$$(5u + 4)(u - 1) = 0$$

$$5u + 4 = 0 \qquad or \quad u - 1 = 0$$

$$u = -\frac{4}{5} \qquad or \qquad u = 1$$

$$y^2 = -\frac{4}{5} \qquad or \qquad y^2 = 1$$

$$y = \pm\frac{2i}{\sqrt{5}} \qquad or \qquad y = \pm 1$$

$$y = \pm\frac{2i\sqrt{5}}{5} \qquad or \qquad y = \pm 1$$

Since $x = 2/y$, if $y = \frac{2i\sqrt{5}}{5}$, $x = \frac{2}{\frac{2i\sqrt{5}}{5}} = \frac{5}{i\sqrt{5}} = \frac{5}{i\sqrt{5}} \cdot \frac{-i\sqrt{5}}{-i\sqrt{5}} = -i\sqrt{5}$; if $y = -\frac{2i\sqrt{5}}{5}$,

$$x = \frac{2}{-\frac{2i\sqrt{5}}{5}} = i\sqrt{5};$$

if $y = 1$, $x = 2/1 = 2$; if $y = -1$, $x = 2/-1 = -2$.

The pairs $\left(-i\sqrt{5}, \dfrac{2i\sqrt{5}}{5}\right)$, $\left(i\sqrt{5}, -\dfrac{2i\sqrt{5}}{5}\right)$, $(2,1)$ and $(-2,-1)$ check. They are the solutions.

55. *Familiarize*. We first make a drawing. We let l and w represent the length and width, respectively.

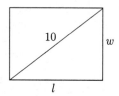

Translate. The perimeter is 28 cm.

$2l + 2w = 28$, or $l + w = 14$

Using the Pythagorean theorem we have another equation.

$l^2 + w^2 = 10^2$, or $l^2 + w^2 = 100$

Carry out. We solve the system:

$l + w = 14$,   (1)

$l^2 + w^2 = 100$ (2)

First solve equation (1) for w.

$w = 14 - l$ (3)

Then substitute $14 - l$ for w in equation (2) and solve for l.

$$l^2 + w^2 = 100$$
$$l^2 + (14 - l)^2 = 100$$
$$l^2 + 196 - 28l + l^2 = 100$$
$$2l^2 - 28l + 96 = 0$$
$$l^2 - 14l + 48 = 0$$
$$(l - 8)(l - 6) = 0$$

$l = 8$ or $l = 6$

If $l = 8$, then $w = 14 - 8$, or 6. If $l = 6$, then $w = 14 - 6$, or 8. Since the length is usually considered to be longer than the width, we have the solution $l = 8$ and $w = 6$, or $(8, 6)$.

Check. If $l = 8$ and $w = 6$, then the perimeter is $2 \cdot 8 + 2 \cdot 6$, or 28. The length of a diagonal is $\sqrt{8^2 + 6^2}$, or $\sqrt{100}$, or 10. The numbers check.

State. The length is 8 cm, and the width is 6 cm.

57. *Familiarize*. We first make a drawing. Let $l =$ the length and $w =$ the width of the rectangle.

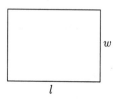

Translate.

Area: $lw = 20$

Perimeter: $2l + 2w = 18$, or $l + w = 9$

Carry out. We solve the system:

Solve the second equation for l: $l = 9 - w$

Substitute $9 - w$ for l in the first equation and solve for w.

$$(9 - w)w = 20$$
$$9w - w^2 = 20$$
$$0 = w^2 - 9w + 20$$
$$0 = (w - 5)(w - 4)$$

$w = 5$ or $w = 4$

If $w = 5$, then $l = 9 - w$, or 4. If $w = 4$, then $l = 9 - 4$, or 5. Since length is usually considered to be longer than width, we have the solution $l = 5$ and $w = 4$, or $(5, 4)$.

Check. If $l = 5$ and $w = 4$, the area is $5 \cdot 4$, or 20. The perimeter is $2 \cdot 5 + 2 \cdot 4$, or 18. The numbers check.

State. The length is 5 in. and the width is 4 in.

59. *Familiarize*. We first make a drawing. Let $l =$ the length and $w =$ the width.

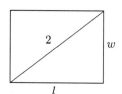

Translate.

Area: $lw = \sqrt{3}$ (1)

From the Pythagorean theorem: $l^2 + w^2 = 2^2$ (2)

Carry out. We solve the system of equations.

We first solve equation (1) for w.

$$lw = \sqrt{3}$$
$$w = \frac{\sqrt{3}}{l}$$

Then we substitute $\dfrac{\sqrt{3}}{l}$ for w in equation 2 and solve for l.

$$l^2 + \left(\frac{\sqrt{3}}{l}\right)^2 = 4$$
$$l^2 + \frac{3}{l^2} = 4$$
$$l^4 + 3 = 4l^2$$
$$l^4 - 4l^2 + 3 = 0$$
$$u^2 - 4u + 3 = 0 \quad \text{Letting } u = l^2$$
$$(u - 3)(u - 1) = 0$$

$u = 3$ or $u = 1$

We now substitute l^2 for u and solve for l.

$l^2 = 3 \quad$ or $\quad l^2 = 1$

$l = \pm\sqrt{3} \quad$ or $\quad l = \pm 1$

Measurements cannot be negative, so we only need to consider $l = \sqrt{3}$ and $l = 1$. Since $w = \sqrt{3}/l$, if $l = \sqrt{3}$, $w = 1$ and if $l = 1$, $w = \sqrt{3}$. Length is usually considered to be longer than width, so we have the solution $l = \sqrt{3}$ and $w = 1$, or $(\sqrt{3}, 1)$.

Check. If $l = \sqrt{3}$ and $w = 1$, the area is $\sqrt{3}\cdot 1 = \sqrt{3}$. Also $(\sqrt{3})^2 + 1^2 = 3 + 1 = 4 = 2^2$. The numbers check.

State. The length is $\sqrt{3}$ m, and the width is 1 m.

61. **Familiarize**. We make a drawing of the dog run. Let $l =$ the length and $w =$ the width.

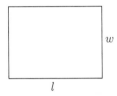

Since it takes 210 yd of fencing to enclose the run, we know that the perimeter is 210 yd.

Translate.

Perimeter: $2l + 2w = 210$, or $l + w = 105$

Area: $lw = 2250$

Carry out. We solve the system:

Solve the first equation for l: $l = 105 - w$

Substitute $105 - w$ for l in the second equation and solve for w.

$(105 - w)w = 2250$

$105w - w^2 = 2250$

$0 = w^2 - 105w + 2250$

$0 = (w - 30)(w - 75)$

$w = 30 \quad$ or $\quad w = 75$

If $w = 30$, then $l = 105 - 30$, or 75. If $w = 75$, then $l = 105 - 75$, or 30. Since length is usually considered to be longer than width, we have the solution $l = 75$ and $w = 30$, or $(75, 30)$.

Check. If $l = 75$ and $w = 30$, the perimeter is $2 \cdot 75 + 2 \cdot 30$, or 210. The area is $75(30)$, or 2250. The numbers check.

State. The length is 75 yd and the width is 30 yd.

63. **Familiarize**. We let $x =$ the length of a side of one test plot and $y =$ the length of a side of the other plot. Make a drawing.

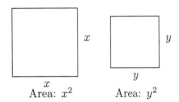

Area: x^2 Area: y^2

Translate.

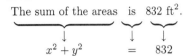

The sum of the areas is 832 ft^2.

$x^2 + y^2 \quad = \quad 832$

The difference of the areas is 320 ft^2.

$x^2 - y^2 \quad = \quad 320$

Carry out. We solve the system of equations.

$$x^2 + y^2 = 832$$
$$\underline{x^2 - y^2 = 320}$$
$$2x^2 \quad\;\; = 1152 \quad \text{Adding}$$
$$x^2 = 576$$
$$x = \pm 24$$

Since measurements cannot be negative, we consider only $x = 24$. Substitute 24 for x in the first equation and solve for y.

$24^2 + y^2 = 832$

$576 + y^2 = 832$

$y^2 = 256$

$y = \pm 16$

Again, we consider only the positive value, 16. The possible solution is $(24, 16)$.

Check. The areas of the test plots are 24^2, or 576, and 16^2, or 256. The sum of the areas is $576 + 256$, or 832. The difference of the areas is $576 - 256$, or 320. The values check.

State. The lengths of the test plots are 24 ft and 16 ft.

65. a) $I(x) = 21.15508287x + 198.9151934$

b) $E(x) = 197.3380928(1.053908223)^x$

c) Find the first coordinate of the point of intersection of the graphs of $y_1 = I(x)$ and $y_2 = E(x)$. It is approximately 25, so income will equal expenditures about 25 years after 1985.

67. Discussion and Writing

69. $2^{3x} = 64$

$2^{3x} = 2^6$

$3x = 6$

$x = 2$

The solution is 2.

71. $\log_3 x = 4$

$x = 3^4$

$x = 81$

The solution is 81.

73. $(x - h)^2 + (y - k)^2 = r^2$

If $(2, 4)$ is a point on the circle, then

$(2 - h)^2 + (4 - k)^2 = r^2$.

If $(3, 3)$ is a point on the circle, then

$(3 - h)^2 + (3 - k)^2 = r^2$.

Thus

$$(2 - h)^2 + (4 - k)^2 = (3 - h)^2 + (3 - k)^2$$

$$4 - 4h + h^2 + 16 - 8k + k^2 =$$
$$9 - 6h + h^2 + 9 - 6k + k^2$$

$$-4h - 8k + 20 = -6h - 6k + 18$$

$$2h - 2k = -2$$

$$h - k = -1$$

If the center (h, k) is on the line $3x - y = 3$, then $3h - k = 3$.

Solving the system

$h - k = -1$,

$3h - k = 3$

we find that $(h, k) = (2, 3)$.

Find r^2, substituting $(2, 3)$ for (h, k) and $(2, 4)$ for (x, y). We could also use $(3, 3)$ for (x, y).

$(x - h)^2 + (y - k)^2 = r^2$

$(2 - 2)^2 + (4 - 3)^2 = r^2$

$0 + 1 = r^2$

$1 = r^2$

The equation of the circle is 1 is $(x-2)^2 + (y-3)^2 = 1$.

75. The equation of the ellipse is of the form $\frac{x^2}{a^2} + \frac{y^2}{b^2} = 1$. Substitute $\left(1, \frac{\sqrt{3}}{2}\right)$ and $\left(\sqrt{3}, \frac{1}{2}\right)$ for (x, y) to get two equations.

$$\frac{1^2}{a^2} + \frac{\left(\frac{\sqrt{3}}{2}\right)^2}{b^2} = 1, \ \ or \ \ \frac{1}{a^2} + \frac{3}{4b^2} = 1$$

$$\frac{(\sqrt{3})^2}{a^2} + \frac{\left(\frac{1}{2}\right)^2}{b^2} = 1, \ \ or \ \ \frac{3}{a^2} + \frac{1}{4b^2} = 1$$

Substitute u for $\frac{1}{a^2}$ and v for $\frac{1}{b^2}$.

$u + \frac{3}{4}v = 1, \qquad\qquad 4u + 3v = 4,$

$\qquad\qquad\qquad or$

$3u + \frac{1}{4}v = 1 \qquad\qquad 12u + v = 4$

Solving for u and v, we get $u = \frac{1}{4}$, $v = 1$. Then $u = \frac{1}{a^2} = \frac{1}{4}$, so $a^2 = 4$; $v = \frac{1}{b^2} = 1$, so $b^2 = 1$.

Then the equation of the ellipse is

$$\frac{x^2}{4} + \frac{y^2}{1} = 1, \text{ or } \frac{x^2}{4} + y^2 = 1.$$

77. $(x - h)^2 + (y - k)^2 = r^2$ Standard form

Substitute $(4, 6)$, $(-6, 2)$, and $(1, -3)$ for (x, y).

$(4 - h)^2 + (6 - k)^2 = r^2$ (1)

$(-6 - h)^2 + (2 - k)^2 = r^2$ (2)

$(1 - h)^2 + (-3 - k)^2 = r^2$ (3)

Thus

$(4 - h)^2 + (6 - k)^2 = (-6 - h)^2 + (2 - k)^2$, or

$5h + 2k = 3$

and

$(4 - h)^2 + (6 - k)^2 = (1 - h)^2 + (-3 - k)^2$, or

$h + 3k = 7$.

We solve the system

$5h + 2k = 3$,

$h + 3k = 7$.

Solving we get $h = -\frac{5}{13}$ and $k = \frac{32}{13}$. Substituting these values in equation (1), (2), or (3), we find that $r^2 = \frac{5365}{169}$.

The equation of the circle is

$$\left(x + \frac{5}{13}\right)^2 + \left(y - \frac{32}{13}\right)^2 = \frac{5365}{169}.$$

79. See the answer section in the text.

81. *Familiarize*. Let x and y represent the numbers.

Translate.

The square of a certain number exceeds twice the square of another number by $\frac{1}{8}$.

$$x^2 = 2y^2 + \frac{1}{8}$$

The sum of the squares is $\frac{5}{16}$.

$$x^2 + y^2 = \frac{5}{16}$$

Carry out. We solve the system.

$$x^2 - 2y^2 = \frac{1}{8}, \quad (1)$$

$$x^2 + y^2 = \frac{5}{16} \quad (2)$$

$$x^2 - 2y^2 = \frac{1}{8},$$

$$\underline{2x^2 + 2y^2 = \frac{5}{8}} \quad \text{Multiplying (2) by 2}$$

$$3x^2 \qquad = \frac{6}{8}$$

$$x^2 = \frac{1}{4}$$

$$x = \pm\frac{1}{2}$$

Substitute $\pm\frac{1}{2}$ for x in (2) and solve for y.

$$\left(\pm\frac{1}{2}\right)^2 + y^2 = \frac{5}{16}$$

$$\frac{1}{4} + y^2 = \frac{5}{16}$$

$$y^2 = \frac{1}{16}$$

$$y = \pm\frac{1}{4}$$

We get $\left(\frac{1}{2}, \frac{1}{4}\right)$, $\left(-\frac{1}{2}, \frac{1}{4}\right)$, $\left(\frac{1}{2}, -\frac{1}{4}\right)$ and $\left(-\frac{1}{2}, -\frac{1}{4}\right)$.

Check. It is true that $\left(\pm\frac{1}{2}\right)^2$ exceeds twice $\left(\pm\frac{1}{4}\right)^2$ by $\frac{1}{8}$: $\frac{1}{4} = 2\left(\frac{1}{16}\right) + \frac{1}{8}$

Also $\left(\pm\frac{1}{2}\right)^2 + \left(\pm\frac{1}{4}\right)^2 = \frac{5}{16}$. The pairs check.

State. The numbers are $\frac{1}{2}$ and $\frac{1}{4}$ or $-\frac{1}{2}$ and $\frac{1}{4}$ or $\frac{1}{2}$ and $-\frac{1}{4}$ or $-\frac{1}{2}$ and $-\frac{1}{4}$.

83. See the answer section in the text.

85. $x^3 + y^3 = 72, \quad (1)$

$x + y = 6 \quad (2)$

Solve equation (2) for y: $y = 6 - x$

Substitute for y in equation (1) and solve for x.

$$x^3 + (6 - x)^3 = 72$$

$$x^3 + 216 - 108x + 18x^2 - x^3 = 72$$

$$18x^2 - 108x + 144 = 0$$

$$x^2 - 6x + 8 = 0 \qquad \text{Multiplying by } \frac{1}{18}$$

$$(x - 4)(x - 2) = 0$$

$x = 4$ or $x = 2$

If $x = 4$, then $y = 6 - 4 = 2$.

If $x = 2$, then $y = 6 - 2 = 4$.

The pairs $(4, 2)$ and $(2, 4)$ check.

87. $p^2 + q^2 = 13, \quad (1)$

$$\frac{1}{pq} = -\frac{1}{6} \quad (2)$$

Solve equation (2) for p.

$$\frac{1}{q} = -\frac{p}{6}$$

$$-\frac{6}{q} = p$$

Substitute $-6/q$ for p in equation (1) and solve for q.

$$\left(-\frac{6}{q}\right)^2 + q^2 = 13$$

$$\frac{36}{q^2} + q^2 = 13$$

$$36 + q^4 = 13q^2$$

$$q^4 - 13q^2 + 36 = 0$$

$$u^2 - 13u + 36 = 0 \qquad \text{Letting } u = q^2$$

$$(u - 9)(u - 4) = 0$$

$u = 9$ or $u = 4$

$x^2 = 9$ or $x^2 = 4$

$x = \pm 3$ or $x = \pm 2$

Since $p = -6/q$, if $q = 3$, $p = -2$; if $q = -3$, $p = 2$; if $q = 2$, $p = -3$; and if $q = -2$, $p = 3$. The pairs $(-2, 3)$, $(2, -3)$, $(-3, 2)$, and $(3, -2)$ check. They are the solutions.

89. $5^{x+y} = 100,$

$3^{2x-y} = 1000$

$(x + y)\log 5 = 2,$ Taking logarithms and

$(2x - y)\log 3 = 3$ simplifying

$x\log 5 + y\log 5 = 2, \quad (1)$

$2x\log 3 - y\log 3 = 3 \quad (2)$

Multiply equation (1) by $\log 3$ and equation (2) by $\log 5$ and add.

$$x \log 3 \cdot \log 5 + y \log 3 \cdot \log 5 = 2 \log 3$$
$$\frac{2x \log 3 \cdot \log 5 - y \log 3 \cdot \log 5 = 3 \log 5}{3x \log 3 \cdot \log 5 \qquad\qquad = 2 \log 3 +}$$
$$\qquad\qquad\qquad\qquad\qquad 3 \log 5$$
$$x = \frac{2 \log 3 + 3 \log 5}{3 \log 3 \cdot \log 5}$$

Substitute in (1) to find y.
$$\frac{2 \log 3 + 3 \log 5}{3 \log 3 \cdot \log 5} \cdot \log 5 + y \log 5 = 2$$
$$y \log 5 = 2 - \frac{2 \log 3 + 3 \log 5}{3 \log 3}$$
$$y \log 5 = \frac{6 \log 3 - 2 \log 3 - 3 \log 5}{3 \log 3}$$
$$y \log 5 = \frac{4 \log 3 - 3 \log 5}{3 \log 3}$$
$$y = \frac{4 \log 3 - 3 \log 5}{3 \log 3 \cdot \log 5}$$

The pair $\left(\dfrac{2 \log 3 + 3 \log 5}{3 \log 3 \cdot \log 5}, \dfrac{4 \log 3 - 3 \log 5}{3 \log 3 \cdot \log 5} \right)$ checks. It is the solution.

91. Find the points of intersection of $y_1 = \ln x + 2$ and $y_2 = x^2$. They are $(1.564, 2.448)$ and $(0.138, 0.019)$.

93. Find the point of intersection of $y_1 = e^x - 1$ and $y_2 = -3x + 4$. It is $(0.871, 1.388)$.

95. Find the points of intersection of $y_1 = e^x$ and $y_2 = x + 2$. They are $(1.146, 3.146)$ and $(-1.841, 0.159)$.

97. Graph $y_1 = \sqrt{19,380,510.36 - x^2}$,

$y_2 = -\sqrt{19,380,510.36 - x^2}$, and

$y_3 = 27,941.25x/6.125$ and find the points of intersection. They are $(0.965, 4402.33)$ and $(-0.965, -4402.33)$.

99. Graph $y_1 = \sqrt{\dfrac{14.5x^2 - 64.5}{13.5}}$,

$y_2 = -\sqrt{\dfrac{14.5x^2 - 64.5}{13.5}}$, and $y_3 = (5.5x - 12.3)/6.3$ and find the points of intersection. They are $(2.112, -0.109)$ and $(-13.041, -13.337)$.

101. Graph $y_1 = \sqrt{\dfrac{56,548 - 0.319x^2}{2688.7}}$,

$y_2 = -\sqrt{\dfrac{56,548 - 0.319x^2}{2688.7}}$,

$y_3 = \sqrt{\dfrac{0.306x^2 - 43,452}{2688.7}}$,

and $y_4 = -\sqrt{\dfrac{0.306x^2 - 43,452}{2688.7}}$ and find the points of intersection. They are $(400, 1.431)$, $(-400, 1.431)$, $(400, -1.431)$, and $(-400, -1.431)$.

Exercise Set 9.5

1. We use the rotation of axes formulas to find x' and y'.
$$\begin{aligned}
x' &= x \cos \theta + y \sin \theta \\
&= \sqrt{2} \cos 45° - \sqrt{2} \sin 45° \\
&= \sqrt{2} \cdot \frac{\sqrt{2}}{2} - \sqrt{2} \cdot \frac{\sqrt{2}}{2} \\
&= 1 - 1 = 0
\end{aligned}$$
$$\begin{aligned}
y' &= -x \sin \theta + y \cos \theta \\
&= -\sqrt{2} \sin 45° - \sqrt{2} \cos 45° \\
&= -\sqrt{2} \cdot \frac{\sqrt{2}}{2} - \sqrt{2} \cdot \frac{\sqrt{2}}{2} \\
&= -1 - 1 = -2
\end{aligned}$$
The coordinates are $(0, -2)$.

3. We use the rotation of axes formulas to find x' and y'.
$$\begin{aligned}
x' &= x \cos \theta + y \sin \theta \\
&= 0 \cdot \cos 30° + 2 \sin 30° \\
&= 0 + 2 \cdot \frac{1}{2} = 1
\end{aligned}$$
$$\begin{aligned}
y' &= -x \sin \theta + y \cos \theta \\
&= -0 \cdot \sin 30° + 2 \cos 30° \\
&= 0 + 2 \cdot \frac{\sqrt{3}}{2} = \sqrt{3}
\end{aligned}$$
The coordinates are $(1, \sqrt{3})$.

5. We use the rotation of axes formulas to find x and y.
$$\begin{aligned}
x &= x' \cos \theta - y' \sin \theta \\
&= 1 \cdot \cos 45° - (-1) \sin 45° \\
&= \frac{\sqrt{2}}{2} + \frac{\sqrt{2}}{2} \\
&= \frac{2\sqrt{2}}{2} = \sqrt{2}
\end{aligned}$$
$$\begin{aligned}
y &= x' \sin \theta + y' \cos \theta \\
&= 1 \cdot \sin 45° - 1 \cdot \cos 45° \\
&= \frac{\sqrt{2}}{2} - \frac{\sqrt{2}}{2} = 0
\end{aligned}$$
The coordinates are $(\sqrt{2}, 0)$.

7. We use the rotation of axes formulas to find x and y.

$$x = x' \cos \theta - y' \sin \theta$$
$$= 2 \cos 30° - 0 \cdot \sin 30°$$
$$= 2 \cdot \frac{\sqrt{3}}{2} - 0 = \sqrt{3}$$

$$y = x' \sin \theta + y' \cos \theta$$
$$= 2 \sin 30° + 0 \cdot \cos 30°$$
$$= 2 \cdot \frac{1}{2} + 0 = 1$$

The coordinates are $(\sqrt{3}, 1)$.

9. $3x^2 - 5xy + 3y^2 - 2x + 7y = 0$

$A = 3$, $B = -5$, $C = 3$

$B^2 - 4AC = (-5)^2 - 4 \cdot 3 \cdot 3 = 25 - 36 = -11$

Since the discriminant is negative, the graph is an ellipse (or circle).

11. $x^2 - 3xy - 2y^2 + 12 = 0$

$A = 1$, $B = -3$, $C = -2$

$B^2 - 4AC = (-3)^2 - 4 \cdot 1 \cdot (-2) = 9 + 8 = 17$

Since the discriminant is positive, the graph is a hyperbola.

13. $4x^2 - 12xy + 9y^2 - 3x + y = 0$

$A = 4$, $B = -12$, $C = 9$

$B^2 - 4AC = (-12)^2 - 4 \cdot 4 \cdot 9 = 144 - 144 = 0$

Since the discriminant is zero, the graph is a parabola.

15. $2x^2 - 8xy + 7y^2 + x - 2y + 1 = 0$

$A = 2$, $B = -8$, $C = 7$

$B^2 - 4AC = (-8)^2 - 4 \cdot 2 \cdot 7 = 64 - 56 = 8$

Since the discriminant is positive, the graph is a hyperbola.

17. $8x^2 - 7xy + 5y^2 - 17 = 0$

$A = 8$, $B = -7$, $C = 5$

$B^2 - 4AC = (-7)^2 - 4 \cdot 8 \cdot 5 = 49 - 160 = -111$

Since the discriminant is negative, the graph is an ellipse (or circle).

19. $3x^2 + 2xy + 3y^2 = 16$

$A = 3$, $B = 2$, $C = 3$

$B^2 - 4AC = 2^2 - 4 \cdot 3 \cdot 3 = 4 - 36 = -32$

Since the discriminant is negative, the graph is an ellipse (or circle). To rotate the axes we first determine θ.

$$\cot 2\theta = \frac{A - C}{B} = \frac{3 - 3}{2} = 0$$

Then $2\theta = 90°$ and $\theta = 45°$, so

$$\sin \theta = \frac{\sqrt{2}}{2} \text{ and } \cos \theta = \frac{\sqrt{2}}{2}.$$

Now substitute in the rotation of axes formulas.

$$x = x' \cos \theta - y' \sin \theta$$
$$= x' \left(\frac{\sqrt{2}}{2} \right) - y' \left(\frac{\sqrt{2}}{2} \right) = \frac{\sqrt{2}}{2}(x' - y')$$

$$y = x' \sin \theta + y' \cos \theta$$
$$= x' \left(\frac{\sqrt{2}}{2} \right) + y' \left(\frac{\sqrt{2}}{2} \right) = \frac{\sqrt{2}}{2}(x' + y')$$

Substitute for x and y in the given equation.

$$3\left[\frac{\sqrt{2}}{2}(x' - y') \right]^2 + 2\left[\frac{\sqrt{2}}{2}(x' - y') \right]\left[\frac{\sqrt{2}}{2}(x' + y') \right] +$$
$$3\left[\frac{\sqrt{2}}{2}(x' + y') \right]^2 = 16$$

After simplifying we have

$$\frac{(x')^2}{4} + \frac{(y')^2}{8} = 1.$$

This is the equation of an ellipse with vertices $(0, -\sqrt{8})$ and $(0, \sqrt{8})$, or $(0, -2\sqrt{2})$ and $(0, 2\sqrt{2})$ on the y'-axis. The x'-intercepts are $(-2, 0)$ and $(2,0)$. We sketch the graph.

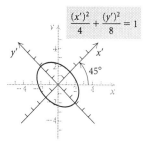

21. $x^2 - 10xy + y^2 + 36 = 0$

$A = 1$, $B = -10$, $C = 1$

$B^2 - 4AC = (-10)^2 - 4 \cdot 1 \cdot 1 = 100 - 4 = 96$

Since the discriminant is positive, the graph is a hyperbola. To rotate the axes we first determine θ.

$$\cot 2\theta = \frac{A - C}{B} = \frac{1 - 1}{-10} = 0$$

Then $2\theta = 90°$ and $\theta = 45°$, so

$$\sin \theta = \frac{\sqrt{2}}{2} \text{ and } \cos \theta = \frac{\sqrt{2}}{2}.$$

Now substitute in the rotation of axes formulas.

$$x = x' \cos \theta - y' \sin \theta$$
$$= x' \left(\frac{\sqrt{2}}{2} \right) - y' \left(\frac{\sqrt{2}}{2} \right) = \frac{\sqrt{2}}{2}(x' - y')$$

$$y = x'\sin\theta + y'\cos\theta$$
$$= x'\left(\frac{\sqrt{2}}{2}\right) + y'\left(\frac{\sqrt{2}}{2}\right) = \frac{\sqrt{2}}{2}(x'+y')$$

Substitute for x and y in the given equation.

$$\left[\frac{\sqrt{2}}{2}(x'-y')\right]^2 - 10\left[\frac{\sqrt{2}}{2}(x'-y')\right]\left[\frac{\sqrt{2}}{2}(x'+y')\right] +$$
$$\left[\frac{\sqrt{2}}{2}(x'+y')\right]^2 + 36 = 0$$

After simplifying we have

$$\frac{(x')^2}{9} - \frac{(y')^2}{6} = 1.$$

This is the equation of a hyperbola with vertices $(-3,0)$ and $(3,0)$ on the x'-axis. The asymptotes are $y' = -\frac{\sqrt{6}}{3}x'$ and $y' = \frac{\sqrt{6}}{3}x'$. We sketch the graph.

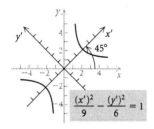

23. $x^2 - 2\sqrt{3}xy + 3y^2 - 12\sqrt{3}x - 12y = 0$

$A = 1$, $B = -2\sqrt{3}$, $C = 3$

$B^2 - 4AC = (-2\sqrt{3})^2 - 4\cdot1\cdot3 = 12 - 12 = 0$

Since the discriminant is zero, the graph is a parabola. To rotate the axes we first determine θ.

$$\cot 2\theta = \frac{A-C}{B} = \frac{1-3}{-2\sqrt{3}} = \frac{-2}{-2\sqrt{3}} = \frac{1}{\sqrt{3}}$$

Then $2\theta = 60°$ and $\theta = 30°$, so

$$\sin\theta = \frac{1}{2} \text{ and } \cos\theta = \frac{\sqrt{3}}{2}.$$

Now substitute in the rotation of axes formulas.

$$x = x'\cos\theta - y'\sin\theta$$
$$= x'\cdot\frac{\sqrt{3}}{2} - y'\cdot\frac{1}{2} = \frac{x'\sqrt{3}}{2} - \frac{y'}{2}$$

$$y = x'\sin\theta + y'\cos\theta$$
$$= x'\cdot\frac{1}{2} + y'\cdot\frac{\sqrt{3}}{2} = \frac{x'}{2} + \frac{y'\sqrt{3}}{2}$$

Substitute for x and y in the given equation.

$$\left(\frac{x'\sqrt{3}}{2} - \frac{y'}{2}\right)^2 - 2\sqrt{3}\left(\frac{x'\sqrt{3}}{2} - \frac{y'}{2}\right)\left(\frac{x'}{2} + \frac{y'\sqrt{3}}{2}\right) +$$
$$3\left(\frac{x'}{2} + \frac{y'\sqrt{3}}{2}\right)^2 - 12\sqrt{3}\left(\frac{x'\sqrt{3}}{2} - \frac{y'}{2}\right) -$$

$$12\left(\frac{x'}{2} + \frac{y'\sqrt{3}}{2}\right) = 0$$

After simplifying we have

$$(y')^2 = 6x'.$$

This is the equation of a parabola with vertex at $(0,0)$ of the $x'y'$-coordinate system and axis of symmetry $y' = 0$. We sketch the graph.

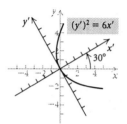

25. $7x^2 + 6\sqrt{3}xy + 13y^2 - 32 = 0$

$A = 7$, $B = 6\sqrt{3}$, $C = 13$

$B^2 - 4AC = (6\sqrt{3})^2 - 4\cdot7\cdot13 = 108 - 364 = -256$

Since the discriminant is negative, the graph is an ellipse or a circle. To rotate the axes we first determine θ.

$$\cot 2\theta = \frac{A-C}{B} = \frac{7-13}{6\sqrt{3}} = \frac{-6}{6\sqrt{3}} = -\frac{1}{\sqrt{3}}$$

Then $2\theta = 120°$ and $\theta = 60°$, so

$$\sin\theta = \frac{\sqrt{3}}{2} \text{ and } \cos\theta = \frac{1}{2}.$$

Now substitute in the rotation of axes formulas.

$$x = x'\cos\theta - y'\sin\theta$$
$$= x'\cdot\frac{1}{2} - y'\cdot\frac{\sqrt{3}}{2} = \frac{x'}{2} - \frac{y'\sqrt{3}}{2}$$

$$y = x'\sin\theta + y'\cos\theta$$
$$= x'\cdot\frac{\sqrt{3}}{2} + y'\cdot\frac{1}{2} = \frac{x'\sqrt{3}}{2} + \frac{y'}{2}$$

Substitute for x and y in the given equation.

$$7\left(\frac{x'}{2} - \frac{y'\sqrt{3}}{2}\right)^2 + 6\sqrt{3}\left(\frac{x'}{2} - \frac{y'\sqrt{3}}{2}\right)\left(\frac{x'\sqrt{3}}{2} + \frac{y'}{2}\right) +$$
$$13\left(\frac{x'\sqrt{3}}{2} + \frac{y'}{2}\right)^2 - 32 = 0$$

After simplifying we have

$$\frac{(x')^2}{2} + \frac{(y')^2}{8} = 1.$$

This is the equation of an ellipse with vertices $(0, -\sqrt{8})$ and $(0, \sqrt{8})$, or $(0, -2\sqrt{2})$ and $(0, 2\sqrt{2})$ on the y'-axis. The x'-intercepts are $(-\sqrt{2}, 0)$ and $(\sqrt{2}, 0)$. We sketch the graph.

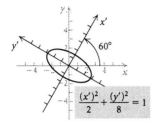

$$\frac{(x')^2}{2} + \frac{(y')^2}{8} = 1$$

27. $11x^2 + 10\sqrt{3}xy + y^2 = 32$

$A = 11$, $B = 10\sqrt{3}$, $C = 1$

$B^2 - 4AC = (10\sqrt{3})^2 - 4 \cdot 11 \cdot 1 = 300 - 44 = 256$

Since the discriminant is positive, the graph is a hyperbola. To rotate the axes we first determine θ.

$$\cot 2\theta = \frac{A - C}{B} = \frac{11 - 1}{10\sqrt{3}} = \frac{10}{10\sqrt{3}} = \frac{1}{\sqrt{3}}$$

Then $2\theta = 60°$ and $\theta = 30°$, so

$$\sin \theta = \frac{1}{2} \text{ and } \cos \theta = \frac{\sqrt{3}}{2}.$$

Now substitute in the rotation of axes formulas.

$$x = x' \cos \theta - y' \sin \theta$$
$$= x' \cdot \frac{\sqrt{3}}{2} - y' \cdot \frac{1}{2} = \frac{x'\sqrt{3}}{2} - \frac{y'}{2}$$

$$y = x' \sin \theta + y' \cos \theta$$
$$= x' \cdot \frac{1}{2} + y' \cdot \frac{\sqrt{3}}{2} = \frac{x'}{2} + \frac{y'\sqrt{3}}{2}$$

Substitute for x and y in the given equation.

$$11\left(\frac{x'\sqrt{3}}{2} - \frac{y'}{2}\right)^2 + 10\sqrt{3}\left(\frac{x'\sqrt{3}}{2} - \frac{y'}{2}\right)\left(\frac{x'}{2} + \frac{y'\sqrt{3}}{2}\right) +$$
$$\left(\frac{x'}{2} + \frac{y'\sqrt{3}}{2}\right)^2 = 32$$

After simplifying we have

$$\frac{(x')^2}{2} - \frac{(y')^2}{8} = 1.$$

This is the equation of a hyperbola with vertices $(-\sqrt{2}, 0)$ and $(\sqrt{2}, 0)$ on the x'-axis. The asymptotes are $y' = -\frac{\sqrt{8}}{\sqrt{2}}x'$ and $y' = \frac{\sqrt{8}}{\sqrt{2}}x'$, or $y' = -2x'$ and $y' = 2x'$. We sketch the graph.

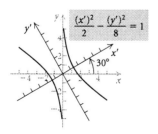

$$\frac{(x')^2}{2} - \frac{(y')^2}{8} = 1$$

29. $\sqrt{2}x^2 + 2\sqrt{2}xy + \sqrt{2}y^2 - 8x + 8y = 0$

$A = \sqrt{2}$, $B = 2\sqrt{2}$, $C = \sqrt{2}$

$B^2 - 4AC = (2\sqrt{2})^2 - 4 \cdot \sqrt{2} \cdot \sqrt{2} = 8 - 8 = 0$

Since the discriminant is zero, the graph is a parabola. To rotate the axes we first determine θ.

$$\cot 2\theta = \frac{A - C}{B} = \frac{\sqrt{2} - \sqrt{2}}{2\sqrt{2}} = 0$$

Then $2\theta = 90°$ and $\theta = 45°$, so

$$\sin \theta = \frac{\sqrt{2}}{2} \text{ and } \cos \theta = \frac{\sqrt{2}}{2}.$$

Now substitute in the rotation of axes formulas.

$$x = x' \cos \theta - y' \sin \theta$$
$$= x'\left(\frac{\sqrt{2}}{2}\right) - y'\left(\frac{\sqrt{2}}{2}\right) = \frac{\sqrt{2}}{2}(x' - y')$$

$$y = x' \sin \theta + y' \cos \theta$$
$$= x'\left(\frac{\sqrt{2}}{2}\right) + y'\left(\frac{\sqrt{2}}{2}\right) = \frac{\sqrt{2}}{2}(x' + y')$$

Substitute for x and y in the given equation.

$$\sqrt{2}\left[\frac{\sqrt{2}}{2}(x'-y')\right]^2 + 2\sqrt{2}\left[\frac{\sqrt{2}}{2}(x'-y')\right]\left[\frac{\sqrt{2}}{2}(x'+y')\right] +$$
$$\sqrt{2}\left[\frac{\sqrt{2}}{2}(x'+y')\right]^2 - 8\cdot\frac{\sqrt{2}}{2}(x'-y') + 8\cdot\frac{\sqrt{2}}{2}(x'+$$
$$y') = 0$$

After simplifying we have

$$y' = -\frac{1}{4}(x')^2.$$

This is the equation of a parabola with vertex at $(0,0)$ of the $x'y'$-coordinate system and axis of symmetry $x' = 0$. We sketch the graph.

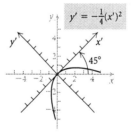

$$y' = -\frac{1}{4}(x')^2$$

31. $x^2 + 6\sqrt{3}xy - 5y^2 + 8x - 8\sqrt{3}y - 48 = 0$

$A = 1$, $B = 6\sqrt{3}$, $C = -5$

$B^2 - 4AC = (6\sqrt{3})^2 - 4 \cdot 1 \cdot (-5) = 108 + 20 = 128$

Since the discriminant is positive, the graph is a hyperbola. To rotate the axes we first determine θ.

$$\cot 2\theta = \frac{A - C}{B} = \frac{1 - (-5)}{6\sqrt{3}} = \frac{6}{6\sqrt{3}} = \frac{1}{\sqrt{3}}$$

Then $2\theta = 60°$ and $\theta = 30°$, so

$$\sin\theta = \frac{1}{2} \text{ and } \cos\theta = \frac{\sqrt{3}}{2}.$$

Now substitute in the rotation of axes formulas.

$$x = x'\cos\theta - y'\sin\theta$$

$$= x' \cdot \frac{\sqrt{3}}{2} - y' \cdot \frac{1}{2} = \frac{x'\sqrt{3}}{2} - \frac{y'}{2}$$

$$y = x'\sin\theta + y'\cos\theta$$

$$= x' \cdot \frac{1}{2} + y' \cdot \frac{\sqrt{3}}{2} = \frac{x'}{2} + \frac{y'\sqrt{3}}{2}$$

Substitute for x and y in the given equation.

$$\left(\frac{x'\sqrt{3}}{2} - \frac{y'}{2}\right)^2 + 6\sqrt{3}\left(\frac{x'\sqrt{3}}{2} - \frac{y'}{2}\right)\left(\frac{x'}{2} + \frac{y'\sqrt{3}}{2}\right) -$$

$$5\left(\frac{x'}{2} + \frac{y'\sqrt{3}}{2}\right)^2 + 8\left(\frac{x'\sqrt{3}}{2} - \frac{y'}{2}\right) -$$

$$8\sqrt{3}\left(\frac{x'}{2} + \frac{y'\sqrt{3}}{2}\right) - 48 = 0$$

After simplifying we have

$$\frac{(x')^2}{10} - \frac{(y'+1)^2}{5} = 1.$$

This is the equation of a hyperbola with vertices $(-\sqrt{10},0)$ and $(\sqrt{10},0)$ and asymptotes $y'+1 = -\frac{\sqrt{5}}{\sqrt{10}}x'$ and $y'+1 = \frac{\sqrt{5}}{\sqrt{10}}x'$, or $y'+1 = -\frac{1}{\sqrt{2}}x'$ and $y'+1 = \frac{1}{\sqrt{2}}x'$ We sketch the graph.

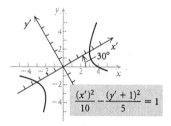

33. $x^2 + xy + y^2 = 24$

$A = 1,\ B = 1,\ C = 1$

$B^2 - 4AC = 1^2 - 4 \cdot 1 \cdot 1 = 1 - 4 = -3$

Since the discriminant is negative, the graph is an ellipse or a circle. To rotate the axes we first determine θ.

$$\cot 2\theta = \frac{A-C}{B} = \frac{1-1}{1} = 0$$

Then $2\theta = 90°$ and $\theta = 45°$, so

$$\sin\theta = \frac{\sqrt{2}}{2} \text{ and } \cos\theta = \frac{\sqrt{2}}{2}.$$

Now substitute in the rotation of axes formulas.

$$x = x'\cos\theta - y'\sin\theta$$

$$= x'\left(\frac{\sqrt{2}}{2}\right) - y'\left(\frac{\sqrt{2}}{2}\right) = \frac{\sqrt{2}}{2}(x' - y')$$

$$y = x'\sin\theta + y'\cos\theta$$

$$= x'\left(\frac{\sqrt{2}}{2}\right) + y'\left(\frac{\sqrt{2}}{2}\right) = \frac{\sqrt{2}}{2}(x' + y')$$

Substitute for x and y in the given equation.

$$\left[\frac{\sqrt{2}}{2}(x' - y')\right]^2 + \left[\frac{\sqrt{2}}{2}(x' - y')\right]\left[\frac{\sqrt{2}}{2}(x' + y')\right] +$$

$$\left[\frac{\sqrt{2}}{2}(x' + y')\right]^2 = 24$$

After simplifying we have

$$\frac{(x')^2}{16} + \frac{(y')^2}{48} = 1.$$

This is the equation of an ellipse with vertices $(0, -\sqrt{48})$ and $(0, \sqrt{48}$, or $(0, -4\sqrt{3})$ and $(0, 4\sqrt{3})$ on the y'-axis. The x'-intercepts are $(-4,0)$ and $(4,0)$. We sketch the graph.

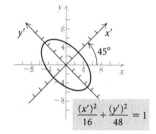

$$\frac{(x')^2}{16} + \frac{(y')^2}{48} = 1$$

35. $4x^2 - 4xy + y^2 - 8\sqrt{5}x - 16\sqrt{5}y = 0$

$A = 4,\ B = -4,\ C = 1$

$B^2 - 4AC = (-4)^2 - 4 \cdot 4 \cdot 1 = 16 - 16 = 0$

Since the discriminant is zero, the graph is a parabola. To rotate the axes we first determine θ.

$$\cot 2\theta = \frac{A-C}{B} = \frac{4-1}{-4} = -\frac{3}{4}$$

Since $\cot 2\theta < 0$, we have $90° < 2\theta < 180°$. We make a sketch.

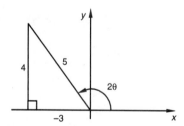

From the sketch we see that $\cos 2\theta = -\frac{3}{5}$. Using half-angle formulas, we have

$$\sin\theta = \sqrt{\frac{1 - \cos 2\theta}{2}} = \sqrt{\frac{1 - \left(-\frac{3}{5}\right)}{2}} = \frac{2}{\sqrt{5}}$$

and

$$\cos\theta = \sqrt{\frac{1+\cos 2\theta}{2}} = \sqrt{\frac{1+\left(-\dfrac{3}{5}\right)}{2}} = \frac{1}{\sqrt5}.$$

Now substitute in the rotation of axes formulas.

$$x = x'\cos\theta - y'\sin\theta$$
$$= x'\cdot\frac{1}{\sqrt5} - y'\cdot\frac{2}{\sqrt5} = \frac{x'}{\sqrt5} - \frac{2y'}{\sqrt5}$$

$$y = x'\sin\theta + y'\cos\theta$$
$$= x'\cdot\frac{2}{\sqrt5} + y'\cdot\frac{1}{\sqrt5} = \frac{2x'}{\sqrt5} + \frac{y'}{\sqrt5}$$

Substitute for x and y in the given equation.

$$4\left(\frac{x'}{\sqrt5} - \frac{2y'}{\sqrt5}\right)^2 - 4\left(\frac{x'}{\sqrt5} - \frac{2y'}{\sqrt5}\right)\left(\frac{2x'}{\sqrt5} + \frac{y'}{\sqrt5}\right) +$$
$$\left(\frac{2x'}{\sqrt5} + \frac{y'}{\sqrt5}\right)^2 - 8\sqrt5\left(\frac{x'}{\sqrt5} - \frac{2y'}{\sqrt5}\right) -$$
$$16\sqrt5\left(\frac{2x'}{\sqrt5} + \frac{y'}{\sqrt5}\right) = 0$$

After simplifying we have

$$(y')^2 = 8x'.$$

This is the equation of a parabola with vertex $(0,0)$ of the $x'y'$-coordinate system and axis of symmetry $y' = 0$. Since we know that $\sin\theta = \dfrac{2}{\sqrt5}$ and $0° < \theta < 90°$, we can use a grapher to find that $\theta \approx 63.4°$. Thus, the xy-axes are rotated through an angle of about $63.4°$ to obtain the $x'y'$-axes. We sketch the graph.

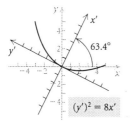

37. $11x^2 + 7xy - 13y^2 = 621$

$A = 11,\ B = 7,\ C = -13$

$B^2 - 4AC = 7^2 - 4\cdot 11\cdot(-13) = 49 + 572 = 621$

Since the discriminant is positive, the graph is a hyperbola. To rotate the axes we first determine θ.

$$\cot 2\theta = \frac{A-C}{B} = \frac{11-(-13)}{7} = \frac{24}{7}$$

Since $\cot 2\theta > 0$, we have $0° < 2\theta < 90°$. We make a sketch.

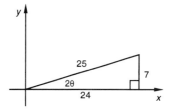

From the sketch we see that $\cos 2\theta = \dfrac{24}{25}$. Using half-angle formulas, we have

$$\sin\theta = \sqrt{\frac{1-\cos 2\theta}{2}} = \sqrt{\frac{1-\dfrac{24}{25}}{2}} = \frac{1}{\sqrt{50}}$$

and

$$\cos\theta = \sqrt{\frac{1+\cos 2\theta}{2}} = \sqrt{\frac{1+\dfrac{24}{25}}{2}} = \frac{7}{\sqrt{50}}.$$

Now substitute in the rotation of axes formulas.

$$x = x'\cos\theta - y'\sin\theta$$
$$= x'\cdot\frac{7}{\sqrt{50}} - y'\cdot\frac{1}{\sqrt{50}} = \frac{7x'}{\sqrt{50}} - \frac{y'}{\sqrt{50}}$$

$$y = x'\sin\theta + y'\cos\theta$$
$$= x'\cdot\frac{1}{\sqrt{50}} + y'\cdot\frac{7}{\sqrt{50}} = \frac{x'}{\sqrt{50}} + \frac{7y'}{\sqrt{50}}$$

Substitute for x and y in the given equation.

$$11\left(\frac{7x'}{\sqrt{50}} - \frac{y'}{\sqrt{50}}\right)^2 + 7\left(\frac{7x'}{\sqrt{50}} - \frac{y'}{\sqrt{50}}\right)\left(\frac{x'}{\sqrt{50}} + \frac{7y'}{\sqrt{50}}\right) -$$
$$13\left(\frac{x'}{\sqrt{50}} + \frac{7y'}{\sqrt{50}}\right)^2 = 621$$

After simplifying we have

$$\frac{(x')^2}{54} - \frac{(y')^2}{46} = 1.$$

This is the equation of a hyperbola with vertices $(-\sqrt{54},0)$ and $(\sqrt{54},0)$, or $(-3\sqrt6,0)$ and $(3\sqrt6,0)$ on the x'-axis. The asymptotes are $y' = -\sqrt{\dfrac{23}{27}}x'$ and $y' = \sqrt{\dfrac{23}{27}}x'$. Since we know that $\sin\theta = \dfrac{1}{\sqrt{50}}$ and $0° < \theta < 90°$, we can use a grapher to find that $\theta \approx 8.1°$. Thus, the xy-axes are rotated through an angle of about $8.1°$ to obtain the $x'y'$-axes. We sketch the graph.

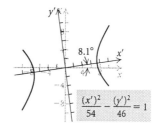

39. Discussion and Writing

41. $120° = 120° \cdot \dfrac{\pi \text{ radians}}{180°}$

$\qquad = \dfrac{120°}{180°}\pi \text{ radians}$

$\qquad = \dfrac{2\pi}{3} \text{ radians}$

43. $\dfrac{\pi}{3} \text{ radians} = \dfrac{\pi}{3} \text{ radians} \cdot \dfrac{180°}{\pi \text{ radians}}$

$\qquad\qquad\quad = \dfrac{\pi}{3\pi} \cdot 180°$

$\qquad\qquad\quad = \dfrac{1}{3} \cdot 180° = 60°$

45. $x' = x \cos\theta + y \sin\theta,$

$\quad\; y' = y \cos\theta - x \sin\theta$

First rewrite the system.

$\quad x \cos\theta + y \sin\theta = x', \quad (1)$

$\quad -x \sin\theta + y \cos\theta = y' \quad (2)$

Multiply Equation (1) by $\sin\theta$ and Equation (2) by $\cos\theta$ and add to eliminate x.

$\quad x \sin\theta \cos\theta + y \sin^2\theta = x' \sin\theta$

$\quad \underline{-x \sin\theta \cos\theta + y \cos^2\theta = y' \cos\theta}$

$\qquad y(\sin^2\theta + \cos^2\theta) = x' \sin\theta + y' \cos\theta$

$\qquad\qquad\qquad\quad y = x' \sin\theta + y' \cos\theta$

Substitute $x' \sin\theta + y' \cos\theta$ for y in Equation (1) and solve for x.

$x \cos\theta + (x' \sin\theta + y' \cos\theta)\sin\theta = x'$

$x \cos\theta + x' \sin^2\theta + y' \sin\theta \cos\theta = x'$

$\qquad\qquad x \cos\theta = x' - x' \sin^2\theta -$

$\qquad\qquad\qquad\qquad y' \sin\theta \cos\theta$

$\qquad\qquad x \cos\theta = x'(1 - \sin^2\theta) -$

$\qquad\qquad\qquad\qquad y' \sin\theta \cos\theta$

$\qquad\qquad x \cos\theta = x' \cos^2\theta -$

$\qquad\qquad\qquad\qquad y' \sin\theta \cos\theta$

$\qquad\qquad\quad x = x' \cos\theta - y' \sin\theta$

Thus, we have $x = x' \cos\theta - y' \sin\theta$ and $y = x' \sin\theta + y' \cos\theta$.

47. $\quad A' + C'$

$= (A \cos^2\theta + B \sin\theta \cos\theta + C \sin^2\theta) +$

$\qquad (A \sin^2\theta - B \sin\theta \cos\theta + C \cos^2\theta)$

$= A(\cos^2\theta + \sin^2\theta) + B(\sin\theta \cos\theta - \sin\theta \cos\theta) +$

$\qquad C(\sin^2\theta + \cos^2\theta)$

$= A \cdot 1 + B \cdot 0 + C \cdot 1$

$= A + C$

Exercise Set 9.6

See the text answer section for odd exercise answers 1-11.

13. Answers may vary.

    A: $(4, 30°)$, $(4, 390°)$, $(-4, 210°)$;

    B: $(5, 300°)$, $(5, -60°)$, $(-5, 120°)$;

    C: $(2, 150°)$, $(2, -210°)$, $(-2, 330°)$;

    D: $(3, 225°)$, $(3, -135°)$, $(-3, 45°)$

15. $(0, -3)$

$r = \sqrt{0^2 + (-3)^2} = \sqrt{9} = 3$

$\tan\theta = \dfrac{-3}{0}$, which is undefined; therefore, since $(0, -3)$ lies on the negative y-axis, $\theta = 270°$, or $\dfrac{3\pi}{2}$.

Thus, $(0, -3) = (3, 270°)$, or $\left(3, \dfrac{3\pi}{2}\right)$.

17. $(3, -3\sqrt{3})$

$r = \sqrt{3^2 + (-3\sqrt{3})^2} = \sqrt{9 + 27} = \sqrt{36} = 6$

$\tan\theta = \dfrac{-3\sqrt{3}}{3} = -\sqrt{3}$, so $\theta = 300°$, or $\dfrac{5\pi}{3}$. (The point is in quadrant IV.)

Thus, $(3, -3\sqrt{3}) = (6, 300°)$, or $\left(6, \dfrac{5\pi}{3}\right)$.

19. $(4\sqrt{3}, -4)$

$r = \sqrt{(4\sqrt{3})^2 + (-4)^2} = \sqrt{48 + 16} = \sqrt{64} = 8$

$\tan\theta = \dfrac{-4}{4\sqrt{3}} = -\dfrac{1}{\sqrt{3}}$, or $-\dfrac{\sqrt{3}}{3}$, so $\theta = 330°$, or $\dfrac{11\pi}{6}$. (The point is in quadrant IV.)

Thus, $(4\sqrt{3}, -4) = (8, 330°)$, or $\left(8, \dfrac{11\pi}{6}\right)$.

21. $(-\sqrt{2}, -\sqrt{2})$

$r = \sqrt{(-\sqrt{2})^2 + (-\sqrt{2})^2} = \sqrt{2 + 2} = \sqrt{4} = 2$

$\tan\theta = \dfrac{-\sqrt{2}}{-\sqrt{2}} = 1$, so $\theta = 225°$, or $\dfrac{5\pi}{4}$. (The point is in quadrant III.)

Thus, $(-\sqrt{2}, -\sqrt{2}) = (2, 225°)$, or $\left(2, \dfrac{5\pi}{4}\right)$.

23. Use the ANGLE feature on a grapher.

$(3, 7) = (7.616, 66.8°)$, or $(7.616, 1.166)$

25. Use the ANGLE feature on a grapher.

$(-\sqrt{10}, 3.4) = (4.643, 132.9°)$, or $(4.643, 2.230)$

27. $(5, 60°)$

$$x = r\cos\theta = 5\cos 60° = 5 \cdot \frac{1}{2} = \frac{5}{2}$$

$$y = r\sin\theta = 5\sin 60° = 5 \cdot \frac{\sqrt{3}}{2} = \frac{5\sqrt{3}}{2}$$

$$(5, 60°) = \left(\frac{5}{2}, \frac{5\sqrt{3}}{2}\right)$$

29. $(-3, 45°)$

$$x = r\cos\theta = -3\cos 45° = -3 \cdot \frac{\sqrt{2}}{2} = -\frac{3\sqrt{2}}{2}$$

$$y = r\sin\theta = -3\sin 45° = -3 \cdot \frac{\sqrt{2}}{2} = -\frac{3\sqrt{2}}{2}$$

$$(-3, 45°) = \left(-\frac{3\sqrt{2}}{2}, -\frac{3\sqrt{2}}{2}\right)$$

31. $(3, -120°)$

$$x = 3\cos(-120°) = 3\left(-\frac{1}{2}\right) = -\frac{3}{2}$$

$$y = 3\sin(-120°) = 3\left(-\frac{\sqrt{3}}{2}\right) = -\frac{3\sqrt{3}}{2}$$

$$(3, -120°) = \left(-\frac{3}{2}, -\frac{3\sqrt{3}}{2}\right)$$

33. $\left(-2, \frac{5\pi}{3}\right)$

$$x = -2\cos\frac{5\pi}{3} = -2 \cdot \frac{1}{2} = -1$$

$$y = -2\sin\frac{5\pi}{3} = -2\left(-\frac{\sqrt{3}}{2}\right) = \sqrt{3}$$

$$\left(-2, \frac{5\pi}{3}\right) = (-1, \sqrt{3})$$

35. Use the ANGLE feature on a grapher.

$(3, -43°) = (2.19, -2.05)$

37. Use the ANGLE feature on a grapher.

$\left(-4.2, \frac{3\pi}{5}\right) = (1.30, -3.99)$

39.
$$3x + 4y = 5$$
$$3r\cos\theta + 4r\sin\theta = 5 \quad (x = r\cos\theta,\ y = r\sin\theta)$$
$$r(3\cos\theta + 4\sin\theta) = 5$$

41.
$$x = 5$$
$$r\cos\theta = 5 \quad (x = r\cos\theta)$$

43.
$$x^2 + y^2 = 36$$
$$(r\cos\theta)^2 + (r\sin\theta)^2 = 36 \quad (x = r\cos\theta,\ y = r\sin\theta)$$
$$r^2\cos^2\theta + r^2\sin^2\theta = 36$$
$$r^2(\cos^2\theta + \sin^2\theta) = 36$$
$$r^2 = 36 \quad (\cos^2\theta + \sin^2\theta = 1)$$
$$r = 6$$

45.
$$x^2 = 25y$$
$$(r\cos\theta)^2 = 25r\sin\theta \quad \text{Substituting for } x \text{ and } y$$
$$r^2\cos^2\theta = 25r\sin\theta$$

47.
$$y^2 - 5x - 25 = 0$$
$$(r\sin\theta)^2 - 5r\cos\theta - 25 = 0 \quad \text{Substituting for } x \text{ and } y$$
$$r^2\sin^2\theta - 5r\cos\theta - 25 = 0$$

49.
$$r = 5$$
$$\pm\sqrt{x^2 + y^2} = 5 \quad \text{Substituting for } r$$
$$x^2 + y^2 = 25 \quad \text{Squaring}$$

51.
$$r\sin\theta = 2$$
$$y = 2 \quad (y = r\sin\theta)$$

53.
$$r + r\cos\theta = 3$$
$$\pm\sqrt{x^2 + y^2} + x = 3$$
$$\pm\sqrt{x^2 + y^2} = 3 - x$$
$$x^2 + y^2 = (3 - x)^2 \quad \text{Squaring both sides}$$
$$x^2 + y^2 = 9 - 6x + x^2$$
$$y^2 = -6x + 9$$

55.
$$r - 9\cos\theta = 7\sin\theta$$
$$r^2 - 9r\cos\theta = 7r\sin\theta \quad \text{Multiplying both sides by } r$$
$$x^2 + y^2 - 9x = 7y$$
$$x^2 - 9x + y^2 - 7y = 0$$

57.
$$r = 5\sec\theta$$
$$r = 5 \cdot \frac{1}{\cos\theta} \quad \left(\sec\theta = \frac{1}{\cos\theta}\right)$$
$$r\cos\theta = 5$$
$$x = 5$$

59. $r = \sin\theta$

Make a table of values. Note that the points begin to repeat at $\theta = 360°$. Plot points and draw the graph.

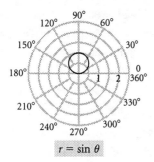

$r = \sin\theta$

61. $r = 4\cos 2\theta$

Make a table of values. Note that the points begin to repeat at $\theta = 180°$. Plot point and draw the graph.

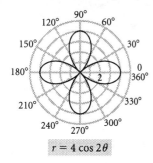

$r = 4\cos 2\theta$

63. Graph (d) is the graph of $r = 3\sin 2\theta$.

65. Graph (g) is the graph of $r = \theta$.

67. Graph (j) is the graph of $r = \dfrac{5}{1 + \cos\theta}$.

69. Graph (b) is the graph of $r = 3\cos 2\theta$.

71. Graph (e) is the graph of $r = 3\sin\theta$.

73. Graph (k) is the graph of $r = 2\sin 3\theta$.

75. Discussion and Writing

77. $2x - 4 = x + 8$
    $x = 12$ Adding 4 and subtracting x
The solution is 12.

79. $y = 2x - 5$

Make a table of values by choosing values for x and finding the corresponding y-values. Plot points and draw the graph.

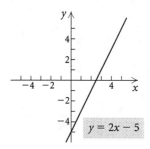

$y = 2x - 5$

81. $x = -3$

Note that any point on the graph has -3 for its first coordinate. Thus, the graph is a vertical line 3 units left of the y-axis.

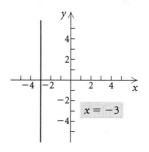

$x = -3$

83.
$$r = \sec^2 \frac{\theta}{2}$$
$$r = \frac{1}{\cos^2 \dfrac{\theta}{2}}$$
$$r = \frac{1}{\dfrac{1 + \cos\theta}{2}}$$
$$r = \frac{2}{1 + \cos\theta}$$
$$r + r\cos\theta = 2$$
$$\pm\sqrt{x^2 + y^2} + x = 2$$
$$x^2 + y^2 = (2 - x)^2$$
$$x^2 + y^2 = 4 - 4x + x^2$$
$$y^2 = -4x + 4$$

85. See the answer section in the text.

87. See the answer section in the text.

89. See the answer section in the text.

91. See the answer section in the text.

Exercise Set 9.7

1. Graph (b) is the graph of $r = \dfrac{3}{1 + \cos\theta}$.

3. Graph (a) is the graph of $r = \dfrac{8}{4 - 2\cos\theta}$.

5. Graph (d) is the graph of $r = \dfrac{5}{3 - 3\sin\theta}$.

7. $r = \dfrac{1}{1 + \cos\theta}$

a) The equation is in the form $r = \dfrac{ep}{1 + e\cos\theta}$ with $e = 1$, so the graph is a parabola.

b) Since $e = 1$ and $ep = 1 \cdot p = 1$, we have $p = 1$. Thus the parabola has a vertical directrix 1 unit to the right of the pole.

c) We find the vertex by letting $\theta = 0$. When $\theta = 0$,
$$r = \frac{1}{1 + \cos 0} = \frac{1}{1 + 1} = \frac{1}{2}.$$
Thus, the vertex is $\left(\dfrac{1}{2}, 0\right)$.

d)

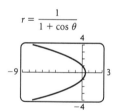

$$r = \frac{1}{1 + \cos\theta}$$

9. $r = \dfrac{15}{5 - 10\sin\theta}$

a) We first divide numerator and denominator by 5:
$$r = \frac{3}{1 - 2\sin\theta}$$
The equation is in the form $r = \dfrac{ep}{1 - e\sin\theta}$ with $e = 2$.

Since $e > 1$, the graph is a hyperbola.

b) Since $e = 2$ and $ep = 2 \cdot p = 3$, we have $p = \dfrac{3}{2}$. Thus the hyperbola has a horizontal directrix $\dfrac{3}{2}$ units below the pole.

c) We find the vertices by letting $\theta = \dfrac{\pi}{2}$ and $\theta = \dfrac{3\pi}{2}$. When $\theta = \dfrac{\pi}{2}$,
$$r = \frac{15}{5 - 10\sin\dfrac{\pi}{2}} = \frac{15}{5 - 10\cdot 1} = \frac{15}{-5} = -3.$$

When $\theta = \dfrac{3\pi}{2}$,
$$r = \frac{15}{5 - 10\sin\dfrac{3\pi}{2}} = \frac{15}{5 - 10(-1)} = \frac{15}{15} = 1.$$

Thus, the vertices are $\left(-3, \dfrac{\pi}{2}\right)$ and $\left(1, \dfrac{3\pi}{2}\right)$.

d)

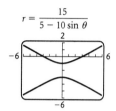

$$r = \frac{15}{5 - 10\sin\theta}$$

11. $r = \dfrac{8}{6 - 3\cos\theta}$

a) We first divide numerator and denominator by 6:
$$r = \frac{4/3}{1 - \dfrac{1}{2}\cos\theta}$$
The equation is in the form $r = \dfrac{ep}{1 - e\cos\theta}$ with $e = \dfrac{1}{2}$.

Since $0 < e < 1$, the graph is an ellipse.

b) Since $e = \dfrac{1}{2}$ and $ep = \dfrac{1}{2} \cdot p = \dfrac{4}{3}$, we have $p = \dfrac{8}{3}$. Thus the ellipse has a vertical directrix $\dfrac{8}{3}$ units to the left of the pole.

c) We find the vertices by letting $\theta = 0$ and $\theta = \pi$. When $\theta = 0$,
$$r = \frac{8}{6 - 3\cos 0} = \frac{8}{6 - 3\cdot 1} = \frac{8}{3}.$$
When $\theta = \pi$,
$$r = \frac{8}{6 - 3\cos\pi} = \frac{8}{6 - 3(-1)} = \frac{8}{9}.$$

Thus, the vertices are $\left(\dfrac{8}{3}, 0\right)$ and $\left(\dfrac{8}{9}, \pi\right)$.

d)

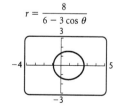

$$r = \frac{8}{6 - 3\cos\theta}$$

13. $r = \dfrac{20}{10 + 15\sin\theta}$

a) We first divide numerator and denominator by 10:

$$r = \dfrac{2}{1 + \dfrac{3}{2}\sin\theta}$$

The equation is in the form $r = \dfrac{ep}{1 + e\sin\theta}$ with $e = \dfrac{3}{2}$.

Since $e > 1$, the graph is a hyperbola.

b) Since $e = \dfrac{3}{2}$ and $ep = \dfrac{3}{2}\cdot p = 2$, we have $p = \dfrac{4}{3}$. Thus the hyperbola has a horizontal directrix $\dfrac{4}{3}$ units above the pole.

c) We find the vertices by letting $\theta = \dfrac{\pi}{2}$ and $\theta = \dfrac{3\pi}{2}$. When $\theta = \dfrac{\pi}{2}$,

$$r = \dfrac{20}{10 + 15\sin\dfrac{\pi}{2}} = \dfrac{20}{10 + 15\cdot 1} = \dfrac{20}{25} = \dfrac{4}{5}.$$

When $\theta = \dfrac{3\pi}{2}$,

$$r = \dfrac{20}{10 + 15\sin\dfrac{\pi}{2}} = \dfrac{20}{10 + 15(-1)} = \dfrac{20}{-5} = -4.$$

Thus, the vertices are $\left(\dfrac{4}{5}, \dfrac{\pi}{2}\right)$ and $\left(-4, \dfrac{3\pi}{2}\right)$.

d)

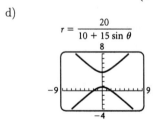

$$r = \dfrac{20}{10 + 15\sin\theta}$$

15. $r = \dfrac{9}{6 + 3\cos\theta}$

a) We first divide numerator and denominator by 6:

$$r = \dfrac{3/2}{1 + \dfrac{1}{2}\cos\theta}$$

The equation is in the form $r = \dfrac{ep}{1 + e\cos\theta}$ with $e = \dfrac{1}{2}$.

Since $0 < e < 1$, the graph is an ellipse.

b) Since $e = \dfrac{1}{2}$ and $ep = \dfrac{1}{2}\cdot p = \dfrac{3}{2}$, we have $p = 3$. Thus the ellipse has a vertical directrix 3 units to the right of the pole.

c) We find the vertices by letting $\theta = 0$ and $\theta = \pi$. When $\theta = 0$,

$$r = \dfrac{9}{6 + 3\cos 0} = \dfrac{9}{6 + 3\cdot 1} = \dfrac{9}{9} = 1.$$

When $\theta = \pi$,

$$r = \dfrac{9}{6 + 3\cos\pi} = \dfrac{9}{6 + 3(-1)} = \dfrac{9}{3} = 3.$$

Thus, the vertices are $(1, 0)$ and $(3, \pi)$.

d)

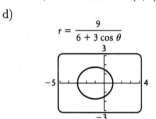

$$r = \dfrac{9}{6 + 3\cos\theta}$$

17. $r = \dfrac{3}{2 - 2\sin\theta}$

a) We first divide numerator and denominator by 2:

$$r = \dfrac{3/2}{1 - \sin\theta}$$

The equation is in the form $r = \dfrac{ep}{1 - e\sin\theta}$ with $e = 1$, so the graph is a parabola.

b) Since $e = 1$ and $ep = 1\cdot p = \dfrac{3}{2}$, we have $p = \dfrac{3}{2}$. Thus the parabola has a horizontal directrix $\dfrac{3}{2}$ units below the pole.

c) We find the vertex by letting $\theta = \dfrac{3\pi}{2}$. When $\theta = \dfrac{3\pi}{2}$,

$$r = \dfrac{3}{2 - 2\sin\dfrac{3\pi}{2}} = \dfrac{3}{2 - 2(-1)} = \dfrac{3}{4}.$$

Thus, the vertex is $\left(\dfrac{3}{4}, \dfrac{3\pi}{2}\right)$.

d)

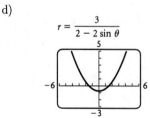

$$r = \dfrac{3}{2 - 2\sin\theta}$$

19. $r = \dfrac{4}{2 - \cos\theta}$

  a) We first divide numerator and denominator by 2:

$$r = \dfrac{2}{1 - \dfrac{1}{2}\cos\theta}$$

  The equation is in the form $r = \dfrac{ep}{1 - e\cos\theta}$ with $e = \dfrac{1}{2}$.

  Since $0 < e < 1$, the graph is an ellipse.

  b) Since $e = \dfrac{1}{2}$ and $ep = \dfrac{1}{2} \cdot p = 2$, we have $p = 4$. Thus the ellipse has a vertical directrix 4 units to the left of the pole.

  c) We find the vertices by letting $\theta = 0$ and $\theta = \pi$. When $\theta = 0$,

$$r = \dfrac{4}{2 - \cos 0} = \dfrac{4}{2 - 1} = 4.$$

  When $\theta = \pi$,

$$r = \dfrac{4}{2 - \cos\pi} = \dfrac{4}{2 - (-1)} = \dfrac{4}{3}.$$

  Thus, the vertices are $(4, 0)$ and $\left(\dfrac{4}{3}, \pi\right)$.

  d)

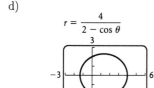

21. $r = \dfrac{7}{2 + 10\sin\theta}$

  a) We first divide numerator and denominator by 2:

$$r = \dfrac{7/2}{1 + 5\sin\theta}$$

  The equation is in the form $r = \dfrac{ep}{1 + e\sin\theta}$ with $e = 5$.

  Since $e > 1$, the graph is a hyperbola.

  b) Since $e = 5$ and $ep = 5 \cdot p = \dfrac{7}{2}$, we have $p = \dfrac{7}{10}$. Thus the hyperbola has a horizontal directrix $\dfrac{7}{10}$ unit above the pole.

  c) We find the vertices by letting $\theta = \dfrac{\pi}{2}$ and $\theta = \dfrac{3\pi}{2}$. When $\theta = \dfrac{\pi}{2}$,

$$r = \dfrac{7}{2 + 10\sin\dfrac{\pi}{2}} = \dfrac{7}{2 + 10 \cdot 1} = \dfrac{7}{12}.$$

  When $\theta = \dfrac{3\pi}{2}$,

$$r = \dfrac{7}{2 + 10\sin\dfrac{3\pi}{2}} = \dfrac{7}{2 + 10(-1)} = -\dfrac{7}{8}.$$

  Thus, the vertices are $\left(\dfrac{7}{12}, \dfrac{\pi}{2}\right)$ and $\left(-\dfrac{7}{8}, \dfrac{3\pi}{2}\right)$.

  d)

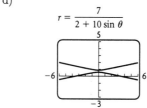

23.
$$r = \dfrac{1}{1 + \cos\theta}$$
$$r + r\cos\theta = 1$$
$$r = 1 - r\cos\theta$$
$$\sqrt{x^2 + y^2} = 1 - x$$
$$x^2 + y^2 = 1 - 2x + x^2$$
$$y^2 = -2x + 1, \text{ or}$$
$$y^2 + 2x - 1 = 0$$

25.
$$r = \dfrac{15}{5 - 10\sin\theta}$$
$$5r - 10r\sin\theta = 15$$
$$5r = 10r\sin\theta + 15$$
$$r = 2r\sin\theta + 3$$
$$\sqrt{x^2 + y^2} = 2y + 3$$
$$x^2 + y^2 = 4y^2 + 12y + 9$$
$$x^2 - 3y^2 - 12y - 9 = 0$$

27.
$$r = \dfrac{8}{6 - 3\cos\theta}$$
$$6r - 3r\cos\theta = 8$$
$$6r = 3r\cos\theta + 8$$
$$6\sqrt{x^2 + y^2} = 3x + 8$$
$$36x^2 + 36y^2 = 9x^2 + 48x + 64$$
$$27x^2 + 36y^2 - 48x - 64 = 0$$

29.
$$r = \frac{20}{10 + 15\sin\theta}$$
$$10r + 15r\sin\theta = 20$$
$$10r = 20 - 15r\sin\theta$$
$$2r = 4 - 3r\sin\theta$$
$$2\sqrt{x^2 + y^2} = 4 - 3y$$
$$4x^2 + 4y^2 = 16 - 24y + 9y^2$$
$$4x^2 - 5y^2 + 24y - 16 = 0$$

31.
$$r = \frac{9}{6 + 3\cos\theta}$$
$$6r + 3r\cos\theta = 9$$
$$6r = 9 - 3r\cos\theta$$
$$2r = 3 - r\cos\theta$$
$$2\sqrt{x^2 + y^2} = 3 - x$$
$$4x^2 + 4y^2 = 9 - 6x + x^2$$
$$3x^2 + 4y^2 + 6x - 9 = 0$$

33.
$$r = \frac{3}{2 - 2\sin\theta}$$
$$2r - 2r\sin\theta = 3$$
$$2r = 2r\sin\theta + 3$$
$$2\sqrt{x^2 + y^2} = 2y + 3$$
$$4x^2 + 4y^2 = 4y^2 + 12y + 9$$
$$4x^2 = 12y + 9, \text{ or}$$
$$4x^2 - 12y - 9 = 0$$

35.
$$r = \frac{4}{2 - \cos\theta}$$
$$2r - r\cos\theta = 4$$
$$2r = r\cos\theta + 4$$
$$2\sqrt{x^2 + y^2} = x + 4$$
$$4x^2 + 4y^2 = x^2 + 8x + 16$$
$$3x^2 + 4y^2 - 8x - 16 = 0$$

37.
$$r = \frac{7}{2 + 10\sin\theta}$$
$$2r + 10r\sin\theta = 7$$
$$2r = 7 - 10r\sin\theta$$
$$2\sqrt{x^2 + y^2} = 7 - 10y$$
$$4x^2 + 4y^2 = 49 - 140y + 100y^2$$
$$4x^2 - 96y^2 + 140y - 49 = 0$$

39. $e = 2,\ r = 3\csc\theta$

The equation of the directrix can be written
$$r = \frac{3}{\sin\theta}, \text{ or } r\sin\theta = 3.$$
This corresponds to the equation $y = 3$ in rectangular coordinates, so the directrix is a horizontal line 3 units above the polar axis. Using the table on page

691 of the text, we see that the equation is of the form
$$r = \frac{ep}{1 + e\sin\theta}.$$
Substituting 2 for e and 3 for p, we have
$$r = \frac{2 \cdot 3}{1 + 2\sin\theta} = \frac{6}{1 + 2\sin\theta}.$$

41. $e = 1,\ r = 4\sec\theta$

The equation of the directrix can be written
$$r = \frac{4}{\cos\theta}, \text{ or } r\cos\theta = 4.$$
This corresponds to the equation $x = 4$ in rectangular coordinates, so the directrix is a vertical line 4 units to the right of the pole. Using the table on page 691 of the text, we see that the equation is of the form
$$r = \frac{ep}{1 + e\cos\theta}.$$
Substituting 1 for e and 4 for p, we have
$$r = \frac{1 \cdot 4}{1 + 1 \cdot \cos\theta} = \frac{4}{1 + \cos\theta}.$$

43. $e = \frac{1}{2},\ r = -2\sec\theta$

The equation of the directrix can be written
$$r = \frac{-2}{\cos\theta}, \text{ or } r\cos\theta = -2.$$
This corresponds to the equation $x = -2$ in rectangular coordinates, so the directrix is a vertical line 2 units to the left of the pole. Using the table on page 691 of the text, we see that the equation is of the form
$$r = \frac{ep}{1 - e\cos\theta}.$$
Substituting $\frac{1}{2}$ for e and 2 for p, we have
$$r = \frac{\frac{1}{2} \cdot 2}{1 - \frac{1}{2}\cos\theta} = \frac{1}{1 - \frac{1}{2}\cos\theta}, \text{ or } \frac{2}{2 - \cos\theta}.$$

45. $e = \frac{3}{4},\ r = 5\csc\theta$

The equation of the directrix can be written
$$r = \frac{5}{\sin\theta}, \text{ or } r\sin\theta = 5.$$
This corresponds to the equation $y = 5$ in rectangular coordinates, so the directrix is a horizontal line 5 units above the polar axis. Using the table on page 691 of the text, we see that the equation is of the form
$$r = \frac{ep}{1 + e\sin\theta}.$$
Substituting $\frac{3}{4}$ for e and 5 for p, we have

$$r = \frac{\frac{3}{4} \cdot 5}{1 + \frac{3}{4}\sin\theta} = \frac{15/4}{1 + \frac{3}{4}\sin\theta} \text{ or } \frac{15}{4 + 3\sin\theta}.$$

47. $e = 4$, $r = -2\csc\theta$

The equation of the directrix can be written

$$r = \frac{-2}{\sin\theta}, \text{ or } r\sin\theta = -2.$$

This corresponds to the equation $y = -2$ in rectangular coordinates, so the directrix is a horizontal line 2 units below the polar axis. Using the table on page 691 of the text, we see that the equation is of the form

$$r = \frac{ep}{1 - e\sin\theta}.$$

Substituting 4 for e and 2 for p, we have

$$r = \frac{4 \cdot 2}{1 - 4\sin\theta} = \frac{8}{1 - 4\sin\theta}.$$

49. Discussion and Writing

51. $f(x) = (x-3)^2 + 4$

$f(t) = (t-3)^2 + 4 = t^2 - 6t + 9 + 4 = t^2 - 6t + 13$

Thus, $f(t) = (t-3)^2 + 4$, or $t^2 - 6t + 13$.

53. $f(x) = (x-3)^2 + 4$

$f(t-1) = (t-1-3)^2 + 4 = (t-4)^2 + 4 = t^2 - 8t + 16 + 4 = t^2 - 8t + 20$

Thus, $f(t-1) = (t-4)^2 + 4$, or $t^2 - 8t + 20$.

55.

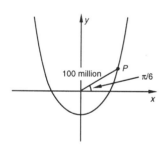

Since the directrix lies above the pole, the equation is of the form $r = \frac{ep}{1 + e\sin\theta}$. The point P on the parabola has coordinates $(1 \times 10^8, \pi/6)$. (Note that 100 million $= 1 \times 10^8$.) Since the conic is a parabola, we know that $e = 1$. We substitute 1×10^8 for r, 1 for e, and $\pi/6$ for θ and then find p.

$$1 \times 10^8 = \frac{1 \cdot p}{1 + 1 \cdot \sin\frac{\pi}{6}}$$

$$1 \times 10^8 = \frac{p}{1 + 0.5}$$

$$1 \times 10^8 = \frac{p}{1.5}$$

$$1.5 \times 10^8 = p$$

The equation of the orbit is $r = \dfrac{1.5 \times 10^8}{1 + \sin\theta}$.

Exercise Set 9.8

1. $x = \frac{1}{2}t$, $y = 6t - 7$; $-1 \le t \le 6$

We graph the equations in the window $[-2, 5, -20, 35]$ with Tstep $= .1$ and Yscl $= 5$.

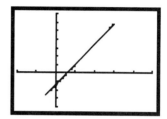

To find an equivalent rectangular equation, we first solve $x = \frac{1}{2}t$ for t.

$$x = \frac{1}{2}t$$

$$2x = t$$

Then we substitute $2x$ for t in $y = 6t - 7$.

$$y = 6(2x) - 7$$

$$y = 12x - 7$$

Given that $-1 \le t \le 6$, we find the corresponding restrictions on x:

For $t = -1$: $x = \frac{1}{2}t = \frac{1}{2}(-1) = -\frac{1}{2}$.

For $t = 6$: $x = \frac{1}{2}t = \frac{1}{2} \cdot 6 = 3$.

Then we have $y = 12x - 7$, $-\frac{1}{2} \le x \le 3$.

3. $x = t^3$, $y = t - 4$; $-1 \le t \le 10$

We graph the equations in the window $[-50, 1050, -10, 10]$ with Tstep $= .1$ and Xscl $= 50$.

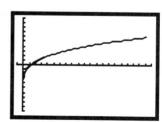

To find an equivalent rectangular equation, we first solve $x = t^3$ for t.

$$x = t^3$$

$$\sqrt[3]{x} = t$$

Then we substitute $\sqrt[3]{x}$ for t in $y = t - 4$: $y = \sqrt[3]{x} - 4$.

Given that $-1 \leq t \leq 10$, we find the corresponding restrictions on x:

For $t = -1$: $x = t^3 = (-1)^3 = -1$.

For $t = 10$: $x = t^3 = 10^3 = 1000$.

Then we have $y = \sqrt[3]{x} - 4$, $-1 \leq x \leq 1000$.

5. $x = t^2$, $y = \sqrt{t}$; $0 \leq t \leq 4$

We graph the equations in the window $[-5, 20, -1, 4]$ with Tstep $= .1$ and Xscl $= 5$.

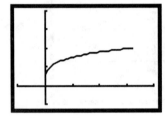

To find an equivalent rectangular equation, we first solve $x = t^2$ for t.

$$x = t^2$$
$$\sqrt{x} = t$$

(We choose the nonnegative square root because $0 \leq t \leq 4$.)

Then we substitute $\sqrt{x}$ for t in $y = \sqrt{t}$:

$$y = \sqrt{\sqrt{x}} = (x^{1/2})^{1/2}$$
$$y = x^{1/4}, \text{ or } \sqrt[4]{x}$$

Given that $0 \leq t \leq 4$, we find the corresponding restrictions on x:

For $t = 0$: $x = t^2 = (0)^2 = 0$.

For $t = 4$: $x = t^2 = 4^2 = 16$.

Then we have $y = \sqrt[4]{x}$, $0 \leq x \leq 16$. (This result could also be expressed as $x = y^4$, $0 \leq x \leq 16$.)

7. $x = t + 3$, $y = \dfrac{1}{t+3}$; $-2 \leq t \leq 2$

We graph the equations in the window $[-1, 7, -1, 3]$ with Tstep $= .1$.

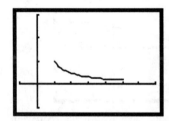

To find an equivalent rectangular equation, we can substitute x for $t + 3$ in $y = \dfrac{1}{t+3}$: $y = \dfrac{1}{x}$.

Given that $-2 \leq t \leq 2$, we find the corresponding restrictions on x:

For $t = -2$: $x = t + 3 = -2 + 3 = 1$.

For $t = 2$: $x = t + 3 = 2 + 3 = 5$.

Then we have $y = \dfrac{1}{x}$, $1 \leq x \leq 5$.

9. $x = 2t - 1$, $y = t^2$; $-3 \leq t \leq 3$

We graph the equations in the window $[-10, 10, -5, 15]$ with Tstep $= .1$ and Yscl $= 5$.

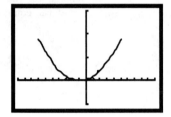

To find an equivalent rectangular equation, we first solve $x = 2t - 1$ for t:

$$x = 2t - 1$$
$$x + 1 = 2t$$
$$\frac{1}{2}(x + 1) = t$$

Then we substitute $\dfrac{1}{2}(x + 1)$ for t in $y = t^2$:

$$y = \left[\frac{1}{2}(x + 1) \right]^2$$
$$y = \frac{1}{4}(x + 1)^2$$

Given that $-3 \leq t \leq 3$, we find the corresponding restrictions on x:

For $t = -3$: $x = 2t - 1 = 2(-3) - 1 = -7$.

For $t = 3$: $x = 2t - 1 = 2 \cdot 3 - 1 = 5$.

Then we have $y = \dfrac{1}{4}(x + 1)^2$, $-7 \leq x \leq 5$.

11. $x = e^{-t}$, $y = e^t$; $-\infty < t < \infty$

We graph the equations in the window $[-1, 10, -1, 10]$ with Tstep $= .1$.

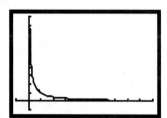

To find an equivalent rectangular equation, we first solve $x = e^{-t}$ for e^t:

$$x = e^{-t}$$
$$x = \frac{1}{e^t}$$
$$e^t = \frac{1}{x}$$

Then we substitute $\frac{1}{x}$ for e^t in $y = e^t$: $y = \frac{1}{x}$.

Given that $-\infty \le t \le \infty$, we find the corresponding restrictions on x:

As t approaches $-\infty$, e^{-t} approaches ∞. As t approaches ∞, e^{-t} approaches 0. Thus, we see that $x > 0$.

Then we have $y = \frac{1}{x}$, $x > 0$.

13. $x = 3\cos t$, $y = 3\sin t$; $0 \le t \le 2\pi$

We graph the equations in the square window $[-6, 6, -4, 4]$ with Tstep $= \pi/24$.

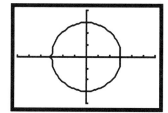

To find an equivalent rectangular equation, we first solve for $\cos t$ and $\sin t$ in the parametric equations:

$$\frac{x}{3} = \cos t, \frac{y}{3} = \sin t$$

Using the identity $\sin^2 \theta + \cos^2 \theta = 1$, we can substitute to eliminate the parameter:

$$\sin^2 t + \cos^2 t = 1$$
$$\left(\frac{y}{3}\right)^2 + \left(\frac{x}{3}\right)^2 = 1$$
$$\frac{x^2}{9} + \frac{y^2}{9} = 1$$
$$x^2 + y^2 = 9$$

For $0 \le t \le 2\pi$, $-3 \le 3\cos t \le 3$.

Then we have $x^2 + y^2 = 9$, $-3 \le x \le 3$.

15. $x = \cos t$, $y = 2\sin t$; $0 \le t \le 2\pi$

We graph the equations in the square window $[-4, 5, -3, 3]$ with Tstep $= \pi/24$.

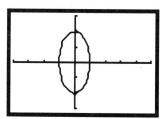

To find an equivalent rectangular equation, we first solve $y = 2\sin t$ for $\sin t$: $\frac{y}{2} = \sin t$.

Using the identity $\sin^2 \theta + \cos^2 \theta = 1$, we can substitute to eliminate the parameter:

$$\sin^2 t + \cos^2 t = 1$$
$$\left(\frac{y}{2}\right)^2 + x^2 = 1$$
$$x^2 + \frac{y^2}{4} = 1$$

For $0 \le t \le 2\pi$, $-1 \le \cos t \le 1$.

Then we have $x^2 + \frac{y^2}{4} = 1$, $-1 \le x \le 1$.

17. $x = \sec t$, $y = \cos t$; $-\frac{\pi}{2} < t < \frac{\pi}{2}$

Using DOT mode we graph the equations in the window $[0, 5, -1, 2]$ with Tstep $= \pi/24$.

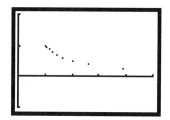

To find an equivalent rectangular equation, we first solve $x = \sec t$ for $\cos t$:

$$x = \sec t$$
$$x = \frac{1}{\cos t}$$
$$\cos t = \frac{1}{x}$$

Then we substitute $\frac{1}{x}$ for $\cos t$ in $y = \cos t$: $y = \frac{1}{x}$.

For $-\frac{\pi}{2} < t < \frac{\pi}{2}$, $1 \le x < \infty$.

Then we have $y = \frac{1}{x}$, $x \ge 1$.

19. $x = 1 + 2\cos t$, $y = 2 + 2\sin t$; $0 \le t \le 2\pi$

We graph the equations in the square window $[-4, 5, -1, 5]$ with Tstep $= \pi/24$.

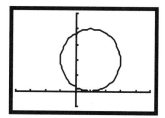

To find an equivalent rectangular equation, we first solve for $\cos t$ and $\sin t$ in the parametric equations:

$$x = 1 + 2\cos t \qquad\qquad y = 2 + 2\sin t$$
$$x - 1 = 2\cos t \qquad\qquad y - 2 = 2\sin t$$
$$\frac{x - 1}{2} = \cos t \qquad\qquad \frac{y - 2}{2} = \sin t$$

Using the identity $\sin^2 \theta + \cos^2 \theta = 1$, we can substitute to eliminate the parameter:

$$\sin^2 t + \cos^2 t = 1$$
$$\left(\frac{y - 2}{2}\right)^2 + \left(\frac{x - 1}{2}\right)^2 = 1$$
$$\frac{(x - 1)^2}{4} + \frac{(y - 2)^2}{4} = 1$$
$$(x - 1)^2 + (y - 2)^2 = 4$$

For $0 \le t \le 2\pi$, $-1 \le 1 + 2\cos t \le 3$.

Then we have $(x - 1)^2 + (y - 2)^2 = 4$, $-1 \le x \le 3$.

21. We can graph the unit circle using the parametric equations $x = \cos t$, $y = \sin t$, $0 \le t \le 2\pi$. Then, using the Value feature, enter $\pi/4$ for t. The value of y, which is $\sin t$, is shown on the screen. It is about 0.7071.

23. We can graph the unit circle using the parametric equations $x = \cos t$, $y = \sin t$, $0 \le t \le 2\pi$. Then, using the Value feature, enter $17\pi/12$ for t. The value of x, which is $\cos t$, is shown on the screen. It is about -0.2588.

25. We can graph the unit circle using the parametric equations $x = \cos t$, $y = \sin t$, $0 \le t \le 2\pi$. Then, using the Value feature, enter $\pi/5$ for t. The values of x and y, which are $\cos t$ and $\sin t$ respectively, are about 0.80901699 and 0.58778525. We find y/x, or $\sin t/\cos t$, which is $\tan t$, by dividing:

$$\frac{0.58778525}{0.80901699} \approx 0.7265.$$

27. We can graph the unit circle using the parametric equations $x = \cos t$, $y = \sin t$, $0 \le t \le 2\pi$. Then, using the Value feature, enter 5.29 for t. The value of x, which is $\cos t$, is shown on the screen. It is about 0.5460.

29. $y = 4x - 3$

Answers may vary.

If $x = t$, then $y = 4t - 3$.

If $x = \dfrac{t}{4}$, then $y = 4 \cdot \dfrac{t}{4} - 3 = t - 3$.

31. $y = (x - 2)^2 - 6x$

Answers may vary.

If $x = t$, then $y = (t - 2)^2 - 6t$.

If $x = t+2$, then $y = (t+2-2)^2 - 6(t+2) = t^2 - 6t - 12$.

33. $x = (80\cos 35°)t$, $y = 6.25 + (80\sin 35°)t - 16t^2$

We first graph the equations in the window [0,225,0,50] with Xscl = 25 and Yscl = 5. We let Tmin = 0, Tmax = 5, and Tstep = .1.

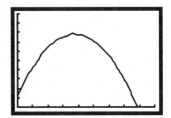

a) To find the height of the ball after 1 sec, we use the Value feature to find the value of y when $t = 1$. We find that the height is about 36.1 ft.

b) To find the total horizontal distance traveled by the ball, we use TRACE and ZOOM to find the first coordinate of the x-intercept of the graph. We find that the total horizontal distance traveled is about 196.3 ft.

c) To find how long it takes the ball to reach the ground, we use TRACE and ZOOM to find the value of t for which $y = 0$. The time is about 3 sec.

35. Graph the equation as in Example 6. Then use TRACE and ZOOM to find the maximum value of y. It is about 73.1, so the ball reaches a maximum height of about 73.1 ft.

37.

$x = 2(t - \sin t)$, $y = 2(1 - \cos t)$; $0 \le t \le 4\pi$

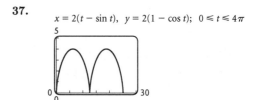

39.

$x = t - \sin t$, $y = 1 - \cos t$; $-2\pi \le t \le 2\pi$

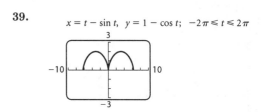

41. Discussion and writing

43. Graph $y = x^3$.

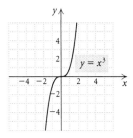

x	y
-2	-8
-1	-1
0	0
1	1
2	8

45. Graph $f(x) = \sqrt{x - 2}$.

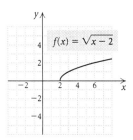

x	$f(x)$
2	0
3	1
6	2
11	3

47. The graph is shown in Exercise 13. The curve is generated clockwise when the equations $x = 3\cos t$, $y = -3\sin t$ are used. Alternatively, the equations $x = 3\sin t$, $y = 3\cos t$ can be used.

Chapter 10

Sequences, Series, and Combinatorics

Exercise Set 10.1

---

1. $a_n = 4n - 1$

$a_1 = 4 \cdot 1 - 1 = 3,$

$a_2 = 4 \cdot 2 - 1 = 7,$

$a_3 = 4 \cdot 3 - 1 = 11,$

$a_4 = 4 \cdot 4 - 1 = 15;$

$a_{10} = 4 \cdot 10 - 1 = 39;$

$a_{15} = 4 \cdot 15 - 1 = 59$

3. $a_n = \dfrac{n}{n-1}, \; n \geq 2$

The first 4 terms are $a_2, a_3, a_4,$ and a_5:

$a_2 = \dfrac{2}{2-1} = 2,$

$a_3 = \dfrac{3}{3-1} = \dfrac{3}{2},$

$a_4 = \dfrac{4}{4-1} = \dfrac{4}{3},$

$a_5 = \dfrac{5}{5-1} = \dfrac{5}{4};$

$a_{10} = \dfrac{10}{10-1} = \dfrac{10}{9};$

$a_{15} = \dfrac{15}{15-1} = \dfrac{15}{14}$

5. $a_n = \dfrac{n^2 - 1}{n^2 + 1},$

$a_1 = \dfrac{1^2 - 1}{1^2 + 1} = 0,$

$a_2 = \dfrac{2^2 - 1}{2^2 + 1} = \dfrac{3}{5},$

$a_3 = \dfrac{3^2 - 1}{3^2 + 1} = \dfrac{8}{10} = \dfrac{4}{5},$

$a_4 = \dfrac{4^2 - 1}{4^2 + 1} = \dfrac{15}{17};$

$a_{10} = \dfrac{10^2 - 1}{10^2 + 1} = \dfrac{99}{101};$

$a_{15} = \dfrac{15^2 - 1}{15^2 + 1} = \dfrac{224}{226} = \dfrac{112}{113}$

7. $a_n = (-1)^n n^2$

$a_1 = (-1)^1 1^2 = -1,$

$a_2 = (-1)^2 2^2 = 4,$

$a_3 = (-1)^3 3^2 = -9,$

$a_4 = (-1)^4 4^2 = 16;$

$a_{10} = (-1)^{10} 10^2 = 100;$

$a_{15} = (-1)^{15} 15^2 = -225$

9. $a_n = 5 + \dfrac{(-2)^{n+1}}{2^n}$

$a_1 = 5 + \dfrac{(-2)^{1+1}}{2^1} = 5 + \dfrac{4}{2} = 7,$

$a_2 = 5 + \dfrac{(-2)^{2+1}}{2^2} = 5 + \dfrac{-8}{4} = 3,$

$a_3 = 5 + \dfrac{(-2)^{3+1}}{2^3} = 5 + \dfrac{16}{8} = 7,$

$a_4 = 5 + \dfrac{(-2)^{4+1}}{2^4} = 5 + \dfrac{-32}{16} = 3;$

$a_{10} = 5 + \dfrac{(-2)^{10+1}}{2^{10}} = 5 + \dfrac{-1 \cdot 2^{11}}{2^{10}} = 3;$

$a_{15} = 5 + \dfrac{(-2)^{15+1}}{2^{15}} = 5 + \dfrac{2^{16}}{2^{15}} = 7$

11. $a_n = 5n - 6$

$a_8 = 5 \cdot 8 - 6 = 40 - 6 = 34$

13. $a_n = (2n + 3)^2$

$a_6 = (2 \cdot 6 + 3)^2 = 225$

15. $a_n = 5n^2(4n - 100)$

$a_{11} = 5(11)^2(4 \cdot 11 - 100) = 5(121)(-56) = -33,880$

17. $a_n = \ln e^n$

$a_{67} = \ln e^{67} = 67$

19. $2, 4, 6, 8, 10, \ldots$

These are the even integers, so the general term might be $2n$.

21. $-2, 6, -18, 54, \ldots$

We can see a pattern if we write the sequence as

$-1 \cdot 2 \cdot 1, \; 1 \cdot 2 \cdot 3, \; -1 \cdot 2 \cdot 9, \; 1 \cdot 2 \cdot 27, \ldots$

The general term might be $(-1)^n 2(3)^{n-1}$.

23. $\dfrac{2}{3}, \dfrac{3}{4}, \dfrac{4}{5}, \dfrac{5}{6}, \dfrac{6}{7}, \ldots$

These are fractions in which the denominator is 1 greater than the numerator. Also, each numerator is 1 greater than the preceding numerator. The general term might be $\dfrac{n+1}{n+2}$.

25. $1 \cdot 2, \, 2 \cdot 3, \, 3 \cdot 4, \, 4 \cdot 5, \ldots$

These are the products of pairs of consecutive natural numbers. The general term might be $n(n+1)$.

27. $0, \log 10, \log 100, \log 1000, \ldots$

We can see a pattern if we write the sequence as

$\log 1, \log 10, \log 100, \log 1000, \ldots$

The general term might be $\log 10^{n-1}$. This is equivalent to $n-1$.

29. $1, 2, 3, 4, 5, 6, 7, \ldots$

$S_3 = 1 + 2 + 3 = 6$

$S_7 = 1 + 2 + 3 + 4 + 5 + 6 + 7 = 28$

31. $2, 4, 6, 8, \ldots$

$S_4 = 2 + 4 + 6 + 8 = 20$

$S_5 = 2 + 4 + 6 + 8 + 10 = 30$

33. $\displaystyle\sum_{k=1}^{5} \dfrac{1}{2k} = \dfrac{1}{2 \cdot 1} + \dfrac{1}{2 \cdot 2} + \dfrac{1}{2 \cdot 3} + \dfrac{1}{2 \cdot 4} + \dfrac{1}{2 \cdot 5}$

$\qquad = \dfrac{1}{2} + \dfrac{1}{4} + \dfrac{1}{6} + \dfrac{1}{8} + \dfrac{1}{10}$

$\qquad = \dfrac{60}{120} + \dfrac{30}{120} + \dfrac{20}{120} + \dfrac{15}{120} + \dfrac{12}{120}$

$\qquad = \dfrac{137}{120}$

35. $\displaystyle\sum_{i=0}^{6} 2^i = 2^0 + 2^1 + 2^2 + 2^3 + 2^4 + 2^5 + 2^6$

$\qquad = 1 + 2 + 4 + 8 + 16 + 32 + 64$

$\qquad = 127$

37. $\displaystyle\sum_{k=7}^{10} \ln k = \ln 7 + \ln 8 + \ln 9 + \ln 10 =$

$\ln(7 \cdot 8 \cdot 9 \cdot 10) = \ln 5040 \approx 8.5252$

39. $\displaystyle\sum_{k=1}^{8} \dfrac{k}{k+1} = \dfrac{1}{1+1} + \dfrac{2}{2+1} + \dfrac{3}{3+1} + \dfrac{4}{4+1} +$

$\qquad \dfrac{5}{5+1} + \dfrac{6}{6+1} + \dfrac{7}{7+1} + \dfrac{8}{8+1}$

$\qquad = \dfrac{1}{2} + \dfrac{2}{3} + \dfrac{3}{4} + \dfrac{4}{5} + \dfrac{5}{6} + \dfrac{6}{7} + \dfrac{7}{8} + \dfrac{8}{9}$

$\qquad = \dfrac{15,551}{2520}$

41. $\displaystyle\sum_{i=1}^{5} (-1)^i$

$= (-1)^1 + (-1)^2 + (-1)^3 + (-1)^4 + (-1)^5$

$= -1 + 1 - 1 + 1 - 1$

$= -1$

43. $\displaystyle\sum_{k=1}^{8} (-1)^{k+1} 3k$

$= (-1)^2 3 \cdot 1 + (-1)^3 3 \cdot 2 + (-1)^4 3 \cdot 3 +$

$\quad (-1)^5 3 \cdot 4 + (-1)^6 3 \cdot 5 + (-1)^7 3 \cdot 6 +$

$\quad (-1)^8 3 \cdot 7 + (-1)^9 3 \cdot 8$

$= 3 - 6 + 9 - 12 + 15 - 18 + 21 - 24$

$= -12$

45. $\displaystyle\sum_{k=0}^{6} \dfrac{2}{k^2+1} = \dfrac{2}{0^2+1} + \dfrac{2}{1^2+1} + \dfrac{2}{2^2+1} + \dfrac{2}{3^2+1} +$

$\qquad \dfrac{2}{4^2+1} + \dfrac{2}{5^2+1} + \dfrac{2}{6^2+1}$

$\qquad = 2 + 1 + \dfrac{2}{5} + \dfrac{2}{10} + \dfrac{2}{17} + \dfrac{2}{26} + \dfrac{2}{37}$

$\qquad = 2 + 1 + \dfrac{2}{5} + \dfrac{1}{5} + \dfrac{2}{17} + \dfrac{1}{13} + \dfrac{2}{37}$

$\qquad = \dfrac{157,351}{40,885}$

47. $\displaystyle\sum_{k=0}^{5} (k^2 - 2k + 3)$

$= (0^2 - 2 \cdot 0 + 3) + (1^2 - 2 \cdot 1 + 3) +$

$\quad (2^2 - 2 \cdot 2 + 3) + (3^2 - 2 \cdot 3 + 3) +$

$\quad (4^2 - 2 \cdot 4 + 3) + (5^2 - 2 \cdot 5 + 3)$

$= 3 + 2 + 3 + 6 + 11 + 18$

$= 43$

49. $\displaystyle\sum_{i=0}^{10} \dfrac{2i}{2^i+1}$

$= \dfrac{2^0}{2^0+1} + \dfrac{2^1}{2^1+1} + \dfrac{2^2}{2^2+1} + \dfrac{2^3}{2^3+1} + \dfrac{2^4}{2^4+1} +$

$\quad \dfrac{2^5}{2^5+1} + \dfrac{2^6}{2^6+1} + \dfrac{2^7}{2^7+1} + \dfrac{2^8}{2^8+1} + \dfrac{2^9}{2^9+1} +$

$\quad \dfrac{2^{10}}{2^{10}+1}$

$= \dfrac{1}{2} + \dfrac{2}{3} + \dfrac{4}{5} + \dfrac{8}{9} + \dfrac{16}{17} + \dfrac{32}{33} + \dfrac{64}{65} + \dfrac{128}{129} +$

$\quad \dfrac{256}{257} + \dfrac{512}{513} + \dfrac{1024}{1025}$

≈ 9.736

51. $5 + 10 + 15 + 20 + 25 + \ldots$

This is a sum of multiples of 5, and it is an infinite series. Sigma notation is

$$\sum_{k=1}^{\infty} 5k.$$

53. $2 - 4 + 8 - 16 + 32 - 64$

This is a sum of powers of 2 with alternating signs. Sigma notation is

$$\sum_{k=1}^{6} (-1)^{k+1} 2k, \text{ or } \sum_{k=1}^{6} (-1)^{k-1} 2k$$

55. $-\dfrac{1}{2} + \dfrac{2}{3} - \dfrac{3}{4} + \dfrac{4}{5} - \dfrac{5}{6} + \dfrac{6}{7}$

This is a sum of fractions in which the denominator is one greater than the numerator. Also, each numerator is 1 greater than the preceding numerator and the signs alternate. Sigma notation is

$$\sum_{k=1}^{6} (-1)^{k} \frac{k}{k+1}.$$

57. $4 - 9 + 16 - 25 + \dots + (-1)^n n^2$

This is a sum of terms of the form $(-1)^k k^2$, beginning with $k = 2$ and continuing through $k = n$. Sigma notation is

$$\sum_{k=2}^{n} (-1)^{k} k^2.$$

59. $\dfrac{1}{1 \cdot 2} + \dfrac{1}{2 \cdot 3} + \dfrac{1}{3 \cdot 4} + \dfrac{1}{4 \cdot 5} + \dots$

This is a sum of fractions in which the numerator is 1 and the denominator is a product of two consecutive integers. The larger integer in each product is the smaller integer in the succeeding product. It is an infinite series. Sigma notation is

$$\sum_{k=1}^{\infty} \frac{1}{k(k+1)}.$$

61. $a_1 = 4, \qquad a_{k+1} = 1 + \dfrac{1}{a_k}$

$a_2 = 1 + \dfrac{1}{4} = 1\dfrac{1}{4}, \text{ or } \dfrac{5}{4}$

$a_3 = 1 + \dfrac{1}{\frac{5}{4}} = 1 + \dfrac{4}{5} = 1\dfrac{4}{5}, \text{ or } \dfrac{9}{5}$

$a_4 = 1 + \dfrac{1}{\frac{9}{5}} = 1 + \dfrac{5}{9} = 1\dfrac{5}{9}, \text{ or } \dfrac{14}{9}$

63. $a_1 = 6561, \qquad a_{k+1} = (-1)^k \sqrt{a_k}$

$a_2 = (-1)^1 \sqrt{6561} = -81$

$a_3 = (-1)^2 \sqrt{-81} = 9i$

$a_4 = (-1)^3 \sqrt{9i} = -3\sqrt{i}$

65. $a_1 = 2, \qquad a_{k+1} = a_k + a_{k-1}$

$a_2 = 3$

$a_3 = 3 + 2 = 5$

$a_4 = 5 + 3 = 8$

67. See the answer section in the text.

69. See the answer section in the text.

71. a) $a_1 = \$1000(1.062)^1 = \$1062$

$a_2 = \$1000(1.062)^2 \approx \$1127.84$

$a_3 = \$1000(1.062)^3 \approx \$1197.77$

$a_4 = \$1000(1.062)^4 \approx \$1272.03$

$a_5 = \$1000(1.062)^5 \approx \$1350.90$

$a_6 = \$1000(1.062)^6 \approx \$1434.65$

$a_7 = \$1000(1.062)^7 \approx \$1523.60$

$a_8 = \$1000(1.062)^8 \approx \$1618.07$

$a_9 = \$1000(1.062)^9 \approx \$1718.39$

$a_{10} = \$1000(1.062)^{10} \approx \$1824.93$

b) $a_{20} = \$1000(1.062)^{20} \approx \$3330.35$

73. Find each term by multiplying the preceding term by 2. Find 17 terms, beginning with $a_1 = 1$, since there are 16 fifteen minute periods in 4 hr.

1, 2, 4, 8, 16, 32, 64, 128, 256, 512, 1024,

2048, 4096, 8192, 16,384, 32,768, 65,536

75. a)

n	Un
15	27.25
16	26.108
17	25.002
18	23.932
19	22.898
20	21.9
21	20.938
22	20.012
23	19.122
24	18.268

b)

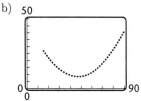

c) From the table in part (a), we know that $a_{16} = 26.108 \approx 26$ and $a_{23} = 19.122 \approx 19$. Using a table set in Ask mode we find that $a_{50} = 8.7 \approx 9$, $a_{75} = 22.45 \approx 22$, and $a_{85} = 34.25 \approx 34$.

77. $a_1 = 1$ (Given)

$a_2 = 1$ (Given)

$a_3 = a_2 + a_1 = 1 + 1 = 2$

$a_4 = a_3 + a_2 = 2 + 1 = 3$

$a_5 = a_4 + a_3 = 3 + 2 = 5$

$a_6 = a_5 + a_4 = 5 + 3 = 8$

$a_7 = a_6 + a_5 = 8 + 5 = 13$

79. Discussion and Writing

81. **Familiarize.** Let $x =$ the amount spent on original research and development and $y =$ the amount spent on purchased research and development.

Translate. A total of \$1.19 billion was spent, so we have one equation:

$x + y = 1.19$

We also know that the amount spent on original research and development is \$0.837 billion more than the amount spent on purchased research and development. This gives us another equation:

$x = y + 0.837$

We have a system of equations:

$x + y = 1.19,$

$x = y + 0.837$

Carry out. We use the substitution method.

$(y + 0.837) + y = 1.19$

$2y + 0.837 = 1.19$

$2y = 0.352$

$y = 0.1765$

Back-substitute to find x.

$x = 0.1765 + 0.837 = 1.0135$

Check. The total spent is \$1.0135 billion + \$0.837 billion, or \$1.19 billion. Also, \$1.0135 billion is \$0.837 billion more than \$0.1765 billion. The answer checks.

State. In 1998 \$1.0135 billion was spent on original research and development and \$0.1765 was spent on purchased research and development.

83. We complete the square twice.

$$x^2 + y^2 + 5x - 8y = 2$$

$$x^2 + 5x + y^2 - 8y = 2$$

$$x^2 + 5x + \frac{25}{4} + y^2 - 8y + 16 = 2 + \frac{25}{4} + 16$$

$$\left(x + \frac{5}{2}\right)^2 + (y - 4)^2 = \frac{97}{4}$$

$$\left[x - \left(-\frac{5}{2}\right)\right]^2 + (y - 4)^2 = \left(\frac{\sqrt{97}}{2}\right)^2$$

The center is $\left(-\frac{5}{2}, 4\right)$ and the radius is $\frac{\sqrt{97}}{2}$.

85. $a_n = i^n$

$a_1 = i$

$a_2 = i^2 = -1$

$a_3 = i^3 = -i$

$a_4 = i^4 = 1$

$a_5 = i^5 = i^4 \cdot i = i$

$S_5 = i - 1 - i + 1 + i = i$

87. $S_n = \ln 1 + \ln 2 \ln 3 + \cdots + \ln n$

$= \ln(1 \cdot 2 \cdot 3 \cdots n)$

Exercise Set 10.2

1. 3, 8, 13, 18, . . .

$a_1 = 3$

$d = 5$ $(8 - 3 = 5,\ 13 - 8 = 5,\ 18 - 13 = 5)$

3. 9, 5, 1, -3, . . .

$a_1 = 9$

$d = -4$ $(5 - 9 = -4,\ 1 - 5 = -4,\ -3 - 1 = -4)$

5. $\dfrac{3}{2}, \dfrac{9}{4}, 3, \dfrac{15}{4}, \ldots$

$a_1 = \dfrac{3}{2}$

$d = \dfrac{3}{4}$ $\left(\dfrac{9}{4} - \dfrac{3}{2} = \dfrac{3}{4}, 3 - \dfrac{9}{4} = \dfrac{3}{4}\right)$

7. $a_1 = \$316$

$d = -\$3$ $(\$313 - \$316 = -\$3,$
$\$310 - \$313 = -\$3,\ \$307 - \$310 = -\$3)$

9. 2, 6, 10, . . .

$a_1 = 2,\ d = 4,$ and $n = 12$

$a_n = a_1 + (n - 1)d$

$a_{12} = 2 + (12 - 1)4 = 2 + 11 \cdot 4 = 2 + 44 = 46$

11. 3, $\dfrac{7}{3}, \dfrac{5}{3}, \ldots$

$a_1 = 3,\ d = -\dfrac{2}{3},$ and $n = 14$

$a_n = a_1 + (n - 1)d$

$a_{14} = 3 + (14 - 1)\left(-\dfrac{2}{3}\right) = 3 - \dfrac{26}{3} = -\dfrac{17}{3}$

13. \$2345.78, \$2967.54, \$3589.30, . . .

$a_1 = \$2345.78,\ d = \$621.76,$ and $n = 10$

$a_n = a_1 + (n - 1)d$

$a_{10} = \$2345.78 + (10 - 1)(\$621.76) = \$7941.62$

15. $a_1 = 0.07$, $d = 0.05$

$a_n = a_1 + (n-1)d$

Let $a_n = 1.67$, and solve for n.

$1.67 = 0.07 + (n-1)(0.05)$

$1.67 = 0.07 + 0.05n - 0.05$

$1.65 = 0.05n$

$33 = n$

The 33rd term is 1.67.

17. $a_1 = 3$, $d = -\dfrac{2}{3}$

$a_n = a_1 + (n-1)d$

Let $a_n = -27$, and solve for n.

$-27 = 3 + (n-1)\left(-\dfrac{2}{3}\right)$

$-81 = 9 + (n-1)(-2)$

$-81 = 9 - 2n + 2$

$-92 = -2n$

$46 = n$

The 46th term is -27.

19. $a_n = a_1 + (n-1)d$

$33 = a_1 + (8-1)4$ Substituting 33 for a_8,
$\qquad\qquad\qquad\qquad$ 8 for n, and 4 for d

$33 = a_1 + 28$

$5 = a_1$

(Note that this procedure is equivalent to subtracting d from a_8 seven times to get a_1: $33 - 7(4) = 33 - 28 = 5$)

21. $a_n = a_1 + (n-1)d$

$-507 = 25 + (n-1)(-14)$

$-507 = 25 - 14n + 14$

$-546 = -14n$

$39 = n$

23. $\dfrac{25}{3} + 15d = \dfrac{95}{6}$

$\qquad 15d = \dfrac{45}{6}$

$\qquad\quad d = \dfrac{1}{2}$

$a_1 = \dfrac{25}{3} - 16\left(\dfrac{1}{2}\right) = \dfrac{25}{3} - 8 = \dfrac{1}{3}$

The first five terms of the sequence are $\dfrac{1}{3}, \dfrac{5}{6}, \dfrac{4}{3}, \dfrac{11}{6}$, $\dfrac{7}{3}$.

25. $5 + 8 + 11 + 14 + \ldots$

Note that $a_1 = 5$, $d = 3$, and $n = 20$. First we find a_{20}:

$a_n = a_1 + (n-1)d$

$a_{20} = 5 + (20-1)3$

$\qquad = 5 + 19 \cdot 3 = 62$

Then

$S_n = \dfrac{n}{2}(a_1 + a_n)$

$S_{20} = \dfrac{20}{2}(5 + 62)$

$\qquad = 10(67) = 670.$

27. The sum is $2 + 4 + 6 + \ldots + 798 + 800$. This is the sum of the arithmetic sequence for which $a_1 = 2$, $a_n = 800$, and $n = 400$.

$S_n = \dfrac{n}{2}(a_1 + a_n)$

$S_{400} = \dfrac{400}{2}(2 + 800) = 200(802) = 160,400$

29. The sum is $7 + 14 + 21 + \ldots + 91 + 98$. This is the sum of the arithmetic sequence for which $a_1 = 7$, $a_n = 98$, and $n = 14$.

$S_n = \dfrac{n}{2}(a_1 + a_n)$

$S_{14} = \dfrac{14}{2}(7 + 98) = 7(105) = 735$

31. First we find a_{20}:

$a_n = a_1 + (n-1)d$

$a_{20} = 2 + (20-1)5$

$\qquad = 2 + 19 \cdot 5 = 97$

Then

$S_n = \dfrac{n}{2}(a_1 + a_n)$

$S_{20} = \dfrac{20}{2}(2 + 97)$

$\qquad = 10(99) = 990.$

33. $\displaystyle\sum_{k=1}^{40}(2k+3)$

Write a few terms of the sum:

$5 + 7 + 9 + \ldots + 83$

This is a series coming from an arithmetic sequence with $a_1 = 5$, $n = 40$, and $a_{40} = 83$. Then

$S_n = \dfrac{n}{2}(a_1 + a_n)$

$S_{40} = \dfrac{40}{2}(5 + 83)$

$\qquad = 20(88) = 1760$

35. $\displaystyle\sum_{k=0}^{19} \frac{k-3}{4}$

Write a few terms of the sum:

$$-\frac{3}{4} - \frac{1}{2} - \frac{1}{4} + 0 + \frac{1}{4} + \ldots + 4$$

Since k goes from 0 through 19, there are 20 terms. Thus, this is equivalent to a series coming from an arithmetic sequence with $a_1 = -\frac{3}{4}$, $n = 20$, and $a_{20} = 4$. Then

$$S_n = \frac{n}{2}(a_1 + a_n)$$

$$S_{20} = \frac{20}{2}\left(-\frac{3}{4} + 4\right)$$

$$= 10 \cdot \frac{13}{4} = \frac{65}{2}.$$

37. $\displaystyle\sum_{k=12}^{57} \frac{7-4k}{13}$

Write a few terms of the sum:

$$-\frac{41}{13} - \frac{45}{13} - \frac{49}{13} - \ldots - \frac{221}{13}$$

Since k goes from 12 through 57, there are 46 terms. Thus, this is equivalent to a series coming from an arithmetic sequence with $a_1 = -\frac{41}{13}$, $n = 46$, and $a_{46} = -\frac{221}{13}$. Then

$$S_n = \frac{n}{2}(a_1 + a_n)$$

$$S_{46} = \frac{46}{2}\left(-\frac{41}{13} - \frac{221}{13}\right)$$

$$= 23\left(-\frac{262}{13}\right) = -\frac{6026}{13}.$$

39. Familiarize. We go from 50 poles in a row, down to six poles in the top row, so there must be 45 rows. We want the sum $50 + 49 + 48 + \ldots + 6$. Thus we want the sum of an arithmetic sequence. We will use the formula $S_n = \frac{n}{2}(a_1 + a_n)$.

Translate. We want to find the sum of the first 45 terms of an arithmetic sequence with $a_1 = 50$ and $a_{45} = 6$.

Carry out. Substituting into the formula, we have

$$S_{45} = \frac{45}{2}(50 + 6)$$

$$= \frac{45}{2} \cdot 56 = 1260$$

Check. We can do the calculation again, or we can do the entire addition:

$50 + 49 + 48 + \ldots + 6$.

State. There will be 1260 poles in the pile.

41. We first find how many plants will be in the last row.

Familiarize. The sequence is 35, 31, 27, ... It is an arithmetic sequence with $a_1 = 35$ and $d = -4$. Since each row must contain a positive number of plants, we must determine how many times we can add -4 to 35 and still have a positive result.

Translate. We find the largest integer x for which $35 + x(-4) > 0$. Then we evaluate the expression $35 - 4x$ for that value of x.

Carry out. We solve the inequality.

$$35 - 4x > 0$$
$$35 > 4x$$
$$\frac{35}{4} > x$$
$$8\frac{3}{4} > x$$

The integer we are looking for is 8. Thus $35 - 4x = 35 - 4(8) = 3$.

Check. If we add -4 to 35 eight times we get 3, a positive number, but if we add -4 to 35 more than eight times we get a negative number.

State. There will be 3 plants in the last row.

Next we find how many plants there are altogether.

Familiarize. We want to find the sum $35 + 31 + 27 + \ldots + 3$. We know $a_1 = 35$ $a_n = 3$, and, since we add -4 to 35 eight times, $n = 9$. (There are 8 terms after a_1, for a total of 9 terms.) We will use the formula $S_n = \frac{n}{2}(a_1 + a_n)$.

Translate. We want to find the sum of the first 9 terms of an arithmetic sequence in which $a_1 = 35$ and $a_9 = 3$.

Carry out. Substituting into the formula, we have

$$S_9 = \frac{9}{2}(35 + 3)$$

$$= \frac{9}{2} \cdot 38 = 171$$

Check. We can check the calculations by doing them again. We could also do the entire addition:

$35 + 31 + 27 + \ldots + 3$.

State. There are 171 plants altogether.

43. Familiarize. We have a sequence 10, 20, 30, . . . It is an arithmetic sequence with $a_1 = 10$, $d = 10$, and $n = 31$.

Translate. We want to find $S_n = \frac{n}{2}(a_1 + a_n)$ where $a_n = a_1 + (n-1)d$, $a_1 = 10$, $d = 10$, and $n = 31$.

Carry out. First we find a_{31}.

$a_{31} = 10 + (31-1)10 = 10 + 30 \cdot 10 = 310$

Then $S_{31} = \frac{31}{2}(10 + 310) = \frac{31}{2} \cdot 320 = 4960.$

Check. We can do the calculation again, or we can do the entire addition:

$$10 + 20 + 30 + \ldots + 310.$$

State. A total of 4960¢, or $49.60 is saved.

45. Familiarize. We have arithmetic sequence with $a_1 = 28$, $d = 4$, and $n = 20$.

Translate. We want to find $S_n = \dfrac{n}{2}(a_1 + a_n)$ where $a_n = a_1 + (n-1)d$, $a_1 = 28$, $d = 4$, and $n = 20$.

Carry out. First we find a_{20}.

$$a_{20} = 28 + (20-1)4 = 104$$

Then $S_{20} = \dfrac{20}{2}(28 + 104) = 10 \cdot 132 = 1320.$

Check. We can do the calculations again, or we can do the entire addition:

$$28 + 32 + 36 + \ldots 104.$$

State. There are 1320 seats in the first balcony.

47. Yes; $d = 6 - 3 = 9 - 6 = 3n - 3(n-1) = 3.$

49. Discussion and Writing

51.
$$\begin{aligned} 2x + y + 3z &= 12 \\ x - 3y - 2z &= -1 \\ 5x + 2y - 4z &= -4 \end{aligned}$$

We will use Gauss-Jordan elimination with matrices. First we write the augmented matrix.

$$\left[\begin{array}{ccc|c} 2 & 1 & 3 & 12 \\ 1 & -3 & -2 & -1 \\ 5 & 2 & -4 & -4 \end{array}\right]$$

Next we interchange the first two rows.

$$\left[\begin{array}{ccc|c} 1 & -3 & -2 & -1 \\ 2 & 1 & 3 & 12 \\ 5 & 2 & -4 & -4 \end{array}\right]$$

Now multiply the first row by -2 and add it to the second row. Also multiply the first row by -5 and add it to the third row.

$$\left[\begin{array}{ccc|c} 1 & -3 & -2 & -1 \\ 0 & 7 & 7 & 14 \\ 0 & 17 & 6 & 1 \end{array}\right]$$

Multiply the second row by $\dfrac{1}{7}$.

$$\left[\begin{array}{ccc|c} 1 & -3 & -2 & -1 \\ 0 & 1 & 1 & 2 \\ 0 & 17 & 6 & 1 \end{array}\right]$$

Multiply the second row by 3 and add it to the first row. Also multiply the second row by -17 and add it to the third row.

$$\left[\begin{array}{ccc|c} 1 & 0 & 1 & 5 \\ 0 & 1 & 1 & 2 \\ 0 & 0 & -11 & -33 \end{array}\right]$$

Multiply the third row by $-\dfrac{1}{11}$.

$$\left[\begin{array}{ccc|c} 1 & 0 & 1 & 5 \\ 0 & 1 & 1 & 2 \\ 0 & 0 & 1 & 3 \end{array}\right]$$

Multiply the third row by -1 and add it to the first row and also to the second row.

$$\left[\begin{array}{ccc|c} 1 & 0 & 0 & 2 \\ 0 & 1 & 0 & -1 \\ 0 & 0 & 1 & 3 \end{array}\right]$$

Now we can read the solution from the matrix. It is $(2, -1, 3)$.

53. The vertices are on the y-axis, so the transverse axis is vertical and $a = 5$. The length of the minor axis is 4, so $b = 4/2 = 2$. The equation is

$$\frac{x^2}{4} + \frac{y^2}{25} = 1.$$

55. $S_n = \dfrac{n}{2}(1 + 2n - 1) = n^2$

57. Let $d =$ the common difference. Then $a_4 = a_2 + 2d$, or

$$\begin{aligned} 10p + q &= 40 - 3q + 2d \\ 10p + 4q - 40 &= 2d \\ 5p + 2q - 20 &= d. \end{aligned}$$

Also, $a_1 = a_2 - d$, so we have

$$\begin{aligned} a_1 &= 40 - 3q - (5p + 2q - 20) \\ &= 40 - 3q - 5p - 2q + 20 \\ &= 60 - 5p - 5q. \end{aligned}$$

59. $4, m_1, m_2, m_3, 1$

We look for m_1, m_2, and m_3 such that $4, m_1, m_2, m_3, 13$ is an arithmetic sequence. In this case, $a_1 = 4$, $n = 5$, and $a_5 = 12$. First we find d:

$$\begin{aligned} a_n &= a_1 + (n-1)d \\ 12 &= 4 + (5-1)d \\ 12 &= 4 + 4d \\ 8 &= 4d \\ 2 &= d \end{aligned}$$

Then we have
$$m_1 = a_1 + d = 4 + 2 = 6$$
$$m_2 = m_1 + d = 6 + 2 = 8$$
$$m_3 = m_2 + d = 8 + 2 = 10$$

61. $-3, m_1, m_2, m_3, 5$

We look for m_1, m_2, and m_3 such that $-3, m_1, m_2, m_3, 5$ is an arithmetic sequence. In this case, $a_1 = -3$, $n = 5$, and $a_5 = 5$. First we find d:
$$a_n = a_1 + (n-1)d$$
$$5 = -3 + (5-1)d$$
$$8 = 4d$$
$$2 = d$$

Then we have
$$m_1 = a_1 + d = -3 + 2 = -1,$$
$$m_2 = m_1 + d = -1 + 2 = 1,$$
$$m_3 = m_2 + d = 1 + 2 = 3.$$

63. $1, 1+d, 1+2d, \ldots, 50$ has n terms and $S_n = 459$.

Find n:
$$459 = \frac{n}{2}(1 + 50)$$
$$18 = n$$

Find d:
$$50 = 1 + (18-1)d$$
$$\frac{49}{17} = d$$

The sequence has a total of 18 terms, so we insert 16 arithmetic means between 1 and 50 with $d = \dfrac{49}{17}$.

65.
$$m = p + d$$
$$\underline{m = q - d}$$
$$2m = p + q \quad \text{Adding}$$
$$m = \frac{p+q}{2}$$

Exercise Set 10.3

1. $2, 4, 8, 16, \ldots$
$$\frac{4}{2} = 2, \frac{8}{4} = 2, \frac{16}{8} = 2$$
$$r = 2$$

3. $1, -1, 1, -1, \ldots$
$$\frac{-1}{1} = -1, \frac{1}{-1} = -1, \frac{-1}{1} = -1$$
$$r = -1$$

5. $\dfrac{2}{3}, -\dfrac{4}{3}, \dfrac{8}{3}, -\dfrac{16}{3}, \ldots$
$$\frac{-\frac{4}{3}}{\frac{2}{3}} = -2, \quad \frac{\frac{8}{3}}{-\frac{4}{3}} = -2, \quad \frac{-\frac{16}{3}}{\frac{8}{3}} = -2$$
$$r = -2$$

7. $\dfrac{0.6275}{6.275} = 0.1, \quad \dfrac{0.06275}{0.6275} = 0.1$
$$r = 0.1$$

9. $\dfrac{\frac{5a}{2}}{5} = \dfrac{a}{2}, \quad \dfrac{\frac{5a^2}{4}}{\frac{5a}{2}} = \dfrac{a}{2}, \quad \dfrac{\frac{5a^3}{8}}{\frac{5a^2}{4}} = \dfrac{a}{2}$
$$r = \frac{a}{2}$$

11. $2, 4, 8, 16, \ldots$

$a_1 = 2$, $n = 7$, and $r = \dfrac{4}{2}$, or 2.

We use the formula $a_n = a_1 r^{n-1}$.
$$a_7 = 2(2)^{7-1} = 2 \cdot 2^6 = 2 \cdot 64 = 128$$

13. $2, 2\sqrt{3}, 6, \ldots$

$a_1 = 2$, $n = 9$, and $r = \dfrac{2\sqrt{3}}{2}$, or $\sqrt{3}$
$$a_n = a_1 r^{n-1}$$
$$a_9 = 2(\sqrt{3})^{9-1} = 2(\sqrt{3})^8 = 2 \cdot 81 = 162$$

15. $\dfrac{7}{625}, -\dfrac{7}{25}, \ldots$

$a_1 = \dfrac{7}{625}$, $n = 23$, and $r = \dfrac{-\frac{7}{25}}{\frac{7}{625}} = -25$.
$$a_n = a_1 r^{n-1}$$
$$a_{23} = \frac{7}{625}(-25)^{23-1} = \frac{7}{625}(-25)^{22}$$
$$= \frac{7}{25^2} \cdot 25^2 \cdot 25^{20} = 7(25)^{20}, \text{ or } 7(5)^{40}$$

17. $1, 3, 9, \ldots$

$a_1 = 1$ and $r = \dfrac{3}{1}$, or 3
$$a_n = a_1 r^{n-1}$$
$$a_n = 1(3)^{n-1} = 3^{n-1}$$

19. $1, -1, 1, -1, \ldots$

$a_1 = 1$ and $r = \dfrac{-1}{1} = -1$
$$a_n = a_1 r^{n-1}$$
$$a_n = 1(-1)^{n-1} = (-1)^{n-1}$$

21. $\frac{1}{x}, \frac{1}{x^2}, \frac{1}{x^2}, \ldots$

$a_1 = \frac{1}{x}$ and $r = \dfrac{\frac{1}{x^2}}{\frac{1}{x}} = \frac{1}{x}$

$a_n = a_1 r^{n-1}$

$a_n = \frac{1}{x}\left(\frac{1}{x}\right)^{n-1} = \frac{1}{x} \cdot \frac{1}{x^{n-1}} = \frac{1}{x^{1+n-1}} = \frac{1}{x^n}$

23. $6 + 12 + 24 + \ldots$

$a_1 = 6$, $n = 7$, and $r = \frac{12}{6}$, or 2

$S_n = \frac{a_1(1-r^n)}{1-r}$

$S_7 = \frac{6(1-2^7)}{1-2} = \frac{6(1-128)}{-1} = \frac{6(-127)}{-1} = 762$

25. $\frac{1}{18} - \frac{1}{6} + \frac{1}{2} - \ldots$

$a_1 = \frac{1}{18}$, $n = 9$, and $r = \dfrac{-\frac{1}{6}}{\frac{1}{18}} = -\frac{1}{6} \cdot \frac{18}{1} = -3$

$S_n = \frac{a_1(1-r^n)}{1-r}$

$S_9 = \dfrac{\frac{1}{18}\left[1-(-3)^9\right]}{1-(-3)} = \dfrac{\frac{1}{18}(1+19{,}683)}{4}$

$\dfrac{\frac{1}{18}(19{,}684)}{4} = \frac{1}{18}(19{,}684)\left(\frac{1}{4}\right) = \frac{4921}{18}$

27. $4 + 2 + 1 + \ldots$

$|r| = \left|\frac{2}{4}\right| = \left|\frac{1}{2}\right| = \frac{1}{2}$, and since $|r| < 1$, the series does have a sum.

$S_\infty = \frac{a_1}{1-r} = \dfrac{4}{1-\frac{1}{2}} = \dfrac{4}{\frac{1}{2}} = 4 \cdot \frac{2}{1} = 8$

29. $25 + 20 + 16 + \ldots$

$|r| = \left|\frac{20}{25}\right| = \left|\frac{4}{5}\right| = \frac{4}{5}$, and since $|r| < 1$, the series does have a sum.

$S_\infty = \frac{a_1}{1-r} = \dfrac{25}{1-\frac{4}{5}} = \dfrac{25}{\frac{1}{5}} = 25 \cdot \frac{5}{1} = 125$

31. $8 + 40 + 200 + \ldots$

$|r| = \left|\frac{40}{8}\right| = |5| = 5$, and since $|r| > 1$ the series does not have a sum.

33. $0.6 + 0.06 + 0.006 + \ldots$

$|r| = \left|\frac{0.06}{0.6}\right| = |0.1| = 0.1$, and since $|r| < 1$, the series does have a sum.

$S_\infty = \frac{a_1}{1-r} = \frac{0.6}{1-0.1} = \frac{0.6}{0.9} = \frac{6}{9} = \frac{2}{3}$

35. $\sum_{k=1}^{11} 15\left(\frac{2}{3}\right)^k$

$a_1 = 15 \cdot \frac{2}{3}$ or 10; $|r| = \left|\frac{2}{3}\right| = \frac{2}{3}$, $n = 11$

$S_{11} = \dfrac{10\left[1-\left(\frac{2}{3}\right)^{11}\right]}{1-\frac{2}{3}} = \dfrac{10\left[1-\frac{2048}{177{,}147}\right]}{\frac{1}{3}}$

$= 10 \cdot \frac{175{,}099}{177{,}147} \cdot 3$

$= \frac{1{,}750{,}990}{59{,}049}$, or $29\frac{38{,}569}{59{,}049}$

37. $\sum_{k=1}^{\infty} \left(\frac{1}{2}\right)^{k-1}$

$a_1 = 1$, $|r| = \left|\frac{1}{2}\right| = \frac{1}{2}$

$S_\infty = \frac{a_1}{1-r} = \dfrac{1}{1-\frac{1}{2}} = \dfrac{1}{\frac{1}{2}} = 2$

39. $\sum_{k=1}^{\infty} 12.5^k$

Since $|r| = 12.5 > 1$, the sum does not exist.

41. $\sum_{k=1}^{\infty} \$500(1.11)^{-k}$

$a_1 = \$500(1.11)^{-1}$, or $\frac{\$500}{1.11}$; $|r| = |1.11^{-1}| = \frac{1}{1.11}$

$S_\infty = \frac{a_1}{1-r} = \dfrac{\frac{\$500}{1.11}}{1-\frac{1}{1.11}} = \dfrac{\frac{\$500}{1.11}}{\frac{0.11}{1.11}} \approx \$4545.\overline{45}$

43. $\sum_{k=1}^{\infty} 16(0.1)^{k-1}$

$a_1 = 16$, $|r| = |0.1| = 0.1$

$S_\infty = \frac{a_1}{1-r} = \frac{16}{1-0.1} = \frac{16}{0.9} = \frac{160}{9}$

45. $0.131313 \ldots = 0.13 + 0.0013 + 0.000013 + \ldots$

This is an infinite geometric series with $a_1 = 0.13$.

$|r| = \left|\frac{0.0013}{0.13}\right| = |0.01| = 0.01 < 1$, so the series has a limit.

$S_\infty = \frac{a_1}{1-r} = \frac{0.13}{1-0.01} = \frac{0.13}{0.99} = \frac{13}{99}$

47. We will find fractional notation for $0.999\overline{9}$ and then add 8.

$$0.999\overline{9} = 0.9 + 0.09 + 0.009 + 0.0009 + \ldots$$

This is an infinite geometric series with $a_1 = 0.9$.

$|r| = \left|\dfrac{0.09}{0.9}\right| = |0.1| = 0.1 < 1$, so the series has a limit.

$$S_\infty = \frac{a_1}{1-r} = \frac{0.9}{1-0.1} = \frac{0.9}{0.9} = 1$$

Then $8.999\overline{9} = 8 + 1 = 9$.

49. $3.4125\overline{125} = 3.4 + 0.0125\overline{125}$

We will find fractional notation for $0.0125\overline{125}$ and then add

3.4, or $\dfrac{34}{10}$, or $\dfrac{17}{5}$.

$$0.0125\overline{125} = 0.0125 + 0.0000125 + \ldots$$

This is an infinite geometric series with $a_1 = 0.0125$.

$|r| = \left|\dfrac{0.0000125}{0.0125}\right| = |0.001| = 0.001 < 1$, so the series has a limit.

$$S_\infty = \frac{a_1}{1-r} = \frac{0.0125}{1-0.001} = \frac{0.0125}{0.999} = \frac{125}{9990}$$

Then $\dfrac{17}{5} + \dfrac{125}{9990} = \dfrac{33,966}{9990} + \dfrac{125}{9990} = \dfrac{34,091}{9990}$

51. a) **Familiarize.** The rebound distances form a geometric sequence:

$$\frac{1}{4} \times 16, \quad \left(\frac{1}{4}\right)^2 \times 16, \quad \left(\frac{1}{4}\right)^3 \times 16, \ldots,$$

or $4, \quad \dfrac{1}{4} \times 4, \quad \left(\dfrac{1}{4}\right)^2 \times 4, \ldots$

The height of the 6th rebound is the 6th term of the sequence.

Translate. We will use the formula $a_n = a_1 r^{n-1}$, with $a_1 = 4$, $r = \dfrac{1}{4}$, and $n = 6$:

$$a_6 = 4\left(\frac{1}{4}\right)^{6-1}$$

Carry out. We calculate to obtain $a_6 = \dfrac{1}{256}$.

Check. We can do the calculation again.

State. It rebounds $\dfrac{1}{256}$ ft the 6th time.

b) $S_\infty = \dfrac{a}{1-r} = \dfrac{4}{1-\frac{1}{4}} = \dfrac{4}{\frac{3}{4}} = \dfrac{16}{3}$ ft, or $5\dfrac{1}{3}$ ft

53. a) **Familiarize.** The rebound distances form a geometric sequence:

$$0.6 \times 200, \quad (0.6)^2 \times 200, \quad (0.6)^3 \times 200, \ldots,$$

or $120, \ 0.6 \times 120, \ (0.6)^2 \times 120, \ldots$

The total rebound distance after 9 rebounds is the sum of the first 9 terms of this sequence.

Translate. We will use the formula

$$S_n = \frac{a_1(1-r^n)}{1-r}$$ with $a_1 = 120$, $r = 0.6$, and $n = 9$.

Carry out.

$$S_9 = \frac{120[1-(0.6)^9]}{1-0.6} \approx 297$$

Check. We repeat the calculation.

State. The bungee jumper has traveled about 297 ft upward after 9 rebounds.

b) $S_\infty = \dfrac{a_1}{1-r} = \dfrac{120}{1-0.6} = 300$ ft

55. **Familiarize.** The amount of the annuity is the geometric series

$\$1000 + \$1000(1.062) + \$1000(1.062)^2 + \ldots + \$1000(1.062)^{17}$, where $a_1 = \$1000$, $r = 1.062$, and $n = 18$.

Translate. Using the formula

$$S_n = \frac{a_1(1-r^n)}{1-r}$$

we have

$$S_{18} = \frac{\$1000[1-(1.062)^{18}]}{1-1.062}$$

Carry out. We carry out the computation and get $S_{18} \approx \$31,497.57$.

Check. Repeat the calculations.

State. The amount of the annuity is $\$31,497.57$.

57. **Familiarize.** The amounts owed at the beginning of successive years form a geometric sequence:

$\$120,000, \ (1.12)\$120,000, \ (1.12)^2\$120,000, \ldots$

The amount to be repaid at the end of 13 years is the amount owed at the beginning of the 14th year.

Translate. Use the formula $a_n = a_1 r^{n-1}$ with $a_1 = 120,000$, $r = 1.12$, and $n = 14$:

$$a_{14} = 120,000(1.12)^{14-1}$$

Carry out. We perform the calculation, obtaining $a_{14} \approx \$523,619.17$.

Check. Repeat the calculation.

State. At the end of 13 years, $\$523,619.17$ will be repaid.

59. **Familiarize.** The total effect on the economy is the sum of an infinite geometric series

$\$13,000,000,000 + \$13,000,000,000(0.85) + \$13,000,000,000(0.85)^2 + \ldots$

with $a_1 = \$13,000,000,000$ and $r = 0.85$.

Translate. Using the formula

$$S_\infty = \frac{a_1}{1-r}$$

we have

$$S_\infty = \frac{\$13,000,000,000}{1-0.85}.$$

Carry out. Perform the calculation:

$S_\infty \approx \$86,666,666,667.$

Check. Repeat the calculation.

State. The total effect on the economy is
$86,666,666,667.

61. Discussion and Writing

63. $f(x) = x^2$, $g(x) = 4x + 5$

$(f \circ g)(x) = f(g(x)) = f(4x + 5) = (4x + 5)^2 = 16x^2 + 40x + 25$

$(g \circ f)(x) = g(f(x)) = g(x^2) = 4x^2 + 5$

65.
$$5^x = 35$$
$$\ln 5^x = \ln 35$$
$$x \ln 5 = \ln 35$$
$$x = \frac{\ln 35}{\ln 5}$$
$$x \approx 2.209$$

67. See the answer section in the text.

69. a) If the sequence is arithmetic, then $a_2 - a_1 = a_3 - a_2$.

$$x + 7 - (x + 3) = 4x - 2 - (x + 7)$$
$$x = \frac{13}{3}$$

The three given terms are $\frac{13}{3} + 3 = \frac{22}{3}$,
$\frac{13}{3} + 7 = \frac{34}{3}$, and $4 \cdot \frac{13}{3} - 2 = \frac{46}{3}$.
Then $d = \frac{12}{3}$, or 4, so the fourth term is
$\frac{46}{3} + \frac{12}{3} = \frac{58}{3}$.

b) If the sequence is geometric, then $a_2/a_1 = a_3/a_2$.

$$\frac{x+7}{x+3} = \frac{4x-2}{x+7}$$
$$x = -\frac{11}{3} \text{ or } x = 5$$

For $x = -\frac{11}{3}$: The three given terms are

$$-\frac{11}{3} + 3 = -\frac{2}{3}, \quad -\frac{11}{3} + 7 = \frac{10}{3}, \text{ and}$$
$$4\left(-\frac{11}{3}\right) - 2 = -\frac{50}{3}.$$

Then $r = -5$, so the fourth term is

$$-\frac{50}{3}(-5) = \frac{250}{3}.$$

For $x = 5$: The three given terms are $5 + 3 = 8$,
$5 + 7 = 12$, and $4 \cdot 5 - 2 = 18$. Then $r = \frac{3}{2}$, so
the fourth term is $18 \cdot \frac{3}{2} = 27$.

71. $x^2 - x^3 + x^4 - x^5 + \cdots$

This is a geometric series with $a_1 = x^2$ and $r = -x$.

$$S_n = \frac{a_1(1 - r^n)}{1 - r} = \frac{x^2(1 - (-x)^n)}{1 - (-x)} = \frac{x^2(1 - (-x)^n)}{1 + x}$$

73. See the answer section in the text.

75. Familiarize. The length of a side of the first square
is 16 cm. The length of a side of the next square is
the length of the hypotenuse of a right triangle with
legs 8 cm and 8 cm, or $8\sqrt{2}$ cm. The length of a side
of the next square is the length of the hypotenuse of
a right triangle with legs $4\sqrt{2}$ cm and $4\sqrt{2}$ cm, or
8 cm. The areas of the squares form a sequence:

$$(16)^2, \quad (8\sqrt{2})^2, \quad (8)^2, \ldots, \text{ or}$$
$$256, \quad 128, \quad 64, \ldots.$$

This is a geometric sequence with $a_1 = 256$ and
$r = \frac{1}{2}$.

Translate. We find the sum of the infinite geometric
series $256 + 128 + 64 + \cdots$.

$$S_\infty = \frac{a_1}{1 - r}$$
$$S_\infty = \frac{256}{1 - \frac{1}{2}}$$

Carry out. We calculate to obtain $S_\infty = 512$.

Check. We can do the calculation again.

State. The sum of the areas is 512 cm^2.

Exercise Set 10.4

---

1. $n^2 < n^3$

$1^2 < 1^3$, $2^2 < 2^3$, $3^2 < 3^3$, $4^2 < 4^3$, $5^2 < 5^3$

The first statement is false, and the others are true.

3. A polygon of n sides has $\frac{n(n-3)}{2}$ diagonals.

A polygon of 3 sides has $\frac{3(3-3)}{2}$ diagonals.

A polygon of 4 sides has $\frac{4(4-3)}{2}$ diagonals.

A polygon of 5 sides has $\frac{5(5-3)}{2}$ diagonals.

A polygon of 6 sides has $\dfrac{6(6-3)}{2}$ diagonals.

A polygon of 7 sides has $\dfrac{7(7-3)}{2}$ diagonals.

Each of these statements is true.

5. - 25. See the answer section in the text.

27. Discussion and Writing

29.
$$x + \ y + \ z = 3, \ (1)$$
$$2x - 3y - 2z = 5, \ (2)$$
$$3x + 2y + 2z = 8 \ (3)$$

We will use Gaussian elimination. First multiply equation (1) by -2 and add it to equation (2). Also multiply equation (1) by -3 and add it to equation (3).

$$x + \ y + z = \ 3$$
$$-5y - 4z = -1$$
$$-y - \ z = -1$$

Now multiply the last equation above by 5 to make the y-coefficient a multiple of the y-coefficient in the equation above it.

$$x + \ y + \ z = \ 3 \ (1)$$
$$- 5y - 4z = -1 \ (4)$$
$$- 5y - 5z = -5 \ (5)$$

Multiply equation (4) by -1 and add it to equation (3).

$$x + \ y + \ z = \ 3 \ (1)$$
$$- 5y - 4z = -1 \ (4)$$
$$- \ z = -4 \ (6)$$

Now solve equation (6) for z.

$$-z = -4$$
$$z = 4$$

Back-substitute 4 for z in equation (4) and solve for y.

$$-5y - 4 \cdot 4 = -1$$
$$-5y - 16 = -1$$
$$-5y = 15$$
$$y = -3$$

Finally, back-substitute -3 for y and 4 for z in equation (1) and solve for x.

$$x - 3 + 4 = 3$$
$$x + 1 = 3$$
$$x = 2$$

The solution is $(2, -3, 4)$.

31. *Familiarize*. Let x, y, and z represent the amounts invested at 6%, 8%, and 10%, respectively.

Translate. We know that simple interest for one year was $376. This gives us one equation:

$$0.06x + 0.08y + 0.1z = 376$$

The amount invested at 8% is twice the amount invested at 6%:

$$y = 2x, \text{ or } -2x + y = 0$$

There is $400 more invested at 10% than at 8%:

$$z = y + 400, \text{ or } -y + z = 400$$

We have a system of equations:

$$0.06x + 0.08y + 0.1z = 376,$$
$$-2x + \ y \ = \ 0$$
$$- \ y + \ z = \ 400$$

Carry out. Solving the system of equations, we get $(800, 1600, 2000)$.

Check. Simple interest for one year would be $0.06(\$800) + 0.08(\$1600) + 0.1(\$2000)$, or $\$48 + \$128 + \$200$, or $\$376$. The amount invested at 8%, $1600, is twice $800, the amount invested at 6%. The amount invested at 10%, $2000, is $400 more than $1600, the amount invested at 8%. The answer checks.

State. Martin invested $800 at 6%, $1600 at 8%, and $2000 at 10%.

33. - 41. See the answer section in the text.

Exercise Set 10.5

1. $_6P_6 = 6! = 6 \cdot 5 \cdot 4 \cdot 3 \cdot 2 \cdot 1 = 720$

3. Using formula (1), we have

$$_{10}P_7 = 10 \cdot 9 \cdot 8 \cdot 7 \cdot 6 \cdot 5 \cdot 4 = 604,800.$$

Using formula (2), we have

$$_{10}P_7 = \frac{10!}{(10-7)!} = \frac{10!}{3!} = \frac{10 \cdot 9 \cdot 8 \cdot 7 \cdot 6 \cdot 5 \cdot 4 \cdot 3!}{3!} =$$

$604,800.$

5. $5! = 5 \cdot 4 \cdot 3 \cdot 2 \cdot 1 = 120$

7. $0!$ is defined to be 1.

9. $\dfrac{9!}{5!} = \dfrac{9 \cdot 8 \cdot 7 \cdot 6 \cdot 5!}{5!} = 9 \cdot 8 \cdot 7 \cdot 6 = 3024$

11. $(8 - 3)! = 5! = 5 \cdot 4 \cdot 3 \cdot 2 \cdot 1 = 120$

13. $\dfrac{10!}{7!3!} = \dfrac{10 \cdot 9 \cdot 8 \cdot 7!}{7!3 \cdot 2 \cdot 1} = \dfrac{10 \cdot 3 \cdot 3 \cdot 4 \cdot 2}{3 \cdot 2 \cdot 1} =$

$10 \cdot 3 \cdot 4 = 120$

15. Using formula (2), we have

$$_8P_0 = \frac{8!}{(8-0)!} = \frac{8!}{8!} = 1.$$

17. Using a grapher, we find $_{52}P_4 = 6,497,400$

19. Using formula (1), we have $_nP_3 = n(n-1)(n-2)$.

Using formula (2), we have

$$_nP_3 = \frac{n!}{(n-3)!} = \frac{n(n-1)(n-2)(n-3)!}{(n-3)!} =$$
$$n(n-1)(n-2).$$

21. Using formula (1), we have $_nP_1 = n$.

Using formula (2), we have

$$_nP_1 = \frac{n!}{(n-1)!} = \frac{n(n-1)!}{(n-1)!} = n.$$

23. $_6P_6 = 6! = 720$

25. $_9P_9 = 9! = 362,880$

27. $_9P_4 = 9 \cdot 8 \cdot 7 \cdot 6 = 3024$

29. Without repetition: $_5P_5 = 5! = 120$

With repetition: $5^5 = 3125$

31. BUSINESS: 1 B, 1 U, 3 S's, 1 I, 1 N, 1 E, a total of 8.

$$= \frac{8!}{1! \cdot 1! \cdot 3! \cdot 1! \cdot 1! \cdot 1!}$$
$$= \frac{8!}{3!} = \frac{8 \cdot 7 \cdot 6 \cdot 5 \cdot 4 \cdot 3!}{3!} = 8 \cdot 7 \cdot 6 \cdot 5 \cdot 4 = 6720$$

BIOLOGY: 1 B, 1 I, 2 0's, 1 L, 1 G, 1 Y, a total of 7.

$$= \frac{7!}{1! \cdot 1! \cdot 2! \cdot 1! \cdot 1! \cdot 1!}$$
$$= \frac{7!}{2!} = \frac{7 \cdot 6 \cdot 5 \cdot 4 \cdot 3 \cdot 2!}{2!} = 7 \cdot 6 \cdot 5 \cdot 4 \cdot 3 = 2520$$

MATHEMATICS: 2 M's, 2 A's, 2 T's, 1 H, 1 E, 1 I, 1 C, 1 S, a total of 11.

$$= \frac{11!}{2! \cdot 2! \cdot 2! \cdot 1! \cdot 1! \cdot 1! \cdot 1! \cdot 1!}$$
$$= \frac{11!}{2! \cdot 2! \cdot 2!} = \frac{11 \cdot 10 \cdot 9 \cdot 8 \cdot 7 \cdot 6 \cdot 5 \cdot 4 \cdot 3 \cdot 2!}{2! \cdot 2! \cdot 2!}$$
$$= \frac{11 \cdot 10 \cdot 9 \cdot 8 \cdot 7 \cdot 6 \cdot 5 \cdot 4 \cdot 3}{2 \cdot 1 \cdot 2 \cdot 1}$$
$$= 4,989,600$$

33. The first number can be any of the eight digits other than 0 and 1. The remaining 6 numbers can each be any of the ten digits 0 through 9. We have

$$8 \cdot 10^6 = 8,000,000$$

Accordingly, there can be 8,000,000 telephone numbers within a given area code before the area needs to be split with a new area code.

35. $a^2b^3c^4 = a \cdot a \cdot b \cdot b \cdot b \cdot c \cdot c \cdot c \cdot c$

There are 2 a's, 3 b's, and 4 c's, for a total of 9. We have

$$\frac{9!}{2! \cdot 3! \cdot 4!}$$
$$= \frac{9 \cdot 8 \cdot 7 \cdot 6 \cdot 5 \cdot 4!}{2 \cdot 1 \cdot 3 \cdot 2 \cdot 1 \cdot 4!} = \frac{9 \cdot 8 \cdot 7 \cdot 6 \cdot 5}{2 \cdot 3 \cdot 2} = 1260.$$

37. a) $_6P_5 = 6 \cdot 5 \cdot 4 \cdot 3 \cdot 2 = 720$

b) $6^5 = 7776$

c) The first letter can only be D. The other four letters are chosen from A, B, C, E, F without repetition. We have

$$1 \cdot_5 P_4 = 1 \cdot 5 \cdot 4 \cdot 3 \cdot 2 = 120.$$

d) The first letter can only be D. The second letter can only be E. The other three letters are chosen from A, B, C, F without repetition. We have

$$1 \cdot 1 \cdot_4 P_3 = 1 \cdot 1 \cdot 4 \cdot 3 \cdot 2 = 24.$$

39. a) Since repetition is allowed, each of the 5 digits can be chosen in 10 ways. The number of zip-codes possible is $10 \cdot 10 \cdot 10 \cdot 10 \cdot 10$, or 100,000.

b) Since there are 100,000 possible zip-codes, there could be 100,000 post offices.

41. a) Since repetition is allowed, each digit can be chosen in 10 ways. There can be

$10 \cdot 10 \cdot 10 \cdot 10 \cdot 10 \cdot 10 \cdot 10 \cdot 10 \cdot 10$, or 1,000,000,000 social security numbers.

b) Since more than 275 million social security numbers are possible, each person can have a social security number.

43. Discussion adn Writing

45.
$$x^2 + x - 6 = 0$$
$$(x+3)(x-2) = 0$$
$$x + 3 = 0 \quad or \quad x - 2 = 0$$
$$x = -3 \quad or \qquad x = 2$$

We could also graph $y = x^2 + x - 6$ and use the Zero feature twice. The solutions are -3 and 2.

47. $f(x) = x^3 - 4x^2 - 7x + 10$

We use synthetic division to find one factor of the polynomial. We try $x - 1$.

$$\begin{array}{r|rrrr} 1 & 1 & -4 & -7 & 10 \\ & & 1 & -3 & -10 \\ \hline & 1 & -3 & -10 & 0 \end{array}$$

$$x^3 - 4x^2 - 7x + 10 = 0$$
$$(x - 1)(x^2 - 3x - 10) = 0$$
$$(x - 1)(x - 5)(x + 2) = 0$$
$$x - 1 = 0 \quad or \quad x - 5 = 0 \quad or \quad x + 2 = 0$$
$$x = 1 \quad or \qquad x = 5 \quad or \qquad x = -2$$

We could also graph $y = x^3 - 4x^2 - 7x + 10$ and use the Zero feature three times. The solutions are -2, 1, and 5.

49.
$$_nP_4 = 8 \cdot {}_{n-1}P_3$$
$$\frac{n!}{(n-4)!} = 8 \cdot \frac{(n-1)!}{(n-1-3)!}$$
$$\frac{n!}{(n-4)!} = 8 \cdot \frac{(n-1)!}{(n-4)!}$$
$$n! = 8 \cdot (n-1)! \qquad \text{Multiplying by } (n-4)!$$
$$n(n-1)! = 8 \cdot (n-1)!$$
$$n = 8 \qquad \text{Dividing by } (n-1)!$$

51.
$$_nP_4 = 8 \cdot {}_nP_3$$
$$\frac{n!}{(n-4)!} = 8 \cdot \frac{n!}{(n-3)!}$$
$$(n-3)! = 8(n-4)! \qquad \begin{array}{l}\text{Multiplying by}\\ \dfrac{(n-4)!(n-3)!}{n!}\end{array}$$
$$(n-3)(n-4)! = 8(n-4)!$$
$$n - 3 = 8 \qquad \text{Dividing by } (n-4)!$$
$$n = 11$$

53. There is one losing team per game. In order to leave one tournament winner there must be $n - 1$ losers produced in $n - 1$ games.

Exercise Set 10.6

---

1. $_{13}C_2 = \dfrac{13!}{2!(13-2)!}$

$= \dfrac{13!}{2!11!} = \dfrac{13 \cdot 12 \cdot 11!}{2 \cdot 1 \cdot 11!}$

$= \dfrac{13 \cdot 12}{2 \cdot 1} = \dfrac{13 \cdot 6 \cdot 2}{2 \cdot 1}$

$= 78$

3. $\dbinom{13}{11} = \dfrac{13!}{11!(13-11)!}$

$= \dfrac{13!}{11!2!}$

$= 78 \qquad \text{(See Exercise 1.)}$

5. $\dbinom{7}{1} = \dfrac{7!}{1!(7-1)!}$

$= \dfrac{7!}{1!6!} = \dfrac{7 \cdot 6!}{1 \cdot 6!}$

$= 7$

7. $\dfrac{_5P_3}{3!} = \dfrac{5 \cdot 4 \cdot 3}{3!}$

$= \dfrac{5 \cdot 4 \cdot 3}{3 \cdot 2 \cdot 1} = \dfrac{5 \cdot 2 \cdot 2 \cdot 3}{3 \cdot 2 \cdot 1}$

$= 5 \cdot 2 = 10$

9. $\dbinom{6}{0} = \dfrac{6!}{0!(6-0)!}$

$= \dfrac{6!}{0!6!} = \dfrac{6!}{6! \cdot 1}$

$= 1$

11. $\dbinom{6}{2} = \dfrac{6 \cdot 5}{2 \cdot 1} = 15$

13. $\dbinom{n}{r} = \dbinom{n}{n-r}$, so

$$\dbinom{7}{0} + \dbinom{7}{1} + \dbinom{7}{2} + \dbinom{7}{3} + \dbinom{7}{4} +$$
$$\dbinom{7}{5} + \dbinom{7}{6} + \dbinom{7}{7}$$
$$= 2\left[\dbinom{7}{0} + \dbinom{7}{1} + \dbinom{7}{2} + \dbinom{7}{3}\right]$$
$$= 2\left[\dfrac{7!}{7!0!} + \dfrac{7!}{6!1!} + \dfrac{7!}{5!2!} + \dfrac{7!}{4!3!}\right]$$
$$= 2(1 + 7 + 21 + 35) = 2 \cdot 64 = 128$$

15. Use a grapher.

$_{52}C_4 = 270,725$

17. Use a grapher.

$\dbinom{27}{11} = {}_{27}C_{11} = 13,037,895$

19. $\dbinom{n}{1} = \dfrac{n!}{1!(n-1)!} = \dfrac{n(n-1)!}{1!(n-1)!} = n$

21. $\dbinom{m}{m} = \dfrac{m!}{m!(m-m)!} = \dfrac{m!}{m!0!} = 1$

23. $_{23}C_4 = \dfrac{23!}{4!(23-4)!}$

$= \dfrac{23!}{4!19!} = \dfrac{23 \cdot 22 \cdot 21 \cdot 20 \cdot 19!}{4 \cdot 3 \cdot 2 \cdot 1 \cdot 19!}$

$= \dfrac{23 \cdot 22 \cdot 21 \cdot 20}{4 \cdot 3 \cdot 2 \cdot 1} = \dfrac{23 \cdot 2 \cdot 11 \cdot 3 \cdot 7 \cdot 4 \cdot 5}{4 \cdot 3 \cdot 2 \cdot 1}$

$= 8855$

25. $\displaystyle {}_{13}C_{10} = \frac{13!}{10!(13-10)!}$

$\displaystyle = \frac{13!}{10!3!} = \frac{13 \cdot 12 \cdot 11 \cdot 10!}{10! \cdot 3 \cdot 2 \cdot 1}$

$\displaystyle = \frac{13 \cdot 12 \cdot 11}{3 \cdot 2 \cdot 1} = \frac{13 \cdot 3 \cdot 2 \cdot 2 \cdot 11}{3 \cdot 2 \cdot 1}$

$= 286$

27. Since two points determine a line and no three of these 8 points are colinear, we need to find the number of combinations of 8 points taken 2 at a time, ${}_8C_2$.

$\displaystyle {}_8C_2 = \binom{8}{2} = \frac{8!}{2!(8-2)!}$

$\displaystyle = \frac{8 \cdot 7 \cdot 6!}{2 \cdot 1 \cdot 6!} = \frac{4 \cdot 2 \cdot 7}{2 \cdot 1}$

$= 28$

Thus 28 lines are determined.

Since three noncolinear points determine a triangle, we need to find the number of combinations of 8 points taken 3 at a time, ${}_8C_3$.

$\displaystyle {}_8C_3 = \binom{8}{3} = \frac{8!}{3!(8-3)!}$

$\displaystyle = \frac{8 \cdot 7 \cdot 6 \cdot 5!}{3 \cdot 2 \cdot 1 \cdot 5!} = \frac{8 \cdot 7 \cdot 3 \cdot 2}{3 \cdot 2 \cdot 1}$

$= 56$

Thus 56 triangles are determined.

29. ${}_{52}C_5 = 2{,}598{,}960$

31. a) ${}_{31}P_2 = 930$

b) $31 \cdot 31 = 961$

c) ${}_{31}C_2 = 465$

33. Discussion and Writing

35. $\qquad 2x^2 - x = 3$

$\qquad 2x^2 - x - 3 = 0$

$\qquad (2x-3)(x+1) = 0$

$\qquad 2x - 3 = 0 \quad or \quad x + 1 = 0$

$\qquad 2x = 3 \quad or \qquad x = -1$

$\qquad x = \dfrac{3}{2} \quad or \qquad x = -1$

We could also graph $y_1 = 2x^2 - x$ and $y_2 = 3$ and use the Intersect feature twice to find the first coordinates of the points of intersection of the graphs. The solutions are $\dfrac{3}{2}$ and -1, or 1.5 and -1.

37. $\qquad x^3 + 3x^2 - 10x = 24$

$\qquad x^3 + 3x^2 - 10x - 24 = 0$

We use synthetic division to find one factor of the polynomial on the left side of the equation. We try $x - 3$.

$$\begin{array}{r|rrrr} 3 & 1 & 3 & -10 & -24 \\ & & 3 & 18 & 24 \\ \hline & 1 & 6 & 8 & 0 \end{array}$$

Now we have:

$\qquad (x-3)(x^2 + 6x + 8) = 0$

$\qquad (x-3)(x+2)(x+4) = 0$

$\quad x - 3 = 0 \quad or \quad x + 2 = 0 \quad or \quad x + 4 = 0$

$\quad x = 3 \quad or \qquad x = -2 \quad or \qquad x = -4$

We could also graph $y_1 = x^3 + 3x^2 - 10x$ and $y_2 = 24$ and use the Intersect feature three times to find the first coordinates of the points of intersection of the graphs. The solutions are -4, -2, and 3.

39. There are 13 diamonds, and we choose 5. We have ${}_{13}C_5 = 1287$.

41. Playing once: ${}_nC_2$

Playing twice: $2 \cdot {}_nC_2$

43. $\qquad\qquad \dbinom{n}{n-2} = 6$

$\qquad \dfrac{n!}{(n-(n-2))!(n-2)!} = 6$

$\qquad\qquad \dfrac{n!}{2!(n-2)!} = 6$

$\qquad\quad \dfrac{n(n-1)(n-2)!}{2 \cdot 1 \cdot (n-2)!} = 6$

$\qquad\qquad \dfrac{n(n-1)}{2} = 6$

$\qquad\qquad n(n-1) = 12$

$\qquad\qquad n^2 - n = 12$

$\qquad\qquad n^2 - n - 12 = 0$

$\qquad\qquad (n-4)(n+3) = 0$

$n = 4 \ \text{ or } \ n = -3$

Only 4 checks. The solution is 4.

45.
$$\binom{n+2}{4} = 6 \cdot \binom{n}{2}$$

$$\frac{(n+2)!}{(n+2-4)!4!} = 6 \cdot \frac{n!}{(n-2)!2!}$$

$$\frac{(n+2)!}{(n-2)!4!} = 6 \cdot \frac{n!}{(n-2)!2!}$$

$$\frac{(n+2)!}{4!} = 6 \cdot \frac{n!}{2!} \quad \text{Multiplying by } (n-2)!$$

$$4! \cdot \frac{(n+2)!}{4!} = 4! \cdot 6 \cdot \frac{n!}{2!}$$

$$(n+2)! = 72 \cdot n!$$

$$(n+2)(n+1)n! = 72 \cdot n!$$

$$(n+2)(n+1) = 72 \quad \text{Dividing by } n!$$

$$n^2 + 3n + 2 = 72$$

$$n^2 + 3n - 70 = 0$$

$$(n+10)(n-7) = 0$$

$$n = -10 \quad \text{or} \quad n = 7$$

Only 7 checks. The solution is 7.

47. See the answer section in the text.

Exercise Set 10.7

1. Expand: $(x+5)^4$.

We have $a = x$, $b = 5$, and $n = 4$.

Pascal's triangle method: Use the fifth row of Pascal's triangle.

$$1 \quad 4 \quad 6 \quad 4 \quad 1$$

$$(x+5)^4$$

$$= 1 \cdot x^4 + 4 \cdot x^3 \cdot 5 + 6 \cdot x^2 \cdot 5^2 +$$

$$\quad 4 \cdot x \cdot 5^3 + 1 \cdot 5^4$$

$$= x^4 + 20x^3 + 150x^2 + 500x + 625$$

Factorial notation method:

$$(x+5)^4$$

$$= \binom{4}{0}x^4 + \binom{4}{1}x^3 \cdot 5 + \binom{4}{2}x^2 \cdot 5^2 +$$

$$\binom{4}{3}x \cdot 5^3 + \binom{4}{4}5^4$$

$$= \frac{4!}{0!4!}x^4 + \frac{4!}{1!3!}x^3 \cdot 5 + \frac{4!}{2!2!}x^2 \cdot 5^2 +$$

$$\frac{4!}{3!1!}x \cdot 5^3 + \frac{4!}{4!0!}5^4$$

$$= x^4 + 20x^3 + 150x^2 + 500x + 625$$

3. Expand: $(x-3)^5$.

We have $a = x$, $b = -3$, and $n = 5$.

Pascal's triangle method: Use the sixth row of Pascal's triangle.

$$1 \quad 5 \quad 10 \quad 10 \quad 5 \quad 1$$

$$(x-3)^5$$

$$= 1 \cdot x^5 + 5x^4(-3) + 10x^3(-3)^2 + 10x^2(-3)^3 +$$

$$\quad 5x(-3)^4 + 1 \cdot (-3)^5$$

$$= x^5 - 15x^4 + 90x^3 - 270x^2 + 405x - 243$$

Factorial notation method:

$$(x-3)^5$$

$$= \binom{5}{0}x^5 + \binom{5}{1}x^4(-3) + \binom{5}{2}x^3(-3)^2 +$$

$$\binom{5}{3}x^2(-3)^3 + \binom{5}{4}x(-3)^4 + \binom{5}{5}(-3)^5$$

$$= \frac{5!}{0!5!}x^5 + \frac{5!}{1!4!}x^4(-3) + \frac{5!}{2!3!}x^3(9) +$$

$$\frac{5!}{3!2!}x^2(-27) + \frac{5!}{4!1!}x(81) + \frac{5!}{5!0!}(-243)$$

$$= x^5 - 15x^4 + 90x^3 - 270x^2 + 405x - 243$$

5. Expand: $(x-y)^5$.

We have $a = x$, $b = -y$, and $n = 5$.

Pascal's triangle method: We use the sixth row of Pascal's triangle.

$$1 \quad 5 \quad 10 \quad 10 \quad 5 \quad 1$$

$$(x-y)^5$$

$$= 1 \cdot x^5 + 5x^4(-y) + 10x^3(-y)^2 + 10x^2(-y)^3 +$$

$$\quad 5x(-y)^4 + 1 \cdot (-y)^5$$

$$= x^5 - 5x^4y + 10x^3y^2 - 10x^2y^3 + 5xy^4 - y^5$$

Factorial notation method:

$$(x-y)^5$$

$$= \binom{5}{0}x^5 + \binom{5}{1}x^4(-y) + \binom{5}{2}x^3(-y)^2 +$$

$$\binom{5}{3}x^2(-y)^3 + \binom{5}{4}x(-y)^4 + \binom{5}{5}(-y)^5$$

$$= \frac{5!}{0!5!}x^5 + \frac{5!}{1!4!}x^4(-y) + \frac{5!}{2!3!}x^3(y^2) +$$

$$\frac{5!}{3!2!}x^2(-y^3) + \frac{5!}{4!1!}x(y^4) + \frac{5!}{5!0!}(-y^5)$$

$$= x^5 - 5x^4y + 10x^3y^2 - 10x^2y^3 + 5xy^4 - y^5$$

7. Expand: $(5x+4y)^6$.

We have $a = 5x$, $b = 4y$, and $n = 6$.

Pascal's triangle method: Use the seventh row of Pascal's triangle.

1   6   15   20   15   6   1

$(5x + 4y)^6$

$= 1 \cdot (5x)^6 + 6 \cdot (5x)^5(4y) + 15(5x)^4(4y)^2 +$

$\quad 20(5x)^3(4y)^3 + 15(5x)^2(4y)^4 + 6(5x)(4y)^5 +$

$\quad 1 \cdot (4y)^6$

$= 15,625x^6 + 75,000x^5y + 150,000x^4y^2 +$

$\quad 160,000x^3y^3 + 96,000x^2y^4 + 30,720xy^5 + 4096y^6$

Factorial notation method:

$(5x + 4y)^6$

$= \begin{pmatrix} 6 \\ 0 \end{pmatrix}(5x)^6 + \begin{pmatrix} 6 \\ 1 \end{pmatrix}(5x)^5(4y) +$

$\quad \begin{pmatrix} 6 \\ 2 \end{pmatrix}(5x)^4(4y)^2 + \begin{pmatrix} 6 \\ 3 \end{pmatrix}(5x)^3(4y)^3 +$

$\quad \begin{pmatrix} 6 \\ 4 \end{pmatrix}(5x)^2(4y)^4 + \begin{pmatrix} 6 \\ 5 \end{pmatrix}(5x)(4y)^5 + \begin{pmatrix} 6 \\ 6 \end{pmatrix}(4y)^6$

$= \frac{6!}{0!6!}(15,625x^6) + \frac{6!}{1!5!}(3125x^5)(4y) +$

$\quad \frac{6!}{2!4!}(625x^4)(16y^2) + \frac{6!}{3!3!}(125x^3)(64y^3) +$

$\quad \frac{6!}{4!2!}(25x^2)(256y^4) + \frac{6!}{5!1!}(5x)(1024y^5) +$

$\quad \frac{6!}{6!0!}(4096y^6)$

$= 15,625x^6 + 75,000x^5y + 150,000x^4y^2 +$

$\quad 160,000x^3y^3 + 96,000x^2y^4 + 30,720xy^5 +$

$\quad 4096y^6$

9. Expand: $\left(2t + \dfrac{1}{t}\right)^7$.

We have $a = 2t$, $b = \dfrac{1}{t}$, and $n = 7$.

Pascal's triangle method: Use the eighth row of Pascal's triangle.

1   7   21   35   35   21   7   1

$\left(2t + \dfrac{1}{t}\right)^7$

$= 1 \cdot (2t)^7 + 7(2t)^6\left(\dfrac{1}{t}\right) + 21(2t)^5\left(\dfrac{1}{t}\right)^2 +$

$\quad 35(2t)^4\left(\dfrac{1}{t}\right)^3 + 35(2t)^3\left(\dfrac{1}{t}\right)^4 + 21(2t)^2\left(\dfrac{1}{t}\right)^5 +$

$\quad 7(2t)\left(\dfrac{1}{t}\right)^6 + 1 \cdot \left(\dfrac{1}{t}\right)^7$

$= 128t^7 + 7 \cdot 64t^6 \cdot \dfrac{1}{t} + 21 \cdot 32t^5 \cdot \dfrac{1}{t^2} +$

$\quad 35 \cdot 16t^4 \cdot \dfrac{1}{t^3} + 35 \cdot 8t^3 \cdot \dfrac{1}{t^4} + 21 \cdot 4t^2 \cdot \dfrac{1}{t^5} +$

$\quad 7 \cdot 2t \cdot \dfrac{1}{t^6} + \dfrac{1}{t^7}$

$= 128t^7 + 448t^5 + 672t^3 + 560t + 280t^{-1} +$

$\quad 84t^{-3} + 14t^{-5} + t^{-7}$

Factorial notation method:

$\left(2t + \dfrac{1}{t}\right)^7$

$= \begin{pmatrix} 7 \\ 0 \end{pmatrix}(2t)^7 + \begin{pmatrix} 7 \\ 1 \end{pmatrix}(2t)^6\left(\dfrac{1}{t}\right) +$

$\quad \begin{pmatrix} 7 \\ 2 \end{pmatrix}(2t)^5\left(\dfrac{1}{t}\right)^2 + \begin{pmatrix} 7 \\ 3 \end{pmatrix}(2t)^4\left(\dfrac{1}{t}\right)^3 +$

$\quad \begin{pmatrix} 7 \\ 4 \end{pmatrix}(2t)^3\left(\dfrac{1}{t}\right)^4 + \begin{pmatrix} 7 \\ 5 \end{pmatrix}(2t)^2\left(\dfrac{1}{t}\right)^5 +$

$\quad \begin{pmatrix} 7 \\ 6 \end{pmatrix}(2t)\left(\dfrac{1}{t}\right)^6 + \begin{pmatrix} 7 \\ 7 \end{pmatrix}\left(\dfrac{1}{t}\right)^7$

$= \frac{7!}{0!7!}(128t^7) + \frac{7!}{1!6!}(64t^6)\left(\dfrac{1}{t}\right) + \frac{7!}{2!5!}(32t^5)\left(\dfrac{1}{t^2}\right) +$

$\quad \frac{7!}{3!4!}(16t^4)\left(\dfrac{1}{t^3}\right) + \frac{7!}{4!3!}(8t^3)\left(\dfrac{1}{t^4}\right) +$

$\quad \frac{7!}{5!2!}(4t^2)\left(\dfrac{1}{t^5}\right) + \frac{7!}{6!1!}(2t)\left(\dfrac{1}{t^6}\right) + \frac{7!}{7!0!}\left(\dfrac{1}{t^7}\right)$

$= 128t^7 + 448t^5 + 672t^3 + 560t + 280t^{-1} +$

$\quad 84t^{-3} + 14t^{-5} + t^{-7}$

11. Expand: $(x^2 - 1)^5$.

We have $a = x^2$, $b = -1$, and $n = 5$.

Pascal's triangle method: Use the sixth row of Pascal's triangle.

1   5   10   10   5   1

$(x^2 - 1)^5$

$= 1 \cdot (x^2)^5 + 5(x^2)^4(-1) + 10(x^2)^3(-1)^2 +$

$\quad 10(x^2)^2(-1)^3 + 5(x^2)(-1)^4 + 1 \cdot (-1)^5$

$= x^{10} - 5x^8 + 10x^6 - 10x^4 + 5x^2 - 1$

Factorial notation method:

$(x^2 - 1)^5$

$= \binom{5}{0}(x^2)^5 + \binom{5}{1}(x^2)^4(-1) +$

$\quad \binom{5}{2}(x^2)^3(-1)^2 + \binom{5}{3}(x^2)^2(-1)^3 +$

$\quad \binom{5}{4}(x^2)(-1)^4 + \binom{5}{5}(-1)^5$

$= \dfrac{5!}{0!5!}(x^{10}) + \dfrac{5!}{1!4!}(x^8)(-1) + \dfrac{5!}{2!3!}(x^6)(1) +$

$\quad \dfrac{5!}{3!2!}(x^4)(-1) + \dfrac{5!}{4!1!}(x^2)(1) + \dfrac{5!}{5!0!}(-1)$

$= x^{10} - 5x^8 + 10x^6 - 10x^4 + 5x^2 - 1$

13. Expand: $(\sqrt{5} + t)^6$.

We have $a = \sqrt{5}$, $b = t$, and $n = 6$.

Pascal's triangle method: We use the seventh row of Pascal's triangle:

$$\begin{array}{ccccccc} 1 & 6 & 15 & 20 & 15 & 6 & 1 \end{array}$$

$(\sqrt{5} + t)^6 = 1 \cdot (\sqrt{5})^6 + 6(\sqrt{5})^5(t) +$

$\quad 15(\sqrt{5})^4(t^2) + 20(\sqrt{5})^3(t^3) +$

$\quad 15(\sqrt{5})^2(t^4) + 6\sqrt{5}t^5 + 1 \cdot t^6$

$\quad = 125 + 150\sqrt{5}\,t + 375t^2 + 100\sqrt{5}\,t^3 +$

$\quad\quad 75t^4 + 6\sqrt{5}\,t^5 + t^6$

Factorial notation method:

$(\sqrt{5} + t)^6 = \binom{6}{0}(\sqrt{5})^6 + \binom{6}{1}(\sqrt{5})^5(t) +$

$\quad \binom{6}{2}(\sqrt{5})^4(t^2) + \binom{6}{3}(\sqrt{5})^3(t^3) +$

$\quad \binom{6}{4}(\sqrt{5})^2(t^4) + \binom{6}{5}(\sqrt{5})(t^5) +$

$\quad \binom{6}{6}(t^6)$

$= \dfrac{6!}{0!6!}(125) + \dfrac{6!}{1!5!}(25\sqrt{5})t + \dfrac{6!}{2!4!}(25)(t^2) +$

$\quad \dfrac{6!}{3!3!}(5\sqrt{5})(t^3) + \dfrac{6!}{4!2!}(5)(t^4) +$

$\quad \dfrac{6!}{5!1!}(\sqrt{5})(t^5) + \dfrac{6!}{6!0!}(t^6)$

$= 125 + 150\sqrt{5}\,t + 375t^2 + 100\sqrt{5}\,t^3 +$

$\quad 75t^4 + 6\sqrt{5}\,t^5 + t^6$

15. Expand: $\left(a - \dfrac{2}{a}\right)^9$.

We have $a = a$, $b = -\dfrac{2}{a}$, and $n = 9$.

Pascal's triangle method: Use the tenth row of Pascal's triangle.

$$\begin{array}{ccccccccccc} 1 & 9 & 36 & 84 & 126 & 126 & 84 & 36 & 9 & 1 \end{array}$$

$\left(a - \dfrac{2}{a}\right)^9 = 1 \cdot a^9 + 9a^8\left(-\dfrac{2}{a}\right) + 36a^7\left(-\dfrac{2}{a}\right)^2 +$

$\quad 84a^6\left(-\dfrac{2}{a}\right)^3 + 126a^5\left(-\dfrac{2}{a}\right)^4 +$

$\quad 126a^4\left(-\dfrac{2}{a}\right)^5 + 84a^3\left(-\dfrac{2}{a}\right)^6 +$

$\quad 36a^2\left(-\dfrac{2}{a}\right)^7 + 9a\left(-\dfrac{2}{a}\right)^8 + 1 \cdot \left(-\dfrac{2}{a}\right)^9$

$= a^9 - 18a^7 + 144a^5 - 672a^3 + 2016a -$

$\quad 4032a^{-1} + 5376a^{-3} - 4608a^{-5} +$

$\quad 2304a^{-7} - 512a^{-9}$

Factorial notation method:

$\left(a - \dfrac{2}{a}\right)^9$

$= \binom{9}{0}a^9 + \binom{9}{1}a^8\left(-\dfrac{2}{a}\right) + \binom{9}{2}a^7\left(-\dfrac{2}{a}\right)^2 +$

$\quad \binom{9}{3}a^6\left(-\dfrac{2}{a}\right)^3 + \binom{9}{4}a^5\left(-\dfrac{2}{a}\right)^4 +$

$\quad \binom{9}{5}a^4\left(-\dfrac{2}{a}\right)^5 + \binom{9}{6}a^3\left(-\dfrac{2}{a}\right)^6 +$

$\quad \binom{9}{7}a^2\left(-\dfrac{2}{a}\right)^7 + \binom{9}{8}a\left(-\dfrac{2}{a}\right)^8 +$

$\quad \binom{9}{9}\left(-\dfrac{2}{a}\right)^9$

$= \dfrac{9!}{9!0!}a^9 + \dfrac{9!}{8!1!}a^8\left(-\dfrac{2}{a}\right) + \dfrac{9!}{7!2!}a^7\left(\dfrac{4}{a^2}\right) +$

$\quad \dfrac{9!}{6!3!}a^6\left(-\dfrac{8}{a^3}\right) + \dfrac{9!}{5!4!}a^5\left(\dfrac{16}{a^4}\right) +$

$\quad \dfrac{9!}{4!5!}a^4\left(-\dfrac{32}{a^5}\right) + \dfrac{9!}{3!6!}a^3\left(\dfrac{64}{a^6}\right) +$

$\quad \dfrac{9!}{2!7!}a^2\left(-\dfrac{128}{a^7}\right) + \dfrac{9!}{1!8!}a\left(\dfrac{256}{a^8}\right) +$

$\quad \dfrac{9!}{0!9!}\left(-\dfrac{512}{a^9}\right)$

$= a^9 - 9(2a^7) + 36(4a^5) - 84(8a^3) + 126(16a) -$

$\quad 126(32a^{-1}) + 84(64a^{-3}) - 36(128a^{-5}) +$

$\quad 9(256a^{-7}) - 512a^{-9}$

$= a^9 - 18a^7 + 144a^5 - 672a^3 + 2016a - 4032a^{-1} +$

$\quad 5376a^{-3} - 4608a^{-5} + 2304a^{-7} - 512a^{-9}$

17. $(\sqrt{2}+1)^6 - (\sqrt{2}-1)^6$

First, expand $(\sqrt{2}+1)^6$.

$$(\sqrt{2}+1)^6 = \binom{6}{0}(\sqrt{2})^6 + \binom{6}{1}(\sqrt{2})^5(1)+$$

$$\binom{6}{2}(\sqrt{2})^4(1)^2 + \binom{6}{3}(\sqrt{2})^3(1)^3+$$

$$\binom{6}{4}(\sqrt{2})^2(1)^4 + \binom{6}{5}(\sqrt{2})(1)^5+$$

$$\binom{6}{6}(1)^6$$

$$= \frac{6!}{6!0!} \cdot 8 + \frac{6!}{5!1!} \cdot 4\sqrt{2} + \frac{6!}{4!2!} \cdot 4+$$

$$\frac{6!}{3!3!} \cdot 2\sqrt{2} + \frac{6!}{2!4!} \cdot 2 + \frac{6!}{1!5!} \cdot \sqrt{2}+\frac{6!}{0!6!}$$

$$= 8 + 24\sqrt{2} + 60 + 40\sqrt{2} + 30 + 6\sqrt{2} + 1$$

$$= 99 + 70\sqrt{2}$$

Next, expand $(\sqrt{2}-1)^6$.

$$(\sqrt{2}-1)^6$$

$$= \binom{6}{0}(\sqrt{2})^6 + \binom{6}{1}(\sqrt{2})^5(-1)+$$

$$\binom{6}{2}(\sqrt{2})^4(-1)^2 + \binom{6}{3}(\sqrt{2})^3(-1)^3+$$

$$\binom{6}{4}(\sqrt{2})^2(-1)^4 + \binom{6}{5}(\sqrt{2})(-1)^5+$$

$$\binom{6}{6}(-1)^6$$

$$= \frac{6!}{6!0!} \cdot 8 - \frac{6!}{5!1!} \cdot 4\sqrt{2} + \frac{6!}{4!2!} \cdot 4 - \frac{6!}{3!3!} \cdot 2\sqrt{2}+$$

$$\frac{6!}{2!4!} \cdot 2 - \frac{6!}{1!5!} \cdot \sqrt{2}+\frac{6!}{0!6!}$$

$$= 8 - 24\sqrt{2} + 60 - 40\sqrt{2} + 30 - 6\sqrt{2} + 1$$

$$= 99 - 70\sqrt{2}$$

$$(\sqrt{2}+1)^6 - (\sqrt{2}-1)^6$$

$$= (99 + 70\sqrt{2}) - (99 - 70\sqrt{2})$$

$$= 99 + 70\sqrt{2} - 99 + 70\sqrt{2}$$

$$= 140\sqrt{2}$$

19. Expand: $(x^{-2} + x^2)^4$.

We have $a = x^{-2}$, $b = x^2$, and $n = 4$.

Pascal's triangle method: Use the fifth row of Pascal's triangle.

$$1 \quad 4 \quad 6 \quad 4 \quad 1.$$
$$(x^{-2} + x^2)^4$$
$$= 1 \cdot (x^{-2})^4 + 4(x^{-2})^3(x^2) + 6(x^{-2})^2(x^2)^2+$$
$$4(x^{-2})(x^2)^3 + 1 \cdot (x^2)^4$$
$$= x^{-8} + 4x^{-4} + 6 + 4x^4 + x^8$$

Factorial notation method:

$$(x^{-2} + x^2)^4$$

$$= \binom{4}{0}(x^{-2})^4 + \binom{4}{1}(x^{-2})^3(x^2)+$$

$$\binom{4}{2}(x^{-2})^2(x^2)^2 + \binom{4}{3}(x^{-2})(x^2)^3+$$

$$\binom{4}{4}(x^2)^4$$

$$= \frac{4!}{4!0!}(x^{-8}) + \frac{4!}{3!1!}(x^{-6})(x^2) + \frac{4!}{2!2!}(x^{-4})(x^4)+$$

$$\frac{4!}{1!3!}(x^{-2})(x^6) + \frac{4!}{0!4!}(x^8)$$

$$= x^{-8} + 4x^{-4} + 6 + 4x^4 + x^8$$

21. Find the 3rd term of $(a + b)^7$.

First, we note that $3 = 2+1$, $a = a$, $b = b$, and $n = 7$. Then the 3rd term of the expansion of $(a + b)^7$ is

$$\binom{7}{2}a^{7-2}b^2, \text{ or } \frac{7!}{2!5!}a^5b^2, \text{ or } 21a^5b^2.$$

23. Find the 6th term of $(x - y)^{10}$.

First, we note that $6 = 5+1$, $a = x$, $b = -y$, and $n = 10$. Then the 6th term of the expansion of $(x - y)^{10}$ is

$$\binom{10}{5}x^5(-y)^5, \text{ or } -252x^5y^5.$$

25. Find the 12th term of $(a - 2)^{14}$.

First, we note that $12 = 11 + 1$, $a = a$, $b = -2$, and $n = 14$. Then the 12th term of the expansion of $(a - 2)^{14}$ is

$$\binom{14}{11}a^{14-11} \cdot (-2)^{11} = \frac{14!}{3!11!}a^3(-2048)$$

$$= 364a^3(-2048)$$

$$= -745,472a^3$$

27. Find the 5th term of $(2x^3 - \sqrt{y})^8$.

First, we note that $5 = 4 + 1$, $a = 2x^3$, $b = -\sqrt{y}$, and $n = 8$. Then the 5th term of the expansion of $(2x^3 - \sqrt{y})^8$ is

$$\binom{8}{4}(2x^3)^{8-4}(-\sqrt{y})^4$$

$$= \frac{8!}{4!4!}(2x^3)^4(-\sqrt{y})^4$$

$$= 70(16x^{12})(y^2)$$

$$= 1120x^{12}y^2$$

29. The expansion of $(2u - 3v^2)^{10}$ has 11 terms so the 6th term is the middle term. Note that $6 = 5 + 1$,

$a = 2u$, $b = -3v^2$, and $n = 10$. Then the 6th term of the expansion of $(2u - 3v^2)^{10}$ is

$$\binom{10}{5}(2u)^{10-5}(-3v^2)^5$$

$$= \frac{10!}{5!5!}(2u)^5(-3v^2)^5$$

$$= 252(32u^5)(-243v^{10})$$

$$= -1,959,552u^5v^{10}$$

31. The number of subsets is 2^7, or 128

33. The number of subsets is 2^{24}, or 16,777,216.

35. The term of highest degree of $(x^5 + 3)^4$ is the first term, or

$$\binom{4}{0}(x^5)^{4-0}3^0 = \frac{4!}{4!0!}x^{20} = x^{20}.$$

Therefore, the degree of $(x^5 + 3)^4$ is 20.

37. We use factorial notation. Note that
$a = 3$, $b = i$, and $n = 5$.

$$(3 + i)^5$$

$$= \binom{5}{0}(3^5) + \binom{5}{1}(3^4)(i) + \binom{5}{2}(3^3)(i^2) +$$

$$\binom{5}{3}(3^2)(i^3) + \binom{5}{4}(3)(i^4) + \binom{5}{5}(i^5)$$

$$= \frac{5!}{0!5!}(243) + \frac{5!}{1!4!}(81)(i) + \frac{5!}{2!3!}(27)(-1) +$$

$$\frac{5!}{3!2!}(9)(-i) + \frac{5!}{4!1!}(3)(1) + \frac{5!}{5!0!}(i)$$

$$= 243 + 405i - 270 - 90i + 15 + i$$

$$= -12 + 316i$$

39. We use factorial notation. Note that
$a = \sqrt{2}$, $b = -i$, and $n = 4$.

$$(\sqrt{2}-i)^4 = \binom{4}{0}(\sqrt{2})^4 + \binom{4}{1}(\sqrt{2})^3(-i) +$$

$$\binom{4}{2}(\sqrt{2})^2(-i)^2 + \binom{4}{3}(\sqrt{2})(-i)^3 +$$

$$\binom{4}{4}(-i)^4$$

$$= \frac{4!}{0!4!}(4) + \frac{4!}{1!3!}(2\sqrt{2})(-i) +$$

$$\frac{4!}{2!2!}(2)(-1) + \frac{4!}{3!1!}(\sqrt{2})(i) +$$

$$\frac{4!}{4!0!}(1)$$

$$= 4 - 8\sqrt{2}i - 12 + 4\sqrt{2}i + 1$$

$$= -7 - 4\sqrt{2}i$$

41. $(a - b)^n = \binom{n}{0}a^n(-b)^0 + \binom{n}{1}a^{n-1}(-b)^1 +$

$$\binom{n}{2}a^{n-2}(-b)^2 + \cdots +$$

$$\binom{n}{n-1}a^1(-b)^{n-1} + \binom{n}{n}a^0(-b)^n$$

$$= \binom{n}{0}(-1)^0a^nb^0 + \binom{n}{1}(-1)^1a^{n-1}b^1 +$$

$$\binom{n}{2}(-1)^2a^{n-2}b^2 + \cdots +$$

$$\binom{n}{n-1}(-1)^{n-1}a^1b^{n-1} +$$

$$\binom{n}{n}(-1)^na^0b^n$$

$$= \sum_{k=0}^{n}\binom{n}{k}(-1)^ka^{n-k}b^k$$

43. $\dfrac{(x + h)^n - x^n}{h}$

$$= \frac{\binom{n}{0}x^n + \binom{n}{1}x^{n-1}h + \cdots + \binom{n}{n}h^n - x^n}{h}$$

$$= \binom{n}{1}x^{n-1} + \binom{n}{2}x^{n-2}h + \cdots + \binom{n}{n}h^{n-1}$$

$$= \sum_{k=1}^{n}\binom{n}{k}x^{n-k}h^{k-1}$$

45. Discussion and Writing

47. $(fg)(x) = f(x)g(x) = (x^2 + 1)(2x - 3) = 2x^3 - 3x^2 + 2x - 3$

49. $(g \circ f)(x) = g(f(x)) = g(x^2 + 1) = 2(x^2 + 1) - 3 = 2x^2 + 2 - 3 = 2x^2 - 1$

51. $\displaystyle\sum_{k=0}^{4}\binom{4}{k}5^{4-k}x^k = 64$

The left side of the equation is sigma notation for $(5 + x)^4$, so we have:

$$(5 + x)^4 = 64$$

$$5 + x = \pm2\sqrt{2} \qquad \text{Taking the 4th root on both sides}$$

$$x = -5 \pm 2\sqrt{2}$$

The real solutions are $-5 \pm 2\sqrt{2}$.

If we also observe that $(2\sqrt{2}i)^4 = 64$, we also find the imaginary solutions $-5 \pm 2\sqrt{2}i$.

53. $\displaystyle\sum_{k=0}^{4}\binom{4}{k}(-1)^kx^{4-k}6^k = \sum_{k=0}^{4}\binom{4}{k}x^{4-k}(-6)^k$, so the left side of the equation is sigma notation for $(x - 6)^4$. We have:

$(x - 6)^4 = 81$

$\quad x - 6 = \pm 3$ Taking the 4th root on both sides

$x - 6 = 3 \quad$ or $\quad x - 6 = -3$

$\quad x = 9 \quad$ or $\qquad x = 3$

The solutions are 9 and 3.

If we also observe that $(3i)^4 = 81$, we also find the imaginary solutions $6 \pm 3i$.

55. The expansion of $(x^2 - 6y^{3/2})^6$ has **7** terms, so the 4th term is the middle term.

$\binom{6}{3}(x^2)^3(-6y^{3/2})^3 = \dfrac{6!}{3!3!}(x^6)(-216y^{9/2}) =$

$-4320x^6y^{9/2}$

57. The $(k + 1)$st term of $\left(\sqrt[3]{x} - \dfrac{1}{\sqrt{x}} \right)^7$ is

$\binom{7}{k}(\sqrt[3]{x})^{7-k}\left(-\dfrac{1}{\sqrt{x}} \right)^k$. The term containing $\dfrac{1}{x^{1/6}}$ is the term in which the sum of the exponents is $-1/6$. That is,

$\left(\dfrac{1}{3} \right)(7 - k) + \left(-\dfrac{1}{2} \right)(k) = -\dfrac{1}{6}$

$\dfrac{7}{3} - \dfrac{k}{3} - \dfrac{k}{2} = -\dfrac{1}{6}$

$-\dfrac{5k}{6} = -\dfrac{15}{6}$

$k = 3$

Find the $(3 + 1)$st, or 4th term.

$\binom{7}{3}(\sqrt[3]{x})^4\left(-\dfrac{1}{\sqrt{x}} \right)^3 = \dfrac{7!}{4!3!}(x^{4/3})(-x^{-3/2}) =$

$-35x^{-1/6}$, or $-\dfrac{35}{x^{1/6}}$.

59. $_{100}C_0 +_{100}C_1 + \cdots +_{100}C_{100}$ is the total number of subsets of a set with 100 members, or 2^{100}.

61. $\displaystyle\sum_{k=0}^{23} \binom{23}{k}(\log_a x)^{23-k}(\log_a t)^k =$

$\qquad (\log_a x + \log_a t)^{23} = [\log_a(xt)]^{23}$

63. See the answer section in the text.

Exercise Set 10.8

1. a) We use Principle P.

For 1: $P = \dfrac{18}{100}$, or 0.18

For 2: $P = \dfrac{24}{100}$, or 0.24

For 3: $P = \dfrac{23}{100}$, or 0.23

For 4: $P = \dfrac{23}{100}$, or 0.23

For 5: $P = \dfrac{12}{100}$, or 0.12

b) Opinions may vary, but it seems that people tend not to select the first or last numbers.

3. The company can expect 78% of the 15,000 pieces of advertising to be opened and read. We have:

$78\%(15,000) = 0.78(15,000) = 11,700$.

5. a) The consonants with the 5 greatest numbers of occurrences are the 5 consonants with the greatest probability of occurring. They are T, S, R, N, and L.

b) E is the vowel with the greatest number of occurrences, so E is the vowel with the greatest probability of occurring.

c) Yes

7. a) Since there are 14 equally likely ways of selecting a marble from a bag containing 4 red marbles and 10 green marbles, we have, by Principle P,

$P(\text{selecting a red marble}) = \dfrac{4}{14} = \dfrac{2}{7}$.

b) Since there are 14 equally likely ways of selecting a marble from a bag containing 4 red marbles and 10 green marbles, we have, by Principle P,

$P(\text{selecting a green marble}) = \dfrac{10}{14} = \dfrac{5}{7}$.

c) Since there are 14 equally likely ways of selecting a marble from a bag containing 4 red marbles and 10 green marbles, we have, by Principle P,

$P(\text{selecting a purple marble}) = \dfrac{0}{14} = 0$.

d) Since there are 14 equally likely ways of selecting a marble from a bag containing 4 red marbles and 10 green marbles, we have, by Principle P,

$P(\text{selecting a red or a green marble}) =$

$\dfrac{4 + 10}{14} = 1$.

9. The total number of coins is $7 + 5 + 10$, or 22 and the total number of coins to be drawn is $4 + 3 + 1$, or 8. The number of ways of selecting 8 coins from a group of 22 is $_{22}C_8$. Four dimes can be selected in $_7C_4$ ways, 3 nickels in $_5C_3$ ways, and 1 quarter in $_{10}C_1$ ways.

P(selecting 4 dimes, 3 nickels, and 1 quarter) =

$$\frac{_7C_4 \cdot _5C_3 \cdot _{10}C_1}{_{22}C_8}, \text{ or } \frac{350}{31,977}$$

11. The number of ways of selecting 5 cards from a deck of 52 cards is $_{52}C_5$. Three sevens can be selected in $_4C_3$ ways and 2 kings in $_4C_2$ ways.

P(drawing 3 sevens and 2 kings) $= \dfrac{_4C_3 \cdot _4C_2}{_{52}C_5}$, or

$\dfrac{1}{108,290}$.

13. The number of ways of selecting 5 cards from a deck of 52 cards is $_{52}C_5$. Since 13 of the cards are spades, then 5 spades can be drawn in $_{13}C_5$ ways

P(drawing 5 spades) $= \dfrac{_{13}C_5}{_{52}C_5} = \dfrac{1287}{2,598,960} =$

$\dfrac{33}{66,640}$

15. a), b) Answers will vary.

17. The roulette wheel contains 38 equally likely slots. Eighteen of the 38 slots are colored black. Thus, by Principle P,

P(the ball falls in a black slot) $= \dfrac{18}{38} = \dfrac{9}{19}$.

19. The roulette wheel contains 38 equally likely slots. Thirty-six of the slots are colored red or black. Then, by Principle P,

P(the ball falls in a red or a black slot) $= \dfrac{36}{38} = \dfrac{18}{19}$.

21. The roulette wheel contains 38 equally likely slots. Only 1 slot is numbered 0. Then, by Principle P,

P(the ball falls in the 0 slot) $= \dfrac{1}{38}$.

23. The roulette wheel contains 38 equally likely slots. Eighteen of the slots are odd-numbered. Then, by Principle P,

P(the ball falls in a an odd-numbered slot) =

$\dfrac{18}{38} = \dfrac{9}{19}$.

25. The dartboard can be thought of as having 18 areas of equal size. Of these, 6 are red, 4 are green, 3 are blue, and 5 are yellow.

$P(\text{red}) = \dfrac{6}{18} = \dfrac{1}{3}$

$P(\text{green}) = \dfrac{4}{18} = \dfrac{2}{9}$

$P(\text{blue}) = \dfrac{3}{18} = \dfrac{1}{6}$

$P(\text{yellow}) = \dfrac{5}{18}$

27. Discussion and Writing

29. $2x^3 + 5x^2 - 4x - 3 = 0$

We use synthetic division to find a factor of the polynomial on the left side of the equation. We try $x - 1$.

$$\begin{array}{r|rrrr} 1 & 2 & 5 & -4 & -3 \\ & & 2 & 7 & 3 \\ \hline & 2 & 7 & 3 & 0 \end{array}$$

Then we have:

$$(x - 1)(2x^2 + 7x + 3) = 0$$
$$(x - 1)(2x + 1)(x + 3) = 0$$

$x - 1 = 0 \quad or \quad 2x + 1 = 0 \quad or \quad x + 3 = 0$

$\quad x = 1 \quad or \quad\quad 2x = -1 \quad or \quad\quad x = -3$

$\quad x = 1 \quad or \quad\quad x = -\dfrac{1}{2} \quad or \quad\quad x = -3$

We could also graph $y = 2x^3 + 5x^2 - 4x - 3$ and use the Zero feature three times. The solutions are -3, $-\dfrac{1}{2}$, and 1, or -3, -0.5, and 1.

31. $\begin{aligned} 2x + \ \ y - 3z &= 5, \\ 3x + 3y - 5z &= 4, \\ x - 2y + 2z &= 11 \end{aligned}$

We use Gauss-Jordan elimination with matrices. First we write the augmented matrix.

$$\left[\begin{array}{rrr|r} 2 & 1 & -3 & 5 \\ 3 & 3 & -5 & 4 \\ 1 & -2 & 2 & 11 \end{array}\right]$$

Now interchange the first and third rows.

$$\left[\begin{array}{rrr|r} 1 & -2 & 2 & 11 \\ 3 & 3 & -5 & 4 \\ 2 & 1 & -3 & 5 \end{array}\right]$$

Multiply the first row by -3 and add it to the second row. Also multiply the first row by -2 and add it to the third row.

$$\left[\begin{array}{rrr|r} 1 & -2 & 2 & 11 \\ 0 & 9 & -11 & -29 \\ 0 & 5 & -7 & -17 \end{array}\right]$$

Multiply the third row by -2 and add it to the second row.

$$\begin{bmatrix} 1 & -2 & 2 & | & 11 \\ 0 & -1 & 3 & | & 5 \\ 0 & 5 & -7 & | & -17 \end{bmatrix}$$

Now multiply the second row by -1.

$$\begin{bmatrix} 1 & -2 & 2 & | & 11 \\ 0 & 1 & -3 & | & -5 \\ 0 & 5 & -7 & | & -17 \end{bmatrix}$$

Multiply the second row by 2 and add it to the first row. Also multiply the second row by -5 and add it to the third row.

$$\begin{bmatrix} 1 & 0 & -4 & | & 1 \\ 0 & 1 & -3 & | & -5 \\ 0 & 0 & 8 & | & 8 \end{bmatrix}$$

Multiply the third row by $\frac{1}{8}$.

$$\begin{bmatrix} 1 & 0 & -4 & | & 1 \\ 0 & 1 & -3 & | & -5 \\ 0 & 0 & 1 & | & 1 \end{bmatrix}$$

Multiply the third row by 4 and add it to the first row. Also multiply the third row by 3 and add it to the second row.

$$\begin{bmatrix} 1 & 0 & 0 & | & 5 \\ 0 & 1 & 0 & | & -2 \\ 0 & 0 & 1 & | & 1 \end{bmatrix}$$

Now we can read the solution from the matrix. We have $x = 5$, $y = -2$, $z = 1$. The solution is $(5, -2, 1)$.

33. Consider a suit

A K Q J 10 9 8 7 6 5 4 3 2

A straight flush can be any of the following combinations in the same suit.

K	Q	J	10	9
Q	J	10	9	8
J	10	9	8	7
10	9	8	7	6
9	8	7	6	5
8	7	6	5	4
7	6	5	4	3
6	5	4	3	2
5	4	3	2	A

Remember a straight flush does not include A K Q J 10 which is a royal flush.

a) Since there are 9 straight flushes per suit, there are 36 straight flushes in all 4 suits.

b) Since 2,598,960, or $_{52}C_5$, poker hands can be dealt from a standard 52-card deck and 36 of those hands are straight flushes, the probability of getting a straight flush is $\frac{36}{2,598,960}$, or about 1.39×10^{-5}.

35. a) There are 13 ways to select a denomination. Then from that denomination there are $_4C_3$ ways to pick 3 of the 4 cards in that denomination. Now there are 12 ways to select any one of the remaining 12 denominations and $_4C_2$ ways to pick 2 cards from the 4 cards in that denomination. Thus the number of full houses is $(13 \cdot_4 C_3) \cdot (12 \cdot_4 C_2)$ or 3744.

b) $\dfrac{3744}{_{52}C_5} = \dfrac{3744}{2,598,960} \approx 0.00144$

37. a) There are 4 ways to select a suit and then $\dbinom{13}{5}$ ways to choose 5 cards from that suit. But these combinations include hands with all the cards in sequence, so we subtract the 4 royal flushes (Exercise 28) and the 36 straight flushes (Exercise 29). Thus there are $4 \cdot \dbinom{13}{5} - 4 - 36$, or 5108 flushes.

b) $\dfrac{5108}{_{52}C_5} = \dfrac{5108}{2,598,960} \approx 0.00197$

39. a) There are 10 sets of 5 consecutive cards:

A	K	Q	J	10
K	Q	J	10	9

.

.

.

5	4	3	2	A

In each of these 10 sets there are 4 ways to choose (from 4 suits) each of the 5 cards. These combinations include the 4 royal flushes and the 36 straight flushes, both of which consist of 5 cards of the *same suit* in sequence.

Thus there are $10 \cdot 4 \cdot 4 \cdot 4 \cdot 4 - 4 - 36$, or 10,200 straights.

b) $\dfrac{10,200}{_{52}C_5} = \dfrac{10,200}{2,598,960} \approx 0.00392$

Sample Chapter Tests

Chapter G Test

1. Make a hand-drawn graph of the set of points. Label each point.

   $\{(-3, 0), (1, 4), (-2, -3), (0, 2), (3, -2), (-1, 5)\}$

2. Determine whether each ordered pair is a solution of the equation $2x - 3y = 7$. Answer yes or no.

   a) $(4, 5)$ b) $(2, -1)$

3. Make a hand-drawn graph of each equation.

   a) $y = 2x - 3$ b) $y = x^2 + 1$.

Match the equations with graphs (a) - (d).

4. $y = |x - 3|$

5. $y = \sqrt{x + 2}$

6. $y = 5 - 3x$

7. $y = x^2 - 4$

   a)

   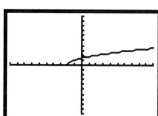

   b)

   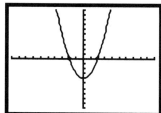

   c)

   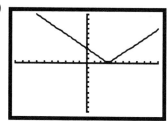

   d)

   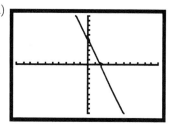

8. Use a grapher to complete the table for $y_1 = \sqrt{9 - x^2}$ and $y_2 = 3 - x$.

   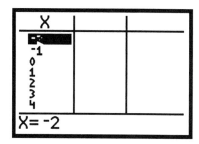

9. Graph the equation $y = x^3 - 2x + 2$ using the window $[-4, 4, -8, 8]$, Xscl $= 1$, Yscl $= 1$.

10. Graph the equation $y = x^2 - 10$ in the standard $[-10, 10, -10, 10]$ window and in the window $[-5, 5, -5, 5]$. Determine which window better shows the shape of the graph and where it crosses the x- and y-axes.

11. a) Graph the equations $y_1 = 2x^3 - 3x^2 + 3$ and $y_2 = 7 - x^2$ in the same viewing window using the dimensions $[-10, 10, -10, 10]$, Xscl $= 1$, Yscl $= 1$.

    b) Find the coordinates of the point of intersection of the graphs in part (a). Round each coordinate to two decimal places.

12. Solve $3.4x^3 - 1.9x = -4$. Round your answer to two decimal places.

Chapter R Test

1. Consider the numbers -8, $\dfrac{11}{3}$, $\sqrt{15}$, 0, -5.49, 36, $\sqrt[3]{7}$, $10\dfrac{1}{6}$.

   a) Which are integers?

   b) Which are rational numbers?

2. Simplify: $|-1.2xy|$.

3. Find the distance between -9 and 5 on a number line.

4. Write interval notation for $\{x|-3 < x \le 6\}$.

5. Compute: $32 \div 2^3 - 12 \div 4 \cdot 3$.

6. Compute and write scientific notation for the answer: $\dfrac{2.7 \times 10^4}{3.6 \times 10^{-3}}$.

Simplify.

7. $(-3a^5b^{-4})(5a^{-1}b^3)$

8. $(3x^4 - 2x^2 + 6x) - (5x^3 - 3x^2 + x)$

9. $(x + 3)(2x - 5)$

10. $(2y - 1)^2$

11. $3\sqrt{75} + 2\sqrt{27}$

12. $\dfrac{\dfrac{x}{y} - \dfrac{y}{x}}{x + y}$

Factor.

13. $2n^2 + 5n - 12$

14. $8x^2 - 18$

15. $m^3 - 8$

16. Use a grapher to graph the equation(s) you would use to check the factorization $x^2 + x - 6 = (x + 3)(x - 2)$. Show your graph in a window and write the equation(s) you graphed above the window.

17. Multiply and simplify: $\dfrac{x^2 + x - 6}{x^2 + 8x + 15} \cdot \dfrac{x^2 - 25}{x^2 - 4x + 4}$.

18. Subtract and simplify: $\dfrac{x}{x^2 - 1} - \dfrac{3}{x^2 + 4x - 5}$.

19. Convert to radical notation: $t^{5/7}$.

20. Rationalize the denominator: $\dfrac{5}{7 - \sqrt{3}}$.

Solve.

21. $6x + 7 = 1$

22. $(2x - 1)(x + 5) = 0$

23. $x^2 - 2x - 3 = 0$

24. $6x^2 - 36 = 0$

25. Solve $V = \dfrac{2}{3}\pi r^2 h$ for h.

26. How long is a guy wire that reaches from the top of a 12-ft pole to a point on the ground 5 feet from the pole?

SYNTHESIS

27. Solve: $(x + 1)^3 = x^3 + 7$

Chapter 1 Test

1. a) Determine whether the relation
$$\{(-4,7),(3,0),(1,5),(0,7)\}$$
is a function. Answer yes or no.
   b) Find the domain of the relation.
   c) Find the range of the relation.

2. Given that $f(x) = 2x^2 - x + 5$, find
   a) $f(-1)$;   b) $f(a+2)$.

3. a) Use a grapher to graph $f(x) = |x - 2| + 3$.
   b) Visually estimate the domain of $f(x)$.
   c) Visually estimate the range of $f(x)$.

4. Determine whether each graph is that of a function. Answer yes or no.
   a)
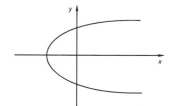
   b)
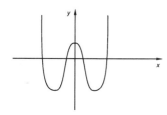

5. For the function $f(x) = x^3 - 4x^2 - 3x + 5$, use a grapher to find
   a) the intervals on which the function is increasing or decreasing;
   b) any relative maxima or minima.

6. Make a hand-drawn graph of $f(x)$.
$$f(x) = \begin{cases} x^2, & \text{for } x < -1, \\ |x|, & \text{for } -1 \le x \le 1, \\ \sqrt{x-1}, & \text{for } x > 1, \end{cases}$$

7. Find the slope and the y-intercept of the graph of $-3x + 2y = 5$.

8. Write an equation for the line that passes through $(-5, 4)$ and $(3, -2)$.

9. Find an equation of the line containing the point $(-1, 3)$ and parallel to the line $x + 2y = -6$.

10. The table below shows the per capita consumption of tea in the United States in several years.

Year, x	Per Capita Consumption of Tea, in gallons
0. '93	8.4
1. '94	8.2
2. '95	8.0
3. '96	8.0

   a) Use a grapher to fit a regression line $y = ax + b$ to the data, where x is the number of years after 1993.
   b) What is the correlation coefficient for the regression line?
   c) Use the regression line in part (a) to predict the per capita consumption of tea in 2005.

11. Find the distance between $(5,8)$ and $(-1,5)$.

12. Find the midpoint of the segment with endpoints $(-2, 6)$ and $(-4, 3)$.

13. Find an equation of the circle with center $(-1, 2)$ and radius $\sqrt{5}$.

14. a) Use a grapher to graph $y = x^4 - 2x^2$ in the window $[-5, 5, -5, 5]$.
    b) Determine whether the graph in part (a) is symmetric with respect to the x-axis, the y-axis, and/or the origin.

15. Test algebraically whether the function $f(x) = \dfrac{2x}{x^2 + 1}$ is even, odd, or neither even nor odd. Show your work.

16. Write an equation for a function that has the shape of $y = x^2$, but shifted right 2 units and down 1 unit.

17. Write an equation for a function that has the shape of $y = x^2$ but shifted left 2 units and down 3 units.

18. The graph of a function $y = f(x)$ is shown below. No formula for f is given. Make a hand-drawn graph of $y = -\dfrac{1}{2}f(x)$.

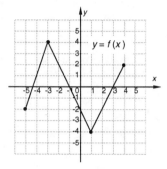

19. For $f(x) = x^2$ and $g(x) = \sqrt{x-3}$, find

a) the domain of f;

b) the domain of g;

c) $(f - g)(x)$;

d) $(fg)(x)$;

e) the domain of $(f/g)(x)$.

20. Find an equation of variation in which y varies inversely as x and $y = 5$ when $x = 6$.

21. The stopping distance d of a car after the brakes have been applied varies directly as the square of the speed r. If a car traveling 60 mph can stop in 200 ft, how long will it take a car traveling 30 mph to stop?

SYNTHESIS

22. If $(-3, 1)$ is a point on the graph of $y = f(x)$, what point do you know is on the graph of $y = f(3x)$?

Chapter 2 Test

Find the zero(s) of each function.

1. $f(x) = 3x + 9$

2. $f(x) = 4x^2 - 11x - 3$

3. $f(x) = 2x^2 - x - 7$

4. Solve $x^2 + 4x = 1$ by completing the square. Find exact solutions. Show your work.

Solve. Find exact solutions.

5. $2t^2 - 3t + 4 = 0$

6. $x + 5\sqrt{x} - 36 = 0$

7. $\dfrac{3}{3x + 4} + \dfrac{2}{x - 1} = 2$

8. $\sqrt{x + 4} - 2 = 1$

9. $|4y - 3| = 5$

10. $-7 < 2x + 3 < 9$

11. $|x + 5| > 2$

12. Solve for n: $R = \sqrt{3np}$.

Express in terms of i.

13. $\sqrt{-43}$

14. $-\sqrt{-25}$

Simplify.

15. $(3 + 4i)(2 - i)$

16. i^{33}

17. For the graph of the function $f(x) = -x^2 + 2x + 8$,
  a) find the vertex;
  b) find the line of symmetry;
  c) state whether there is a maximum or minimum value and find that value;
  d) find the range.

18. *Maximum Area* A homeowner wants to fence a rectangular play yard using 60 ft of fencing. The side of the house will be used as one side of the rectangle. Find the dimensions for which the area is a maximum.

19. Use the regression feature on a grapher to fit a quadratic function to the data in the table.

x	y
-3	300
-1	167
1	98
3	136
5	310

SYNTHESIS

20. Find a such that $f(x) = ax^2 - 4x + 3$ has a maximum value of 12.

Chapter 3 Test

1. Use a grapher to graph the function $f(x) = 2x^3 + 6x^2 - x - 5$. Then estimate the function's

   a) real zeros;

   b) relative maxima;

   c) relative minima;

   d) domain;

   e) range.

2. *Interest compounded annually* When P dollars is invested at interest rate i, compounded annually, for t years, the investment grows to A dollars, where
$$A = P(1+i)^t.$$
   Find the interest rate i if $1500 grows to $1858.24 in 4 years.

3. a) Use the regression feature on a grapher to fit a cubic function $y = ax^3 + bx^2 + cx + d$ to the data in the table.

x	y
-5	-200
-3	-66
0	-4
2	10
6	175

   b) Use your answer to part (a) to find the function value for $x = 9$.

4. Use long division to find the quotient and remainder. Show your work.
$$(x^4 + 3x^3 + 2x - 5) \div (x^2 - 1)$$

5. Use synthetic division to find the quotient and remainder. Show your work.
$$(3x^3 - 12x + 7) \div (x - 5)$$

6. Use synthetic division to determine whether -2 is a zero of $f(x) = x^3 + 4x^2 + x - 6$. Answer yes or no. Show your work.

7. Use synthetic division to find $P(-3)$ for $P(x) = 2x^3 - 6x^2 + x - 4$. Show your work.

8. Suppose that a polynomial function of degree 5 with rational coefficients has 1, $\sqrt{3}$, and $2 - i$ as zeros. Find the other zeros.

9. For the polynomial function $P(x) = x^4 + 2x^3 - 4x^2 - 5x + 6$,

   a) use a grapher to graph $P(x)$;

   b) solve $P(x) = 0$;

   c) express $P(x)$ as a product of linear factors.

10. Make a hand-drawn graph of $f(x) = \dfrac{2}{(x-3)^2}$. Label all the asymptotes.

11. Find a rational function that has vertical asymptotes $x = -1$ and $x = 2$ and x-intercept $(-4, 0)$.

Solve.

12. $2x^2 > 5x + 3$

13. $\dfrac{x+1}{x-4} \le 3$

14. The function $S(t) = -16t^2 + 64t + 192$ gives the height S, in feet, of a model rocket launched from a hill that is 192 ft high with a velocity of 64 ft/sec.

   a) Determine how long it will take the rocket to reach the ground.

   b) Find the interval on which the height of the rocket is greater than 240 ft.

SYNTHESIS

15. Find the domain of $f(x) = \sqrt{x^2 + x - 12}$.

Chapter 4 Test

1. For $f(x) = x - 5$ and $g(x) = x^2 + 1$, find $(f \circ g)(x)$ and $(g \circ f)(x)$.

2. Find the inverse of the relation
$$\{(-2, 5), \ (4, 3), \ (0, -1), \ (-6, -3)\}.$$

3. Determine whether the function shown below is one-to-one. Answer yes or no.

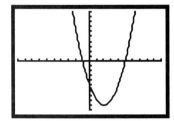

4. Find a formula for the inverse of the function $f(x) = x^3 + 1$.

Graph each of the following functions using a grapher.

5. $f(x) = e^x - 3$

6. $f(x) = \ln(x + 2)$

7. Convert to an exponential equation: $\ln x = 4$.

8. Convert to a logarithmic equation: $3^x = 5.4$.

Solve.

9. $\log_{25} 5 = x$

10. $\log_3 x + \log_3(x + 8) = 2$

11. $3^{4-x} = 27^x$

12. $e^x = 65$

13. Express in terms of sums and differences of logarithms: $\ln \sqrt[5]{x^2 y}$.

14. Given that $\log_a 2 = 0.328$ and $\log_a 8 = 0.984$, find $\log_a 4$.

15. Simplify: $\ln e^{-4t}$.

16. *Growth rate* The population of a country doubled in 45 yr. What was the exponential growth rate?

17. *Compound interest* Suppose $1000 is invested at interest rate k, compounded continuously, and grows to $1144.54 in 3 yr.

   a) Find the interest rate.

   b) Find the exponential growth function.

   c) Find the balance after 8 yr.

   d) Find the doubling time.

18. The following table contains data regarding the sales of a small business.

Year, x	Sales, y (in millions of dollars)
0. 1995	2.8
1. 1996	4.4
2. 1997	6.5
3. 1998	10.1
4. 1999	15.4
5. 2000	23.0

   a) Create a scatterplot of the data.

   b) Use regression to fit an exponential function $y = ab^x$ or $y = ae^{kx}$ to the data.

   c) Use the function to predict the sales in 2003.

   d) After how long will the sales be $50 million?

SYNTHESIS

19. Solve: $4^{\sqrt[3]{x}} = 8$.

Chapter 5 Test

Find the exact function value, if it exists.

1. $\cos 3\pi$

2. $\sec \dfrac{5\pi}{4}$

3. $\sin 120°$

4. $\tan(-45°)$

Find the function values. Round to four decimal places.

5. $\sec \dfrac{5\pi}{9}$

6. $\cos 76.07$

7. $\tan 526.4°$

8. $\sin(-12°)$

9. Find a positive angle and a negative angle coterminal with a $112°$ angle.

10. Convert $38°27'56"$ to decimal degree notation. Round to two decimal places.

11. Find the supplement of $\dfrac{5\pi}{6}$.

12. Convert $210°$ to radian measure in terms of π.

13. Convert $\dfrac{3\pi}{4}$ to degree measure.

14. Find the length of an arc of a circle given a central angle of $\pi/3$ and a radius of 16 cm.

15. Given that $\sin \theta = -4/\sqrt{41}$ and that the terminal side is in quadrant IV, find the other five trigonometric function values.

16. Find the six trigonometric functions values of θ.

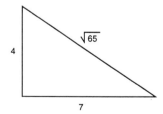

17. Find the exact acute angle θ, in degrees, for which $\sin \theta = \dfrac{1}{2}$.

18. Given that $\sin 28.4° \approx 0.4756, \cos 28.4° \approx 0.8796$, and $\tan 28.4° \approx 0.5407$, find the six trigonometric function values of $61.6°$.

Consider the function $y = -\sin(x - \pi/2) + 1$.

19. Find the amplitude.

20. Find the period.

21. Find the phase shift.

22. Which is the graph of the function?

a)

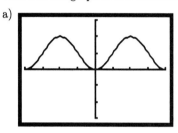

b)

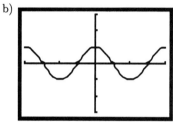

c)

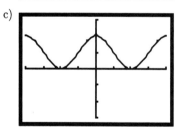

d)

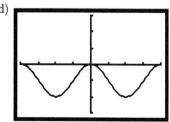

23. *Linear Speed* A ferris wheel has a radius of 6 m and revolves at 1.5 rpm. What is the linear speed in meters per minute?

24. *Height of a Kite* The angle of elevation of a kite is 65° with 490 ft of string out. Assuming the string is taut, how high is the kite?

SYNTHESIS

25. Determine the domain of $f(x) = \dfrac{-3}{\sqrt{\cos x}}$.

Chapter 6 Test

1. Simplify: $\dfrac{2\cos^2 x - \cos x - 1}{\cos x - 1}$.

2. Given that $x = 2\sin\theta$, express $\sqrt{4 - x^2}$ as a trigonometric function without radicals. Assume $0 < \theta < \pi/2$.

3. Use a sum or difference identity to find $\sin 75°$ exactly.

4. Given that $\cos\theta = -\dfrac{2}{3}$ and that the terminal side is in quadrant II, find $\cos\left(\dfrac{\pi}{2} - \theta\right)$.

5. Given that $\sin\theta = -\dfrac{4}{5}$ and θ is in quadrant III, find $\sin 2\theta$.

6. Use a half-angle identity to evaluate $\cos\dfrac{\pi}{12}$ exactly.

7. Use a grapher to determine which expression can be used to complete an identity $\dfrac{\cos x}{1 - \sin x} = \cdots$

a) $y = \tan(x/2)$

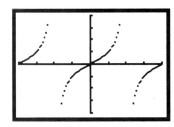

b) $y = \cos x - \sin x$

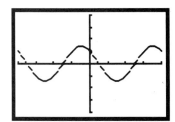

c) $y = \sin 2x$

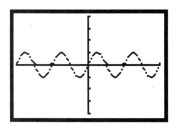

d) $y = \dfrac{1 + \sin x}{\cos x}$

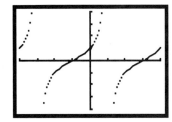

Prove each of the following identities.

8. $\csc x - \cos x \cot x = \sin x$.

9. $(\sin x + \cos x)^2 = 1 + \sin 2x$

10. Find $\sin^{-1}\left(-\dfrac{\sqrt{2}}{2}\right)$ exactly in degrees.

11. Find $\arctan\sqrt{3}$ exactly in radians.

12. Use a grapher to find $\cos^{-1}(-0.6716)$ exactly in radians, rounded to three decimal places.

13. Evaluate $\cos\left(\sin^{-1}\dfrac{1}{2}\right)$.

Solve, finding all solutions in $[0, 2\pi)$.

14. $4\cos^2 x = 3$

15. $2\sin^2 x = \sqrt{2}\sin x$

16. Solve with a grapher finding all solutions in $[0, 2\pi)$: $5\sin x = \cos 2x + 3$.

SYNTHESIS

17. Find $\cos\theta$ given that $\cos 2\theta = \dfrac{5}{6}$, $\dfrac{3\pi}{2} < \theta < 2\pi$.

Chapter 7 Test

1. Solve the right triangle with $b = 45.1$ and $A = 35.9°$. Standard lettering has been used.

2. *Location* A pickup-truck camper travels at 50 mph for 6 hr in a direction of 115° from Buffalo, Wyoming. At the end of that time, how far east of Buffalo is the camper?

Solve $\triangle ABC$, if possible.

3. $a = 18$ ft, $B = 54°$, $C = 43°$

4. $b = 8$ m, $c = 5$ m, $C = 36°$

5. $a = 16.1$ in., $b = 9.8$ in., $c = 11.2$ in.

6. Find the area of $\triangle ABC$ if $C = 106.4°$, $a = 7$ cm, and $b = 13$ cm.

7. *Distance Across a Lake* Points A and B are on opposite sides of a lake. Point C is 52 m from A. The measure of $\angle BAC$ is determined to be 108° and the measure of $\angle ACB$ is determined to be 44°. What is the distance from A to B?

8. *Location of Airplanes* Two airplanes leave an airport at the same time. The first flies 210 km/h in a direction of 290°. The second flies 180 km/h in a direction of 185°. After 3 hr, how far apart are the planes?

9. Graph $-4 + i$.

10. Find the absolute value of $2 - 3i$.

11. Find trigonometric notation for $3 - 3i$.

12. Find standard notation, $a + bi$, for $2(\cos 90° + i \sin 90°)$.

13. Divide and express the result in standard notation $a + bi$:
$$\frac{2\left(\cos \dfrac{2\pi}{3} + i\sin \dfrac{2\pi}{3}\right)}{8\left(\cos \dfrac{\pi}{6} + i\sin \dfrac{\pi}{6}\right)}.$$

14. Find $(1 - i)^8$ and write standard notation for the answer.

15. Find the cube roots of -1.

16. For vectors $\mathbf{u}$ and $\mathbf{v}$, $|\mathbf{u}| = 8$, $|\mathbf{v}| = 5$, and the angle between the vectors is 63°. Find $\mathbf{u} + \mathbf{v}$. Give the magnitude to the nearest tenth and give the direction by specifying the angle the resultant makes with $\mathbf{u}$, to the nearest degree.

17. For $\mathbf{u} = 2i - 7j$ and $\mathbf{v} = 5i + j$, find $2\mathbf{u} - 3\mathbf{v}$.

18. Find a unit vector in the same direction as $-4i + 3j$.

SYNTHESIS

19. A parallelogram has sides of length 15.4 and 9.8. Its area is 72.9. Find the measures of the angles.

Chapter 8 Test

1. Graph the system of equations in the window $[-10, 10, -10, 10]$.
$$x - y = 5,$$
$$2x + 3y = 5.$$

Solve. Use any method.

2. $3x + 2y = 1,$
$2x - y = -11$

3. $2x - 3y = 8,$
$5x - 2y = 9$

4. $4x + 2y + z = 4,$
$3x - y + 5z = 4$
$5x + 3y - 3z = -2$

5. Classify the system of equations in Exercise 2 as consistent or inconsistent

6. Classify the system of equations in Exercise 3 as dependent or independent.

7. *Ticket Sales* One evening 750 tickets were sold for Shortridge Community College's spring musical. Tickets cost $3 for students and $5 for non-students. Total receipts were $3066. How many of each type of ticket were sold?

For Exercises 8 - 13, let

$$A = \begin{bmatrix} 1 & -1 & 3 \\ -2 & 5 & 2 \end{bmatrix}, B = \begin{bmatrix} -5 & 1 \\ -2 & 4 \end{bmatrix}, \text{ and } C = \begin{bmatrix} 3 & -4 \\ -1 & 0 \end{bmatrix}.$$

Find each of the following, if possible.

8. $B + C$

9. $A - C$

10. CB

11. AB

12. $2A$

13. C^{-1}

14. *Food Service Management* The table below shows the cost per serving, in cents, for items on three lunch menus served at a senior-citizens' center.

Menu	Main Dish	Side Dish	Dessert
1	49	10	13
2	43	12	11
3	51	8	12

On a particular day 26 Menu 1 meals, 18 Menu 2 meals, and 23 Menu 3 meals are served.

a) Write the information in the table as a 3×3 matrix M.

b) Write a row matrix N that represents the number of each menu served.

c) Find the product NM.

d) State what the entries of NM represent.

15. Write a matrix equation equivalent to the system of equations
$$3x - 4y + 2z = -8,$$
$$2x + 3y + z = 7,$$
$$x - 5y - 3z = 3.$$

16. Make a hand-drawn graph of $3x + 4y \leq -12$.

17. Find the maximum and minimum values of $Q = 2x + 3y$ subject to
$$x + y \leq 6,$$
$$2x - 3y \geq -3,$$
$$x \geq 1,$$
$$y \geq 0.$$

18. *Maximizing Profit* Casey's Cakes prepares pound cakes and carrot cakes. In a given week, at most 100 cakes can be prepared of which 25 pound cakes and 15 carrot cakes are required by regular customers. The profit from each pound cake is $3 and the profit from each carrot cake is $4. How many of each type of cake should be prepared in order to maximize the profit?

19. Decompose into partial fractions: $\dfrac{3x - 11}{x^2 + 2x - 3}$.

SYNTHESIS

20. Three solutions of the equation $Ax - By = Cz - 8$ are $(2, -2, 2)$, $(-3, -1, 1)$, and $(4, 2, 9)$. Find A, B, and C.

Chapter 9 Test

Match each of the following with its graph.

1. $4x^2 - y^2 = 4$

2. $x^2 - 2x - 3y = 5$

3. $x^2 + 4x + y^2 - 2y - 4 = 0$

4. $9x^2 + 4y^2 = 36$

a)
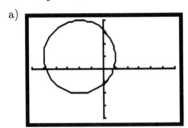

b)
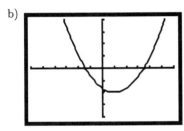

c)
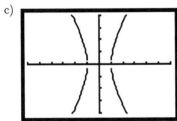

d)
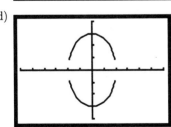

5. Find an equation of the parabola with focus $(0,2)$ and directrix $y = -2$.

6. Find the vertex of the parabola given by $y^2 - 8y + 2x - 4 = 0$.

7. Find the center and the radius of the circle given by $x^2 + y^2 + 2x - 6y - 15 = 0$.

8. Find an equation of the ellipse having vertices $(0, -5)$ and $(0,5)$ with minor axis of length 4.

9. Find the asymptotes of the hyperbola given by $2y^2 - x^2 = 18$.

10. *Satellite Dish* A satellite dish has a parabolic cross-section that is 18 in. wide at the opening and 6 in. deep at the vertex. How far from the vertex is the focus?

Solve.

11. $2x^2 - 3y^2 = -10,$
     $x^2 + 2y^2 = 9$

12. $x^2 + y^2 = 13,$
     $x + y = 1$

13. $x + y = 5,$
     $xy = 6$

14. *Landscaping* Leisurescape is planting a rectangular flower garden with a perimeter of 18 ft and a diagonal of $\sqrt{41}$ ft. Find the dimensions of the garden.

15. Graph: $5x^2 - 8xy + 5y^2 = 9$.

16. Find the polar coordinates of $(-1, \sqrt{3})$. Express the angle in degrees using the smallest possible positive angle.

17. Use a grapher to convert $(-1, 4.8)$ to rectangular coordinates. Round the coordinates to the nearest hundredth.

18. Convert to a polar equation: $x^2 + y^2 = 10$.

19. Graph $r = 1 - \cos\theta$.

20. Graph $r = \dfrac{2}{1 - \sin\theta}$. State whether the directrix is vertical or horizontal, describe its location in relation to the pole, and find the vertex or vertices.

21. Find a polar equation of the conic with a focus at the pole, eccentricity 2, and directrix $r = 3\sec\theta$.

22. Graph the plane curve given by the parametric equations $x = \sqrt{t}$, $y = t + 2$, $0 \le t \le 16$.

23. Find a rectangular equation equivalent to $x = 3\cos\theta$, $y = 3\sin\theta$, $0 \leq \theta \leq 2\pi$.

24. Find two sets of parametric equations for the rectangular equation $y = x - 5$.

SYNTHESIS

25. Find an equation of a circle that passes through the points $(1,1)$ and $(5, -3)$ and whose center is on the line $x - y = 4$.

Chapter 10 Test

1. For the sequence whose nth term is $a_n = (-1)^n(2n + 1)$, find a_{21}.

2. Find and evaluate: $\displaystyle\sum_{k=1}^{4}(k^2 + 1)$.

3. Use a grapher to graph the first 10 terms of the sequence with general term $a_n = \dfrac{n+1}{n+2}$.

4. Find the 15th term of the arithmetic sequence $2, 5, 8, \ldots$.

5. The 1st term of an arithmetic sequence is 8 and the 21st term is 108. Find the 7th term.

6. Find the sum of the first 20 terms of the series $17 + 13 + 9 + \ldots$.

7. For a geometric sequence, $r = 0.2$ and $S_4 = 1248$. Find a_1.

8. Find the sum, if it exists: $18 + 6 + 2 \cdots$.

9. Find fractional notation for $0.\overline{56}$.

10. *Amount of an Annuity* To create a college fund, a parent makes a sequence of 18 yearly deposits of $2500 each in a savings account on which interest is compounded annually at 5.6%. Find the amount of the annuity.

11. Use mathematical induction to prove that, for every natural number n, $2 + 5 + 8 + \cdots + (3n - 1) = \dfrac{n(3n + 1)}{2}$.

Evaluate.

12. $_{15}P_6$

13. $_{21}C_{10}$

14. $\dbinom{n}{4}$

15. How many 4-digit numbers can be formed using the digits 1, 3, 5, 6, 7, and 9 without repetition?

16. *Test Options* On a test with 20 questions, a student must answer 8 of the first 12 questions and 4 of the last 8. In how many ways can this be done?

17. Expand $(x + 1)^5$.

18. Determine the number of subsets of a set containing 9 members.

19. *Marbles* Suppose we select, without looking, one marble from a bag containing 6 red marbles and 8 blue marbles. What is the probability of selecting a blue marble?

SYNTHESIS

20. Solve for n: $_{n}P_7 = 9 \cdot _{n}P_6$.

Final Examination

1. Make a hand-drawn graph of the equation $y = 4 - 2x$.

2. Find the coordinates of the point of intersection of $y_1 = x^3 - 5x + 7$ and $y_2 = -2x + 1$. (Round to three decimal places.)

3. Compute: $100 - 80 \div 2^2 \cdot 5 + 3$.

4. Find an equation of variation in which y varies directly as x and $y = 30$ when $x = 45$.

5. Solve. Write the answer in interval notation.
$$-16 \le 3x + 2 < 5$$

6. Given that $f(x) = 5x^3 + 4x^2 - 6x - 8$, find $f(-1)$.

7. Use a grapher to find the zeros of the function $f(x) = -x^3 + 2x^2 + 5x - 4$. (Round to three decimal places.)

8. Write an equation for the line that passes through $(1, -4)$ and $(3, -6)$.

9. The table below shows the per capita consumption of flour and cereal products in the United States in recent years.

Year, x	Per Capita Consumption of Flour and Cereal Products, in pounds
0. '93	191.0
1. '94	194.1
2. '95	192.5
3. '96	198.5

   a) Use a grapher to fit a regression line to the data.

   b) What is the correlation coefficient for the regression line?

10. a) Use a grapher to graph $y = 3x^3 - 2x$ in the window $[-5, 5, -5, 5]$.

    b) Determine whether the graph in part (a) is symmetric with respect to the x-axis, the y-axis, and/or the origin.

11. Solve $4x^2 - 3x + 2 = 0$.

12. For the graph of the function $f(x) = -x^2 + 3x + 5$,

    a) find the vertex;

    b) find the line of symmetry;

    c) state whether there is a maximum or minimum value and find that value;

    d) find the range.

13. For the polynomial function $P(x) = x^4 + 5x^3 + 3x^2 - 5x - 4$, solve $P(x) = 0$.

14. Solve $\dfrac{x-2}{x+3} < 2$.

15. Use the regression feature on a grapher to fit a quadratic function to the data in the table. (Round a, b, and c to three decimal places.)

x	y
0	252
1	216
3	198
7	243
9	278

16. Find a formula for the inverse of the function $f(x) = 3x - 5$.

17. Use a grapher to graph $f(x) = e^{x-1}$ in the window $[-10, 10, -10, 10]$.

18. Solve $16^x = 2^{3x-1}$.

19. Express in terms of sums and differences of logarithms of x, y, and z: $\log \dfrac{x^3 y}{z^2}$.

20. Suppose \$3000 is invested at interest rate k, compounded continuously, and grows to \$3635 in 4 years.

    a) Find the exponential growth function.

    b) Find the doubling time.

21. Find the six trigonometric function values of θ.

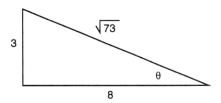

22. Solve the right triangle with $c = 36$ and $B = 42°$. Standard lettering has been used.

23. Convert $\dfrac{7\pi}{9}$ to degree measure.

24. Graph the function $y = -\cos(x + \pi/2) - 1$ in the window $[-2\pi, 2\pi, -4, 1]$.

25. *Distance* A moving van travels at 50 mph for 4 hr in a direction of 145° from Boise, Idaho. At the end of that time, how far east of Boise is the van?

26. Use a sum, difference, or half-angle identity to find $\cos 105°$.

27. Use a grapher to determine which expression can be used to complete an identity $\tan \dfrac{x}{2} = \cdots$.

a) $y = \dfrac{\tan x}{2}$

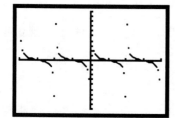

b) $y = \sin x - \cos x$

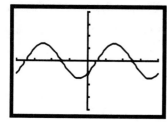

c) $y = \sin^2 x - \cos^2 x$

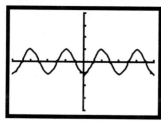

d) $y = \dfrac{1 - \cos x}{\sin x}$

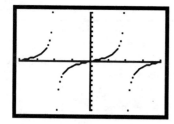

28. Prove the identity: $\dfrac{2 \tan x}{1 + \tan^2 x} = \sin 2x$.

29. Find $\cos^{-1}\left(-\dfrac{1}{2}\right)$ exactly, in radians.

30. Solve, finding all solutions in $[0, 2\pi)$:
$$\cos^2 x = \cos x.$$

31. Solve $\triangle ABC$, if possible.
$$b = 15 \text{ in.}, \ A = 43°, \ C = 99°$$

32. *Distance* Two airplanes leave an airport at the same time. The first flies 225 km/h in a direction of 220°. The second flies 200 km/h in a direction of 100°. After 2 hr, how far apart are the planes?

33. Find trigonometric notation for $2 - 2\sqrt{3}i$.

34. Graph $r = 2 \sin \theta$.

35. For $\mathbf{u} = 3\mathbf{i} + 4\mathbf{j}$ and $\mathbf{v} = \mathbf{i} - 6\mathbf{j}$, find $3\mathbf{u} - 4\mathbf{v}$.

Solve. Use any method.

36. $2x + 3y = 1,$
$\quad\ 3x - 2y = 21$

37. $2x - 3y + \ z = 11,$
$\quad\ 3x + 2y + 2z = -1,$
$\quad\ 4x - 5y - 3z = -1$

38. For $A = \begin{bmatrix} 2 & -5 & 4 \\ -3 & 1 & -1 \end{bmatrix}$ and $B = \begin{bmatrix} 6 & -4 \\ -1 & 2 \\ -3 & 5 \end{bmatrix}$, find AB, if possible.

39. Make a hand-drawn graph of $3x - 2y > 6$.

40. *Maximizing Profit* Henry's Bakery produces blueberry muffins and bran muffins. Each day at most 60 dozen muffins can be produced of which 12 dozen blueberry muffins and 6 dozen bran muffins are required by regular customers. The profit from 1 dozen blueberry muffins is \$3.50 and the profit from 1 dozen bran muffins is \$2.50. How many dozen of each type of muffin should be prepared in order to maximize the profit?

Match each of the following with its graph.

41. $2x^2 + 3x - 4y = 1$

42. $2y^2 - 3x^2 = 6$

43. $4x^2 + y^2 = 16$

44. $x^2 - 2x + y^2 - 4y - 4 = 0$

a)

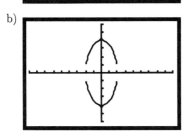

b)

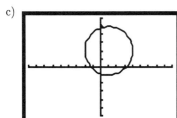

c)

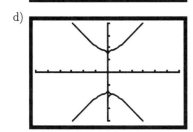

d)

45. Solve: $x^2 - y^2 = 24,,$
$$x + y = 2$$

46. Graph $x^2 + 2\sqrt{3}xy + 3y^2 - 12\sqrt{3}x + 12y = 0$.

47. Find the polar coordinates of $(-2, 2)$. Express the angle in degrees using the smallest possible positive angle.

48. Graph $r = 2\sin\theta$.

49. Graph $r = \dfrac{6}{3 + 3\cos\theta}$. State whether the directrix is vertical or horizontal, describe its location in relation to the pole, and find the vertex or vertices.

50. Graph the plane curve given by the parametric equations $x = 3\sin\theta$, $y = \cos\theta$, $0 \le \theta \le 2\pi$.

51. Find the 18th term of the arithmetic sequence $-7, -4, -1, \ldots$.

52. Find the sum, if it exists: $48 + 12 + 3 + \ldots$.

53. Use mathematical induction to prove that, for every natural number n, $2 + 7 + 12 + \cdots + (5n - 3) = \dfrac{n(5n - 1)}{2}$.

54. In how many ways can 5 different books be arranged on a shelf?

55. From a bag containing 3 red marbles, 8 blue marbles, and 5 green marbles, 4 marbles are drawn all at once. What is the probability of getting 2 blue marbles and 2 green marbles?

SYNTHESIS

56. If $(5, -2)$ is a point on the graph of $y = f(x)$, what point do you know is on the graph of $y = f(x + 3)$?

57. Find the domain of $f(x) = \log_3(2x + 1)$.

CHAPTER TEST ANSWERS

Chapter G

1.

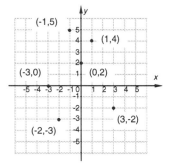

2. a) No; b) yes

3. a)

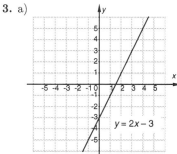

b)

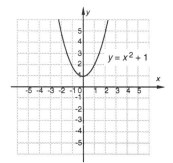

4. c   **5.** a   **6.** d   **7.** b

8.

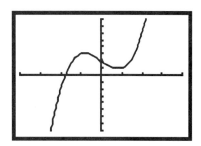

9.

10. Standard window

11. a)

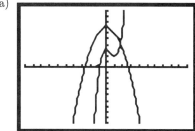

b) (1.70,4.12)   **12.** -1.23

Chapter R

1. a) $-8, 0, 36$; b) $-8, \dfrac{11}{3}, 0, -5.49, 36, 10\dfrac{1}{6}$ **2.** $1.2|xy|$, or $1.2|x| \cdot |y|$ **3.** 14 **4.** $(-3, 6]$ **5.** -5 **6.** 7.5×10^6

7. $-15a^4 b^{-1}$, or $-\dfrac{15a^4}{b}$ **8.** $3x^4 - 5x^3 + x^2 + 5x$

9. $2x^2 + x - 15$ **10.** $4y^2 - 4y + 1$ **11.** $21\sqrt{3}$

12. $\dfrac{x - y}{xy}$ **13.** $(2n - 3)(n + 4)$ **14.** $2(2x + 3)(2x - 3)$

15. $(m - 2)(m^2 + 2m + 4)$

16. $y_1 = x^2 + x - 6, y_2 = (x + 3)(x - 2)$

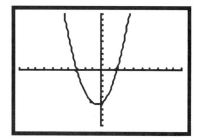

17. $\dfrac{x - 5}{x - 2}$ **18.** $\dfrac{x + 3}{(x + 1)(x + 5)}$ **19.** $\sqrt[7]{t^5}$

20. $\dfrac{35 + 5\sqrt{3}}{46}$ **21.** -1 **22.** $-5, \dfrac{1}{2}$ **23.** $-1, 3$

24. $\pm\sqrt{6}$ **25.** $h = \dfrac{3V}{2\pi r^2}$ **26.** 13 ft   **27.** $-2, 1$

Chapter 1

1. a) Yes; b) $\{-4, 3, 1, 0\}$; c) $\{7, 0, 5\}$ **2.** a) 8;

b) $2a^2 + 7a + 11$

3. a)

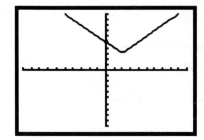

b) $(-\infty, \infty)$; c) $[3, \infty)$ **4.** a) No; b) yes

5. a) Increasing: $(-\infty, -0.33), (3, \infty)$, decreasing:

$(-0.33, 3)$; b) Maximum: 5.52 at $x = -0.33$, minimum:

-13 at $x = 3$

6.

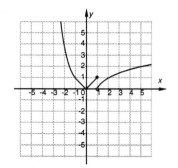

7. Slope: $\dfrac{3}{2}$; y-intercept: $\left(0, \dfrac{5}{2}\right)$

8. $y - 4 = -\dfrac{3}{4}(x - (-5))$, or $y - (-2) = -\dfrac{3}{4}(x - 3)$, or

$y = -\dfrac{3}{4}x + \dfrac{1}{4}$ **9.** $y - 3 = -\dfrac{1}{2}(x + 1)$, or $y = -\dfrac{1}{2}x + \dfrac{5}{2}$

10. a) $y = -0.14x + 8.36$; b) $r \approx -0.9439$; c) 6.7 gal

11. $\sqrt{45} \approx 6.708$ **12.** $\left(-3, \dfrac{9}{2}\right)$

13. $(x + 1)^2 + (y - 2)^2 = 5$

14. a)

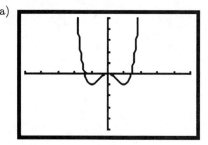

b) x-axis: no, y-axis: yes, origin: no   **15.** Odd

16. $f(x) = (x - 2)^2 - 1$ **17.** $f(x) = (x + 2)^2 - 3$

18.

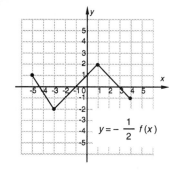

19. a) $(-\infty, \infty)$; b) $[3, \infty)$; c) $x^2 - \sqrt{x - 3}$;

d) $x^2\sqrt{x - 3}$; e) $(3, \infty)$ **20.** $y = \dfrac{30}{x}$ **21.** 50 ft

22. $(-1, 1)$

Chapter 2

1. -3 **2.** $-\dfrac{1}{4}$ and 3, or -0.25 and 3   **3.** $\dfrac{1 \pm \sqrt{57}}{4}$, or

approximately -1.637 and 2.137 **4.** $-2 \pm \sqrt{5}$

5. $\dfrac{3}{4} \pm \dfrac{\sqrt{23}}{4}i$ **6.** 16   **7.** $-1, \dfrac{13}{6}$ **8.** 5   **9.** $-\dfrac{1}{2}, 2$

10. $(-5, 3)$ **11.** $(-\infty, -7) \cup (-3, \infty)$ **12.** $n = \dfrac{R^2}{3p}$

13. $\sqrt{43}i$ **14.** $-5i$ **15.** $10 + 5i$ **16.** i **17.** a) $(1, 9)$;

b) $x = 1$; c) maximum: 9; d) $(-\infty, 9]$ **18.** 15 ft by 30

ft   **19.** $y = 12.875x^2 - 26.3x + 112.625$ **20.** $\dfrac{1}{6}$

Chapter 3

1. a) $-2.87, -1, 0.87$; b) 5.04 at $x = -2.08$; c) -5.04 at

$x = 0.08$; d) $(-\infty, \infty)$; e) $(-\infty, \infty)$ **2.** 5.5%

3. a) $y = 0.9360221882x^3 - 1.786434004x^2 +$

$6.861222601x - 4.039496246$; **b)** approximately 595

4. $x^2 + 3x + 1 + \dfrac{5x - 4}{x^2 - 1}$ **5.** $3x^2 + 15x + 63$, R 322

6. Yes **7.** -115 **8.** $-\sqrt{3}, 2 + i$

9. a)

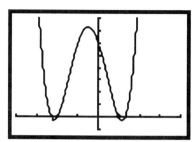

b) $1, -2, \dfrac{-1 \pm \sqrt{13}}{2}$;

c) $P(x) = (x-1)(x+2)\left(x - \dfrac{-1+\sqrt{13}}{2}\right)\left(x - \dfrac{-1-\sqrt{13}}{2}\right)$

10.

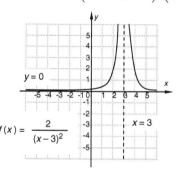

$f(x) = \dfrac{2}{(x-3)^2}$

11. Answers may vary; $f(x) = \dfrac{x+4}{x^2 - x - 2}$

12. $\left(-\infty, -\dfrac{1}{2}\right) \cup (3, \infty)$ **13.** $(-\infty, 4) \cup \left[\dfrac{13}{2}, \infty\right)$

14. a) 6 sec; **b)** (1,3) **15.** $(-\infty, -4] \cup [3, \infty)$

Chapter 4

1. $(f \circ g)(x) = x^2 - 4$; $(g \circ f)(x) = x^2 - 10x + 26$

2. $\{(5, -2), (3, 4), (-1, 0), (-3, -6)\}$ **3.** No

4. $f^{-1}(x) = \sqrt[3]{x - 1}$

5.

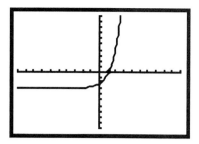

6.

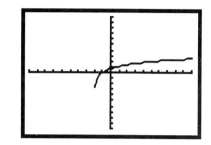

7. $x = e^4$ **8.** $x = \log_3 5.4$ **9.** $\dfrac{1}{2}$ **10.** 1 **11.** 1

12. 4.174 **13.** $\dfrac{2}{5}\ln x + \dfrac{1}{5}\ln y$ **14.** 0.656 **15.** $-4t$

16. 0.0154 **17. a)** 4.5%; **b)** $P(t) = 1000e^{0.045t}$; **c)**

$1433.33; **d)** 15.4 yr

18.

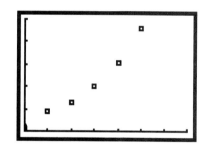

b) $y = 2.83547343(1.523196852)^x$, or

$y = 2.83547343e^{0.4208113184x}$; **c)** $82.2 million; **d)** about

6.8 yr **19.** $\dfrac{27}{8}$

Chapter 5

1. -1 **2.** $-\sqrt{2}$ **3.** $\dfrac{\sqrt{3}}{2}$ **4.** -1 **5.** -5.7588

6. 0.7827 **7.** -0.2419 **8.** -0.2079 **9.** Answers may

vary; $472°, -248°$ **10.** $38.47°$ **11.** $\dfrac{\pi}{6}$ **12.** $\dfrac{7\pi}{6}$

13. $135°$ **14.** $\dfrac{16\pi}{3} \approx 16.755$ cm

15. $\cos\theta = 5/\sqrt{41}$; $\tan\theta = -4/5$; $\csc\theta =$

$-\sqrt{41}/4; \sec\theta = \sqrt{41}/5; \cot\theta = -5/4$

16. $\sin\theta = \dfrac{4}{\sqrt{65}}$, or $\dfrac{4\sqrt{65}}{65}; \cos\theta = \dfrac{7}{\sqrt{65}}$, or $\dfrac{7\sqrt{65}}{65}; \tan\theta = \dfrac{4}{7}; \csc\theta = \dfrac{\sqrt{65}}{4}; \sec\theta = \dfrac{\sqrt{65}}{7}; \cot\theta = \dfrac{7}{4}$

17. $30°$ **18.** $\sin 61.6° \approx 0.8796; \cos 61.6° \approx$

$0.4756; \tan 61.6° \approx 1.8495; \csc 61.6° \approx$

$1.1369; \sec 61.6° \approx 2.1026; \cot 61.6° \approx 0.5407$ **19.** 1

20. 2π **21.** $\pi/2$ **22.** c **23.** $18\pi \approx 56.55$ m/min

24. About 444 ft

25. $\left\{x \,\middle|\, -\dfrac{\pi}{2} + 2k\pi < x < \dfrac{\pi}{2} + 2k\pi, k \text{ an integer}\right\}$

Chapter 6

1. $2\cos x + 1$ **2.** $2\cos\theta$ **3.** $\dfrac{\sqrt{2}+\sqrt{6}}{4}$ **4.** $\sqrt{5}/3$

5. $\dfrac{24}{25}$ **6.** $\dfrac{\sqrt{2}+\sqrt{3}}{2}$ **7.** d

8. $\csc x - \cos x \cot x$

$= \dfrac{1}{\sin x} - \cos x \cdot \dfrac{\cos x}{\sin x}$

$= \dfrac{1 - \cos^2 x}{\sin x}$

$= \dfrac{\sin^2 x}{\sin x}$

$= \sin x$

9. $(\sin x + \cos x)^2$

$= \sin^2 x + 2\sin x \cos x + \cos^2 x$

$= 1 + 2\sin x \cos x$

$= 1 + \sin 2x$

10. $-45°$ **11.** $\dfrac{\pi}{3}$ **12.** 2.307 **13.** $\dfrac{\sqrt{3}}{2}$

14. $\dfrac{\pi}{6}, \dfrac{5\pi}{6}, \dfrac{7\pi}{6}, \dfrac{11\pi}{6}$ **15.** $0, \dfrac{\pi}{4}, \dfrac{3\pi}{4}, \pi$ **16.** $0.69, 2.45$

17. $-\sqrt{\dfrac{11}{12}}$

Chapter 7

1. $B = 54.1°, a \approx 32.6, c \approx 55.7$ **2.** About 272 mi

3. $A = 83°, b \approx 14.7$ ft, $c \approx 12.4$ ft **4.** $A \approx 73.9°$,

$B \approx 70.1°, a \approx 8.2$ m, or $A \approx 34.1°, B \approx 109.9°$,

$a \approx 4.8$ m **5.** $A \approx 99.9°, B \approx 36.8°, C \approx 43.3°$

6. About 43.6 cm^2 **7.** About 77 m **8.** About 930 km

9.

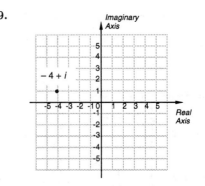

10. $\sqrt{13}$ **11.** $3\sqrt{2}(\cos 315° + i\sin 315°)$ **12.** $2i$

13. $\dfrac{1}{4}i$ **14.** 16 **15.** $\dfrac{1}{2} + i\dfrac{\sqrt{3}}{4}, -1, \dfrac{1}{2} - i\dfrac{\sqrt{3}}{4}$

16. Magnitude: 11.2, direction: 23.4° **17.** $-11i - 17j$

18. $-\dfrac{4}{5}i + \dfrac{3}{5}j$ **19.** $28.9°, 151.1°$

Chapter 8

1.

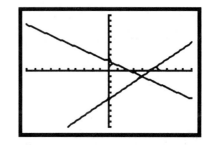

2. $(-3, 5)$ **3.** $(1, -2)$ **4.** $(-1, 3, 2)$ **5.** Consistent

6. Independent **7.** Student: 342, non-student: 408

8. $\begin{bmatrix} -2 & -3 \\ -3 & 4 \end{bmatrix}$ **9.** Not defined **10.** $\begin{bmatrix} -7 & -13 \\ 5 & -1 \end{bmatrix}$

11. Not defined **12.** $\begin{bmatrix} 2 & -2 & 6 \\ -4 & 10 & 4 \end{bmatrix}$

13. $\begin{bmatrix} 0 & -1 \\ -0.25 & -0.75 \end{bmatrix}$ **14.** a) $\begin{bmatrix} 49 & 10 & 13 \\ 43 & 12 & 11 \\ 51 & 8 & 12 \end{bmatrix}$;

b) $[26 \quad 18 \quad 23]$; c) $[3221 \quad 660 \quad 812]$; d) the total

cost, in cents, for each type of menu item served on the

given day **15.** $\begin{bmatrix} 3 & -4 & 2 \\ 2 & 3 & 1 \\ 1 & -5 & -3 \end{bmatrix}\begin{bmatrix} x \\ y \\ z \end{bmatrix} = \begin{bmatrix} -8 \\ 7 \\ 3 \end{bmatrix}$

16.

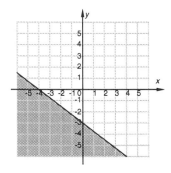

17. Maximum: 15 at (3,3), minimum: 2 at (1,0)

18. 25 pound cakes, 75 carrot cakes

19. $-\dfrac{2}{x-1} + \dfrac{5}{x+3}$ **20.** $A = 1, B = -3, C = 2$

Chapter 9

1. c   **2.** b   **3.** a   **4.** d   **5.** $x^2 = 8y$ **6.** (10,4)

7. Center: $(-1, 3)$, radius: 5   **8.** $\dfrac{y^2}{25} + \dfrac{x^2}{4} = 1$

9. $y = \dfrac{\sqrt{2}}{2}x$, $y = -\dfrac{\sqrt{2}}{2}x$ **10.** $\dfrac{27}{8}$ in.

11. $(1, 2)$, $(1, -2)$, $(-1, 2)$, $(-1, -2)$

12. $(3, -2)$, $(-2, 3)$ **13.** $(2,3)$, $(3,2)$ **14.** 5 ft by 4 ft

15. After using the rotation of axes formulas with

$\theta = 45°$, we have $\dfrac{(x')^2}{9} + (y')^2 = 1$.

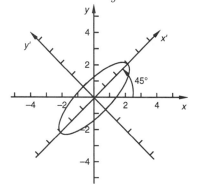

16. $2(\cos 120° + i \sin 120°)$ **17.** $(-0.09, 1.00)$

18. $r = \sqrt{10}$

19.

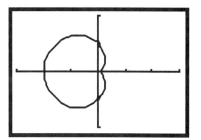

20. Horizontal directrix 2 units below the pole; vertex:

$(1, 3\pi/2)$;

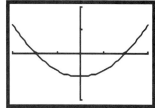

21. $r = \dfrac{6}{1 + 2\cos\theta}$

22.

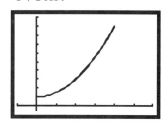

23. $x^2 + y^2 = 9$, $-3 \le x \le 3$ **24.** Answers may vary;

$x = t$, $y = t - 5$; $x = t^2$, $y = t^2 - 5$

25. $(x - 1)^2 + (y + 3)^2 = 16$

Chapter 10

1. -43 **2.** $2 + 5 + 10 + 17 = 34$

3.

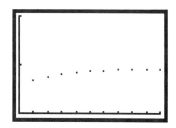

4. 44   **5.** 38   **6.** -420 **7.** 156,000   **8.** 27   **9.** $\dfrac{56}{99}$

10. \$74,399.77

11. $S_1 : 2 = \dfrac{1(3 \cdot 1 + 1)}{2}$

$S_k : 2 + 5 + 8 + \cdots + (3k - 1) = \dfrac{k(3k+1)}{2}$

$S_{k+1} : 2 + 5 + 8 + \cdots + (3k - 1) + [3(k+1) - 1] = \dfrac{(k+1)[3(k+1)+1]}{2}$

Basis step: $\dfrac{1(3 \cdot 1 + 1)}{2} = \dfrac{1 \cdot 4}{2} = 2$, so S_1 is true.

Induction step:
$2 + 5 + 8 + \cdots + (3k - 1) + [3(k+1) - 1]$

$= \dfrac{k(3k+1)}{2} + [3k + 3 - 1]$

$= \dfrac{3k^2}{2} + \dfrac{k}{2} + 3k + 2$

$= \dfrac{3k^2}{2} + \dfrac{7k}{2} + 2$

$= \dfrac{3k^2 + 7k + 4}{2}$

$= \dfrac{(k+1)(3k+4)}{2}$

$= \dfrac{(k+1)[3(k+1)+1]}{2}$

12. 3,603,600 **13.** 352,716

14. $\dfrac{n(n-1)(n-2)(n-3)}{24}$ **15.** 360 **16.** 34,650

17. $x^5 + 5x^4 + 10x^3 + 10x^2 + 5x + 1$ **18.** $2^9 = 512$

19. $\dfrac{4}{7}$ **20.** 15

Final Examination

1.

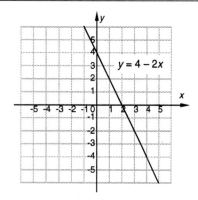

2. $(-2.355, 5.711)$ **3.** 3 **4.** $y = \dfrac{2}{3}x$ **5.** $[-6, 1)$ **6.** -3

7. $-1.856, 0.678, 3.177$ **8.** $y - (-4) = -1(x - 1)$, or

$y - (-6) = -1(x - 3)$, or $y = -x - 3$

9. a) $y = 2.09x + 190.89$; **b)** $r \approx 0.8326$

10. a)

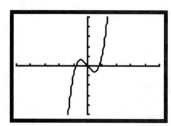

b) x-axis: no; y-axis: no; origin: yes **11.** $\dfrac{3}{8} \pm \dfrac{\sqrt{23}}{8}i$

12. a) $\left(\dfrac{3}{2}, \dfrac{29}{4}\right)$; **b)** $x = \dfrac{3}{2}$; **c)** maximum: $\dfrac{29}{4}$ at $x = \dfrac{3}{2}$;

d) $\left(-\infty, \dfrac{29}{4}\right]$ **13.** $-4, -1, 1$

14. $(-\infty, -8) \cup (-3, \infty)$

15. $y = 3.038x^2 - 22.924x + 244.036$

16. $f^{-1}(x) = \dfrac{x + 5}{3}$

17.

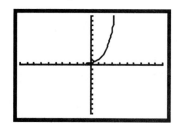

18. $-1s$ **19.** $3 \log x + \log y - 2 \log z$

20. a) $P(t) = 3000e^{0.048t}$; **b)** about 14.4 yr

21. $\sin \theta = \dfrac{3}{\sqrt{73}}$, or $\dfrac{3\sqrt{73}}{73}$; $\cos \theta = \dfrac{8}{\sqrt{73}}$, or

$\dfrac{8\sqrt{73}}{73}$; $\tan \theta = \dfrac{3}{8}$; $\csc \theta = \dfrac{\sqrt{73}}{3}$; $\sec \theta = \dfrac{\sqrt{73}}{8}$; $\cot \theta = \dfrac{8}{3}$

22. $A = 48°$, $a \approx 26.8$, $b \approx 24.1$ **23.** $140°$

24.

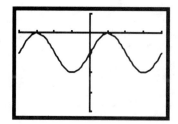

25. About 114.7 mi **26.** $\dfrac{\sqrt{2} - \sqrt{6}}{4}$ **27.** d

28. $\dfrac{2\tan x}{1 + \tan^2 x}$

$= \dfrac{2 \cdot \dfrac{\sin x}{\cos x}}{\sec^2 x}$

$= \dfrac{\dfrac{2\sin x}{\cos x}}{\dfrac{1}{\cos^2 x}}$

$= 2\sin x \cos x$

$= \sin 2x$

29. $\dfrac{2\pi}{3}$ **30.** $0,\ \dfrac{\pi}{2},\ \dfrac{3\pi}{2}$ **31.** $B = 38°$,

$a \approx 16.6,\ c \approx 24.1$ **32.** About 736.5 km

33. $4(\cos 300° + i\sin 300°)$

34.

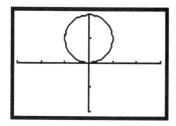

35. $5i + 36j$ **36.** $(5, -3)$ **37.** $(-1, -3, 4)$

38. $\begin{bmatrix} 5 & 2 \\ -16 & 9 \end{bmatrix}$

39.

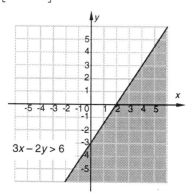

$3x - 2y > 6$

40. 54 dozen blueberry, 6 dozen bran   **41.** a   **42.** d

43. b   **44.** c   **45.** $(7, -5)$ **46.** After using the

rotation of axes formulas with $\theta = 60°$, we have

$(x')^2 = -6y'$.

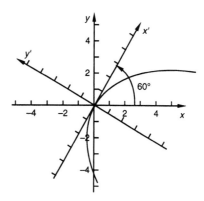

47. $(2\sqrt{2}, 135°)$, or $(2\sqrt{2}, 3\pi/4)$

48.

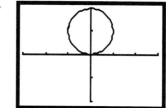

49. Vertical directrix 2 units to the right of the pole;

vertex: $(1,0)$;

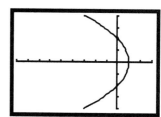

50.

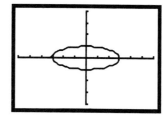

51. 44   **52.** 64

453. $S_1 : 2 = \dfrac{1(5 \cdot 1 - 1)}{2}$

$S_k : 2 + 7 + 12 + \cdots + (5k - 3) = \dfrac{k(5k - 1)}{2}$

$S_{k+1} : 2 + 7 + 12 + \cdots + (5k - 3) + [5(k+1) - 3] =$
$\dfrac{(k+1)[5(k+1) - 1]}{2}$

Basis step: $\dfrac{1(5 \cdot 1 - 1)}{2} = \dfrac{1 \cdot 4}{2} = 2$, so S_1 is true.

Induction step:

$$2 + 7 + 12 + \cdots + (5k - 3) + [5(k + 1) - 3]$$

$$= \frac{k(5k - 1)}{2} + [5k + 5 - 3]$$

$$= \frac{5k^2}{2} - \frac{k}{2} + 5k + 2$$

$$= \frac{5k^2 + 9k + 4}{2}$$

$$= \frac{(k + 1)(5k + 4)}{2}$$

$$= \frac{(k + 1)[5(k + 1) - 1]}{2}$$

454. 120   **55.** $\dfrac{2}{13}$ **56.** $(2, -2)$ **57.** $\left(-\dfrac{1}{2}, \infty\right)$